Springer Series in Surface Sciences

Volume 64

AF823418

This series covers the whole spectrum of surface sciences, including structure and dynamics of clean and adsorbate-covered surfaces, thin films, basic surface effects, analytical methods and also the physics and chemistry of interfaces. Written by leading researchers in the field, the books are intended primarily for researchers in academia and industry and for graduate students.

More information about this series at http://www.springer.com/series/409

Olaf Stenzel · Miloslav Ohlídal
Editors

Optical Characterization of Thin Solid Films

Springer

Editors
Olaf Stenzel
Abbe School of Photonics
Friedrich-Schiller-University Jena
Jena, Thüringen
Germany

and

Fraunhofer Institut für Angewandte Optik
und Feinmechanik IOF
Jena
Germany

Miloslav Ohlídal
Institute of Physical Engineering
Brno University of Technology
Brno
Czech Republic

ISSN 0931-5195 ISSN 2198-4743 (electronic)
Springer Series in Surface Sciences
ISBN 978-3-030-09200-9 ISBN 978-3-319-75325-6 (eBook)
https://doi.org/10.1007/978-3-319-75325-6

Printed on acid-free paper

This Springer imprint is published by the registered company Springer International Publishing AG part of Springer Nature.
The registered company address is: Gewerbestrasse 11, 6330 Cham, Switzerland

Fluorite Crystal

Detail of a painting by Vera Andrianova Forwerk, a Bulgarian master student of photonics at Abbe School of Photonics, Friedrich Schiller University Jena, Germany. Printed with permission.

Fluorite is a naturally occurring crystalline modification of calcium fluoride, a widely used optical material for applications as windows, lenses and the like from the middle infrared up to the vacuum ultraviolet spectral regions. See Sect. 8.2.2 for optical constants.

Foreword

The present book provides the reader with top-notch reports from the workbenches of renowned colleagues in the field of thin film characterization.

The field of thin films is an ever-expanding area with optical coatings representing the oldest, yet perhaps the fastest, still developing area among the various, multidisciplinary applications of thin films.

A most current snapshot of the vast expanse and speed by which the field of thin films is moving is provided in this book.

Thin films are abundant in modern science and engineering owing to sophisticated and continuously improved techniques for their preparation, their usefulness in basic and applied research, and their suitability for mass production in a sheer limitless variety of new and advancing device structures. Perhaps one of the most intriguing peculiarities of thin films is inscribed within their optical properties, which depend in most complex ways on the structural arrangements of matter from atomic to macroscopic scales.

The authors in this book provide the reader with a set of chapters to gain a fast foothold in understanding of the topics needed to engage in optical thin film characterization at the present level of the state of the art.

The electromagnetic spectrum offers a broadest interval over which the response of thin films can be interrogated. Advanced optical and structural investigations techniques are required for characterization of thin films. Likewise, intricate theoretical tools are needed when the attempt is made to relate the observations made to the real structure of the thin films. The tools for calculations descent from quantum mechanics as well as continuum physics concepts, and their implementations can be made easily on personal computers or may reach the limits of contemporary computation capabilities.

The book reflects the complexity of this topic from a multitude of perspectives.

Introductory portions provide the reader with the information on techniques for optical and structural characterization of advanced optical thin films. Examples are presented where the reader is walked through the complexity and challenges which one faces when simply questions are related to complex structured real samples. Rather than attempting to cover the vastness of all possible situations, the book focuses on selected examples, such as microstructure gradients, porosity, damages, defects, thin films composed of regular nano- and microstructures, and challenges in instrumentation and their circumvention.

The book is not the first in this field and will not be the last. It is the current best step stone to the next level of understanding.

I found the book a very interesting read and recommend it highly to beginners and experts alike, both in the field of optical characterization of thin films as well as to anyone in our expanding, multidisciplinary scientific community who is working with thin films.

Lincoln, NE, USA
July 2017

Mathias Schubert

Preface

The idea for this book arose at a bilateral workshop held by colleagues from Czech Republic and Germany on optical coating characterization in Brno, Czech Republic, October 11–13, 2016. The workshop dealt with different aspects of thin film optical characterization, including modeling, spectrophotometric approaches, spectroellipsometric approaches, specifics of characterization of defected or corrugated coatings, as well as the measurement of smallest optical losses with laser-based experimental equipment. It thus provided a snapshot on the current state of the art in diverse branches of optical coating characterization.

In a narrower sense, all speakers at the workshop stemmed from research facilities located either in Brno (Czech Republic) or in Jena (Germany), so that the broad circle of characterization approaches presented there also served as a demonstration of the analytic possibilities located in these two cities.

We strongly believe that the material discussed and presented at the workshop may be of interest to a broader audience. Therefore, each of the speakers was asked to extend the content of his or her presentation in order to supply a chapter that became part of the present book. It was our first intention to provide a guide to modern and powerful thin film characterization techniques while emphasizing the benefits of their coherent interaction, even when including non-optical techniques into the characterization approach. Thus, atomic force microscopy combined with measurements of the elastically scattered light appears to be a powerful tool for surface and film roughness investigation. Information on film porosity may be gained from simple spectrophotometric (or spectroellipsometric) measurements performed in different ambient conditions and spectral ranges, combined with an elemental analysis performed, for example, by energy-dispersive X-ray spectroscopy.

Our second intention was to emphasize the role of modeling. Practically throughout the whole book text, emphasis is placed on the development and presentation of manageable theoretical models with highest predictive power, and with clearly defined interfaces to interact with experimentally available input data. All presented models are applicable to broadest classes of characterization problems currently needed in industrial and partially also in academic research. In this context, characterization examples are presented, which are in most cases of high practical relevance today. They stem from real characterization problems and illustrate the feasibility of the coherent interaction of robust theoretical approaches and modern commercially available measurement facilities.

Concerning the target audience of this book, it is our feeling that every scientist or engineer having finished a master study in a physical, chemical, or engineering discipline and being involved in thin film characterization tasks can benefit from reading this book.

The editors are extremely grateful to all authors for their effort to provide high-quality contributions in a very short time—it took less than nine months of the workshop to the submittal of the book manuscript. Many thanks are also to Professor Mathias Schubert, Electrical & Computer Engineering, University of Nebraska–Lincoln, for critical remarks to the manuscript and, in particular, for contributing a foreword to this book.

All German authors acknowledge the financial support to the workshop provided by the Deutsche Forschungsgemeinschaft DFG. All authors thank to the Brno University of Technology and Masaryk University, Brno, for organizing the workshop.

Jena, Germany Olaf Stenzel
Brno, Czech Republic Miloslav Ohlídal

Contents

Contributors

Martin Čermák Department of Physical Electronics, Faculty of Science, Masaryk University, Brno, Czech Republic

Daniel Franta Faculty of Science, Department of Physical Electronics, Masaryk University, Brno, Czech Republic

David Holec Department of Physical Metallurgy and Materials Testing, Montanuniversität Leoben, Leoben, Austria

Christian Karras Leibniz-Institute of Photonic Technology IPHT, Jena, Germany

Petr Klapetek Czech Metrology Institute, Brno, Czech Republic

Stefanie Kroker LENA Laboratory for Emerging Nanometrology, Technische Universität Braunschweig, Braunschweig, Germany; Physikalisch-Technische Bundesanstalt, Braunschweig, Germany

Christian Mühlig Leibniz-Institute of Photonic Technology IPHT, Jena, Germany

David Nečas Plasma Technologies, CEITEC, Masaryk University, Brno, Czech Republic

Miloslav Ohlídal Institute of Physical Engineering, Brno University of Technology, Brno, Czech Republic; Institute of Physics, Faculty of Mining and Geology, VŠB-Technical University of Ostrava, Ostrava, Czech Republic

Pavel Ondračka Plasma Technologies, CEITEC, Masaryk University, Brno, Czech Republic; Faculty of Science, Department of Physical Electronics, Masaryk University, Brno, Czech Republic

Ivan Ohlídal Faculty of Science, Department of Physical Electronics, Masaryk University, Brno, Czech Republic

S. Schröder Fraunhofer Institute for Applied Optics and Precision Engineering IOF, Jena, Germany

Thomas Siefke Physikalisch-Technische Bundesanstalt, Braunschweig, Germany; Friedrich-Schiller-University Jena, Institute of Applied Physics, Jena, Germany

O. Stenzel Fraunhofer Institut für Angewandte Optik und Feinmechanik IOF, Jena, Germany; Abbe School of Photonics, Friedrich-Schiller-University Jena, Jena, Germany

M. Trost Fraunhofer Institute for Applied Optics and Precision Engineering IOF, Jena, Germany

Jiří Vodák Institute of Physical Engineering, Brno University of Technology, Brno, Czech Republic; Department of Physical Electronics, Masaryk University, Brno, Czech Republic; Czech Metrology Institute, Brno, Czech Republic

Jiří Vohánka Faculty of Science, Department of Physical Electronics, Masaryk University, Brno, Czechia

Steffen Wilbrandt Fraunhofer Institute for Applied Optics and Precision Engineering IOF, Jena, Germany

Lenka Zajíčková Plasma Technologies, CEITEC, Masaryk University, Brno, Czech Republic; Faculty of Science, Department of Physical Electronics, Masaryk University, Brno, Czech Republic

Symbols and Abbreviations

a_j	Parameter “a” in the *j*-th layer of a layer stack
A	Absorptance
α	Absorption coefficient
β	Roughening exponent (in Chap. 14)
β	Two-photon absorption coefficient (in Chap. 16)
c	Velocity of light in vacuum
DF	Discrepancy function
δ	Phase, phase shift
ΔP_s	Scattered power
$\Delta\Omega_s$	Detector solid angle
$\boldsymbol{E}$, E	Electric field strength
E_0	Field amplitude
E_g	Band gap
E_n	Energy level (in quantum mechanics)
e	Euler’s number (2.71828182…)
e	Elementary charge ($1.60217662 \times 10^{-19}$ C)
ε	Complex dielectric function
ε_r	Real part of the dielectric function
ε_i	Imaginary part of the dielectric function
ε_{xx}	Diagonal element of the dielectric tensor
$\hat{\varepsilon}$	Dielectric function
ε_0	Vacuum permittivity ($8.854187817 \times 10^{-12}\,\mathrm{F \cdot m^{-1}}$)
f_l	Relative strength of the absorption lines
f_{lm}	Oscillator strength in quantum mechanics
φ	Incident angle
Φ_a	Azimuth angle
φ_B	Brewster’s angle
γ	Damping constant
Γ	Homogeneous linewidth
h	Physical film thickness

h	Thickness
$\hbar$	Reduced Planck constant ($1.0545718 \times 10^{-34}$ J · s)
h_{sub}	Physical substrate thickness
$\boldsymbol{H}$, H	Magnetic field strength
$\boldsymbol{H}$	Hamilton operator, Hamiltonian
I	Intensity
i	Imaginary unit
$\hat{\boldsymbol{I}}$	Characteristic matrix
$\boldsymbol{j}$	Electric current density
k	Extinction coefficient
$\boldsymbol{k}$	Wave vector
k_0	Size of wave vector in vacuum $k_0 = 2\pi/\lambda$
k_B	Boltzmann's constant$1.380648 \times 10^{-23}\,\mathrm{J}\cdot\mathrm{K}^{-1}$
L	Depolarization factor (Chaps. 2, 7and 10)
L	Optical loss
λ	Wavelength of light in vacuum
Λ	Spatial period
m	Mass
m_e	Electron rest mass (9.109383×10^{-31} Kg)
μ_0	Permeability of free space
N	Concentration
N	Number (where specified)
n	Refractive index
n_{sub}	Substrate refractive index
$\hat{n}$	Complex index of refraction
ν	Wavenumber
π	The ratio of a circle's circumference to its diameter (3.14159265…)
p	Dipole moment
$\boldsymbol{P}$, P	Polarization
P_i	Incident power (in Chap. 14)
$P^{(l)}$	Polarization of l-th order
$\boldsymbol{p}_{ml}$	Matrix element of the dipole operator
ψ	Refraction angle, propagation angle
q	Charge
R	Radius
R	Reflectance
R_p	Reflectance of p-polarized light
R_s	Reflectance of s-polarized light
$\boldsymbol{r}$	Position vector with $\mathbf{r} = (x, y, z)^T$
$r_{(s,p)}$	Field reflection coefficient (for s- or p-polarized light)
ρ	Mass density
S	RMS of thickness variation
S	Surface area
S	Signal (in Chap. 16)

σ	RMS surface roughness
σ	RMS value of the heights of irregularities
σ_{stat}	Static value of the electrical conductivity (in Chap. 7)
T	Absolute temperature
T	Transmittance
$t_{(s,p)}$	Field transmission coefficient (for s- or p-polarized light)
t	Time
τ	Time constant, relaxation time
θ	Heaviside step function
θ_s	(polar) scatter angle
V	Volume
V_j	Fraction volume
v	Velocity
v	Wavenumber
W	Probability
ω	Angular frequency
ω_0	Eigenfrequency, resonance frequency
ω_p	Plasma frequency
ω_{nm}	Transition frequency, resonance frequency in quantum mechanics
$\Delta\omega$	Spectral bandwidth
χ	Linear dielectric susceptibility
ACF	Autocorrelation function
AFM	Atomic force microscopy
ARS	Angular-resolved scattering
BSE	Bethe–Salpeter equation
CCD	Charge-coupled device
CMOS	Complementary metal–oxide–semiconductor
DFT	Density functional theory
DOS	Density of states
DHO	Damped Harmonic Oscillators
EMA	Effective medium approximation
FIR	Far-infrared spectral region
FWHM	Full width half maximum
GGA	Generalized gradient approximation
GWS	Grating waveguide structure
HW	Halfwave
IR	Infrared
IR	Infrared spectral region
ISR	Imaging spectroscopic reflectometry
ITF	Identical thin film
JDOS	Joint density of states
KK	Kramers–Kronig
KS	Kohn–Sham
LDA	Local density approximation
LIDT	Laser-induced damage threshold

LSM	Least squares method
MIR	Middle-infrared spectral region
NIR	Near-infrared spectral region
NIR	Near-infrared
PBC	Periodic boundary conditions
PCSA	Polarizer–compensator–sample–analyzer
PIAD	Plasma-ion-assisted electron beam evaporation (a film preparation technique)
PJDOS	Parameterization of the joint density of states
PSCA	Polarizer–sample–compensator–analyzer
PSDF	Power spectral density function
QP	Quasiparticle
QW	Quarterwave
RGB	Red–Green–Blue
RMS	Root-mean-square
RPA	Random phase approximation
RRT	Rayleigh–Rice theory
SDT	Scalar diffraction theory
SPM	Scanning probe microscopy
SQS	Special quasirandom structure
SR	Spectroscopic reflectometry
TB-mBJ	Tran–Blaha modified Becke–Johnson
TRK	Thomas–Reiche–Kuhn
TS	Total scattering
UDM	Universal dispersion model
UV	Ultraviolet
UV	Ultraviolet spectral region
VASE	Variable angle spectroscopic ellipsometry
VIS	Visible
VIS	Visible spectral region

Part I
Introduction and Modelling

Chapter 1
Introduction

O. Stenzel and Miloslav Ohlídal

Abstract Optical thin film characterization represents the total of all theoretical and experimental activities, which pursue determination of various thin film construction parameters. The purpose of optical characterization is usually in the determination of film thickness, refractive index and extinction coefficient, porosity, surface roughness, film stoichiometry, film density and the like.

1.1 First Considerations

The interaction of electromagnetic radiation with matter is in the basis of any optical phenomenon. Whenever we are looking into our surrounding, our eyes receive light from various objects (material systems). This light may by emitted from an object itself, but it may also result from specular as well as diffuse reflections of light originating from a remote light source that is used to illuminate the object. At the moment, while you are reading this text, both of these situations may be relevant. In the case that you read it from the computer screen, your eyes receive photons actively

O. Stenzel (✉)
Fraunhofer Institut für Angewandte Optik und Feinmechanik IOF, Albert-Einstein-Str. 7, 07745 Jena, Germany
e-mail: olaf.stenzel@iof.fraunhofer.de; optikbuch@optimon.de

O. Stenzel
Abbe School of Photonics, Friedrich-Schiller-University Jena, Albert-Einstein-Straße 6, 07745 Jena, Germany

M. Ohlídal
Institute of Physical Engineering, Brno University of Technology, Technická 2, 61669 Brno, Czech Republic
e-mail: ohlidal@fme.vutbr.cz

M. Ohlídal
Institute of Physics, Faculty of Mining and Geology, VŠB-Technical University of Ostrava, Ostrava, Czech Republic

O. Stenzel and M. Ohlídal (eds.), *Optical Characterization of Thin Solid Films*, Springer Series in Surface Sciences 64,
https://doi.org/10.1007/978-3-319-75325-6_1

emitted by the screen. In this case, any external light may be strongly disturbing. However, if you are reading this text in a printed book, you certainly make use of an external light source, and your eyes receive photons that are (in this particular case diffusely) reflected from the book page.

Of course, the particular properties of any reflected radiation are to a certain extent defined by the process of interaction of the incident light with the corresponding material system. Hence, the reflected light carries some information about the objects properties. Understanding the mechanisms of the interaction of light with matter is absolutely necessary to disclose this information in a quantitative manner. Consequently, we may gain information on the specifics of a material system by analyzing the properties of light that has interacted with that system. Essentially, this is the idea of optical characterization.

Optical characterization techniques are applied to many different types of samples, among them thin films. Thin films enter into our daily lives in many different situations. For us the most significant are those thin films we prepare for improvement of quality of our lives by means of various technologies. The ranges of utilization of thin film optical properties, of technologies for thin film preparation, and of techniques which determine parameters of thin film optical properties are huge.

Thin films are used in many different areas of technology, including semiconductor technology, optoelectronics, mechanical or chemical surface conditioning, or optical surface functionalization. All of these films may, in principle, be characterized by optical techniques.

If we focus on the measuring techniques of optical properties of thin films, which are currently used, the number of them is of the order of one hundred. This number is related to the fact that thin film is a specific object with various properties (microstructure, variations in stoichiometry, inhomogeneity, anisotropy), which may differ significantly from the properties of the corresponding bulk substance. These properties typically depend on the conditions of the thin film preparation. The production of thin films or film systems with new and more challenging properties fulfilling requirements of optical practice require more and more advanced techniques measuring these properties.

Once optical characterization is based on the interaction of light with matter, a basic understanding of the mechanisms of that interaction is absolutely essential for performing an optical sample characterization. In a narrow sense, light may be understood as electromagnetic radiation in the visible spectral range only. When further speaking on optical characterization, we will nevertheless use a somewhat broader understanding of the terminus "light", including at least the infrared and ultraviolet spectral regions into the discussion. The actual spectral range included into a particular optical characterization may differ from case to case, but this is in fact not so essential. The more important point is that any electromagnetic wave is characterized by electric and magnetic field strength vectors, which interact with different degrees of freedom within the material system. Fields cause forces acting for example on a test charge, and the natural writing of the electric and magnetic fields in a light wave would operate with real (in the mathematical sense) functions and coefficients only. Thus, for such real electric fields and the particular case of a plane wave, a description of the type:

$$E_{real}(t, \boldsymbol{r}) = E_{0,real} \cos(\omega t - \boldsymbol{kr} + \delta_0) \tag{1.1}$$

is often used in order to quantify the electric field strength as a function of time and space coordinates.

The plane wave expression for the electric field as introduced in (1.1) is applicable when the light emitter is placed "sufficiently far" from the sample, when comparing with the wavelength of the light. This is guaranteed in a multiplicity of optical thin film characterization techniques today. On the other hand, for thin film samples, interference effects within the film are of crucial importance. Therefore we really have to consider electric fields, which carry phase information, while it would be insufficient to work with light rays characterized by certain intensity.

However, a more convenient mathematical treatment is achieved when rewriting (1.1) in the following manner:

$$E_{real}(t, \boldsymbol{r}) = \frac{1}{2}\left[E_{0,real}\mathrm{e}^{-\mathrm{i}(\omega t - \boldsymbol{kr})}e^{-\mathrm{i}\delta_0} + E_{0,real}\mathrm{e}^{\mathrm{i}(\omega t - \boldsymbol{kr})}e^{\mathrm{i}\delta_0}\right] \equiv E_0\mathrm{e}^{-\mathrm{i}(\omega t - \boldsymbol{kr})} + c.c \tag{1.2}$$

Here "c.c." denotes the conjugate complex to the preceding expression. We recognize that the *real* electric field (1.1) may be expressed as the superposition of a *complex* field and its conjugate complex counterpart. Clearly, the latter does not contain any new physical information. The complex field amplitude E_0 is defined as:

$$E_0 \equiv \frac{E_{0,real}\mathrm{e}^{-\mathrm{i}\delta_0}}{2} \tag{1.3}$$

In our further treatment we will therefore make use of the complex field defined by (1.4):

$$E(t, \boldsymbol{r}) = E_0\mathrm{e}^{-\mathrm{i}(\omega t - \boldsymbol{kr})} \tag{1.4}$$

Then, for the real field we have:

$$E_{real}(t, \boldsymbol{r}) = 2\mathrm{Re}E(t, \boldsymbol{r}) \tag{1.5}$$

The choice of (1.4) for the complex writing of the electric field defines a particular convention, which will be used throughout this book. When recalling (1.2), it becomes evident there exists an alternative writing:

$$E_{real}(t, \boldsymbol{r}) = \frac{1}{2}\left[E_{0,real}\mathrm{e}^{+\mathrm{i}(\omega t - \boldsymbol{kr})}e^{+\mathrm{i}\delta_0} + E_{0,real}\mathrm{e}^{-\mathrm{i}(\omega t - \boldsymbol{kr})}e^{-\mathrm{i}\delta_0}\right] \equiv E_0\mathrm{e}^{+\mathrm{i}(\omega t - \boldsymbol{kr})} + c.c;$$

$$E_0 \equiv \frac{E_{0,real}\mathrm{e}^{+\mathrm{i}\delta_0}}{2}$$

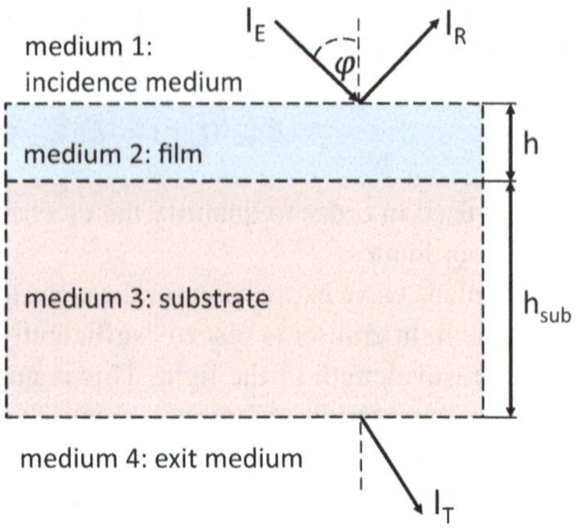

Fig. 1.1 Thin film on a thick substrate, irradiated by light under an incidence angle φ. For details see text

In this case, instead of (1.4) we would have:

$$E(t, \boldsymbol{r}) = E_0 e^{+i(\omega t - \boldsymbol{kr})} \tag{1.6}$$

For our further purposes, it makes absolutely no physical difference, which of the (1.4) or (1.6) is used for building the theory. In our particular treatment, we will use the minus-sign as fixed in (1.4). Let us remark that the writing of the expression for the complex index of refraction (n + ik or n – ik) depends on the choice of either (1.4) or (1.6). In our further treatment, in consistency with the choice of the time-dependence (1.4), the complex index of reraction will be given as n + ik.

Let us now assume a thin film, deposited on a thick substrate as shown in Fig. 1.1.

The intensity I of the light can be understood as the amount of light energy penetrating a unit surface area per unit time interval. After interaction with the sample, a part of the intensity of the incident light (I_E) may be transmitted into the exit medium with the intensity I_T, while another part is reflected back into the incident medium with the intensity I_R. Note that in the case of oblique incidence of the light, both transmission an reflection are strongly dependent on the polarization of the incident light.

As it follows from Fig. 1.1, a light wave which has penetrated a thin film sample, will carry information about the *materials* which form the sample (i.e. about both film and substrate material constants), as well as about its *geometry* (here the thicknesses of film h and substrate h_{sub}). Generally, the same will be true for the reflected wave, because all interfaces can principally contribute to the reflectance spectrum. So that we have to expect, that both I_T and I_R will be rather complicated functions of all the mentioned construction parameters. Thus, measured transmission and reflection spectra can be used to gain information about both material properties and sample geometry.

In the model case of optically homogeneous, isotropic, and non-magnetic media, the linear optical *material* properties may be expressed in terms of a scalar complex

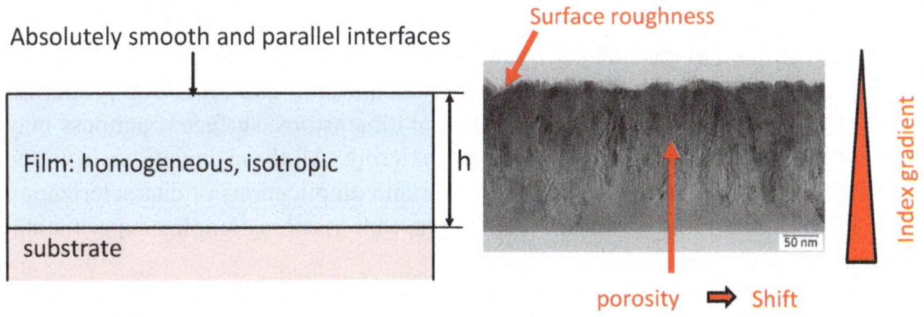

Fig. 1.2 *On left*: ideal thin film sample; *on right*: TEM cross-section of a real hafnia film

dielectric function. When further neglecting all effects related to spatial dispersion, this dielectric function appears to be a function of the angular frequency ω only [1, 2]. When assuming (1.4), $\varepsilon(\omega)$ is related to the optical constants (refractive index n, which defines the phase velocity of light in the medium, and extinction coefficient k, which it defines the light amplitude damping in the medium) through the relationship:

$$n(\omega) + \mathrm{i}k(\omega) = \sqrt{\varepsilon(\omega)} \equiv \hat{n}(\omega) \tag{1.7}$$

Here $\hat{n}$ is the so-called complex index of refraction; its frequency-dependence is called *dispersion*. The absorption coefficient α is defined as:

$$\alpha(\omega) = 2\frac{\omega}{c}k(\omega) \tag{1.8}$$

c is the velocity of light in vacuum. Note that use of the alternative convention according to (1.6) would result in the appearance of the minus-sign in the left-hand term in (1.7). Let us also mention that a positive imaginary part of the dielectric function will result in energy dissipation in a medium. Whenever the dielectric function is purely real, no energy will be dissipated.

When further assuming absolutely smooth and parallel surfaces and interfaces in the system shown in Fig. 1.1, then the transmission and reflection of a plane incident light wave will be performed only into well-defined directions, which are related to the propagation direction of the incident light by Snells law of refraction. In other words, the idealized thin film model as shown in Fig. 1.1 will not allow for elastic light scatter losses.

In any practical situation, the real film geometry may show more or less significant deviations from that model case, so that more complicated models may be necessary in order to reproduce measured spectral sample features in a consistent quantitative manner. This is visualized in Fig. 1.2.

It is straightforward to give a theoretical description of the optical properties of a film that is close to the idealized model sketched in Fig. 1.2 on the left. In reality, as it is exemplified in Fig. 1.2 on the right, the ideal model assumptions are not fulfilled. Columnar film morphology may violate homogeneity and isotropy of the film material. Samples may be porous, which tends to reduce the refractive index and results in vacuum and thermal shift phenomena. Refractive index gradients as well as surface and interface roughness alter the transmission and reflection properties of the film. Depending on characteristic lateral dimensions, surface roughness may also provide a mechanism for diffuse light scattering. All these deviations from the ideal model assumptions will result in significant complications of characterization tasks. They require more effort in modelling as well as more complex experimental equipment.

1.2 Coating Characterization and Quality Control

Many analytical tools such as spectrophotometers, interferometers, or scanning probe microscopes may be used in coating characterization as well as in coating quality control. Let us therefore clearly distinguish between these two different tasks:

- The terminus "quality control" will be used in our treatment in order to define a set of measurements, performed with the only goal to check the degree of adherence of sample properties with some pre-defined target specification.
- The terminus "sample characterization" will be used to define the total of all theoretical and experimental investigations, performed with the purpose to determine general construction parameters of the considered sample.

1.3 Organisation of the Book

The book is subdivided into four parts. The first part is of introductory character with certain emphasis on modelling.

Chapter 2 provides a short overview of characterization techniques relevant in the context of the present book. The mutual interplay of selected characterization techniques is exemplified in application to the analysis of the properties of a porous zirconia thin film.

In Chap. 3, a dielectric response of a material is modeled by a universal dispersion model. This model consists of a collection of dispersion models which describe individual elementary excitation in solids while respecting the basic conditions that follow from the theory of dispersion (time reversal symmetry, Kramers–Kronig consistency and finite sum rule integral). The combination of these models can be used for the description of complete dielectric response of a wide range of materials in the spectral range from far infrared to X-ray.

Chapter 4 deals with ab initio modeling techniques that can be used for the quantum-mechanical calculation of electronic band structure of solids within the framework of density functional theory and many body perturbation theory, with a particular focus on the evaluation of optical properties of these solids. A reader of the chapter should gain a general idea about the possibilities of the standard ab initio techniques, not only in the case of interpreting experimental results, but also in the case of predicting optical properties of even not-yet-known materials.

The second part is focused on spectrophotometric and spectro-ellipsometric thin film characterization of different real thin film samples.

Chapter 5 is devoted to the optical characterization of thin films, which exhibit area non-uniformity in their optical properties, by means of the non-microscopic imaging spectroscopic reflectometry. Essential features and implementation of this relatively new technique are given here together with the basic experimental setup of imaging spectroscopic reflectometers. The classification of the experimental data processing methods is also performed from the point of view of the way in which the information contained in the image of the film measured is exploited. Potential of imaging spectroscopic reflectometry utilization in the field of optical characterization of thin films is demonstrated by examples of results achieved.

Data processing approaches and algorithms for imaging optical techniques are discussed in Chap. 6, with focus on imaging spectrophotometry in the visible and ultraviolet regions. It therefore follows on from the issue of Chap. 5. Efficient fitting methods are developed for huge number of experimental data. This is achieved by splitting the least squares problem in different ways in order to extract useful information and by constructing effective models for individual phenomena and types of samples. Efficient computation of optical quantities is discussed for ideal, non-uniform, and rough films. Modelling of optical constants and spectral or angular averaging are discussed too.

The Chaps. 7 and 8 introduce the reader to basic skills of modern spectrophotometric substrate and coating characterization. Both ex situ and in situ spectrophotometry versions are presented. Examples demonstrate the application of spectrophotometry to the characterization of dielectric, semiconductor, and metal films. The Drude model, the Lorentzian oscillator model, as well as the newly developed Beta-distributed oscillator (β_do) model are applied to fit the experimental data.

Chapter 9 focuses on a brief description of the principles of ellipsometry from theoretical and experimental points of view. The theoretical results are formulated using Jones, Stokes–Mueller, Yeh matrix formalisms, and matrix formalism for optically isotropic layered systems. The basic principles of conventional, generalized and Mueller-matrix ellipsometry together with the most common ellipsometric techniques and model approaches are also described.

The third part concentrates on samples that show rather strong deviations from the ideal case postulated in Fig. 1.1.

Chapter 10 introduces theoretical approaches usable for describing the most common defects in thin film systems. Specific topics treated are random roughness of boundaries, thickness non-uniformity, optical inhomogeneity corresponding to refractive index profiles, overlayers, and transition layers. The influence of defects is demonstrated by means of several numerical and experimental examples.

Chapter 11 focuses on the use of Scanning Probe Microscopy in the optical thin film analysis. Even if this is not an optical technique, it is often used to get surface roughness characteristics as important input to the models of the optical response of thin film surfaces. Basic measurement principles, data processing, and roughness statistical quantities most frequently used in the field of thin film optics are listed in the reader's reference.

Chapters 12 and 13 deal with periodically structured films, while structuring is performed along lateral directions. Thus, in Chap. 12, resonant Grating Waveguide structures are introduced as candidates for narrowline high reflectors. Chapter 13 is focused on design, preparation, and characterization of grid polarizer devices used for application in the deep ultraviolet region.

And finally, the fourth part of the book is dedicated to the characterization of close-to-ideal films that show smallest absorption or scattering losses.

Here, small scattering losses are in the focus of Chap. 14. The chapter provides an introduction to both theoretical background and experimental techniques necessary for measurement and interpretation of scattered light on surfaces, single films, and multilayer coatings.

Similarly, in Chap. 15, the limits of smallest absorption loss measurements are discussed. A short overview is provided on the state-of-the-art in absorption measurement. Then, focus is given on photothermal deflection techniques and measurements of the Laser Induced Fluorescence (LIF).

The forth part is finished with Chap. 16, which explains the application of Cavity Ringdown Decay (CRD) techniques to the accurate measurement of highest reflectance, and thus smallest reflection losses. The techniques are demonstrated in application to high reflector coatings.

Chapter 2
Characterization of Porous Zirconia Samples as an Example of the Interplay Between Optical and Non-optical Characterization Methods

O. Stenzel

Abstract Optical characterization appears particularly strong when it is combined with suitable non-optical characterization techniques, as well as theoretical modelling efforts. As an example, we provide results on the characterization of porous zirconia samples by a combination of spectrophotometry and non-optical characterization techniques like transmission electron microscopy, energy dispersive x-ray spectroscopy and x-ray reflection.

2.1 Optical and Non-optical Coating Characterization

In the context of the Sect. 1.2, we now define *optical characterization* as a particular case of sample characterization, where the experimental input data stem from any kind of optical measurements. Optical coating characterization is thus based on the interaction of electromagnetic radiation (light) with a specific kind of matter (the *sample*, which is in our case a thin film system). After interaction, certain parameters characterizing the properties of the electromagnetic radiation will have changed in a specific manner – we will use them to judge the particular properties of the sample. In many cases this is possible without any destruction of the sample.

O. Stenzel (✉)
Fraunhofer Institut für Angewandte Optik und Feinmechanik IOF, Albert-Einstein-Str. 7, 07745 Jena, Germany
e-mail: olaf.stenzel@iof.fraunhofer.de; optikbuch@optimon.de

O. Stenzel
Abbe School of Photonics, Friedrich-Schiller-University Jena, Albert-Einstein-Straße 6, 07745 Jena, Germany

O. Stenzel and M. Ohlídal (eds.), *Optical Characterization of Thin Solid Films*, Springer Series in Surface Sciences 64,
https://doi.org/10.1007/978-3-319-75325-6_2

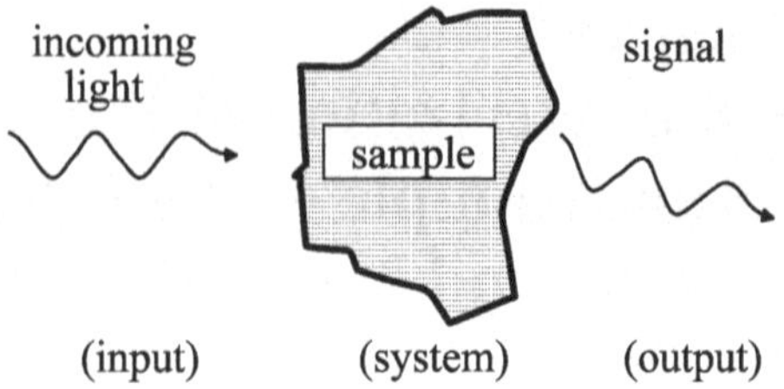

Fig. 2.1 Optical signal as the result of interaction of an electromagnetic wave with the sample

Let us start our discussion from a monochromatic plane light wave incident on a sample. In a complex notation (compare (1.4)), the electric field strength of the wave is given by (Fig. 2.1):

$$\boldsymbol{E} = \boldsymbol{E}(t, \boldsymbol{r}) = \boldsymbol{E}_0 \mathrm{e}^{-i(\omega t - \boldsymbol{kr})} \tag{2.1}$$

In usual characterization practice, the parameters characterizing the incident light may be supposed to be known. For example, we may illuminate the sample with monochromatic light of a well-defined frequency. But the signal may contain new frequencies, obtained as the result of the light-matter interaction. These new frequencies may arise from luminescence, from spontaneous Raman scattering, or from various nonlinear optical processes. But anyway they contain valuable information about the specifics of the sample.

The same applies to other parameters of the incident light. The intensity of the light may change, for examples as a result of absorption processes. Intensity changes are detected by means of photometric techniques. The interaction processes may further result in changes of the polarization state of the light, as detected by ellipsometry and used for further sample characterization. The simple refraction process at the interface between two transparent media is a manifestation of changes in the propagation direction of the light and provides access to differences in the refractive index.

Thus, the diversity of parameters which define the properties of light (in reality they are more than those mentioned here for the model case of an incident plane monochromatic wave) gives rise to quite diverse measurement techniques in optical coating characterization. Some major optical techniques, which are in the focus of this book, are summarized in Table 2.1.

After having clarified what is meant with the terminus “optical characterization”, it is clear that “non-optical characterization” may be defined as a sample characterization process based on non-optical measurement techniques. Although the non-optical techniques definitely constitute a powerful set of individual characterization techniques, we will mainly use them as a source of side information for use in optical characterization tasks.

Clearly, the distinction between optical and non-optical measurement techniques is of relative nature. For example, in a strong sense, X-ray-reflectometry XRR should be regarded as an optical technique as well, because it is entirely based on the interaction of (X-ray) photons with matter. Nevertheless, when primarily being used

Tab.2.1 Survey of selected experimental techniques, used in optical characterization and discussed further in this book

Measurement technique	Chapter
Ex situ spectrophotometry (MIR/NIR/VIS/UV): measurement of transmittance and reflectance at different angles of incidence	2, 5–8, 10
In situ spectrophotometry (NIR/VIS): measurement of transmittance and/or reflectance (usually at a fixed angle of incidence)	7, 8
Cavity ring-down decay CRD: accurate measurement of highest transmittance or reflectance values	16
Absorption measurements, Laser induced fluorescence LIF: Determination of small absorption losses	15
Elastic light scattering: determination of small amounts of scattered light	14
Variable angle spectroscopic ellipsometry VASE	9, 10

Tab.2.2 Survey of selected non-optical characterization techniques

Technique	**Information gained**	**Chapter**
X-ray reflectometry XRR	Density, surface and interface roughness, layer thickness	2
Scanning electron microscopy SEM	Surface topology	12, 13
Energy dispersive X-ray spectroscopy EDX	Atomic composition	2
Scanning probe microscopy	Surface profile, Power spectral density, surface roughness	11, 14
Transmission electron microscopy TEM	Film Morphology	2, 13

as a source of side information in optical characterization focused on other spectral regions, it is often tackled as a non-optical technique.

Table 2.2 summarizes important classes of non-optical coating characterization techniques, with relevance to the topic of this book.

As an example for the value of non-optical side information, Fig. 2.2 shows the cross-sectional transmission electron microscopic (TEM) image of an approximately 200 nm thick zirconium oxide thin film [2]. Note that the morphology of that real film is quite different from the idealized picture initially developed in Fig. 1.1. In Fig. 2.2, one can directly recognize small elongated pores in the film (these are the bright features, the solid fraction appears grey). These pores do not only result in a reduction of the mean density of the film, but also cause certain surface roughness (Chap. 14) as well as a gradient in the refractive index (Chap. 10). Such side information is utmost important in coating characterization practice, because it is extremely helpful for the correct interpretation of the measured optical spectra. Therefore, spectrophotometric characterization always benefits from accompanying non-optical characterization. We will return to this later in Sect. 2.3.

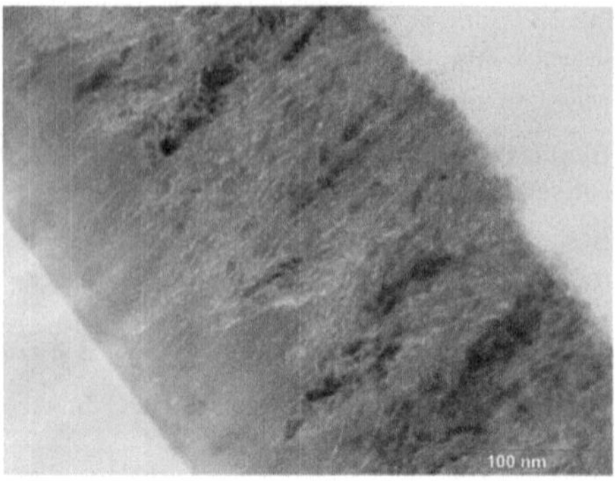

Fig. 2.2 TEM cross sectional image of a porous zirconium oxide thin film. The pores can be well identified as elongated bright structures. Left: substrate side. The image is courtesy of Johannes Biskupek and Ute Kaiser, Ulm University

2.2 Optical Characterization Based on Intensity Measurements

We will proceed with a characterization example based on spectrophotometric measurements. In this case, it is the intensity of the light that has to be measured and discussed in order to gain information on the sample properties. The intensity I of the light is defined as the amount of light energy penetrating a unit surface area per unit time interval. In the case that the electric field of the wave is written in the complex notation according to (2.1), the expression for the light intensity in SI units is obtained as given in (2.2) [1, 3, 4]

$$I = 2\frac{n}{c\mu_0}|E_0|^2 \tag{2.2}$$

Note that in the real notation (compare (1.1)), we have $E_{real}(t, \boldsymbol{r}) = E_{0,real}\cos(\omega t - \boldsymbol{kr} + \delta_0)$, and the expression for the intensity changes to

$$I = \frac{n}{2\mu_0 c}\left|E_{0,real}\right|^2 \tag{2.2a}$$

The *transmittance* T and *reflectance* R of the light are defined through the directed transmitted (I_T) or specularly reflected (I_R) light intensities, divided by the intensity of the incident light (I_E):

$$T \equiv \frac{I_T}{I_E}$$
$$R \equiv \frac{I_R}{I_E} \tag{2.3}$$

As soon as the thin film has been prepared on a transparent substrate, the spectrally resolved measurement of T and R (at any chosen incidence angle φ and any required polarization state of the incident light) appears as a widely used straightforward

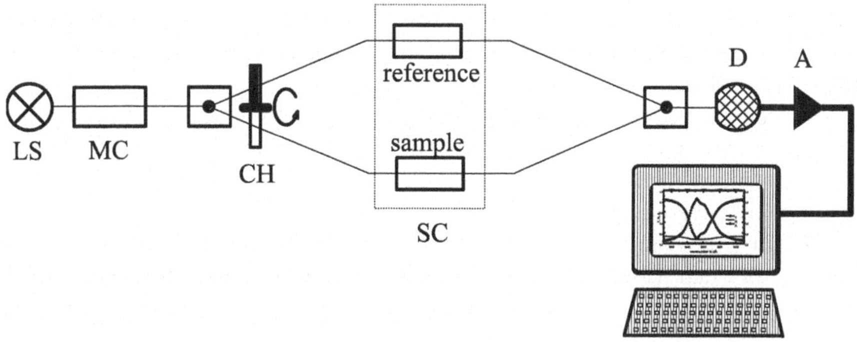

Fig. 2.3 Principle scheme of a double beam dispersive spectrophotometer; (LS) light source, (MC) monochromator, (CH) chopper, (SC) sample compartment, (D) detector, (A) amplifier

characterization tool (Chaps. 7 and 8). Alternatively, spectrally resolved ellipsometric measurements become more and more frequently used in coating characterization practice (Chaps. 9 and 10).

It is a relative advantage of spectrophotometry, that the measurement of both T and R under rather identical conditions gives direct access to the optical loss L, which is composed from total scatter TS and absorptance A. Thus, we find from energy conservation:

$$1 - T - R = L = TS + A \tag{2.4}$$

Details concerning sophisticated measurements of T, R, A, and TS form the content of the later book Chaps. 8 and 14–16. Let us mention here that measurements of the transmittance are most easily accessible, because transmission spectrophotometers belong to the commercially available standard equipment in many labs today. The construction principle of a typical dispersive spectrophotometer is shown in Fig. 2.3.

For many spectrophotometers, suitable specular reflectance accessories are optionally available, so that the measurement of T and R is in most cases accessible. From (2.4), the full optical loss may be determined straightforwardly.

Discrimination between absorption and scatter losses needs additional effort. Nevertheless, some *indications* of surface scatter may already be drawn from the T-and R-spectra: If the first surface of a dielectric sample is rough, both the directed transmittance and the specular reflectance gradually decrease down to zero with increasing frequency. Absorption, on the contrary, tends to show more complicated frequency behaviour (absorption lines, absorption edges). Moreover, absorption in a smooth sample suppresses the transmittance, while the reflectance remains finite because of the Fresnel reflection at the first sample surface.

In the case of small optical losses (typically below 0.01 or 1%), the application of (2.4) is no more suitable for loss determination because of the inaccuracies in T- and R-measurements. It is then preferred to measure scatter and absorptance directly.

Scatter losses are usually measured by means of integrating spheres, and the state of the art in the measurement and interpretation of light scattering will form the topic of Chap. 14.

The accurate measurement of smallest *absorption losses* is based on the principle, that any energy portion initially accumulated in the sample as the result of light absorption must either enhance the sample temperature or leave the sample in order to re-establish thermodynamic equilibrium conditions. So that the idea is to make use of energy relaxation processes in order to detect the amount of absorbed light intensity. In Chap. 15, absorption measurements will be discussed in detail.

2.3 Characterization Example: Interplay Between Optical and Non-optical Methods

2.3.1 Optical Constants

We will now exemplify the interplay between spectrophotometry and selected non-optical characterization techniques in application to porous zirconia thin films, such as shown in Fig. 2.2. Let it be the task to estimate the packing density or alternatively the degree of porosity in such a film. What we will present here is a rather condensed version of a treatment that is explained in full detail in [2]. In principle contrast to [2], we will emphasize the interaction between optical and non-optical characterization methods, instead of emphasizing the interplay between deposition parameters and layer properties.

In the course of optical thin film characterization, the first step may be to fit experimental transmission and reflection spectra such as shown in Fig. 2.4 (left) in order to obtain the linear optical constants n and k as shown in Fig. 2.4 (right).

Figure 2.4 presents experimental spectra which stem from a real characterization experiment. The zirconium oxide thin films have been prepared by plasma ion assisted electron beam evaporation on two different substrates, namely fused silica and both-side polished silicon wafers [5]. (Near) normal incidence VIS/UV T- and R-spectra of samples deposited on fused silica have been measured with a dispersive spectrophotometer. In addition, the MIR T- and R-spectra of the samples on silicon have been recorded by means of an FTIR spectrometer. A representative set of such spectra is shown in Fig. 2.4 on left.

This is a typical experimental input a researcher might have in order to perform an optical characterization of the film(s). Assuming for simplicity that the films optical constants do not depend on the substrate (that is a severe simplification!), and that there are no prominent spectral features in the films optical response in the NIR (a rather feasible assumption), the optical constants of our zirconia samples may be determined by means of a spectra fitting procedure assuming a multi oscillator model for describing the dispersion of the optical constants [1]. As a result of the fit in terms of the LCalc software [6], we obtain the optical constants as shown in Fig. 2.4 on

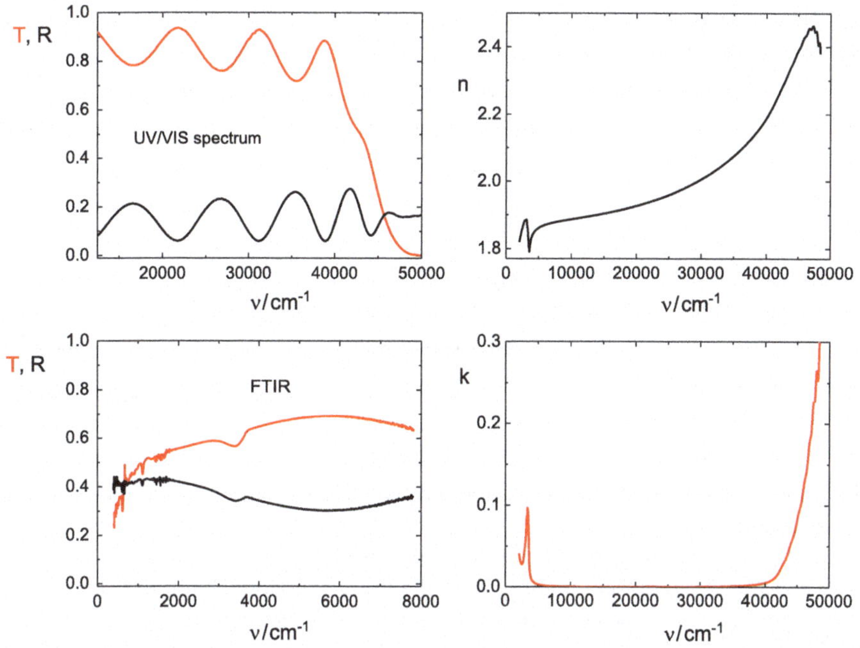

Fig. 2.4 Measured spectra and optical constants of zirconia (ZrO_2) films: left on top: T- and R-spectra of a film deposited on fused silica; left on bottom: the same as deposited on a silicon wafer; right on top: refractive index n, right on bottom: Extinction coefficient k

right, and a film thickness of approximately 240 nm. Let us note at this point that the oscillator model provides a robust but sometimes inconvenient tool for modelling the dispersion of optical constants of solids, some more refined dispersion models such as the Universal Dispersion Model UDM or beta distributed oscillator model (β_do model) will be introduced and discussed in the Chaps. 3, 7 and 8 of this book.

The obtained optical constants show a rather typical behaviour. The extinction coefficient reveals a broad transparency region, where the refractive index shows normal dispersion. In the UV, the extinction increases because of the optical excitation of valence electrons, which results in anomalous dispersion in the refractive index.

The absorption feature in the MIR is of obvious extrinsic nature. As a particular conclusion from the TEM image shown in Fig. 2.2, it makes sense to tackle the real film as a binary mixture, composed from a solid and a pore fraction. Stored in normal ambient, the pores are expected to be filled with water, and indeed, a comparison of the extinction shown in Fig. 2.4 with the optical constants of water (Fig. 2.5) [7] suggests, that there is a water contribution to the optical response of the real film. Clearly, the binary mixture model is the utmost simplest model assumption, because neither the possible co-existence of different solid phases nor a possibly different filling state of the pores can be taken into account explicitly this way. But

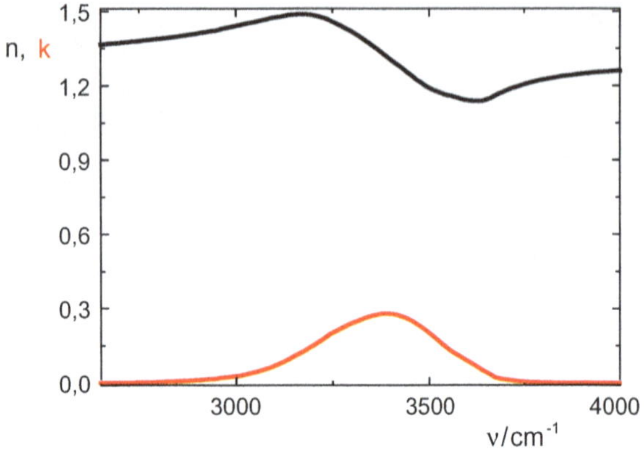

Fig. 2.5 Optical constants of water from 2650 to 4000 cm^{-1} [7]

this simplest model will give access to understanding major features in the behavior of the coatings.

So that as a working hypothesis, we will postulate that the optical constants as shown in Fig. 2.5 represent the response of a mixture coating, built up from a solid and a pore fractions. It is therefore worth having a look on frequently applied mixing models.

2.3.2 *Relation of Optical Constants to Porosity: Mixing Models*

Let us start from the general mixing formula [8]:

$$\frac{(\varepsilon_{eff} - \varepsilon_h)}{\varepsilon_h + (\varepsilon_{eff} - \varepsilon_h)L} = \sum_j p_j \frac{(\varepsilon_j - \varepsilon_h)}{\varepsilon_h + (\varepsilon_j - \varepsilon_h)L} \tag{2.5}$$

Here L is the so-called depolarization factor, which depends on the shape and orientation of the inclusions. p_j is the volume filling factor of the j-th type of inclusion with dielectric function ε_j, and ε_h being the dielectric function of the material which is supposed to act as the host for the inclusions. ε_{eff} denotes the effective dielectric function of the mixture [1, 8]. From the effective dielectric function, the effective optical constants n_{eff} and k_{eff} are calculated in the usual way by means of (2.6):

$$n_{eff} + \mathrm{i}k_{eff} = \sqrt{\varepsilon_{eff}} \tag{2.6}$$

When regarding the real film shown in Fig. 2.2 as a mixture of pores and a solid fraction, (2.5) provides a vehicle for understanding the impact of pores on the films optical constants in a quantitative manner.

For spherical inclusions, $L = 1/3$. Note that $\varepsilon_h = 1$ results in the Lorentz–Lorenz mixing approach, while for $\varepsilon_h = \varepsilon_{eff}$, from (2.5) the Bruggeman approach (also called EMT = Effective Medium Theory or EMA = effective medium approximation) is obtained. In the case that one of the mixing partners itself acts as the host medium (guest-host-system), we arrive at the Maxwell Garnett approach.

Note that physically reasonable values of L are confined by the bounds $L = 0$ and $L = 1$. Regardless of the choice of ε_h, in these cases, (2.5) results in the simpler relationships

$$L = 0 \Rightarrow \varepsilon_{eff} = \sum_j p_j \varepsilon_j \tag{2.7}$$

$$L = 1 \Rightarrow \varepsilon_{eff}^{-1} = \sum_j p_j {\varepsilon_j}^{-1} \tag{2.8}$$

In the case of a binary mixture with real and positive dielectric functions, (2.7) and (2.8) form the Wiener bounds to the dielectric function of the binary mixture.

On the other hand, expressions (2.7) and (2.8) may be regarded as the limiting cases of a class of more empirically formulated mixing models, as described by (2.9):

$$\varepsilon_{eff}^{\beta} = \sum_j p_j \varepsilon_j^{\beta}; \qquad (-1 \le \beta \le 1) \tag{2.9}$$

Equation (2.9) is a general writing of the Lichtenecker mixing formula [9], for $\beta = \pm 1$ we again obtain (2.7) and (2.8). $\beta = ½$ corresponds to a mixing model where the refractive indices of the mixing partners are linearly superimposed; such an approach has proven useful in the design of rugate filters built up from mixtures of oxide coatings [10]. For $\beta = 1/3$, (2.9) results in the Looyenga mixing formula, which may be exactly derived for isotropic inclusions and the assumption of only small differences between the individual values of ε_j [11].

2.3.3 *Application to a Porous Film*

When assuming a columnar film structure built up from free-standing cylinders, the mixing model of Bragg and Pippard may be a good choice for modelling the complex index of refraction $\hat{n}$ of the mixture. This model is obtained from (2.5) when considering a binary mixture where the pore fraction forms the host medium, and the solid fraction the inclusion (guest). In films with columnar structure, it is reasonable setting $L = ½$. Then from (2.5) we obtain

$$\hat{n}^2 = \frac{(1 - p_s)\hat{n}_v^4 + (1 + p_s)\hat{n}_v^2\hat{n}_s^2}{(1 + p_s)\hat{n}_v^2 + (1 - p_s)\hat{n}_s^2} \tag{2.10}$$

Here, p_s denotes the volume filling factor of the solid fraction (or packing density), while $\hat{n}_s$ is the complex refractive index of the solid fraction. $\hat{n}_v$ is the complex refractive index of the pore or void fraction, the corresponding porosity is 1-p_s. Knowledge of $\hat{n}_v$ and $\hat{n}_s$ thus allows determining the packing density from measured values of the complex refractive index of the real film, hence it provides access to information about the microstructure of the film.

We will present three different optical approaches for estimating the porosity:

(a) ***Estimation of the packing density from the refractive index in the visible spectral region***

In this case, all optical constants may be regarded as real, and the packing density may be directly determined from inverting (2.10):

$$p_s = \frac{(n_v^2 - n^2)(n_v^2 + n_s^2)}{(n_v^2 - n_s^2)(n_v^2 + n^2)} \tag{2.11}$$

When n was determined at atmosphere ($n = n_{atm}$), it is reasonable assuming that the pores are filled with water, i.e. $n_v = 1.33$. As an approximation for n_s, one can take the refractive index of a rather dense film.

(b) ***Estimation of the packing density from the Thermal and Vacuum spectral Shift***

Generally, when a porous coating is stored in atmospheric conditions, after a certain time most of its pores are filled with water. Shift measurements [12] may be used to judge this amount of water. The basic idea of our shift measurements is to compare the transmission spectrum of a sample measured first in atmospheric conditions at room temperature with that measured after evacuation of the sample chamber in vacuum at 100 °C. Figure 2.6 shows the result of such a shift measurement performed with a zirconia film on fused silica. For details of the measurement equipment, the reader is referred to Chap. 8.

Both water migration and sample heating cause a change in the coatings optical thickness. This may be quantified in terms of the shift of the wavelength λ_m of a selected interference structure in the normal incidence spectrum according to:

$$\begin{aligned} shift &\equiv \frac{\Delta(nh)}{nh} * 100\% = \frac{\lambda_{m,vacuum,100^\circ C} - \lambda_{m,atmosphere,roomtemperature}}{\lambda_{m,atmosphere,roomtemperature}} * 100\% \\ &= \frac{\Delta n}{n} * 100\% + \frac{\Delta h}{h} * 100\% \end{aligned} \tag{2.12}$$

Here Δn is the change in refractive index mainly caused by water migration from the coating. When further setting $\Delta h/h$*100% $\approx$ 0.2% in order to account for the thermal expansion, we find the film refractive index in vacuum conditions (at 100 °C) n_{vac} according to:

$$n_{vac} = n_{atm} + \Delta n = n_{atm}\left(1 + \frac{shift - 0.2\%}{100\%}\right) \tag{2.13}$$

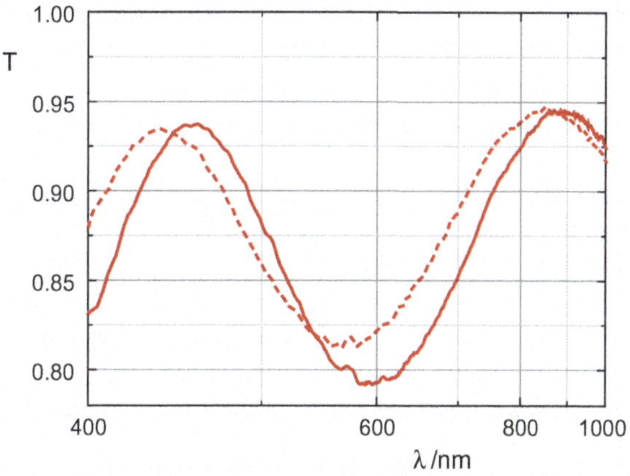

Fig. 2.6 Shift measurement: transmittance of a zirconia film at room temperature at atmosphere (solid line), and at 100 °C in vacuum (dashed line)

n_{atm} is the refractive index as determined in atmosphere at room temperature. In fact we assume, that during the short evacuation time relevant for the shift measurement, not all the water will leave the coating. Instead, it may be only a part of the pores which lose their water, and we will call these pores for simplicity the "larger pores". Other pores remain filled with water, and we well call them the "smaller pores". The solid fraction of the film and the smaller pores thus form a sub-system that practically does not change its refractive index n_O during the shift measurement.

In terms of (2.10) and (2.13), the following expression is obtained for n_O:

$$n_0^2 = \frac{(n_v^2-1)(n_{atm}^2 n_{vac}^2 - n_v^2)}{2(n_v^2 n_{vac}^2 - n_{atm}^2)} \pm \sqrt{\left[\frac{(n_v^2-1)(n_{atm}^2 n_{vac}^2 - n_v^2)}{2(n_v^2 n_{vac}^2 - n_{atm}^2)}\right]^2 + n_v^2}\,\Bigg|_{n_v=1.33} \tag{2.14}$$

We then find for the "large pores" (compare (2.11)):

$$p_{''large\ pores''} = 1 - \frac{(n_v^2 - n_{atm}^2)(n_v^2 + n_0^2)}{(n_v^2 - n_0^2)(n_v^2 + n_{atm}^2)}\Bigg|_{n_v=1.33} = 1 - \frac{(1 - n_{vac}^2)(1 + n_0^2)}{(1 - n_0^2)(1 + n_{vac}^2)} \tag{2.15}$$

According to (2.11), the subsystem formed from the solid fraction and the smaller pores has a packing density given by:

$$p_{subsystem} = \frac{(n_v^2 - n_0^2)(n_v^2 + n_s^2)}{(n_v^2 - n_s^2)(n_v^2 + n_0^2)}\Bigg|_{n_v=1.33} \tag{2.16}$$

So that for the full packing density p_s as defined through (2.10) or (2.11) we find:

$$p_s = (1 - p_{''large\ pores''})p_{subsystem} \tag{2.17}$$

Finally we calculate the volume factor corresponding to the fraction of "smaller pores" from (2.15) and (2.17):

$$p_{''small\ pores''} = 1 - p_{''large\ pores''} - p_s = (1 - p_{''large\ pores''})(1 - p_{subsystem}) \quad (2.18)$$

In our terminology, the value of the combined discussion of shift and refractive index is in the possibility to discriminate between rather large and rather small pores. Of course, the full porosity is:

$$p_{''pores''} = p_{''small\ pores''} + p_{''\text{large pores}''} \quad (2.19)$$

(c) ***Estimation of the water content from Infrared Spectroscopy***

Water shows characteristic absorption bands in the infrared, and we will focus on the features caused by O-H stretching vibrations around the wavenumber of 3400 cm^{-1}. From the infrared spectra, we will therefore estimate the water content in the films again in terms of (2.10), but now making use of the *imaginary part* of the corresponding optical constants. In this case, we explore the pronounced absorption structure which is shown in Fig. 2.4 right on bottom in the MIR spectral region.

The estimation of the water content is then based on an integral approach according to [12]:

$$p_{H_2O,FTIR} \propto \int_{\nu_1}^{\nu_2} n(\nu)\alpha(\nu)d\nu \quad (2.20)$$

with $\alpha = 4\pi\nu k$ (compare (1.8)) and $\nu_1 = 2650\ cm^{-1}$; $\nu_2 = 4000\ cm^{-1}$. The method is calibrated by simulations in terms of (2.10). That leads to the following picture for the integrand in relation (2.20) (Fig. 2.7):

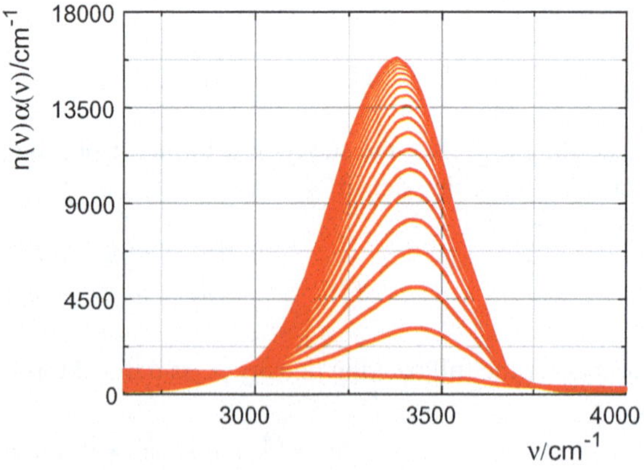

Fig. 2.7 Simulation: the product n(ν)α(ν) of the densest zirconia film (the line closest to the bottom of the figure) intermixed with optical constants of water (increment 5%)

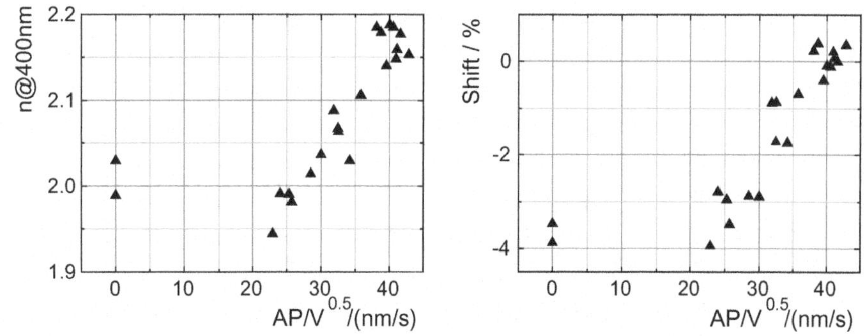

Fig. 2.8 Refractive index and shift of zirconia films, deposited at different levels of assistance

After having performed that calibration, the water content in the film may be estimated from experimental optical constants according to (2.20).

Because of possible other contributions to the MIR absorption (any OH-, NH-, or CH-infrared active vibrations originating from plenty of possible contaminations), our FTIR treatment rather serves as an estimation of the upper limit of the water content. Hence, $p_{H_2O,FTIR} > 1 - p_s$ when being determined from (2.20) at atmospheric conditions.

Results

Figure 2.8 presents experimental results on the refractive index at a wavelength of 400 nm, as well as the measured shift. On the abscissa, an assistance parameter AP is used for quantifying the effect of plasma assistance during film growth. It is calculated as the square root of the BIAS voltage [5], divided by the average growth rate [12].

From Fig. 2.8 it is evident, that both refractive index and shift are strongly influenced by the level of assistance. Strongly assisted films appear to have highest refractive indices and a negligible shift; hence we expect that their porosity is lowest. The application of our theoretical apparatus leads to the following results for the porosity (Fig. 2.9):

We notice that both the treatments (a) and (b) (Fig. 2.9 on left) deliver a dependence of the full porosity on the assistance parameter that is in good qualitative agreement to what is obtained from method c) (FTIR). Nevertheless, as expected, the FTIR "porosity" values are found to be larger than the porosity values obtained from the VIS refractive indices. This is not only caused by the mentioned sensitivity of method c) to different kinds of contaminations; in fact a more refined treatment shows that the samples on silicon (used for the FTIR investigations) indeed have another (in most cases a higher) porosity than those on fused silica [2]. So that our initial assumption of optical film constants that do not depend on the substrate was helpful for getting a first understanding of rough trends concerning the relation between assistance and porosity; a finer quantitative treatment would require discussing samples on fused silica and silicon separately. It turns out that optical thin film characterization is

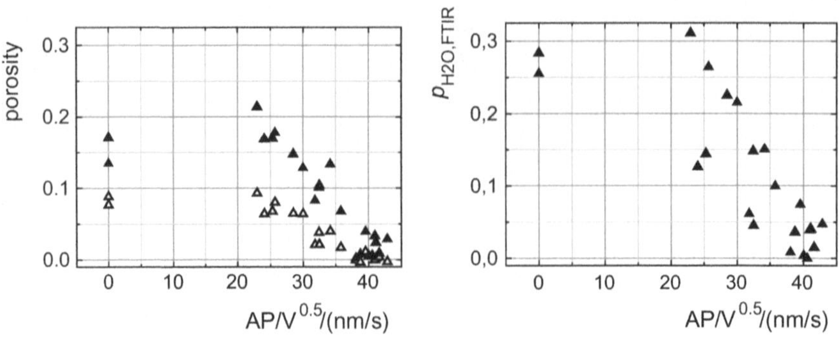

Fig. 2.9 Porosity as obtained from optical measurements; on left: full triangles: full porosity from (2.11) or (2.19), open triangles: amount of small pores according to (2.18); on right: $p_{H2O,FTIR}$ as calculated from (2.20)

often an iterative procedure: simplest initial models reveal robust general trends, while remaining inconsistencies indicate the necessity to use more refined models at a later stage in order to understand the nature of quantitative details.

What is most astonishing, however, is the rather large amount of pores that are too small to contribute to the shift measurement or contain contaminations with a rather low electronic polarizability that are chemically bound to the zirconia network (Fig. 2.9 on left, open triangles). From the FTIR data, but also from the deposition history of the coatings, one should expect that these small pores contain some kind of OH groups, but our optical measurements cannot validate this assumption. This is the point where additional, non-optical characterization tools become useful. In our case it turned out to be helpful investigating the stoichiometry of the samples by Energy Dispersive X-ray spectroscopy EDX [13].

Stoichiometric zirconia films contain 66.7at.% oxygen. But oxygen or water molecules incorporated into the pores, as well as chemically bound hydroxyl groups will result in deviations of the oxygen content from the value of 66.7at.%. And this is really observed (Fig. 2.10). Once EDX measurements are usually performed in vacuum conditions, water molecules contained in large pores are not expected to contribute to the EDX signal. The excess oxygen found by EDX (Fig. 2.10) should therefore stem from smaller pores or diverse chemically bound contaminations.

We note that an increase in AP results in a reduction in that excess oxygen contribution (Fig. 2.10), in good accordance to the results calculated from the shift measurements as shown in Fig. 2.9 (open triangles). Note that excess oxygen in zirconia films has been reported in earlier studies too, compare for example [14], where the oxygen concentration has been determined by Rutherford backscattering.

We come to the result, that all the data concerning optical constants, shift, as well as the oxygen concentration reveal a consistent picture of the water content in the films and their packing density.

The later statement may be validated by an independent measurement as well. In that connection, Fig. 2.11 shows the packing density of zirconia films as determined

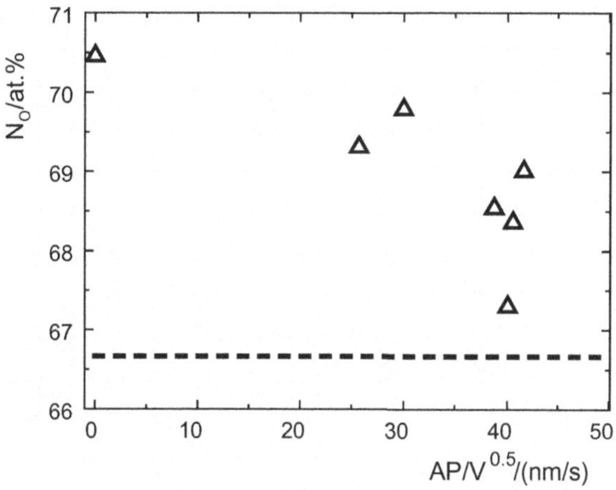

Fig. 2.10 Oxygen atomic concentration inside selected zirconia layers measured by EDX depending on the associated assistance parameter

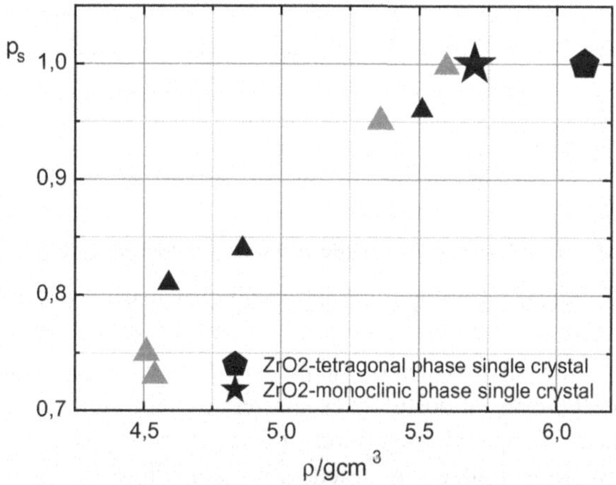

Fig. 2.11 Relation between zirconia packing density (2.11) compared to the mass density ρ obtained by XRR. Grey triangles: samples on silicon; black triangles: samples on fused silica. Crystal data are indicated for reference purposes

by (2.11) opposed to the films mass density ρ as determined by X-Ray-Reflectometry XRR.

From Fig. 2.11 we recognize the excellent agreement between the mass density and the packing density as obtained from the refractive index in the VIS. The experimental dependence even converges to the crystalline value relevant for the monoclinic zirconia crystal phase.

Of course, the distinction between small and large pores as introduced in our treatment is of relative nature. It is clearly the duration of the shift measurement that does matter when discriminating between "large" and "small" pores: When waiting long enough, even smaller pores may be evacuated when the sample is held in vacuum conditions. A few quantitative considerations on this subject will define the content of the next section.

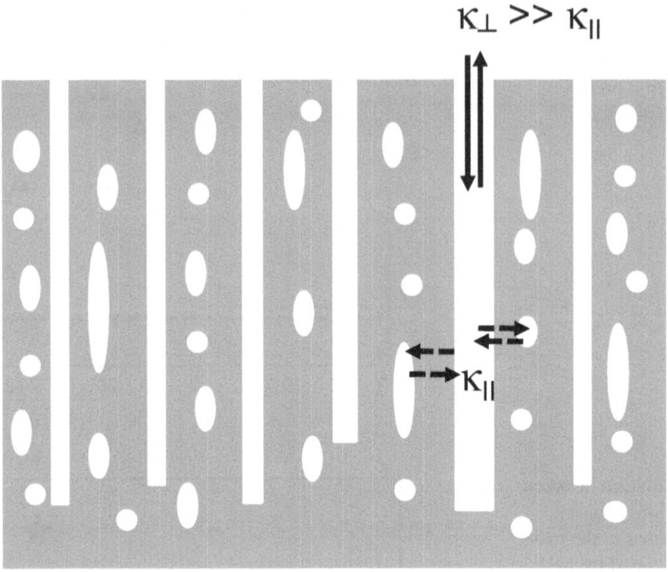

Fig. 2.12 Model of a porous coating

(d) ***Discussion: A simple model of water migration in a porous coating***

Let us now develop a simple model idea on the water migration kinetics in our porous film.

The general idea of the proposed model is schematically presented in Fig. 2.12. As before, we postulate co-existence of different kinds of pores in the film, namely large elongated open pores and small, rather isolated and (nearly) closed pores. In principle, both can be filled with water, while the filling rates κ are quite different for the different types of pores. Indeed, let us make the following model assumptions for the filling kinetics:

(i) "Large pores" interchange water directly with the ambient with a filling or evacuation rate $\kappa_{\perp}$ (solid arrows in Fig. 2.12). We apply this terminus predominantly to elongated cylindrical pores which typically have diameters in the region 1–3 nm and are open to the film surface.

(ii) "Small pores" interchange water with large pores with a filling or evacuation rate κ_{II} (dashed arrows in Fig. 2.12).

(iii) While $\kappa_{\perp}$ appears to be determined by the interplay of capillary and liquid friction forces [15, 16] within the rather cylindrical large pores, κ_{II} is rather defined by diffusion of O, H, OH, or H_2O species through a defected atomic network. We therefore make the reasonable assumption $\kappa_{\perp} >> \kappa_{\text{II}}$.

Let us assume that the sample was held at atmosphere for a time long enough so that all pores are essentially filled with water. Let y express the full water content in the film, y_l the degree of filling of the large pores, and y_s that of the small pores.

Tab.2.3 Water content in pores at different evacuation times

Time regime	Condition	Water content $y(t)$ according to (2.22)
0	$t = 0$	$y(t) = p_{''\text{small pores}''} + p_{''\text{large pores}''}$
I	$t << \kappa_{\perp}^{-1}$	$y(t) \approx p_{''\text{small pores}''} + p_{''\text{large pores}''}$
II	$\kappa_{\perp}^{-1} << t << (p_{''\text{large pores}''}\kappa_{\text{II}})^{-1}$	$y(t) \approx p_{''\text{small pores}''}$
III	$(p_{''\text{large pores}''}\kappa_{\text{II}})^{-1} << t$	$y(t) \rightarrow 0$

Let us further assume, that at the moment $t = 0$, the film is brought into vacuum. So the surrounding is practically free of water, which is controlled by an ambient parameter y_0 that is set equal to zero. Hence, the evacuation kinetics of the film may be described by the following system of rate equations:

$$\begin{aligned} &p_{''\text{large pores}''}\frac{dy_l}{dt} = p_{''\text{large pores}''}\kappa_{\perp}(y_0 - y_l) - p_{''\text{small pores}''}\frac{dy_s}{dt} \\ &\frac{dy_s}{dt} = p_{''\text{large pores}''}\kappa_{\text{II}}(y_l - y_s) \\ &y_0 = 0;\ y_l(t = 0) = 1;\ y_s(t = 0) = 1 \end{aligned} \tag{2.21}$$

Here, $p_{''\text{large pores}''}$ is the volume fraction occupied by the large pores, and $p_{''\text{small pores}''}$ that of the small pores. As a solution of (2.21), the full amount of water $y(t)$ is then obtained as given in (2.22) (compare Chap. 8):

$$\begin{aligned} y(t) &\equiv p_{''\text{small pores}''}y_s(t) + p_{''\text{large pores}''}y_l(t) \\ &\approx p_{''\text{small pores}''}e^{-p_{''\text{large pores}''}\kappa_{\text{II}}t} + p_{''\text{large pores}''}e^{-\kappa_{\perp}t} \end{aligned} \tag{2.22}$$

In a moderately porous layer, large rather cylindrical pores are likely to exist (compare Figs. 1.2 and 2.2). To our experience, in suchlike coatings $\kappa_{\perp}^{-1}$ is typically of the order of a few minutes, while $(p_{''\text{large pores}''}\kappa_{\text{II}})^{-1}$ corresponds to several ours or even days. Therefore, a typical shift measurement rather detects effects that are connected to the evacuation of the „large open pores“. Some more detailed information on the output of (2.22) in different time regimes is condensed in Table 2.3.

FTIR and refractive index measurements of samples that have been stored at atmosphere correspond to measurements in the time regime 0 or I, hence they give access to the full porosity $p = p_{''\text{small pores}''} + p_{''\text{large pores}''}$ (Fig. 2.9). Our shift measurements detect a change in water content between the time regimes 0 and II, so that it is practically the volume fraction of large pores that is quantified by the shift measurement. The comparison with the full porosity allows then estimating the amount of small pores (Fig. 2.9. on left). Our EDX measurements present a snapshot of the oxygen concentration in regime II, so it is the fraction of smaller pores that is accessible this way (Fig. 2.10). Time regime III is rather responsible for remaining storage and aging effects.

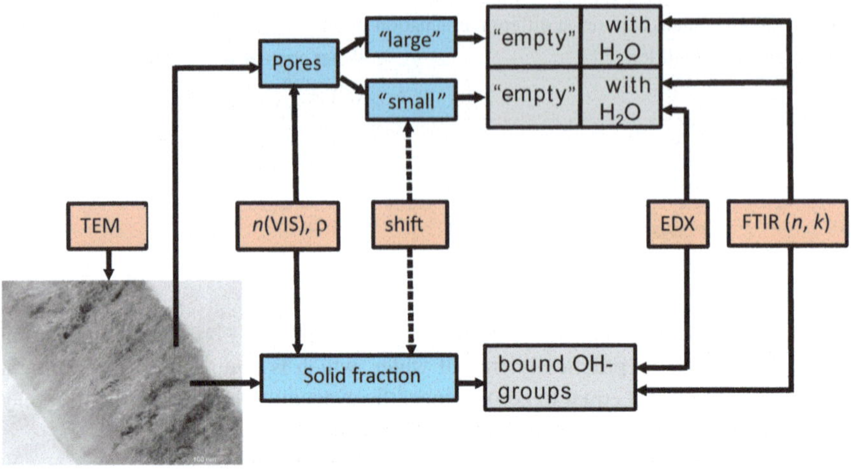

Fig. 2.13 Relation of the discussed characterization tools to the porosity of the zirconia films

2.4 Conclusions

Before concluding this chapter let us remark, that the combination of optical spectroscopy (VIS, MIR) with EDX and XRR reveals a consistent picture of the porosity of zirconia films with a microstructure that is far from the ideal homogeneous and isotropic layer model. In summarizing, Fig. 2.13 shows the interaction of the individual characterization tools as exemplified in this chapter in a schematic manner.

Figure 2.13 demonstrates the relative information obtained from the different characterization techniques. While the VIS refractive index and the mass density obtained by XRR may give, in principle, access to the full porosity, the FTIR method detects only those pores which contain any kind of hydroxyl groups. Shift measurements allow discriminating between pores which allow water migration (large or open pores) and pores which do not (small or closed pores). EDX as performed in vacuum conditions gives a rough measure of any hydroxyl groups that do not migrate during evacuation. So neither of the methods may be called a favorite one, it is in fact their combination which allows estimating amount and kind of porosity of a real sample.

Acknowledgements The author wants to express his gratitude to the PhD student Christian Franke, who performed a significant amount of porosity calculations, and prepared Fig. 2.7 in the course of his PhD thesis work.

References

1. O. Stenzel, The Physics of Thin Film Optical Spectra: An Introduction. Springer Series in Surface Sciences, vol. 44, 2. edn (Springer, Berlin, 2015)
2. C. Franke, O. Stenzel, S. Wilbrandt, S. Schröder, L. Coriand, N. Felde, A. Tünnermann, Porosity and optical properties of zirconia films prepared by plasma ion assisted deposition. Appl. Opt. **56**, 3913–3922 (2017)
3. M. Born, E. Wolf, *Principles of Optics* (Pergamon Press, Oxford, 1968)
4. H. Paul, Photonen. Eine Einführung in die Quantenoptik (Teubner Stuttgart, 1995)
5. S. Pongratz, A. Zöller, Plasma ion assisted deposition: a promising technique for optical coatings. J. Vac. Sci. Techn. A **10**, 1897–1904 (1992)
6. O. Stenzel, S. Wilbrandt, K. Friedrich, N. Kaiser, Realistische Modellierung der NIR/VIS/UV-optischen Konstanten dünner optischer Schichten im Rahmen des Oszillatormodells. Vak. Forsch. Prax. **21**(5), 15–23 (2009)
7. H. Downing, D. Williams, Optical Constants of Water in the Infrared. J. Geogr. Res. **80**, 1656–1661 (1975)
8. D.E. Aspnes, J.B. Theeten, F. Hottier, Investigation of effective-medium models of microscopic surface roughness by spectroscopic ellipsometry. Phys. Rev. B **20**, 3292–3302 (1979)
9. T.P. Leão, E. Perfect, J.S. Tyner, Evaluation of Lichtenecker's mixing model for predicting effective permittivity of soils at 50 MHz. Trans. ASABE **58**(1), 83–91 (2015)
10. A.V. Tikhonravov, M.K. Trubetskov, T.V. Amotchkina, M.A. Kokarev, N. Kaiser, O. Stenzel, S. Wilbrandt, D. Gäbler, New optimization algorithm for the synthesis of rugate optical coat-ings. Appl. Opt. **45**, 1515–1524 (2006)
11. L.D. Landau, E.M. Lifschitz, in, Lehrbuch der theoretischen Physik, Bd. VIII: Elektrodynamik der Kontinua [engl.: Textbook of theoretical physics, vol. VIII: Electrodynamics of continuous media] (Akademie-Verlag, Berlin, 1985)
12. O. Stenzel, Optical Coatings Material Aspects in Theory and Practice (Springer, Berlin, 2014)
13. O. Stenzel, S. Wilbrandt, Optical characterization of high index metal oxide films for UV/VIS applications, prepared by Plasma Ion Assisted Deposition, in *Proceedings of SPIE,* vol. 9627, (2015),pp. 962709-1–923709-12
14. E.E. Khawaja, F. Bouamrane, A.B. Hallak, M.A. Daous, M.A. Salim, Observation of oxygen enrichment in zirconium oxide films. J. Vac. Sci. Technol., A **11**, 580–587 (1993)
15. P. Meakin, A.M. Tartakovsky, Modelling and simulation of pore-scale multiphase fluid flow and reactive transport in fractured and porous media. Rev. Geophys. **47**, (2009), https://doi.org/10.1029/2008rg000263
16. K.P. Hapgood, J.D. Litster, S.R. Biggs, T. Howes, Drop penetration into porous powder beds. J. Colloid Interface Sci. **253**, 353–366 (2002)

Chapter 3
Universal Dispersion Model for Characterization of Thin Films Over Wide Spectral Range

Daniel Franta, Jiří Vohánka and Martin Čermák

Abstract The universal dispersion model is a collection of dispersion models (contributions to the dielectric response) describing individual elementary excitation in solids. All contributions presented in this chapter satisfy the basic conditions that follow from the theory of dispersion (time reversal symmetry, Kramers–Kronig consistency and finite sum rule integral). The individual contributions are presented in an unified formalism. In this formalism the spectral distributions of the contributions are parameterized using dispersion functions normalized with respect to the sum rule. These normalized dispersion functions must be multiplied by the transition strengths parameters which can be related to the density of charged particles. The separation of contributions into the transitions strengths and normalized spectral distributions is beneficial since it allows us to elegantly introduce the temperature dependencies into these models.

3.1 Introduction

The dispersion models are used to describe dielectric response of materials. From the theory of dispersion it follows that the dielectric response must fulfill three fundamental conditions [1–7] and one supplementary condition. For isotropic materials without spatial dispersion in thermodynamic equilibrium these conditions can be written for the dielectric function as follows:

1. *Time-reversal symmetry*

$$\hat{\varepsilon}(\omega) = \hat{\varepsilon}^*(-\omega)\,, \tag{3.1}$$

where the star denotes the complex conjugation. This condition ensures that the response function is a real function in time domain.

D. Franta (✉) · J. Vohánka · M. Čermák
Faculty of Science, Department of Physical Electronics,
Masaryk University, Kotlářská 2, 611 37 Brno, Czechia
e-mail: franta@physics.muni.cz

O. Stenzel and M. Ohlídal (eds.), *Optical Characterization of Thin Solid Films*, Springer Series in Surface Sciences 64,
https://doi.org/10.1007/978-3-319-75325-6_3

2. *Kramers–Kronig (KK) relations*

$$\varepsilon_\mathrm{r}(\omega) = 1 + \frac{1}{\pi} \int_{-\infty}^{\infty} \frac{\varepsilon_\mathrm{i}(\xi)}{\xi - \omega} \, \mathrm{d}\xi \,, \tag{3.2}$$

where the subscripts 'r' and 'i' denote the real and imaginary parts and the integral must be understood as the Cauchy principal value. The condition (3.1) allows us to write the KK relation (3.2) as

$$\varepsilon_\mathrm{r}(\omega) = 1 + \frac{2}{\pi} \int_{0}^{\infty} \frac{\xi \, \varepsilon_\mathrm{i}(\xi)}{\xi^2 - \omega^2} \mathrm{d}\xi \,. \tag{3.3}$$

The KK relations follow from the causality of the response function.

3. *Sum rule*

$$\int_0^\infty \omega \, \varepsilon_\mathrm{i}(\omega) \mathrm{d}\omega = \frac{\pi}{2} \omega_\mathrm{p}^2 \,, \tag{3.4}$$

where the constant ω_p is called the plasma frequency and it is related to the electron density $\mathcal{N}_\mathrm{e}$ as follows [7]

$$\omega_\mathrm{p}^2 = \frac{e^2 \mathcal{N}_\mathrm{e} \mathcal{U}}{\epsilon_0 m_\mathrm{e}} \,. \tag{3.5}$$

The dimensionless correction factor $\mathcal{U}$ compensates the influence of the positive nuclei. If the ratio of nucleon number to proton number is assumed be 2:1 then the theoretical value of this factor is $\mathcal{U} = 1.000274$ [7–11]. The sum rule can be derived with the help of the superconvergence theorem, assuming that at high frequencies, the dielectric function is the same for all materials as

$$\hat{\varepsilon}(\omega) \approx 1 - \frac{\omega_\mathrm{p}^2}{\omega^2} \,. \tag{3.6}$$

The term plasma frequency is motivated by the equation above since it corresponds to the model of sparse noninteracting plasma. The sum rule can be also derived on the basis of either classical or quantum laws of motion.

4. *Dissipative system condition*

$$\varepsilon_\mathrm{i}(\omega) \geq 0 \quad \text{for} \quad \omega > 0 \,. \tag{3.7}$$

This condition corresponds to systems in thermodynamic equilibrium.

This chapter will discuss only models of linear dielectric response of nonmagnetic media without spatial dispersion, which satisfy the three fundamental conditions. In principle the tensor of dielectric functions should be used instead of the scalar dielectric function. For simplicity, only results for the scalar dielectric function will be presented, i. e. for isotropic materials. It is, however, not difficult to generalize these models to anisotropic media.

In this chapter, the collection of models suitable for describing various elementary excitations will be presented. A combination of these models can be used for the description of complete dielectric response of a wide range of materials in the spectral range form far IR to X-ray. Therefore, the collection of these models will be further called the universal dispersion model (UDM). The emphasis will be placed on models for which the dielectric function can be calculated analytically. Although this chapter aims to present the dispersion models from the practical point of view, basic knowledge of condensed matter physics will be necessary in order to understand the principles behind the presented models. Thus, in this chapter, the isotropic version of the UDM will be presented.

The Kramers–Kronig relations (3.2) and (3.3) allow us to calculate the real part of the dielectric function if the imaginary part is known. It is even possible to understand these relations in the sense of generalized dielectric functions, i. e. the imaginary part of the dielectric function can contain delta functions or discontinuities. The situation is more complicated for the inverse Kramers–Kronig relation expressing the imaginary part of the dielectric function. The general form of this Kramers–Kronig relation is as follows

$$\varepsilon_{\mathrm{i}}(\omega) = -\frac{1}{\pi} \int_{-\infty}^{\infty} \frac{\varepsilon_{\mathrm{r}}(\xi) - 1}{\xi - \omega} \, \mathrm{d}\xi + \frac{\lim_{\xi \to 0} \xi \varepsilon_{\mathrm{i}}(\xi)}{\omega} \tag{3.8}$$

or

$$\varepsilon_{\mathrm{i}}(\omega) = -\frac{2\omega}{\pi} \int_{0}^{\infty} \frac{\varepsilon_{\mathrm{r}}(\xi) - 1}{\xi^2 - \omega^2} \, \mathrm{d}\xi + \frac{\lim_{\xi \to 0} \xi \varepsilon_{\mathrm{i}}(\xi)}{\omega} . \tag{3.9}$$

The second term in (3.2) and the first term in (3.8) represent the Hilbert transform and the inverse Hilbert transform of the imaginary and real parts of the susceptibility $\hat{\chi}(\omega) = \hat{\varepsilon}(\omega) - 1$. The second term in (3.8) is zero for dielectrics, but it is nonzero for conducting materials. Therefore, for conducting materials it is not possible to calculate the imaginary part of the susceptibility function simply as the inverse Hilbert transform of its real part [6]. This problem can be avoided if the Kramers–Kronig relations are written for the quantity proportional to $\omega\hat{\chi}(\omega)$ (e. g. complex optical conductivity $\hat{\sigma}(\omega) = -\mathrm{i}\omega\epsilon_0\hat{\chi}(\omega)$). For such quantity the Kramers–Kronig relations are given only by the Hilbert transforms.

3.2 Theoretical Background

3.2.1 *Classical Model*

In the beginning, there was the classical model, and the classical model was used for modeling of everything. Within the frame of the classical model, the medium is described as a system of massive charged particles forming quasi-neutral environment. Two types of forces are considered, the restoring forces which act to return the particles to the equilibrium positions and damping forces slowing down the particles. It is not necessary to consider all the particles forming the system. Since identical particles in equivalent configurations exhibit identical behavior, it is possible to consider an equivalently behaving system consisting of m nonequivalent coupled particles. Taking into account the above considerations (in the isotropic case) the problem leads to m linear ordinary differential equations which can be easily solved [12]. The dielectric function of such a system can be written in a closed matrix form

$$\hat{\varepsilon}(\omega) = 1 + \boldsymbol{\omega}_{\rm p}^{\rm T}[\mathbf{S} - \omega^2\mathbf{I} - {\rm i}\omega\mathbf{G}]^{-1}\boldsymbol{\omega}_{\rm p}\,, \tag{3.10}$$

where the column vector $\boldsymbol{\omega}_{\rm p}$ contains plasma frequencies. These are proportional to square roots of the effective densities of individual modes:

$$\boldsymbol{\omega}_{\rm p}^{\rm T} = (\omega_{{\rm p},1}, \omega_{{\rm p},2}, \cdots, \omega_{{\rm p},m})\,, \tag{3.11}$$

where the symbol $^{\rm T}$ denotes transposition. The matrix $\mathbf{S}$ is diagonal with squares of central frequencies on the diagonal (zero frequencies for the description of free carriers are possible):

$$\mathbf{S} = \begin{pmatrix} \omega_{{\rm c},1}^2 & 0 & \cdots & 0 \\ 0 & \omega_{{\rm c},2}^2 & \cdots & 0 \\ \vdots & \vdots & \ddots & \vdots \\ 0 & 0 & \cdots & \omega_{{\rm c},m}^2 \end{pmatrix}, \tag{3.12}$$

$\mathbf{I}$ denotes the unit matrix and $\mathbf{G}$ is real symmetric positive definite matrix

$$\mathbf{G} = \begin{pmatrix} \gamma_{11} & \gamma_{12} & \cdots & \gamma_{1m} \\ \gamma_{12} & \gamma_{22} & \cdots & \gamma_{2m} \\ \vdots & \vdots & \ddots & \vdots \\ \gamma_{1m} & \gamma_{2m} & \cdots & \gamma_{mm} \end{pmatrix}, \tag{3.13}$$

where γ_{lk} are the damping parameters. The positive definiteness of the matrix $\mathbf{G}$ ensures that system is dissipative. The matrix $\mathbf{G}$ can be conveniently parameterized using Cholesky decomposition as

$$\mathbf{G} = \mathbf{B}\mathbf{B}^{\rm T}\,, \tag{3.14}$$

where $\mathbf{B}$ is a lower triangular matrix with positive entries on the diagonal. The classical model contains 3 parameters for each mode determining the frequency, strength and damping plus additional $m(m-1)/2$ parameters determining the coupling between modes, therefore, the total number of dispersion parameters is $3m + m(m-1)/2$.

The dielectric response (3.10) is a general result for the system of damped harmonic oscillators (DHO) with coupled modes. In practice, it is often possible to assume that the oscillators are independent. If this is the case, the matrix **G** is diagonal and the resulting model is known as the Drude–Lorentz model. The dielectric function of the Drude–Lorentz model is then given as [13]

$$\hat{\varepsilon}(\omega) = 1 + \sum_k \frac{\omega_{\mathrm{p},k}^2}{\omega_{\mathrm{c},k}^2 - \omega^2 - \mathrm{i}\gamma_k \omega}, \tag{3.15}$$

where $\gamma_k \equiv \gamma_{kk}$. The terms for which $\omega_{\mathrm{c},k} = 0$ corresponds to the Drude model describing the dielectric response of free charges. The terms with nonzero $\omega_{\mathrm{c},k}$ correspond to the Lorentz model describing dielectric response of bound charges.

The classical model is derived from equations of motion which ensures that it fulfills the time reversal symmetry and it is KK consistent. From the asymptotic behavior it is possible to prove that the sum rule integral depends only on the plasma frequency parameters as

$$\int_0^\infty \omega\, \varepsilon_\mathrm{i}(\omega)\, \mathrm{d}\omega = \frac{\pi}{2} \sum_k \omega_{\mathrm{p},k}^2 . \tag{3.16}$$

The classical model presented above is expressed in the basis of eigenmode vibrations, i. e. $\omega_{\mathrm{p},k}^2$ describes the strength of the vibration mode, but the sum rule takes the same form also in the basis of individual particles. Thus, the sum of the plasma frequencies over all the vibration modes is

$$\sum_k \omega_{\mathrm{p},k}^2 = \omega_\mathrm{p}^2 , \tag{3.17}$$

where ω_p is the plasma frequency related to the density of the particles in the system by (3.5).

The classical model is universal in the sense that it can be used to model a wide range of absorption phenomena in solids. For some phenomena the accuracy of the classical model is sufficient but for many others, it is not possible to achieve the desired accuracy with a reasonable number of terms. This is not surprising since a number of quantum mechanical effects cannot be described by models based purely on classical mechanics (e. g. bandgap). Therefore, one should look for other models which would be more appropriate for the description of such phenomena.

3.2.2 Models Based on Quantum Mechanics

In principle it is possible to describe solids as many body systems and use the laws of quantum mechanics to derive the dielectric response. This approach, known as *ab initio* calculations, is very important for the study of the properties of solids since they does not need any experimental input. It is, however, numerically intensive and moreover, the current approximate approaches are not accurate or fast enough to be utilized as dispersion models in practice (see Chap. 4 for details). The practical dispersion models are based on the results known from the quantum mechanical theory but apart from that they also require some empirical knowledge.

A good starting point for expressing the dielectric function is using the Fermi golden rule (or alternatively Kubo formula) [10, 13–16] for an open quantum mechanical system at temperature T. The real and imaginary parts of the dielectric function in the frame of the dipole approximation are then expressed as

$$\varepsilon_{\mathrm{r}}(E) = 1 + \frac{2}{\epsilon_0 V} \sum_{i,f}^{f \neq i} P_i(T) |\langle f | \hat{d}_x | i \rangle|^2 \frac{E_f - E_i}{(E_f - E_i)^2 - E^2}, \tag{3.18}$$

$$\varepsilon_{\mathrm{i}}(E) = \frac{\pi}{\epsilon_0 V} \sum_{i,f}^{f \neq i} P_i(T) |\langle f | \hat{d}_x | i \rangle|^2 \left[\delta(E_f - E_i - E) - \delta(E_i - E_f - E) \right], \tag{3.19}$$

where the summation runs over all possible initial states $|i\rangle$ and final states $|f\rangle$. The symbol V denotes the volume of the system and the symbol E denotes the photon energy, which is related to frequency as $E = \hbar\omega$. The symbols E_i and E_f are the initial and final state energies of the system. The interaction of light with the quantum system is represented by stimulated absorption or emission of the photons with simultaneous transitions of the system from the initial to the final state. In (3.19) the absorption and emission processes are represented by delta functions in the imaginary part. The use of the delta functions is correct in the approximation of infinite lifetime of the states. The finite lifetime of the states leads to fluctuation of the energies of the states and the delta functions must be replaced by functions with finite width. This will be discussed in detail in Sect. 3.2.3. The real part (3.18) is calculated using the KK relation (3.2) which leads to the poles. The signs of these functions depend on whether an absorption process is taking place $E_i < E_f$ or an emission process is taking place $E_i > E_f$. The probability of the transition is given by the square of the size of dipole matrix element $\langle f | \hat{d}_x | i \rangle$, where $\hat{d}_x$ is the dipole operator along the coordinate x parallel to the external electric field

$$\hat{d}_x = \sum_j q_j \hat{x}_j \,. \tag{3.20}$$

In the expression above, the summation over j runs over all particles with the charges q_j and position operator $\hat{x}_j$. Therefore, for each term in (3.19) representing absorption process, there is a term representing a reverse emission process with the same transition probability. These terms have opposite signs but the probability $P_i(T)$ of finding the system in the initial state $|i\rangle$ is different for the absorption processes $E_i < E_f$ and for the emission processes $E_i > E_f$. For this reason, the net contribution is nonzero. In equilibrium, the system is described by the canonical distribution with the probability of finding the system in state $|i\rangle$ given by:

$$P_i(T) = \exp\left(\frac{\Omega - E_i}{k_\mathrm{B}T}\right), \tag{3.21}$$

with the canonical potential Ω determined from the normalization condition

$$\sum_i P_i(T) = \sum_i \exp\left(\frac{\Omega - E_i}{k_\mathrm{B}T}\right) = 1\,. \tag{3.22}$$

Note that for systems in equilibrium, the probability for the absorption process is higher than for the reverse emission process. Therefore, the contribution to the imaginary part of the dielectric function in (3.19) is positive for all positive photon energies E. The condition that the imaginary part of the dielectric function is positive is the fourth supplementary condition for dissipative systems that must be fulfilled by dispersion models. In the classical model the fulfillment of this condition is ensured by the positive definiteness of the matrix $\mathbf{G}$ composed of damping coefficients (3.13).

The dipole matrix elements in (3.18) and (3.19) can be rewritten using the current matrix elements as

$$|\langle f|\hat{d}_x|i\rangle|^2 = \frac{\hbar^2|\langle f|\hat{j}_x|i\rangle|^2}{(E_f - E_i)^2}, \tag{3.23}$$

where $\hat{j}_x$ is the volume integrated current density operator (or simply current operator [16]) related to the momentum operators of individual particles $\hat{p}_{x,j}$ as

$$\hat{j}_x = \sum_j \frac{q_j}{m_j}\hat{p}_{x,j}\,. \tag{3.24}$$

The imaginary part of the dielectric function is then expressed as

$$\varepsilon_\mathrm{i}(E) = \frac{\pi\hbar^2}{\epsilon_0 V}\sum_{i,f}^{f\neq i} P_i(T)\frac{|\langle f|\hat{j}_x|i\rangle|^2}{(E_f - E_i)^2}\left[\delta(E_f - E_i - E) - \delta(E_i - E_f - E)\right], \tag{3.25}$$

Since the delta functions ensure that $|E_f - E_i| = E$, the factor $(E_f - E_i)^2$ in the denominator can be written as E^2 and placed outside of the sum. Therefore, if we define the joint density of states (JDOS) function as

$$J(E) = \frac{\pi\hbar^2}{\epsilon_0 V}\sum_{i,f}^{f\neq i} P_i(T)|\langle f|\hat{j}_x|i\rangle|^2\left[\delta(E_f - E_i - E) - \delta(E_i - E_f - E)\right], \tag{3.26}$$

then the imaginary part of the dielectric function reduces to

$$\varepsilon_\mathrm{i}(E) = \frac{J(E)}{E^2}\,. \tag{3.27}$$

In some cases, the JDOS function is defined without the matrix elements inside the sum

$$\mathcal{J}(E) = \frac{1}{V}\sum_{i,f}^{f\neq i} P_i(T)\left[\delta(E_f - E_i - E) - \delta(E_i - E_f - E)\right]. \tag{3.28}$$

This function coincides, up to a constant factor, with the JDOS function (3.26) only if the matrix elements are independent on i and f. The approximation of constant matrix elements is commonly used in the case of transitions of the same type occurring in the narrow energy range. In general, however, the matrix elements cannot be assumed to be constant.

Apart from the functions $\varepsilon_{\rm i}(E)$ and $J(E)$ it is convenient to also introduce the function $F(E)$ given as

$$F(E) = \frac{\pi\hbar^2}{\epsilon_0 V}\sum_{i,f}^{f\neq i} P_i(T)\frac{|\langle f|\hat{j}_x|i\rangle|^2}{E_f - E_i}\left[\delta(E_f - E_i - E) + \delta(E_i - E_f - E)\right]. \tag{3.29}$$

This function will be called the transition strength function and it is related to the imaginary part of the dielectric function as

$$\varepsilon_{\rm i}(E) = \frac{F(E)}{E}. \tag{3.30}$$

Note that, the transition strength function in the frame of the dipole approximation is, up to the constant factor, equivalent to the real part of the complex optical conductivity $\sigma_{\rm r}(E) = F(E)\epsilon_0/\hbar$. The function $F(E)$ is important because it provides a bridge between the classical sum rule and the quantum mechanical Thomas–Reiche–Kuhn (TRK) sum rule [13, 17]. The TRK sum rule can be applied separately to each type of the charged particles (electrons, different nuclei) existing in the system. The TRK sum rule further says that the sum of dimensionless oscillator strengths f_{if}^k over the final states f is independent on the initial states i and is equal to the number of particles of type k in the system:

$$\frac{1}{V}\sum_{f}^{f\neq i} f_{if}^k = \mathcal{N}_k. \tag{3.31}$$

In the formula above, V is the volume of the system and $\mathcal{N}_k$ is the density of the particles of type k. The oscillator strength can be defined with the help of the matrix elements of the dipole $\hat{d}_{xk}$ or current $\hat{j}_{xk}$ operators for particles of type k as

$$f_{if}^k = \frac{2m_k}{\hbar^2 q_k^2}(E_f - E_i)|\langle f|\hat{d}_{xk}|i\rangle|^2 = \frac{2m_k}{q_k^2}\frac{|\langle f|\hat{j}_{xk}|i\rangle|^2}{E_f - E_i}, \tag{3.32}$$

where q_k and m_k are the charge and mass of the particles of type k, respectively. Note that the TRK sum rule (3.31) is derived from the general principles of quantum mechanics and the assumption that the Hamiltonian is a sum of kinetic energy, which depends only on momentum operators, and the potential energy, which depends only on position operators. In the frame of the dipole approximation, the TRK sum rule can be written using of the transition strength function as [7]

$$\int_0^\infty F(E)\,\mathrm{d}E = \sum_k \frac{\pi}{2}\frac{(q_k\hbar)^2}{\epsilon_0 m_k}N_k = \frac{\pi}{2}\frac{(e\hbar)^2}{\epsilon_0 m_\mathrm{e}}N_\mathrm{e}\mathcal{U} = N\,, \tag{3.33}$$

which is equivalent to the classical sum rule (3.16). The quantity N introduced in the above equation will be called the total transition strength. In theory, the quantity N can be determined from the known density of charged particles.

Although the right side of (3.19) is expressed as a sum of delta functions, the number of energy levels in the system is sufficiently high so that it is possible to use the continuous JDOS functions. In practice, it is convenient to express the JDOS function as a sum of contributions corresponding to individual elementary absorption processes. The JDOS function corresponding to the individual elementary absorption processes are modeled on the basis of both the theoretical and empirical knowledge. Although these contributions can be expressed using the transition strength function or the imaginary part of the dielectric function, this method is always referred to as parameterization of the JDOS function (PJDOS). It is beneficial to express the individual contributions using the terms normalized with respect to the sum rule and the transition strength parameters corresponding to the individual absorption processes. The total transition strength function is then given as

$$F(E) = \sum_t N_t F_t^0(E)\,, \tag{3.34}$$

where the index t distinguishes individual contributions (terms), N_t are the transitions strengths parameters and $F_t^0(E)$ are normalized transition strength functions. The dispersion parameters of the whole model consist of the transition strength parameters N_t and parameters of the functions $F_t^0(E)$ which determine spectral distributions of individual contributions (e. g. bandgap energy). The dielectric function is calculated as

$$\varepsilon_\mathrm{r}(E) = 1 + \sum_t N_t \frac{2}{\pi}\int_0^\infty \frac{F_t^0(x)}{x^2 - E^2}\,\mathrm{d}x = 1 + \sum_t N_t \varepsilon_{\mathrm{r},t}^0(E)\,, \tag{3.35}$$

$$\varepsilon_\mathrm{i}(E) = \sum_t N_t \frac{F_t^0(E)}{E} = \sum_t N_t \varepsilon_{\mathrm{i},t}^0(E)\,. \tag{3.36}$$

The expression for the imaginary part follows from (3.30) and the real part is calculated from the KK relation (3.3). In the adopted formalism, the quantities $\hat{\varepsilon}_t^0(E)$ are normalized susceptibilities of individual contributions but they will be called normalized contributions to the dielectric function or just normalized dielectric functions.

Note that in the derivation of the above expressions it was assumed that the eigenstates of the Hamiltonian are known. For the description of specific elementary excitations of the dielectric response we often utilize the second quantization, i. e. describing the system using one-particle states and quasiparticles. It should be further emphasized that the total transition strength N can be expressed on the basis of the real number of particles in the system but the transition strengths of individual elementary excitations cannot be used to calculate the densities of particles involved in these processes. These transition strengths must be understood as strengths corresponding to the effective number of particles or, alternatively, particles with effective masses or effective charges must be considered. For example, the transition strength of valence electron excitations is usually described using the effective number of valence electrons per atom n_{ve} [4]

$$N_{\mathrm{ve}} = \frac{2}{\pi}\frac{(e\hbar)^2}{\epsilon_0 m_{\mathrm{e}}} N_{\mathrm{a}} n_{\mathrm{ve}} , \tag{3.37}$$

where N_{a} is the density of atoms. The transition strength of free carriers is usually described using the effective mass $m^* m_{\mathrm{e}}$

$$N_{\mathrm{fc}} = \frac{2}{\pi}\frac{(e\hbar)^2}{\epsilon_0 m^* m_{\mathrm{e}}} N_{\mathrm{d}} , \tag{3.38}$$

where N_{d} is the density of dopants. Finally the transition strength of phonon excitations is usually expressed using the effective charge $q^* e$

$$N_{\mathrm{ph}} = \frac{2}{\pi}\frac{(q^* e\hbar)^2}{\epsilon_0 m_{\mathrm{n}}} N_{\mathrm{n}} , \tag{3.39}$$

where N_{n} and m_{n} are density and mass of the nuclei.

3.2.3 *Broadening*

As mentioned above, it is not possible to obtain an exact quantum mechanical solution for the complex systems encountered in the solid state physics. In practice, different approximate models are employed for individual elementary excitations. For example, the approximation of non-interacting particles applied to crystalline materials often leads to discrete spectrum or discontinuities in the JDOS (or its first derivative) known as Van Hove singularities [18], which are, however, usually not observed in the real absorption spectra. In reality, the sharp structures are blurred as a consequence of different effects neglected in the approximate models. The effects that contribute to the blurring (broadening) of the sharp structures in the absorption spectra are the finite lifetime of the states (temperature dependent, intrinsic for open systems) or various irregularities in the lattice (usually temperature independent). An empirical broadening procedure is then used to model the blurring of the sharp structures [19–30].

In practice, the unbroadened JDOS is determined based on an approximate model and the broadened dielectric response is then calculated as a convolution of the function representing this dielectric response with appropriately chosen normalized broadening function $\beta(x)$. Since the broadening procedure is an empirical approach, it is not evident on which function, representing the dielectric response, should the broadening procedure be applied. For example, the broadening procedure can be applied to functions $\varepsilon_{\mathrm{i}}(E)$, $F(E)$ or $J(E)$ [31]:

$$\tilde{\varepsilon}_{\mathrm{i}}(E) = \int_{-\infty}^{\infty} \beta(E-t)\,\varepsilon_{\mathrm{i}}(t)\,\mathrm{d}t \qquad \varepsilon - \text{broadening}, \tag{3.40}$$

$$\tilde{\varepsilon}_{\mathrm{i}}(E) = \frac{\tilde{F}(E)}{E} = \frac{1}{E}\int_{-\infty}^{\infty} \beta(E-t)\,F(t)\,\mathrm{d}t \qquad F - \text{broadening}, \tag{3.41}$$

$$\tilde{\varepsilon}_{\mathrm{i}}(E) = \frac{\tilde{J}(E)}{E^2} = \frac{1}{E^2}\int_{-\infty}^{\infty} \beta(E-t)\,J(t)\,\mathrm{d}t \qquad J - \text{broadening}, \tag{3.42}$$

where the tilde denotes the broadened functions. Two one-parametric normalized distribution functions are used for the broadening, the first being the Gaussian function

$$\beta_{\mathrm{G}}(x) = \frac{1}{\sqrt{2\pi}\,B}\exp\left(-\frac{x^2}{2B^2}\right), \tag{3.43}$$

where B is the broadening parameter representing the root mean square (RMS) value. The second distribution function is the Lorentzian function

$$\beta_{\mathrm{L}}(x) = \frac{1}{\pi}\frac{B/2}{x^2 + B^2/4}, \tag{3.44}$$

where the broadening parameter B represents the full width half maximum (FWHM) value. The Gaussian distribution function is more universal (thanks to central limit theorem) than the Lorentzian distribution function but the Lorentzian distribution function often leads to expressions that can be calculated analytically.

It was previously shown that the J-broadening or F-broadening procedure can give nonzero static conductivity [21, 31] even if the unbroadened dielectric response had zero static conductivity ($F(0) = 0$). On the other hand the ε-broadening always gives a dielectric response with zero static conductivity [31]. Therefore, the ε-broadening is appropriate for dispersion models of contributions representing bounded charges (interband electronic transitions, phonon absorption, etc.) and J-broadening or F-broadening is appropriate for dispersion models of contributions having nonzero static conductivity (e. g. contributions of free charges or direct electronic transitions in 2D graphene sheet). Furthermore, it was shown that only the ε-broadening and F-broadening preserve the sum rule integral [31] and, therefore, they preserve the normalization if applied to the normalization contributions.

The ε-broadening and the F-broadening can be written in a concise form with the help of the symbol $*$ denoting the operation of convolution as

$$\tilde{\varepsilon_{\rm i}}^0 = \beta * \varepsilon_{\rm i}^0 \quad \text{and} \quad \tilde{\varepsilon_{\rm i}}^0 = \frac{1}{E}(\beta * F^0)\,. \tag{3.45}$$

The KK relation for the real part of the normalized dielectric function can be written with the help of the Hilbert transform H as [32]

$$\varepsilon_{\rm r}^0(E) = \frac{1}{\pi} \int_{-\infty}^{\infty} \frac{\varepsilon_{\rm i}^0(x)}{x - E}\, {\rm d}x \equiv H[\varepsilon_{\rm i}^0](E)\,. \tag{3.46}$$

Since the order of the Hilbert transform and the convolution can be exchanged [31], the real part of the ε-broadened normalized dielectric function can be written as

$$\tilde{\varepsilon_{\rm r}}^0 = H[\tilde{\varepsilon_{\rm i}}^0] = H[\beta * \varepsilon_{\rm i}^0] = H[\beta] * \varepsilon_{\rm i}^0\,. \tag{3.47}$$

A similar expression can be derived for the real part of the F-broadened normalized dielectric function

$$\tilde{\varepsilon_{\rm r}}^0 = H\left[\frac{\tilde{F}^0}{E}\right] = \frac{1}{E} H[\beta * F^0] = \frac{1}{E}\left(H[\beta] * F^0\right). \tag{3.48}$$

Both the ε-broadened and F-broadened normalized dielectric functions can be written very compactly as

$$\tilde{\varepsilon}^0 = \hat{\beta} * \varepsilon_{\rm i}^0 \quad \text{and} \quad \tilde{\varepsilon}^0 = \frac{1}{E}(\hat{\beta} * F^0)\,. \tag{3.49}$$

where the complex function $\hat{\beta}$ is defined as

$$\hat{\beta} = H[\beta] + {\rm i}\beta\,. \tag{3.50}$$

The fact that the real part of the broadened dielectric function can be calculated with the help of only a single integral instead of double integral (convolution followed by the Hilbert transform) is important if this function is calculated numerically.

The Hilbert transform of the Gaussian distribution function (3.43) is given by

$$H[\beta_{\rm G}](x) = -\frac{\sqrt{2}}{\pi B} {\rm D}\left(\frac{x}{\sqrt{2}B}\right), \tag{3.51}$$

where $D(x)$ denotes the Dawson function (integral) [33–35] defined as

$$\mathrm{D}(x) = \exp(-x^2) \int_0^x \exp(t^2)\,\mathrm{d}t\,. \tag{3.52}$$

The Dawson function can be efficiently evaluated in numerical calculations. The Hilbert transform of the Lorentzian distribution function (3.44) is given by

$$H[\beta_\mathrm{L}](x) = -\frac{1}{\pi}\frac{x}{x^2 + B^2/4}\,. \tag{3.53}$$

The complex Lorentzian broadening function can be elegantly written as

$$\hat{\beta}_\mathrm{L}(x) = -\frac{1}{\pi}\frac{1}{x + \mathrm{i}B/2}\,. \tag{3.54}$$

The Lorentzian ε-broadened dielectric function $\hat{\beta}_\mathrm{L} * \varepsilon_\mathrm{i}^0$ can be calculated as [30]

$$\tilde{\hat{\varepsilon}}^0(E) = \frac{1}{\pi}\int_{-\infty}^{\infty} \frac{\varepsilon_\mathrm{i}^0(x)}{x - \hat{E}}\,\mathrm{d}x \quad \text{where} \quad \hat{E} = E + \mathrm{i}B/2\,. \tag{3.55}$$

The Lorentzian ε-broadening provides a bridge between the quantum and classical dispersion models. The imaginary part of the normalized dielectric function of a discrete transition with the energy difference E_r in the two-level system is given by

$$\varepsilon_\mathrm{i}^0(E) = \frac{1}{E_\mathrm{r}}\Big(\delta(E - E_\mathrm{r}) - \delta(E + E_\mathrm{r})\Big)\,. \tag{3.56}$$

The Lorentzian ε-broadening of this response function gives

$$\tilde{\varepsilon_\mathrm{r}}^0(E) = H[\beta_\mathrm{L}] * \varepsilon_\mathrm{i}^0 = -\frac{1}{\pi E_\mathrm{r}}\left(\frac{E - E_\mathrm{r}}{(E - E_\mathrm{r})^2 + B^2/4} - \frac{E + E_\mathrm{r}}{(E + E_\mathrm{r})^2 + B^2/4}\right), \tag{3.57}$$

$$\tilde{\varepsilon_\mathrm{i}}^0(E) = \beta_\mathrm{L} * \varepsilon_\mathrm{i}^0 = \frac{B}{2\pi E_\mathrm{r}}\left(\frac{1}{(E - E_\mathrm{r})^2 + B^2/4} - \frac{1}{(E + E_\mathrm{r})^2 + B^2/4}\right), \tag{3.58}$$

which is equivalent to the Lorentz term in the classical model because the real and imaginary part of the normalized dielectric function (3.15) are expressed as

$$\varepsilon_\mathrm{r}^0(E) = \frac{2}{\pi}\frac{E_\mathrm{c}^2 - E^2}{(E_\mathrm{c}^2 - E^2)^2 + B^2E^2}\,, \quad \varepsilon_\mathrm{i}^0(E) = \frac{2}{\pi}\frac{BE}{(E_\mathrm{c}^2 - E^2)^2 + B^2E^2}\,. \tag{3.59}$$

The broadened functions (3.57) and (3.58) can be written as (3.59) with the central energy equal to

$$E_\mathrm{c} = \sqrt{E_\mathrm{r}^2 + B^2/4}\,. \tag{3.60}$$

Note that that the Lorentz terms corresponding to critically damped ($E_c = B/2$) or overdamped ($E_c < B/2$) harmonic oscillator cannot be expressed as Lorentzian ε-broadened discrete transitions.

3.3 Dispersion Models of Elementary Excitations

Linear dielectric response corresponds to the transitions of the system between two states, which are accompanied by absorption or emission of photons. The absorption (emission) processes involving electron excitation (recombination) are called electronic transition (absorption) processes. If the phonon occupation numbers are not changed in the electronic transition processes, they are referred to as direct electronic transition processes while if the phonon occupation numbers are changed, they are referred to as indirect electronic transition processes. Phonon absorption (emission) processes, on the other hand, are those without electron excitation (recombination). The electronic transitions cover the whole spectral range from zero frequency (photon energy) up to X-ray. If the Fermi energy lies inside the band of valence electrons or close to it, the states in this band are partially filled and indirect intraband transitions are possible. Indirect intraband transitions are referred to as free carrier contributions. The schematic diagram of the band structure and classification of electronic transitions is depicted in Fig. 3.1. In the materials with the filled valence band which is separated from the empty conduction band, both the direct and indirect interband transitions can exist. The origin of the valence and conduction band is in the splitting of the partially occupied atomic valence orbital states. The band above the conduction band which originates from the orbital states above the valence shell is called higher energy excitation band. The excitations of the valence electrons into these states are called higher energy excitations (see Fig. 3.1). These higher energy excitations can

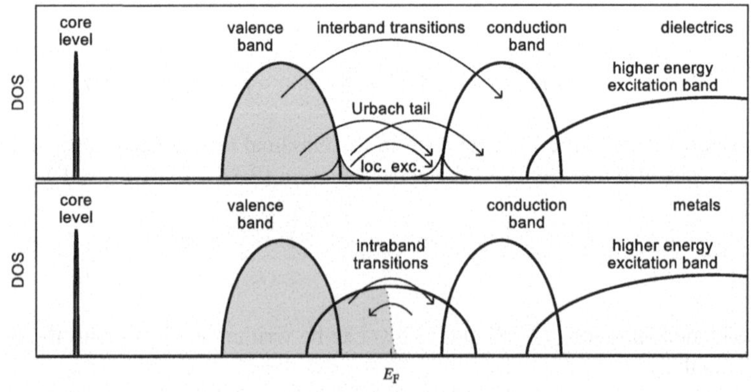

Fig. 3.1 Schematic diagram of the electronic density of states (DOS) distribution function and possible interband an intraband transitions in dielectrics and metals

also be understood as the scattering processes. The orbitals below the valence shell result in discrete core level states and excitations from these states are called core electron excitations.

3.3.1 Phonon Absorption

The phonon absorption exhibits different behavior in different types of materials (crystals vs. disordered, homopolar vs. ionic, etc.). In the following subsections, several dispersion models describing phonon absorption will be discussed.

3.3.1.1 One-Phonon Absorption in Disordered Materials

From the point of view of vibrational modes, disordered solids can be viewed as a collection of a very large number of atoms having independent vibrational modes with the total number of vibrational modes given by the number of degrees of freedom minus three. In disordered solids the vibrational modes are localized, i. e. the eigenvectors describing the amplitudes of to the vibrational modes have non-negligible values concentrated in blocks corresponding to small volumes in the solid. Each of these blocks can be viewed as a molecule with a relatively small number of vibrational modes. The transition strength function of such a molecule is given as a sum of delta functions

$$F_{\mathrm{1ph}}(E) = \sum_p N_p \left[\delta(E - E_p) + \delta(E + E_p)\right] , \tag{3.61}$$

where the index p distinguishes individual vibrational modes (peaks in the spectra). The symbols E_p and N_p denote the energy and transition strength of individual modes, respectively. The imaginary part of the dielectric function is then calculated as (3.30)

$$\varepsilon_{\mathrm{i,1ph}}(E) = \sum_p \frac{N_p}{E_p} \left[\delta(E - E_p) - \delta(E + E_p)\right] . \tag{3.62}$$

This is the same result that one would obtain for non-interacting molecular gases. For this reason the terminology used in the absorption spectroscopy of gases is often adopted also in the study of phonon absorption in disordered solids.

The atoms, which are lying outside of the blocks forming the above mentioned molecules lead to the broadening of the delta functions, forming the absorption spectra. If the molecules are weakly interacting (which is true for gases) then the Lorentzian ε-broadening is appropriate. This is because the interactions (collisions) of molecules can be described by the relaxation time and, therefore, the Lorentz model provides a suitable description.

In disordered solids, the assumption of weak interactions is not valid. The interactions are so strong that they cannot be described by the relaxation time and it is more correct to imagine the interactions as something that distorts (deforms) the molecules. The random deformations of the molecules cause changes in the frequencies of the localized vibrational modes. According to the central limit theorem, the Gaussian distribution provides accurate distribution of the changes in these frequencies [34–37]. Therefore, the Gaussian ε-broadening is more appropriate in this case.

The ε-broadened contribution to the dielectric function (susceptibility) describing one-phonon absorption in disordered solids is then

$$\hat{\varepsilon}_{\rm 1ph}(E) = \sum_p \frac{N_p}{E_p} \hat{\beta}_{{\rm G},p} * \left[\delta(E - E_p) - \delta(E + E_p)\right] , \tag{3.63}$$

where the complex Gaussian broadening functions $\hat{\beta}_{{\rm G},p}$ are given by (3.50), (3.43) and (3.51). The spectral dependency of the normalized Gaussian ε-broadened dielectric function is compared with different ε-broadened dielectric functions in Fig. 3.2. The explicit expression for the normalized dielectric function corresponding to one mode is given by

$$\varepsilon^0_{{\rm r},p}(E) = \frac{-\sqrt{2}}{\pi B_p E_p} \left[{\rm D}\left(\frac{E - E_p}{\sqrt{2} B_p}\right) - {\rm D}\left(\frac{E + E_p}{\sqrt{2} B_p}\right) \right] , \tag{3.64}$$

$$\varepsilon^0_{{\rm i},p}(E) = \frac{1}{\sqrt{2\pi} B_p E_p} \left[\exp\left(-\frac{(E - E_p)^2}{2B_p^2}\right) - \exp\left(-\frac{(E + E_p)^2}{2B_p^2}\right) \right] , \tag{3.65}$$

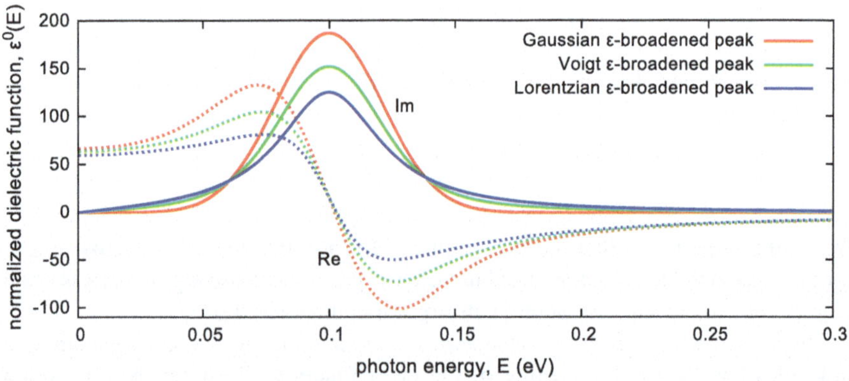

Fig. 3.2 Spectral dependencies of the normalized dielectric functions of ε-broadened discrete spectrum calculated using three different broadening functions: $E_p = 0.1\,$eV (peak energy), $B_p = 0.05\,$eV (FWHM), $L_p = 0.5$ (Lorentzian fraction in Voigt profile)

where B_p is the broadening parameter which may be different for individual vibrational modes. The real part is calculated using the the Dawson integral $\mathrm{D}(x)$ defined in (3.52). The broadening factor B_p in (3.64) and (3.65) gives the RMS value. Since we usually work with FWHM values in the IR spectroscopy (the RMS value is not defined for the Lorentzian distribution), it is useful to point out that in the case of the Gaussian distribution, the RMS and FWHM values are related as

$$\sqrt{2}B^{\mathrm{RMS}} = \frac{B^{\mathrm{FWHM}}}{2\sqrt{\ln 2}} . \tag{3.66}$$

3.3.1.2 One-Phonon Absorption in Crystalline Materials

The one-phonon absorption in ideal crystalline materials is represented by a discrete spectrum of vibrational modes. These modes correspond to the transverse optical (TO) phonons with zero momentum (i.e. in the Γ point). The number of TO modes depends on the number of atoms in the primitive cell. Only the optically active modes, i.e. those with a non-vanishing dipole matrix element, appear in the absorption spectra. The appropriate model in this case is the classical model of independent underdamped harmonic oscillator, which is equivalent to the model of the Lorentzian ε-broadened discrete transitions. The normalized dielectric function corresponding to the TO mode p is then given by

$$\hat{\varepsilon}^0_p(E) = \frac{1}{E_p}\hat{\beta}_{\mathrm{L},p} * \left[\delta(E - E_p) - \delta(E + E_p)\right] = \frac{2/\pi}{E_p^2 + B_p^2/4 - E^2 - \mathrm{i}B_p E} , \tag{3.67}$$

where the complex Lorentzian broadening function $\hat{\beta}_{\mathrm{L},p}$ is given by (3.54). The broadening parameter B_p is the FWHM of the absorption peak and E_p is the resonant energy. The expression $E_p^2 + B_p^2/4$ in the denominator is often written with the help of the central energy defined as $E_{\mathrm{c},p}^2 = E_p^2 + B_p^2/4$. The form with $E_p^2 + B_p^2/4$ is, however, more convenient since the condition that the DHO is underdamped can then be written as $E_p > 0$.

In some cases the phonon absorption cannot be described by independent DHO (Lorentz) model and the model of DHO with coupled modes must be used. The model of DHO with coupled modes was already discussed in Sect. 3.2.1. In this model the contribution to the dielectric function (susceptibility) can be expressed using matrices as

$$\hat{\varepsilon}_{1\mathrm{ph}}(E) = \frac{2}{\pi}\mathbf{N}^{\mathrm{T}}[\mathbf{S} - E^2\mathbf{I} - \mathrm{i}E\mathbf{B}\mathbf{B}^{\mathrm{T}}]^{-1}\mathbf{N} , \tag{3.68}$$

where $\mathbf{N}$ is a real vector formed by the square roots of transition strengths of individual TO modes N_p, $p = 1, \ldots, m$:

$$\mathbf{N}^{\mathrm{T}} = (\sqrt{N_1}, \sqrt{N_2}, \ldots, \sqrt{N_m}) . \tag{3.69}$$

The matrix $\mathbf{S}$ is a real diagonal matrix with squares of central frequencies $E_{c,p}$ on the diagonal

$$\mathbf{S} = \begin{pmatrix} E_{c,1}^2 & 0 & \cdots & 0 \\ 0 & E_{c,2}^2 & \cdots & 0 \\ \vdots & \vdots & \ddots & \vdots \\ 0 & 0 & \cdots & E_{c,m}^2 \end{pmatrix}, \tag{3.70}$$

the symbol $\mathbf{I}$ denotes the unit matrix and the matrix $\mathbf{B}$ is a real lower triangular matrix with positive values on the diagonal

$$\mathbf{B} = \begin{pmatrix} \sqrt{B_1} & 0 & \cdots & 0 \\ \nu_{12}\, B_{12}/\sqrt{|B_{12}|} & \sqrt{B_2} & \cdots & 0 \\ \vdots & \vdots & \ddots & \vdots \\ \nu_{1m}\, B_{1m}/\sqrt{|B_{1m}|} & \nu_{2m}\, B_{2m}/\sqrt{|B_{2m}|} & \cdots & \sqrt{B_m} \end{pmatrix}, \tag{3.71}$$

where the constants B_p and B_{kl} determine the broadening and coupling of modes. The constants ν_{kl} are defined as follows

$$\nu_{kl} = \frac{\sqrt{N_k N_l}}{\mathbf{N}^{\mathrm{T}}\mathbf{N}}. \tag{3.72}$$

The constants ν_{kl} are introduced in order to ensure that the corresponding elements in the matrix $\mathbf{B}$ vanish if N_k or N_l is zero.

This model of DHO with coupled modes produces asymmetrically shaped absorption peaks (see Fig. 3.3). In contrast to the simpler model of independent DHO, the

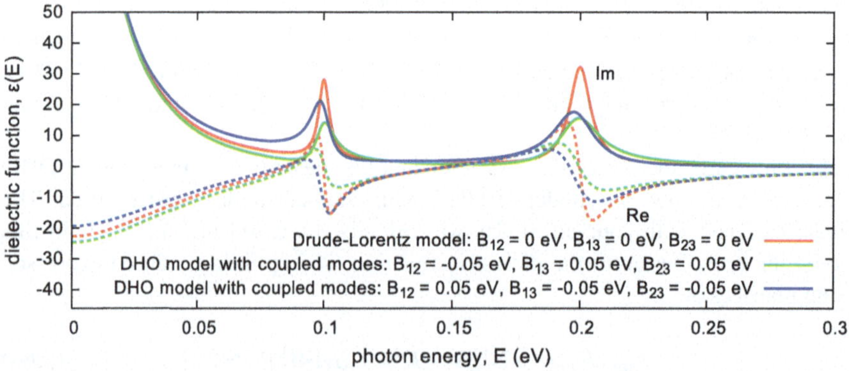

Fig. 3.3 Contributions to the dielectric function of DHO model with three independent (Drude-Lorentz model) and three coupled modes: $N_1 = 0.02\,\mathrm{eV}^2$, $E_{c,1} = 0.1\,\mathrm{eV}$, $B_1 = 0.005\,\mathrm{eV}$, $N_2 = 0.1\,\mathrm{eV}^2$, $E_{c,2} = 0.2\,\mathrm{eV}$, $B_2 = 0.01\,\mathrm{eV}$, $N_3 = 0.1\,\mathrm{eV}^2$, $E_{c,3} = 0\,\mathrm{eV}$, $B_3 = 0.05\,\mathrm{eV}$. The third mode with zero central frequency describes contribution from free charges. The values of the parameters B_{12}, B_{13} and B_{23} are different for each curves

parameters B_p cannot be interpreted as FWHMs of the peaks in the model. The relation between the FWHM of the peaks and parameters B_p and B_{kl} is, in this case, very complicated. Furthermore, the relation between the central energies $E_{c,p}$, which are used to parameterize the model, and resonant energies E_p, which correspond to the photon energies of the peaks, is also very complicated. If the broadening and coupling factors are small then the resonant energies differ only slightly from the central energies.

3.3.1.3 One-Phonon Absorption in Partially Disordered Materials

In some cases the phonon absorption peaks are properly described neither by the Lorentzian nor the Gaussian ε-broadened discrete spectrum and both the broadening procedures must be combined. The resulting broadening procedure is described by a two parameter broadening function known as the Voigt profile corresponding to the convolution of the Lorentzian and Gaussian distribution

$$\hat{\beta}_{\mathrm{V},p} = \hat{\beta}_{\mathrm{G},p} * \beta_{\mathrm{L},p} = \hat{\beta}_{\mathrm{L},p} * \beta_{\mathrm{G},p}\,, \tag{3.73}$$

or with the help of the complex Faddeeva function $\mathrm{W}(\hat{z})$ as [38–42]

$$\hat{\beta}_{\mathrm{V},p}(x) = \frac{\mathrm{i}}{\sqrt{2\pi}\,B_{\mathrm{G},p}} \mathrm{W}\left(\frac{x + \mathrm{i}B_{\mathrm{L},p}/2}{\sqrt{2}B_{\mathrm{G},p}}\right), \tag{3.74}$$

where the Faddeeva function is given by the following integral

$$\mathrm{W}(\hat{z}) = \frac{\mathrm{i}}{\pi}\int_{-\infty}^{\infty} \frac{\exp(-t^2)}{\hat{z} - t}\mathrm{d}t\,. \tag{3.75}$$

The normalized dielectric function corresponding to one mode could, in this case, be expressed as

$$\hat{\varepsilon}_p^0(E) = \frac{\mathrm{i}}{\sqrt{2\pi}\,B_{\mathrm{G},p}E_p}\left[\mathrm{W}\left(\frac{E - E_p + \mathrm{i}B_{\mathrm{L},p}/2}{\sqrt{2}B_{\mathrm{G},p}}\right) - \mathrm{W}\left(\frac{E + E_p + \mathrm{i}B_{\mathrm{L},p}/2}{\sqrt{2}B_{\mathrm{G},p}}\right)\right]. \tag{3.76}$$

The exact dependence of the FWHM of the Voigt profile on parameters $B_{\mathrm{L},p}$ (FWHM of the Lorentzian part) and $B_{\mathrm{G},p}$ (RMS of the Gaussian part) is quite complicated. However, the dependencies of these broadening parameters on the FWHM of the Voigt profile B_p can be approximated with sufficient accuracy using the following formulas [43]

$$B_{\mathrm{L},p} \approx B_p L_p\,, \qquad B_{\mathrm{G},p} \approx \frac{B_p}{2\sqrt{2\ln 2}}\sqrt{(1 - aL_p)^2 - (1 - a)^2 L_p^2}\,, \tag{3.77}$$

where $a = 0.5346$ and L_p is the parameter mixing the Gaussian and Lorentzian part. The comparison of the ε-broadened normalized dielectric functions calculated using the Voigt, Gaussian and Lorentzian broadening functions is in Fig. 3.2.

Since the evaluation of the complex Faddeeva function is more complicated than the evaluation of the real Dawson function, it is convenient to use an approximation in which the Voigt profile is replaced by the linear combination of the Gaussian and Lorentzian distributions. There are several approaches to define this linear combination and they differ in the choice of broadening factors for the Gaussian and Lorentz distributions. The simplest approach is to combine the Gaussian and Lorentzian functions with the same FWHM [44]:

$$\hat{\beta}_{\mathrm{V},p} \approx (1 - L_p)\hat{\beta}_{\mathrm{G},p} + L_p\hat{\beta}_{\mathrm{L},p}\,, \tag{3.78}$$

where L_p is the parameter mixing the Gaussian and Lorentzian parts. Note that this mixing parameter has different meaning than the mixing parameter introduced in (3.77). The deviation from the exact values of the Voigt profile in this approximation is smaller than the usual experimental errors (absolute error lower than 10^{-4} peak value).

3.3.1.4 One-Phonon Absorption with Asymmetric Absorption Peaks

The classical model of DHO with coupled modes, which gives asymmetric peaks, can be approximated by the Lorentz model with complex transition strengths [45]. As mentioned above, the underdamped Lorentz model can be written as the Lorentzian ε-broadened discrete spectrum. The contribution to the dielectric function (susceptibility) can be then expressed as

$$\hat{\varepsilon}_{\mathrm{1ph}}(E) = \frac{1}{C_{\mathrm{N}}} \sum_p \left(N_p + \mathrm{i} M_p \frac{E}{E_p} \right) \frac{1}{E_p} \hat{\beta}_p * \left[\delta(E - E_p) - \delta(E + E_p) \right]\,, \tag{3.79}$$

where M_p determines the asymmetric part of the transition strength of the absorption peak and C_{N} is a normalization constant. In the case of underdamped Lorentz model, the complex broadening function $\hat{\beta}_p$ is given by (3.54) but the model can be easily generalized using different types of broadening functions. While the real parts of transition strengths N_p are always positive, the imaginary parts of transition strengths M_p can be chosen arbitrarily. Such a model fulfills the first two fundamental conditions for the dielectric response (i. e. the time reversal symmetry (3.1) and KK consistency (3.2) and (3.8)). The dimensionless factor E/E_p in (3.79) ensures the time reversal symmetry of the model. Using the superconvergence theorem it is easy to prove that the sum rule integral (3.4) is infinite for all individual asymmetric terms in the sum (3.79). In order to ensure the third fundamental condition (i. e. the convergence of the sum rule integral), the following sum must be equal to zero [45]:

$$\sum_p \frac{M_p}{E_p} = 0 \,. \tag{3.80}$$

The asymmetric part contributes to the sum rule for a Lorentzian broadening function but not for Gaussian broadening. The normalization constant is then given from the superconvergence theorem as

$$C_{\mathrm{N}} = \frac{1}{\sum_p N_p} \sum_p \left(N_p + \frac{M_p B_{\mathrm{L},p}}{E_p} \right) , \tag{3.81}$$

where $B_{\mathrm{L},p}$ is the Lorentzian part of the broadening of the p-peak ($B_{\mathrm{L},p} = L_p B_p$ for both the Voigt and linear combination profile).

The advantage of this model is that for three and more modes the asymmetric part is determined by a smaller number of independent dispersion parameters $(m - 1)$ than in the DHO model with coupled modes $(m(m - 1)/2)$, where m denotes the number of modes. On the other hand, the disadvantage of this model is that the fourth auxiliary condition (i. e. $\varepsilon_{\mathrm{i}} \geq 0$ in the whole spectral range) is difficult to satisfy.

For sufficiently small values of the parameters M_p, the model satisfies the fourth condition, though it is difficult to formulate the general limits under which this condition is satisfied. The asymmetric peaks in Fig. 3.4 do not fulfill this condition. For the selected parameters the fourth condition is fulfilled for $|M_2| < 0.002767$. Thus, this model could be used only for modeling peaks which are only slightly asymmetric. For strongly asymmetric peaks, the model of the DHO with coupled modes must be used or the model presented in (3.79) has to be considered together

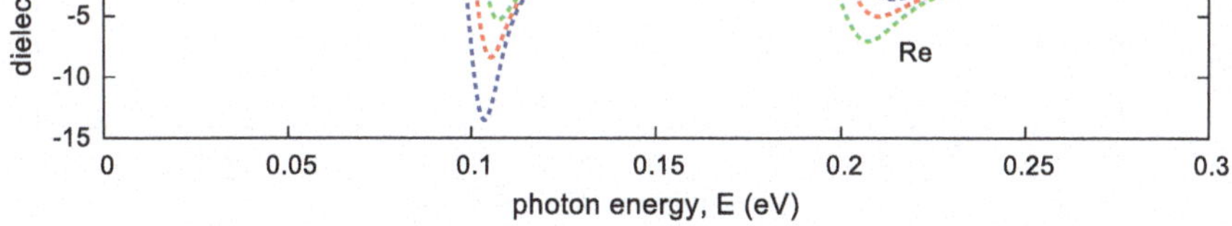

Fig. 3.4 Contributions to the dielectric function of two asymmetric Voigt ε-broadened peaks. Asymmetry represents the redistribution of transition strength due to coupling between phonon modes. The dielectric function without coupling is added for comparison. The parameters were chosen as follows: $N_1 = 0.02\,\mathrm{eV}^2$, $E_1 = 0.1\,\mathrm{eV}$, $B_1 = 0.01\,\mathrm{eV}$, $L_1 = 0.3$, $N_2 = 0.05\,\mathrm{eV}^2$, $E_2 = 0.2\,\mathrm{eV}$, $B_1 = 0.02\,\mathrm{eV}$, $L_2 = 0.7$. The values of the parameters M_1 and M_2 are different for each curves with $M_1 = -(E_1/E_2)M_2$

with some other contribution to the dielectric function that ensures that the imaginary part of the resulting dielectric function is positive in the whole spectral range.

It is necessary to note that it is possible to prove [45] that the model (3.79) with Lorentzian broadening is equivalent to the the model known as 'factorized Lorentz oscillators' introduced in [46].

In the conducting materials (e. g. graphite [47–49]) there are one or more phonon absorption peaks in the IR region which have asymmetric profile. The asymmetry of the profile of these phonon absorption peaks is caused by quantum mechanical interference between the phonons and free charges, called the Fano resonance [50]. In this case, the asymmetric peaks cannot be described by the model based on the ε-broadened discrete transitions with the imaginary parts of transition strengths satisfying the condition (3.80). The problem can be solved if the factor E_p/E is used instead of the factor E/E_p used in (3.79)

$$\hat{\varepsilon}_{1\text{ph}}(E) = \sum_p \left(N_p + \mathrm{i} M_p \frac{E_p}{E} \right) \frac{1}{E_p} \hat{\beta}_p * \left[\delta(E - E_p) - \delta(E + E_p) \right] . \qquad (3.82)$$

In this case, the sum rule integral is always zero for the individual asymmetric parts. Therefore, this model satisfies all three fundamental conditions even if only one phonon vibrational mode is considered. The fourth auxiliary condition can be satisfied if the above dielectric function is used as a part of the UDM because it also contains the contribution from free carriers (see Fig. 3.5). The asymmetric part of the absorption peak should be understood as a redistribution of the transition strength function of free carriers. The asymmetric part of the model changes value of

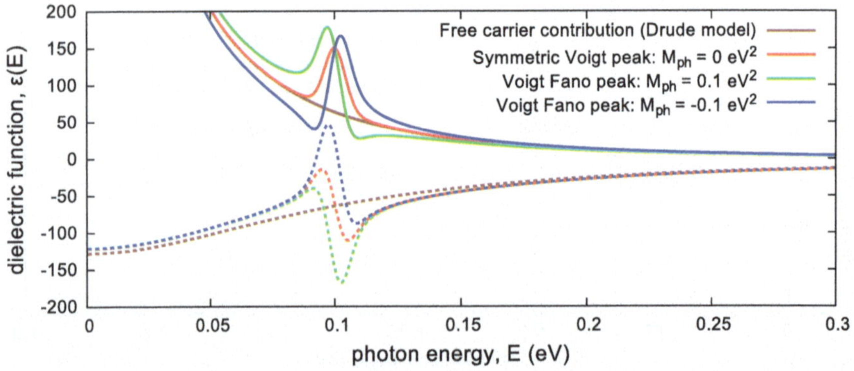

Fig. 3.5 Contributions to the dielectric function of asymmetric Voigt Fano ε-broadened peaks representing redistribution of transition strength due to Fano resonance effect between phonon peak and free carrier contribution. The dielectric functions with symmetrical peak (without Fano resonance effect) and without phonon peak (Drude contribution) are added for comparison. The parameters corresponding to Drude model and phonon peak were chosen as follows: $N_\text{D} = 2\,\text{eV}^2$, $B_\text{D} = 0.1\,\text{eV}$, $N_\text{ph} = 0.1\,\text{eV}^2$, $E_\text{ph} = 0.1\,\text{eV}$, $B_\text{ph} = 0.01\,\text{eV}$ (FWHM), $L_\text{ph} = 0.2$. The value of the parameter M_ph is different for each curves

the imaginary part of the dielectric function in zero energy, i. e. static conductivity. Therefore, using this model as a part of a model without free carrier contribution leads to an artificial positive or negative Drude-like singularity in the dielectric function.

3.3.1.5 Temperature Dependence of One-Phonon Absorption

In the equilibrium the mean values of the phonon occupation numbers are determined by the Bose–Einstein statistics

$$f^{\mathrm{BE}}(E_p, T) = \frac{1}{\exp(E_p/(k_{\mathrm{B}}T)) - 1} . \tag{3.83}$$

This statistical factor can be derived in the quasiparticle approximation from the probability factor of open systems (3.21). In the quasiparticle approximation the probability of the one-phonon absorption process is proportional to the occupation number plus one while the probability of the one-phonon emission process is proportional to the occupation number. The resulting transition strength of the one-phonon absorption peak at energy E_p is a sum of the transition strength of the absorption process $N_{+p}(T)$, which is positive, and the transition strength of the emission process $N_{-p}(T)$, which is negative,

$$N_p(T) = N_{+p}(T) + N_{-p}(T) \sim f^{\mathrm{BE}}(E_p, T) + 1 - f^{\mathrm{BE}}(E_p, T) = 1 . \tag{3.84}$$

This equation says that the strength of the one-phonon absorption is not temperature-dependent through the statistical factor. However, the resonant frequency of phonons E_p and broadening B_p are temperature dependent. Moreover, the phonon matrix elements are slightly temperature-dependent, meaning that the transition strength N_p could also be slightly temperature-dependent. The temperature dependencies of the phonon frequencies are mostly a consequence of the thermal expansion and could be modeled by the average Bose–Einstein statistical factor with dependency given by the average phonon energy [51–53]. In this case, the temperature dependence can be parameterized using the following three parametric formula

$$E_p(T) = E_p^{0K} + \left(E_p^{300K} - E_p^{0K}\right) \frac{\exp(\Theta/300K) - 1}{\exp(\Theta/T) - 1} , \tag{3.85}$$

where E_p^{0K} and E_p^{300K} are the phonon energies for zero and room temperature. Note that the $E_p(T)$ decreases with temperature ($E_p^{300K} < E_p^{0K}$). The parameter Θ is the average phonon energy in Kelvins. This parameter is common to all phonon modes. It represents the contribution from all the phonons which participate in the thermal expansion (i. e. also the optically inactive phonons). Therefore, $k_{\mathrm{B}}\Theta$ is a completely different quantity than the discrete energies E_p of the optically active phonons.

The same type of temperature dependency could be also used for the parts of the broadening factors B_p representing the finite lifetime of phonons [54]. In this case, the broadening factors increase with temperature ($B_p^{300K} > B_p^{0K}$).

3.3.1.6 Multi-phonon Absorption

Multi-phonon absorption is very weak effect compared to the one-phonon absorption, with the exception of homopolar crystalline materials, such as diamond, silicon or germanium. In homopolar materials the multi-phonon absorption is the dominant effect in the IR region because it is not masked by the one-phonon absorption processes which are mostly prohibited due to the selection rules (vanishing dipole matrix elements).

In materials with non-negligible charge transfer (e. g. polar crystals, disordered solids) where the one-phonon absorption is dominant, it is difficult to separate the absorption spectra into parts corresponding to one-phonon and multi-phonon absorption. The magnitude of multi-phonon absorption effects often lies close to the limit of experimental accuracy. Therefore, the choice of the model used for the multi-phonon absorption is not usually very important and the same model as for one-phonon absorption but with broader peaks and smaller transition strength can be utilized. We will not discuss this model in this section and rather, we will focus on the models needed to describe the multi-phonon absorption in the homopolar crystalline materials.

In one-phonon absorption processes, only the phonons with zero momentum (i. e. at the Γ point) contribute to the absorption processes. This is a consequence of the fact that the total momentum in the absorption process must be conserved and the momentum of photons is negligible. In multi-phonon processes the situation is more complicated since we must take into account all the absorption and emission processes for which the sum of momenta of all phonons participating in these processes is zero. In the case of two-phonon processes, the JDOS can be calculated using an integral over the Brillouin zone. In addition to the integration over the Brillouin zone, the summation over all the optical and acoustical phonon branches must be performed. For three and more phonon processes, the situation is even more complicated since multiple integrals over the Brillouin zone are needed and the summation over phonon branches is more complicated. In this section, we will concentrate on models describing the two-phonon absorption.

It is known from the theory of crystalline solids that the integration over the Brillouin zone mentioned above leads to singularities in the derivative of the JDOS function called Van Hove singularities [15, 18]. The Van Hove singularities correspond to critical points in the Brillouin zone which are located at the points of high symmetry or near the points of lower symmetry. In the three-dimensional space (3D) there are four types of the Van Hove singularities. The M_0 type singularity corresponds to critical points representing the minima of energy of absorption processes, the M_1 and M_2 types correspond to saddle points and the M_3 type corresponds to maxima. From the theory, it follows that each type of critical point must occur at

Table 3.1 Shape functions L_i modeling 3D Van Hove singularities

	3D (E_0, E_1, E_2, E_3)		
	$E_0 \le E \le E_1$	$E_1 \le E \le E_2$	$E_2 \le E \le E_3$
M_0	$L_0(E) = \sqrt{X_{\mathrm{I}}(E)}$	$Y_{\mathrm{II}}(E)$	0
M_1	$L_1(E) = 1 - \sqrt{Y_{\mathrm{I}}(E)}$	$Y_{\mathrm{II}}(E)$	0
M_2	$L_2(E) = 0$	$X_{\mathrm{II}}(E)$	$1 - \sqrt{X_{\mathrm{III}}(E)}$
M_3	$L_3(E) = 0$	$X_{\mathrm{II}}(E)$	$\sqrt{Y_{\mathrm{III}}(E)}$

least once for each combination of phonon branches but some of them cannot be distinguished in the spectra due to the degeneracy of phonon branches in the points of symmetry. Note that even in the situation with many critical points the JDOS function can be expressed as a sum of contributions corresponding to basic sequences M_0–M_1–M_2–M_3 of critical points. Moreover, some Van Hove singularities may not be visible in the absorption spectra due to vanishing dipole matrix element. Therefore, the multi-phonon absorption for a single combination of phonon branches can be modeled using the ε-broadened piecewise continuous JDOS functions within the energy interval between E_0–E_3 and discontinuities in the energies E_1 and E_2. The number of absorption bands corresponding to combination of phonon branches is specific for each material. Although we will not discuss the individual materials here, the general aspects of the model of two-phonon absorption will be presented.

The JDOS function for the 3D isotropic Van Hove singularities can be modeled by the shape functions L_i introduced in Table 3.1. The functions X_{I}, X_{II}, X_{III} and Y_{I}, Y_{II}, Y_{III} are defined as follows

$$X_l(E) = \frac{E - E_{l-1}}{E_l - E_{l-1}}, \quad Y_l(E) = \frac{E_l - E}{E_l - E_{l-1}} \quad \text{for} \quad E_{l-1} \le E \le E_l\,, \tag{3.86}$$

where index $l = 1(\mathrm{I}), 2(\mathrm{II}), 3(\mathrm{III})$. The schematic diagram of the JDOS function corresponding to the basic sequence of the critical points is depicted in the upper panel of Fig. 3.6. The behavior around the isotropic critical points is described by the quadratic form with eigenvalues having the same size with the number of negative eigenvalues determining the type of the critical point. In reality, most of the critical points, especially at the boundary of the Brillouin zone, are anisotropic, i. e. with differently sized eigenvalues of the quadratic form. In the extreme cases, the anisotropic 3D critical points may appear as points in lower dimensions, i. e. as 2D (one eigenvalue is zero) or 1D (two eigenvalues are zero) critical points. The JDOS function for the anisotropic critical points can be approximated by the linear combination of contributions corresponding to 3D, 2D and in very rare cases also 1D Van Hove singularities. The functions corresponding to 2D and 1D singularities are introduced in Tables 3.2 and 3.3. The columns in these tables correspond to different alignment of 2D and 1D critical points with respect to 3D critical points. The

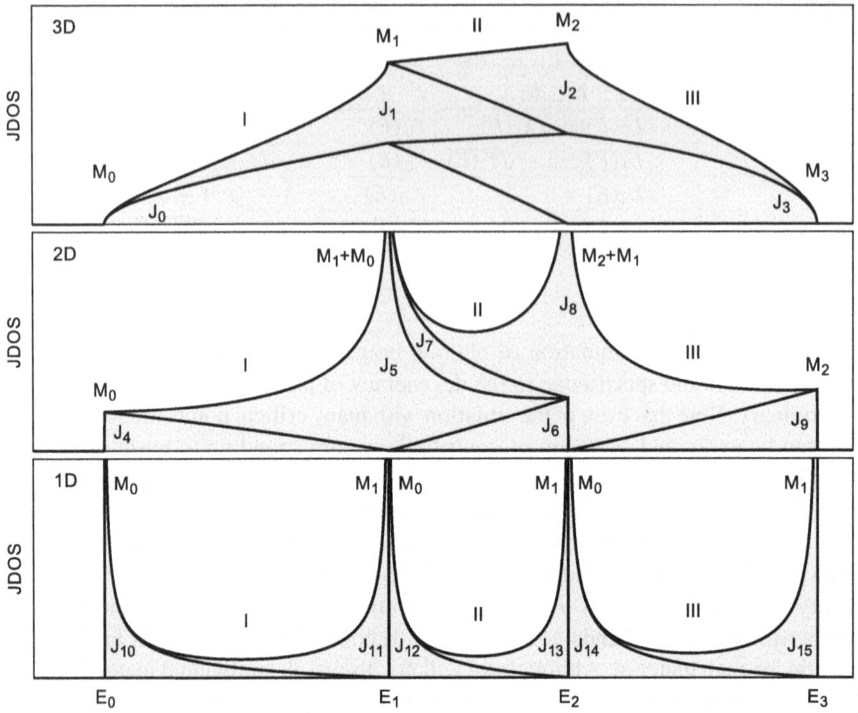

Fig. 3.6 Schematic diagram of joint density of states (JDOS) of two-phonon absorption processes corresponding to 3D, 2D and 1D Van Hove singularities. The individual contributions are calculated as $J_i(E) = E^2 L_i(E)$ with assumptions $E_0 = 0.8E_3$, $E_1 = 0.88E_3$ and $E_2 = 0.93E_3$

Table 3.2 Shape functions L_i modeling 2D Van Hove singularities

	2D (E_0, E_1, E_2)		2D (E_1, E_2, E_3)	
	$E_0 \le E \le E_1$	$E_1 \le E \le E_2$	$E_1 \le E \le E_2$	$E_2 \le E \le E_3$
M_0	$L_4(E) = Y_{\rm I}(E)$	0	$L_7(E) = Y_{\rm II}(E)$	0
M_1	$L_5(E) = -\ln Y_{\rm I}(E)$	$-\ln X_{\rm II}(E)$	$L_8(E) = -\ln Y_{\rm II}(E)$	$-\ln X_{\rm III}(E)$
M_2	$L_6(E) = 0$	$X_{\rm II}(E)$	$L_9(E) = 0$	$X_{\rm III}(E)$

schematic diagrams of the contributions to the JDOS function corresponding to the 2D and 1D singularities are shown in the middle and bottom panels of Fig. 3.6.

The resulting normalized imaginary part of the dielectric function describing the absorption band of two-phonon absorption combining the phonon branches A and B can be calculated as follows

$$\varepsilon^0_{\rm i,2ph}(E) = \frac{f_{\rm A\pm B}(E,T)}{C_{\rm N}} H(E) \sum_{i=0}^{15} A_i L_i(E)\,, \tag{3.87}$$

Table 3.3 Shape functions L_i modeling 1D Van Hove singularities

	1D (E_0, E_1)	1D (E_1, E_2)	1D (E_2, E_3)
	$E_0 \le E \le E_1$	$E_1 \le E \le E_2$	$E_2 \le E \le E_3$
M_0	$L_{10}(E) = 1/\sqrt{X_{\mathrm{I}}(E)} - 1$	$L_{12}(E) = 1/\sqrt{X_{\mathrm{II}}(E)} - 1$	$L_{14}(E) = 1/\sqrt{X_{\mathrm{III}}(E)} - 1$
M_1	$L_{11}(E) = 1/\sqrt{Y_{\mathrm{I}}(E)} - 1$	$L_{13}(E) = 1/\sqrt{Y_{\mathrm{II}}(E)} - 1$	$L_{15}(E) = 1/\sqrt{Y_{\mathrm{III}}(E)} - 1$

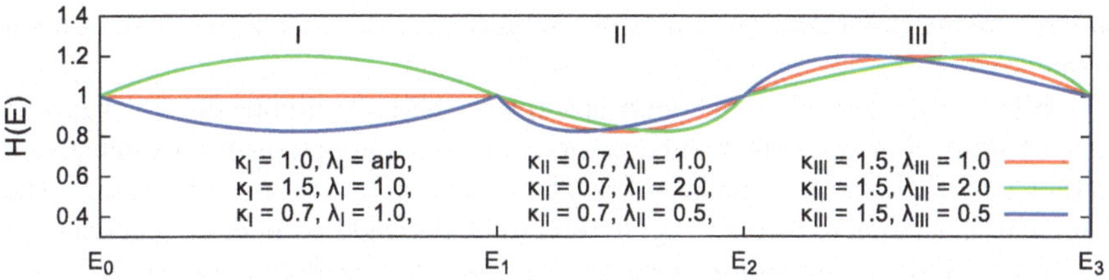

Fig. 3.7 Examples of the shape modifying function $H(E)$ for different values of parameters

where C_{N} is the normalization constant, A_i are weights describing the strength of the individual L_i contributions and $f_{\mathrm{A\pm B}}(E, T)$ is the temperature dependent factor. The index A $\pm$ B distinguishes absorption bands corresponding to different combinations of A and B branches.

The function $H(E)$ is introduced in order to change the shape of the dielectric function in the intervals between the critical points. This function is important because the linear combination of the L_i functions provides correct description only in the vicinity of the Van Hove singularities, meaning that additional parameters influencing the shape of the dielectric function between these singularities are needed. The function $H(E)$ can be, for example, defined as follows

$$H(E) = \frac{\kappa_l \lambda_l Y_l(E)}{(\kappa_l \lambda_l - 1)Y_l(E) + 1} + \frac{\kappa_l/\lambda_l X_l(E)}{(\kappa_l/\lambda_l - 1)X_l(E) + 1} \quad \text{for} \quad E_{l-1} \le E \le E_l\,, \tag{3.88}$$

where κ_l and λ_l are non-zero positive parameters regulating the shape in the corresponding intervals. This definition ensures that $H(E)$ is continuous in the interval between E_0 and E_3 and it is equal to unity at the critical point energies. This function is smooth on the intervals between the critical point energies but its derivatives may exhibit jumps at E_1 and E_2. Note that if the shape modifying parameters κ_l is equal to unity then the function $H(E)$ is constant in the corresponding interval (see Fig. 3.7).

If the structure of the critical points is known for the studied material, it is convenient to express the critical point energies E_0–E_3, which are used to model the Van Hove singularities, using energies of individual phonons participating in the two-phonon absorption process. These energies are then given as a sum (A+B) or difference (A−B) of phonon energies at critical points

$$E_0 = E_{\mathrm{A}}^{M_0} \pm E_{\mathrm{B}}^{M_0}, \quad \ldots \quad E_3 = E_{\mathrm{A}}^{M_3} \pm E_{\mathrm{B}}^{M_3}\,. \tag{3.89}$$

This procedure reduces the number of independent parameters in the model. For example, the two-phonon absorption in crystalline silicon was modeled in [55] by 15 absorption bands with critical energies determined by the set of only 17 independent phonon frequencies.

The temperature dependencies of the critical point energies are modeled by the same average Bose–Einstein statistical factor as in the case of one-phonon absorption, i. e. the temperature dependency is described by the formula (3.85). The same average phonon energy parameter Θ is used for temperature dependencies of all phonon frequencies.

While in the case of one-phonon absorption the temperature dependencies of the probabilities of absorption and emission did not result in temperature dependent factor (see (3.84)) the situation is different for two-phonon absorption processes. The temperature dependent factor $f_{\mathrm{A\pm B}}(E,T)$ can be derived by the same procedure as for the one-phonon absorption in Sect. 3.3.1.5 but the creation/annihilation of two phonons must be taken into account. In the case of simultaneous creation of two phonons or simultaneous annihilation of two phonons temperature dependent factor is given by

$$f_{\mathrm{A+B}}(E,T) = 1 + f^{\mathrm{BE}}(E_{\mathrm{A}}(E),T) + f^{\mathrm{BE}}(E_{\mathrm{B}}(E),T)\,, \tag{3.90}$$

where $E_{\mathrm{A}}(E)$ and $E_{\mathrm{B}}(E)$ describe the phonon energies in A and B branches participating in the absorption or emission processes of photon with energy E. A similar calculation gives the temperature dependent factor for processes, in which one phonon is created and one phonon is annihilated

$$f_{\mathrm{A-B}}(E,T) = f^{\mathrm{BE}}(E_{\mathrm{B}}(E),T) - f^{\mathrm{BE}}(E_{\mathrm{A}}(E),T)\,. \tag{3.91}$$

The values of the functions $E_{\mathrm{A}}(E)$ and $E_{\mathrm{B}}(E)$ at the critical point energies are known from the known structure of phonon branches. In the regions between these energies they can be approximated with the help of linear interpolation:

$$E_{\mathrm{A}}(E) = E_{\mathrm{A}}^{M_{l-1}} Y_l(E) + E_{\mathrm{A}}^{M_l} X_l(E) \quad \text{for} \quad E_{l-1} \le E \le E_l\,. \tag{3.92}$$

The same linear interpolation is used to approximate $E_{\mathrm{B}}(E)$. We should emphasize that the temperature-dependent factors (3.90) and (3.91) do not require the introduction of any new parameters.

The complex dielectric function for two-phonon absorption is calculated from the unbroadened imaginary part of the dielectric function calculated with (3.87) by the ε-broadening procedure described in Sect. 3.2.3. This step can be performed only numerically, i. e. the convolution integrals in (3.49) must be performed numerically. Alternatively, we can split the intervals between the critical point energies into sufficiently small subintervals (E_{j-1}, E_j), use the polynomial approximation of the function (3.87) on these subintervals (e.g. cubic splines) and then use the analytical results for broadened polynomials. On the j-th interval we can write

$$\varepsilon^0_{\mathrm{i},j}(E) = \sum_{n=0}^{m} a_{j,n} E^n \qquad \text{for} \qquad E_{j-1} \le E \le E_j\,, \tag{3.93}$$

where m is the degree of the approximating polynomial and $a_{j,n}$ are its coefficients. Thus, the contribution of this subinterval to the broadened dielectric function is expressed as follows:

$$\tilde{\hat{\varepsilon}}^0_{j+}(E) = \int_{E_{j-1}}^{E_j} \hat{\beta}(E-t) \sum_{n=0}^{m} a_{j,n}\, t^n \,\mathrm{d}t = \int_{E-E_{j-1}}^{E-E_j} \hat{\beta}(x) \sum_{n=0}^{m} A_{j,n}(E)\, x^n \,\mathrm{d}x\,, \tag{3.94}$$

where

$$A_{j,n}(E) = (-1)^n \sum_{k=n}^{m} a_{j,k} \binom{k}{n} E^{k-n}\,. \tag{3.95}$$

The contribution $\tilde{\hat{\varepsilon}}^0_{j+}(E)$ is calculated for the interval (E_{j-1}, E_j) lying in positive values of energy. Because of the time reversal symmetry relating the dielectric function for positive and negative energies we can write the result that includes the contributions from both intervals (E_{j-1}, E_j) and $(-E_j, -E_{j-1})$ as

$$\tilde{\hat{\varepsilon}}^0_j(E) = \sum_{n=0}^{m}\sum_{k=n}^{m} a_{j,k} \binom{k}{n} (-E)^{k-n} \Big[(-1)^k \Big(\hat{B}_n(E-E_{j-1}) - \hat{B}_n(E-E_j) \Big) + \hat{B}_n(E+E_{j-1}) - \hat{B}_n(E+E_j) \Big]\,, \tag{3.96}$$

where $\hat{B}_n(x)$ are the functions defined by the following integrals

$$\hat{B}_n(x) = \int x^n\, \hat{\beta}(x)\,\mathrm{d}x\,. \tag{3.97}$$

For the Lorentzian broadening, the results of the integrals can be expressed analytically

$$\hat{B}_{\mathrm{L},n}(x) = -\frac{1}{\pi}\left(\frac{-\mathrm{i}B}{2}\right)^n \left[\ln\left(1 - \frac{\mathrm{i}2x}{B}\right) + \sum_{k=0}^{n-1} \binom{n}{k} \frac{1}{n-k} \left(\frac{\mathrm{i}2x}{B} - 1\right)^{n-k} \right]. \tag{3.98}$$

In the case of the Gaussian broadening the integrals can be expressed with the help of the following recurrent formulas

$$\hat{B}_{\mathrm{G},n}(x) = D_n(x) + \mathrm{i}G_n(x)\,, \tag{3.99}$$

where

$$D_0(x) = -\frac{2}{\pi}\mathrm{D_i}\left(\frac{x}{\sqrt{2}B}\right), \qquad G_0(x) = \frac{1}{2}\mathrm{erf}\left(\frac{x}{\sqrt{2}B}\right), \tag{3.100}$$

$$D_1(x) = \frac{\sqrt{2}B}{\pi}\mathrm{D}\left(\frac{x}{\sqrt{2}B}\right) - \frac{x}{\pi}, \qquad G_1(x) = \frac{B}{\sqrt{2\pi}}\left(1 - \exp\left(-\frac{x^2}{2B^2}\right)\right), \tag{3.101}$$

$$D_n(x) = (n-1)B^2 D_{n-2}(x) + \frac{\sqrt{2}B}{\pi}x^{n-1}\mathrm{D}\left(\frac{x}{\sqrt{2}B}\right) - \frac{1}{n\pi}x^n, \tag{3.102}$$

$$G_n(x) = (n-1)B^2 G_{n-2}(x) - \frac{B}{\sqrt{2\pi}}x^{n-1}\exp\left(-\frac{x^2}{2B^2}\right). \tag{3.103}$$

The recurrent formulas use three special functions, the Dawson function $\mathrm{D}(x)$ (3.52), its integral $\mathrm{D_i}(x)$ and the error function $\mathrm{erf}(x)$ defined as

$$\mathrm{D_i}(x) = \int_0^x \mathrm{D}(t)\,\mathrm{d}t, \qquad \mathrm{erf}(x) = \frac{2}{\sqrt{\pi}}\int_0^x \exp(-t^2)\,\mathrm{d}t. \tag{3.104}$$

All three functions can be efficiently calculated in computer.

In the case of the Voigt broadening (3.74), it should be possible to write the result using complex special functions but in practice the approximation of the Voigt broadening by the linear combination of the Lorentzian and Gaussian broadening is sufficient (see Sect. 3.3.1.4).

In the presented model of the two-phonon absorption, the functions describing the Van Hove singularities L_0–L_{15} are used to model the imaginary part of the dielectric function. Similar model of the two-phonon absorption in crystalline silicon was presented in [55]. The functions describing the Van Hove singularities were used to model the transition strength function instead of the imaginary part of the dielectric function. This required a different definition of the shape function $H(E)$ because if the critical point energy E_0 is zero then the imaginary part of the dielectric function must behave as $\varepsilon_i(E) \propto E$ for small energies.

3.3.2 Valence Electron Excitations

The valence electron excitations correspond to transition of electrons from the occupied valence band to the empty conduction band or to empty higher energy excitation band. In the crystalline solids in the quasiparticle approximation the one-particle states have well-defined momentum. Due to the momentum conservation it is possible to distinguish between direct and indirect valence-to-conduction interband electron transitions. This classification cannot, however, be used in the case of amorphous materials. Therefore, the valence electron excitations must be studied separately for crystalline and amorphous materials.

3.3.2.1 Direct Valence Electron Excitations in Crystalline Solids

In the frame of one-electron approximation the dielectric function can be modeled with the help of the 3D and 2D Van Hove singularities [13, 15, 18, 21, 56, 57] described in detail in Sect. 3.3.1.6. The imaginary part of the normalized dielectric function is then given by the formula:

$$\varepsilon^{0}_{\mathrm{i,dt}}(E) = \frac{f_{\mathrm{VC}}(E,T)}{C_{\mathrm{N}}E^2} H(E) \sum_{i=0}^{9} A_i J_i(E), \tag{3.105}$$

where $f_{\mathrm{VC}}(E,T)$ is a temperature dependent factor, C_{N} is the normalization constant, the shape modifying function $H(E)$ is defined by (3.88) and the weights A_i describe the strength of individual contributions $J_i(E)$. In the absence of many-body effects, the contributions $J_i(E)$ are given as $J_i(E) = L_i(E)$ with functions $L_i(E)$ given in Tables 3.1 and 3.2. While in the case of two-phonon absorption described in Sect. 3.3.1.6 the functions $L_i(E)$ modeled the imaginary part of the dielectric function, the situation is different in the case of direct transitions (DT). More specifically, these functions are used to model the JDOS function (notice the factor E^2 in the denominator in (3.105)).

In contrast to phonons which could be, for most purposes, considered independent quasiparticles, the interactions between electrons cannot be neglected. The electron–hole interactions are manifested via two phenomena in the absorption spectra. The first phenomenon is the appearance of the discrete transitions, called excitons, in the region below the minimal energy of interband electronic transitions E_0. These discrete transitions correspond to bonded states between the electron–hole pairs. The second phenomenon is the redistribution of the transition strength (probability of transition) from higher energies to lower energies. This redistribution represents the effect of the continuum part of the spectra in the electron–hole system. In the frame of the Elliott theory [15, 24, 57–59], the contribution describing the discrete part $A_0 J^{\mathrm{3D}}_{0\mathrm{ex}}(E)$ must be added as an extra term in the sum in (3.105) and the square root parts of the contributions $J_0(E)$ and $J_3(E)$ must be modified. The schematic diagram of the JDOS function with and without the many-body effects is in the upper panel in Fig. 3.8. The function $J^{\mathrm{3D}}_{0\mathrm{ex}}(E)$ is defined as

$$J^{\mathrm{3D}}_{0\mathrm{ex}}(E) = S^{\mathrm{3D}}_{0}(E_1) \sum_{n=1}^{\infty} \frac{2R}{n^3}\, \delta\!\left(E - E_0 + \frac{R}{n^2}\right), \tag{3.106}$$

where the parameter R is the Rydberg energy describing the strength of the electron–hole interaction. Since the discrete part lies below the critical point energy E_0 we

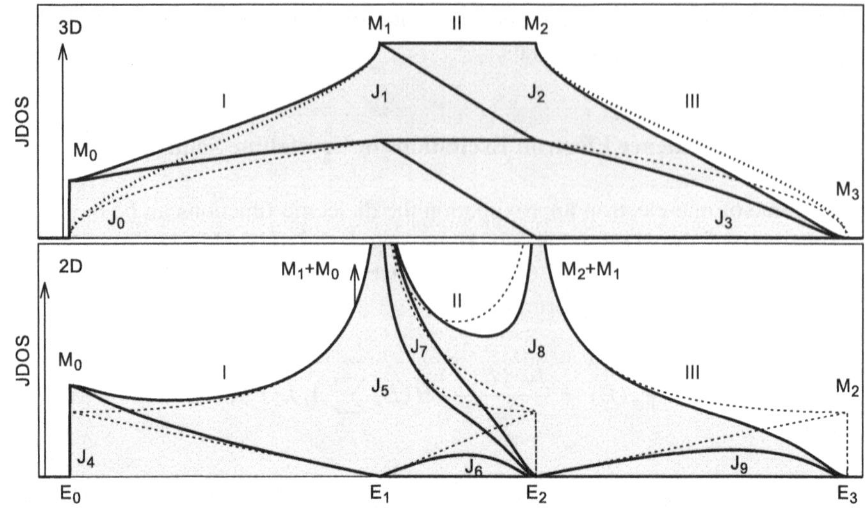

Fig. 3.8 Schematic diagram of joint density of states (JDOS) of direct valence electron excitations corresponding to 3D and 2D Van Hove singularities, assuming $R = 0.004E_3$, $E_0 = 0.5E_3$, $E_1 = 0.7E_3$ and $E_2 = 0.8E_3$. The dotted lines correspond to individual contributions without many-body effects while the solid lines correspond to contributions that include excitonic effects, i.e. corrected by the Sommerfield factors. The arrows indicate the positions of the first excitons ($n = 1$) in the series of discrete transitions

must extend the definition of the function $H(E)$ in such a way that it is equal to unity for $E < E_0$. The functions $J_0(E)$ and $J_3(E)$ are changed to

$$J_0(E) = \frac{S_0^{3D}(E_1)}{S_0^{3D}(E)} \qquad \text{for} \quad E_0 \le E \le E_1\,, \tag{3.107}$$

$$J_3(E) = \frac{S_3^{3D}(E_2)}{S_3^{3D}(E)} \qquad \text{for} \quad E_2 \le E \le E_3\,, \tag{3.108}$$

The functions $S_0^{3D}(E)$ and $S_3^{3D}(E)$, which are called the Sommerfield factors, are defined as

$$S_0^{3D}(E) = 1 - \exp\left(-2\pi\sqrt{\frac{R}{E - E_0}}\right), \quad S_3^{3D}(E) = \exp\left(2\pi\sqrt{\frac{R}{E_3 - E}}\right) - 1\,. \tag{3.109}$$

In the limit of small R, the functions $J_0(E)$ and $J_3(E)$ are identical with $L_0(E)$ and $L_3(E)$.

In the case of the 2D Van Hove singularities the Elliott theory gives two series of discrete transitions, one below the critical point energy E_0 and the other below the E_1:

$$J_{0\text{ex}}^{2\text{D}}(E) = S_0^{2\text{D}}(E_1) \sum_{n=1}^{\infty} \frac{32R}{(2n-1)^3}\, \delta\left(E - E_0 + \frac{4R}{(2n-1)^2}\right), \tag{3.110}$$

$$J_{1\text{ex}}^{2\text{D}}(E) = S_1^{2\text{D}}(E_2) \sum_{n=1}^{\infty} \frac{32R}{(2n-1)^3}\, \delta\left(E - E_1 + \frac{4R}{(2n-1)^2}\right). \tag{3.111}$$

These functions must be added to the sum in (3.105) as two extra terms $A_4 J_{0\text{ex}}^{2\text{D}}(E) + A_7 J_{1\text{ex}}^{2\text{D}}(E)$. Moreover, the contributions 4, 6, 7 and 9 must be changed to:

$$J_4(E) = Y_\text{I}(E) \frac{S_0^{2\text{D}}(E_1)}{S_0^{2\text{D}}(E)} \qquad \text{for} \quad E_0 \le E \le E_1, \tag{3.112}$$

$$J_6(E) = X_\text{II}(E) \frac{S_2^{2\text{D}}(E_1)}{S_2^{2\text{D}}(E)} \qquad \text{for} \quad E_1 \le E \le E_2, \tag{3.113}$$

$$J_7(E) = Y_\text{II}(E) \frac{S_1^{2\text{D}}(E_2)}{S_1^{2\text{D}}(E)} \qquad \text{for} \quad E_1 \le E \le E_2, \tag{3.114}$$

$$J_9(E) = X_\text{III}(E) \frac{S_3^{2\text{D}}(E_2)}{S_3^{2\text{D}}(E)} \qquad \text{for} \quad E_2 \le E \le E_3, \tag{3.115}$$

where the 2D Sommerfield factors are defined as

$$S_0^{2\text{D}}(E) = 1 + \exp\left(-2\pi\sqrt{\frac{R}{E - E_0}}\right), \quad S_2^{2\text{D}}(E) = 1 + \exp\left(2\pi\sqrt{\frac{R}{E_2 - E}}\right), \tag{3.116}$$

$$S_1^{2\text{D}}(E) = 1 + \exp\left(-2\pi\sqrt{\frac{R}{E - E_1}}\right), \quad S_3^{2\text{D}}(E) = 1 + \exp\left(2\pi\sqrt{\frac{R}{E_3 - E}}\right). \tag{3.117}$$

The models of the discrete transitions below E_0, which were introduced above are correct as long as the critical point M_0 is purely 3D, purely 2D, or if the Rydberg energy R is smaller than the broadening parameter and the discrete transitions appear as a single structure (peak) in the absorption spectra. In the case of anisotropic critical point M_0 and Rydberg energy larger than the broadening parameter, it must be ensured that the series of discrete transitions is not doubled. A useful usable approximation is the formula with weighted averages for the amplitudes and positions of discrete transitions. The imaginary part of the normalized dielectric function is in this case given as

$$\varepsilon^0_{\mathrm{i,dt}}(E) = \frac{f_{\mathrm{VC}}(E,T)}{C_{\mathrm{N}}E^2} H(E) \left[\sum_{n=1}^{\infty} A_{0\mathrm{ex},n}\, \delta(E - E_{0\mathrm{ex},n}) + \sum_{n=1}^{\infty} A_{1\mathrm{ex},n}\, \delta(E - E_{1\mathrm{ex},n}) + \sum_{i=0}^{9} A_i J_i(E) \right], \tag{3.118}$$

where the amplitudes and energies of excitons are expressed as

$$A_{0\mathrm{ex},n} = \frac{R}{A_0 + A_4} \left(\frac{2A_0 S_0^{\mathrm{3D}}(E_1)}{n^3} + \frac{32 A_4 S_0^{\mathrm{2D}}(E_1)}{(2n-1)^3} \right), \tag{3.119}$$

$$E_{0\mathrm{ex},n} = E_0 - \frac{R}{A_0 + A_4} \left(\frac{A_0}{n^2} + \frac{4A_4}{(2n-1)^2} \right), \tag{3.120}$$

$$A_{1\mathrm{ex},n} = \frac{32 R A_7 S_1^{\mathrm{2D}}(E_2)}{(2n-1)^3}, \qquad E_{1\mathrm{ex},n} = E_1 - \frac{4R}{(2n-1)^2}. \tag{3.121}$$

The temperature-dependent factor $f_{\mathrm{VC}}(E,T)$ can be derived using the Fermi–Dirac statistics. If the Fermi energy lies in the band gap and both the valence and conduction bands are sufficiently far from the Fermi energy, then the value of this factor can be set to unity. Therefore, temperature dependency in dielectrics and semiconductors introduced by this factor is negligible. The temperature dependencies of the critical point energies could be described by the same formula as in the case of phonons, i. e. with the help of the average Bose–Einstein statistical factor (3.85). In contrast to phonons, separate average phonon energy parameters Θ_0–Θ_3 must be used for individual critical point energies E_0–E_3 [52, 53].

The presented model uses the functions J_0–J_9 describing the Van Hove singularities to model the JDOS function. In materials where all the critical point energies E_0 are greater than zero, i. e. dielectrics and semiconductors, it is not specifically important that the functions describing the Van Hove singularities are used to model the JDOS function, the transition strength function or the imaginary part of the dielectric function. Therefore, the factor E^2 in the denominator in (3.118) could be replaced by E or 1, it is even possible to use a generic factor E^κ with κ being the parameter of the model. The ε-broadening should be used if DT in dielectrics or semiconductors are modeled. In contrast to phonons, where only a single broadening parameter is typically used, in the case of the DT it is necessary to use several broadening parameters. It is often necessary to use different broadening parameters for structures belonging to individual critical points. For example, the structure around M_0 (contributions $J_{0\mathrm{ex}}$, J_0, J_4) are broadened using B_0, the structure around M_1 (contributions $J_{1\mathrm{ex}}$, J_1, J_5) by B_1 etc.

Only the 3D and 2D Van Hove singularities were considered in this section. In principle the model can be extended to include also 1D singularities with corresponding 1D excitonic effects.

The slightly modified variety of the model of DT discussed in this section was used in the temperature-dependent dispersion model of crystalline silicon [60]. It was shown that the presented model could be understood as the extension of the models of Adachi [19, 61], Kim et al. [21, 22], Tanguy [23, 24] and Herzinger and Johs [25, 26].

In metals, the parabolic band approximation cannot be used to describe the behavior around the minimal energy of DT if the initial or final band is not fully filled. Therefore, the JDOS around this minimal energy cannot be described as a Van Hove M_0 singularity and a special model must be devised. Moreover, the initial or final energy is given by the Fermi energy and, therefore, the thermal dependent factor cannot be approximated by unity. We will not discuss the models of DT in metals in this chapter.

Another special case is graphene. The graphene forms a two-dimensional structure in which the valence and conduction bands touch at the K points of the hexagonal Brillouin zone. Moreover, the bands around the points of contact form the cones with the Fermi energy lying exactly at the level where the vertexes of the cones touch. This band structure together with the assumption of the constant momentum (current) matrix element results in constant conductivity (constant transition strength function), known as the universal conductivity. In this case the unbroadened transition strength function of the ordinary part of the dielectric response can be modeled with the help of the functions $F_4(E) \equiv L_4(E)$, $F_5(E) \equiv L_5(E)$ and $F_6(E) \equiv J_6(E)$ describing the 2D Van Hove singularities with critical point energy $E_0 = 0$. In this special case the F-broadening must be used instead of ε-broadening and, furthermore, it must be performed prior to the multiplication by the the temperature dependent factor

$$\hat{\varepsilon}^0_{\mathrm{dt}}(E) = \frac{f_{\mathrm{VC}}(E,T)}{C_{\mathrm{N}} E} \left[\hat{\beta} * \left(\sum_{i=4,5,6} A_i F_i(E) \right) \right], \tag{3.122}$$

where the temperature dependent factor $f_{\mathrm{VC}}(E, T)$ is given as

$$f_{\mathrm{VC}}(E,T) = f^{\mathrm{FD}}(-E/2,T) - f^{\mathrm{FD}}(E/2,T), \tag{3.123}$$

where $f^{\mathrm{FD}}(E, T)$ is the Fermi–Dirac statistical factor with the Fermi energy equal to zero

$$f^{\mathrm{FD}}(E,T) = \frac{1}{\exp(E/(k_{\mathrm{B}}T)) + 1}. \tag{3.124}$$

The similar absorption structure appears also in graphite [48] which is a 3D structure with strongly anisotropic critical points.

3.3.2.2 Indirect Valence Electron Excitations in Crystalline Solids

Apart from the direct electronic transitions there are also indirect electronic transitions in which the transitions of electrons between the valence and conduction bands are accompanied by a simultaneous change in phonon occupation numbers. The dominant absorption processes in this case are the processes involving creation or annihilation of one phonon. The processes in which multiple phonon occupation numbers are changed are much weaker and they can be neglected. Thus, the temperature-dependent imaginary part of the dielectric function can be separated into two parts

$$\varepsilon_{\mathrm{i,idt}}(E) = \sum_p \frac{N_p}{E} \frac{\left(f^{\mathrm{BE}}(E_p,T)+1\right) F^0_{\mathrm{ab}}(E-E_p) + f^{\mathrm{BE}}(E_p,T) F^0_{\mathrm{ab}}(E+E_p)}{2 f^{\mathrm{BE}}(E_p, 300K)+1}, \tag{3.125}$$

where the summation over p is performed for phonon branches corresponding to possible energies of phonons E_p ensuring momentum conservation in indirect absorption processes. The part with $E - E_p$ corresponds to processes in which a phonon is created and the part with $E + E_p$ corresponds to processes in which a phonon is annihilated. The term in the denominator ensures proper normalization of the terms at 300 K. The parameters N_p are the transition strengths for the corresponding phonon branches at 300 K. The Bose–Einstein statistical factors f^{BE} (3.83) determine the phonon occupation numbers, i. e. probabilities of the processes [7, 15, 62]. In principle we should also include the temperature dependent factor $f_{\mathrm{VC}}(E,T)$ but its effect can be usually neglected. The normalized transition strength function of indirect transitions (IDT) corresponding to one branch when the phonon is created or annihilated is calculated as

$$F^0_{\mathrm{ab}}(E) = \frac{(|E|-E_{\mathrm{g}})^2 (E_{\mathrm{h}}-|E|)^2}{C_{\mathrm{N}}\,|E|^{\kappa}\left((|E|-E_{\mathrm{r}})^2 + B^2/4\right)^{\nu}}\, \Pi_{E_{\mathrm{g}},E_{\mathrm{h}}}(|E|), \tag{3.126}$$

where the symbols E_{g} and E_{h} denote the minimal and maximal energy needed for excitation of electrons from the valence band to the conduction band. The parameter ν could be 0 or 1. If $\nu = 0$, the formula represents the broad absorption band between E_{g} and E_{h} with quadratic behavior in neighborhoods above E_{g} and below E_{h}. In the band gap region this behavior is known as the Tauc's law discussed in Sect. 3.3.2.4. The parameter κ regulates the asymmetry of this absorption band. If $\nu = 1$, the shape of the absorption band is modified by the Lorentzian term and the symbols E_{r} and B denote the resonant energy and broadening, respectively. The function $\Pi_{E_{\mathrm{g}},E_{\mathrm{h}}}(E)$ ensures that the transition strength function is zero below E_{g} and above E_{h}

$$\Pi_{E_{\mathrm{g}},E_{\mathrm{h}}}(E) = \Theta(E-E_{\mathrm{g}})\,\Theta(E_{\mathrm{h}}-E). \tag{3.127}$$

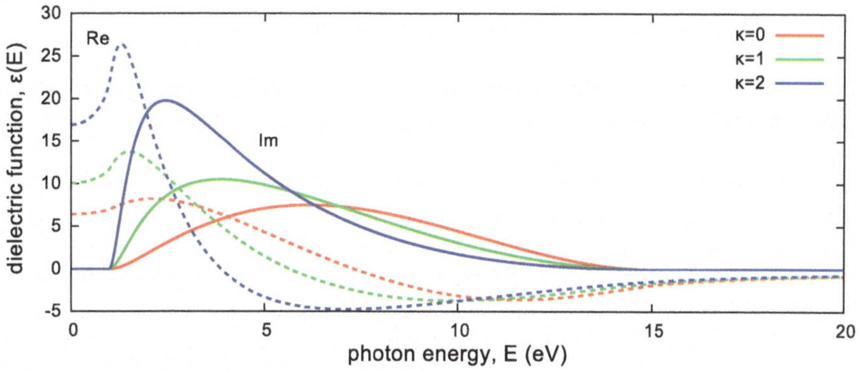

Fig. 3.9 Contributions to the dielectric function of the IDT and IBT model (3.126) or (3.133) for $\nu = 0$ and selected values of κ: $E_g = 1\,\text{eV}$, $E_h = 15\,\text{eV}$ and $N = 400\,\text{eV}^2$

In many cases it is not possible to model the shape of the absorption band by using only one function (3.126) and several terms must be used. Note that separate parameters E_r and B should be used for each absorption band but it impossible in practice to distinguish individual absorption branches with the exception of the region around the band gap energy E_g. Therefore, only one set of parameters E_r and B could be used for all phonon branches.

The real part of the dielectric function and normalization constant C_N must be calculated using the KK relations and normalization integral. The result can be expressed in an analytic form for integer values of the parameter κ (for $\kappa = 1$ and $\nu = 0, 1$ see [63, 64]). For $\kappa = 0, 1, 2$ and $\nu = 0$, the dielectric function is plotted in Fig. 3.9. In order to avoid the numerical integration when the parameter κ has non-integer values, we can use the approximation in which the linear interpolation between integer values of κ is used

$$F^0_{\text{ab}}(E, \kappa) \approx (\kappa - \lfloor\kappa\rfloor)\, F^0_{\text{ab}}(E, 1 + \lfloor\kappa\rfloor) + (1 - \kappa + \lfloor\kappa\rfloor) F^0_{\text{ab}}(E, \lfloor\kappa\rfloor) \quad (3.128)$$

where $\lfloor\kappa\rfloor$ is the floor function.

It is apparent that the total transition strength of IDT is strongly temperature-dependent because the probabilities of IDT depend on the occupation numbers of phonons which are determined by the Bose–Einstein statistics. The transitions strength of the whole system, which includes both the DT and IDT, is only weakly temperature-dependent due to the thermal expansion of the system. Therefore, the increase of the transition strength of IDT with temperature is at the expense of the transition strength of DT.

3.3.2.3 High Energy Valence Electron Excitations

Above a certain energy level the absorption spectra of crystalline solids are smooth without visible structures in the region of interband transitions. This can be explained by the fact that at higher energies, the electrons behave more like free electrons than bonded electrons. Although there is no clearly defined energy above which the transitions from the valence band lead to these structureless absorption spectra it is convenient to separate the excitations of valence electrons into those to conduction band (direct and indirect) and those to the band located above the conduction band (see Fig. 3.1). The transitions of the valence electrons to the states above the conduction band are called high energy valence electron excitations. The transition strength of these electron excitations is only weakly temperature dependent due to the thermal expansion of the system.

The simplest model of high energy transitions (HET) is a one-parametric model with the normalized dielectric function given by

$$\varepsilon^0_{\mathrm{r,het}}(E) = \frac{3E_\mathrm{x}}{\pi E^2}\left(-\frac{(E_\mathrm{x}-E)^2}{E^3}\ln\left|1-\frac{E}{E_\mathrm{x}}\right| + \frac{(E_\mathrm{x}+E)^2}{E^3}\ln\left|1+\frac{E}{E_\mathrm{x}}\right| - \frac{2}{3E_\mathrm{x}} - \frac{2E_\mathrm{x}}{E^2}\right),$$

$$\varepsilon^0_{\mathrm{i,het}}(E) = \frac{3E_\mathrm{x}(|E|-E_\mathrm{x})^2}{E^5}\Theta(|E|-E_\mathrm{x}), \tag{3.129}$$

where the parameter E_x is the energy threshold of the higher energy valence electron excitations. The value of this parameter must lie above the band gap energy E_g and is usually closer to E_g than to the maximal energy of valence-to-conduction interband transitions E_h. The imaginary part of the dielectric function has quadratic behavior near the threshold energy E_x thus there is no pronounced structure associated with this energy in the absorption spectra.

The imaginary part of the dielectric function of the above models falls as $1/E^3$ at high energies. The classical models (see Sect. 3.2.1) exhibit the same behavior and for this reason it is called the classical asymptotic behavior of the dispersion model. From the experiment it is known that for energies in the X-ray region the imaginary part of the dielectric function given mainly by the elastic scattering losses falls faster than $1/E^3$. If we want to model the dielectric response in this region then we can include the term $(|E|+E_\mathrm{a})$ ensuring faster decay above the energy E_a in the denominator of (3.129)

$$\varepsilon^0_{\mathrm{i,het}}(E) = \frac{(|E|-E_\mathrm{x})^2}{C_\mathrm{N}E^5(|E|+E_\mathrm{a})}\Theta(|E|-E_\mathrm{x}). \tag{3.130}$$

For both the presented models of high energy valence electron excitations, it is possible to express the real parts $\varepsilon^0_{\mathrm{r,het}}(E)$ and normalization constants C_N analytically. The imaginary parts of the dielectric functions corresponding to HET models are compared in Fig. 3.10.

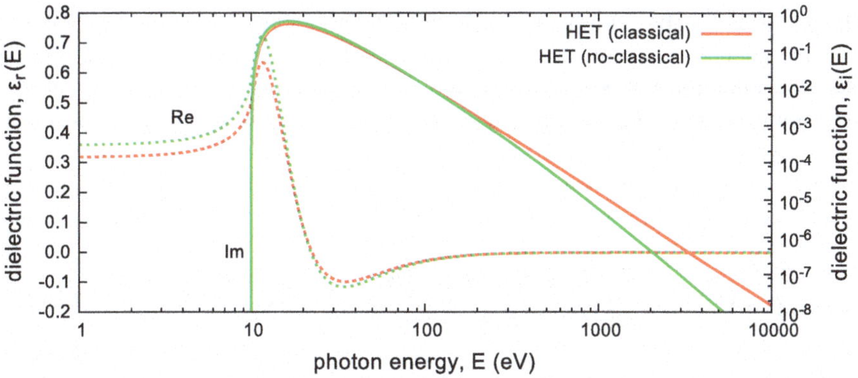

Fig. 3.10 Contributions to the dielectric function corresponding to HET models (3.129) and (3.130): $E_x = 10\,\mathrm{eV}$, $E_a = 500\,\mathrm{eV}$, $N_{het} = 500\,\mathrm{eV}^2$

3.3.2.4 Valence Electron Excitations in Amorphous Materials

In disordered materials, it is not possible to distinguish between valence-to-conduction interband transitions that involve changes in phonon occupation numbers and those in which phonons do not participate. Therefore, the interband transitions can be described by a single broad absorption band between E_g (band gap energy) and E_h (maximal energy of interband transitions). The situation is similar to the case of IDT in the crystalline solids, but without the temperature dependency introduced by the average Bose–Einstein statistical factor for phonons (3.126). This is because the phonon assisted absorption processes have practically the same spectral distribution as the processes in which phonons do not participate and the sum of transitions strengths of both of these processes is only weakly temperature dependent (it is given mostly by the thermal expansion).

The behavior of the JDOS function in the vicinity of the bandgap can be derived using the one-particle approximation and quadratic band approximation. The JDOS function is then given as the correlation between the initial states in the valence band and final states in the conduction band

$$J(E) = \int_{-\infty}^{\infty} f_{VC}(E, T)\, D_V(S)\, D_C(S + E)\, dS\,, \qquad (3.131)$$

where the integration is performed over the energies of initial states S. The symbols D_V and D_C denote the densities of states (DOS) of initial and final states, respectively. If the momentum (current) matrix element is assumed to be constant, the probability of transition is given only by the temperature dependent factor $f_{VC}(E, T)$ which is determined by the Fermi–Dirac distribution as

$$f_{VC}(E, T) = f^{FD}(S, T)\left(1 - f^{FD}(S + E, T)\right)\,. \qquad (3.132)$$

This factor is usually set to unity because the Fermi energy lies in the bandgap far from the valence and conduction bands (see Fig. 3.1). The result, known as the Tauc's law, states that the JDOS function should be quadratic in the neighborhood above the band gap energy E_g and zero below E_g. The similar result can be obtained for the behavior of the JDOS in the vicinity of the maximal energy of transitions E_h.

The model of the interband transitions (IBT) which fulfills the quadratic dependencies above E_g and below E_h can be constructed as a linear combination of terms

$$\varepsilon^0_{\mathrm{i,ibt}}(E) = \frac{(|E| - E_g)^2 (E_h - |E|)^2}{C_N\, E\, |E|^\kappa \left((|E| - E_r)^2 + B^2/4\right)^\nu}\, \Pi_{E_g, E_h}(|E|)\,. \tag{3.133}$$

with the same values of E_g, E_h and κ but with different values of parameters ν, E_r and B for each term. The real part of the dielectric function and normalization constant C_N can be expressed in an analytic form for integer values of the parameter κ (for $\kappa = 1$ and $\nu = 0, 1$ see [63, 64]). For $\kappa = 0, 1, 2$ and $\nu = 0$, the dielectric function is plotted in Fig. 3.9.

The high energy excitations in amorphous materials are modeled in the same way as in crystalline solids. Therefore, the dielectric response of the total valence electron excitations can be expressed as a sum of IBT and HET contributions

$$\hat{\varepsilon}_{\mathrm{vee}}(E) = N_{\mathrm{ibt}} \hat{\varepsilon}^0_{\mathrm{ibt}}(E) + N_{\mathrm{het}} \hat{\varepsilon}^0_{\mathrm{het}}(E)\,. \tag{3.134}$$

The threshold energies E_x and E_h appearing in equations for the HET and IBT cannot be identified by any visible structures in the absorption spectra. Therefore, the HET and IBT are often described by the model which does not contain the parameters E_x and E_h. The Tauc–Lorentz (TL) models combining the Tauc's law valid in the vicinity of the bandgap with the Lorentz model (or Lorentz function) are utilized for this purpose.

The first physically correct model combining the Lorentz model with the Tauc's law was suggested by Campi and Coriasso [65, 66]. The Campi–Coriasso (CC) model is based on the fact that the transition strength function of the Lorentz model (3.59) has quadratic form in the vicinity of zero energy

$$F^0_{\mathrm{LM}}(E) = \frac{2}{\pi} \frac{B E^2}{(E_c^2 - E^2)^2 + B^2 E^2}\,. \tag{3.135}$$

The Campi and Coriasso shifted the Lorentz model by the bandgap energy E_g, i.e. they performed the substitution $E \rightarrow E - E_g$, and for energies below the E_g they set the transition strength to zero. The normalized transition strength function is then given by

$$F^0_{\mathrm{CC}}(E) = F^0_{\mathrm{LM}}(E - E_\mathrm{g})\,\Theta(E - E_\mathrm{g})$$
$$= \frac{2}{\pi} \frac{B\,(E - E_\mathrm{g})^2\,\Theta(E - E_\mathrm{g})}{\left((E_\mathrm{c} - E_\mathrm{g})^2 - (E - E_\mathrm{g})^2\right)^2 + B^2\,(E - E_\mathrm{g})^2}\,, \tag{3.136}$$

where the parameter E_c must fulfill the inequality $E_\mathrm{c} > E_\mathrm{g}$. The dielectric function is calculated using (3.30) and the KK relations, the analytic expressions can be found in [67].

Another approach to constructing the models combining the Tauc's law with the Lorentz model includes multiplying the transition strength function of the Lorentz model by a truncate function that is zero below the bandgap, quadratic in the neighborhood above the bandgap and tends to unity at high energies. Two such models are widely used, the Jellison–Modine (JM) model [68] which uses the one-parametric truncate function

$$T_1(E) = \frac{(E - E_\mathrm{g})^2}{E^2}\,\Theta(E - E_\mathrm{g})\,, \tag{3.137}$$

and the A. S. Ferlauto et al.(ASF) model [69] which uses the two-parametric truncate function

$$T_2(E) = \frac{(E - E_\mathrm{g})^2}{(E - E_\mathrm{g})^2 + E_\mathrm{q}^2}\,\Theta(E - E_\mathrm{g})\,. \tag{3.138}$$

In the literature the JM model is usually referred as the Tauc–Lorentz model whereas ASF model is known as the Cody–Lorentz model. Two models that use the same truncate functions as the JM and ASF models but with the transition strength of the Lorentz model (LM) replaced by the Lorentz function (LF) were suggested in [67]. We will refer to these models as to truncated Lorentz function models TLF1 and TLF2.

The imaginary part of the dielectric function of the JM, ASF, TLF1 and TLF2 models can be expressed with the help of the generic formula

$$\varepsilon^0_{\mathrm{i},m} = \frac{T_a(E)F_b(E)}{C_\mathrm{N}E}\,, \tag{3.139}$$

where C_N is the normalization constant, the truncate function $T_a(E)$ is either (3.137) or (3.138) and $F_b(E)$ is one of

$$F_{\mathrm{LM}}(E) = \frac{E^2}{(E_\mathrm{c}^2 - E^2)^2 + B^2E^2}\,, \qquad F_{\mathrm{LF}}(E) = \frac{1}{(E - E_\mathrm{r})^2 + B^2/4}\,, \tag{3.140}$$

where the energies E_c and E_r are related as (3.60). The summary of these models is in Table 3.4.

Each version of the TL model introduced above has a slightly different shape. In order to compare these models the dielectric function of the CC model calculated in the spectral range 0–10 eV for the selected dispersion parameters was fitted by

Table 3.4 The summary of the truncate functions and transition strength functions used in formula (3.139). The dispersion parameters and the rms value of differences $\delta\varepsilon$ corresponding to the fit to the CC model calculated with $N = 100\,\mathrm{eV}^2$, $E_\mathrm{c} = 3\,\mathrm{eV}$, $B = 3\,\mathrm{eV}$ and $E_\mathrm{g} = 1\,\mathrm{eV}$ are also introduced. Transition strength N for the IBT+HET model (3.134) is $N_\mathrm{ibt} + N_\mathrm{het}$

Model m	Truncate f. $T_a(E)$	Base f. $F_b(E)$	N eV^2	E_g eV	E_c eV	B eV	E_q eV	E_h eV	E_x eV	$\delta\varepsilon$ eV
JM	$T_1(E)$	$F_\mathrm{LM}(E)$	107.1	0.85	2.71	2.56	–	–	–	0.0192
TLF1	$T_1(E)$	$F_\mathrm{LF}(E)$	97.8	1.13	2.93	3.34	–	–	–	0.0123
ASF	$T_2(E)$	$F_\mathrm{LM}(E)$	102.0	0.96	2.85	3.02	0.79	–	–	0.0032
TLF2	$T_2(E)$	$F_\mathrm{LF}(E)$	97.5	0.93	2.59	2.92	2.58	–	–	0.0058
IBT+HET	n.a.		82.2+20.3	1.00	2.71	2.85	–	19.6	6.10	0.0024

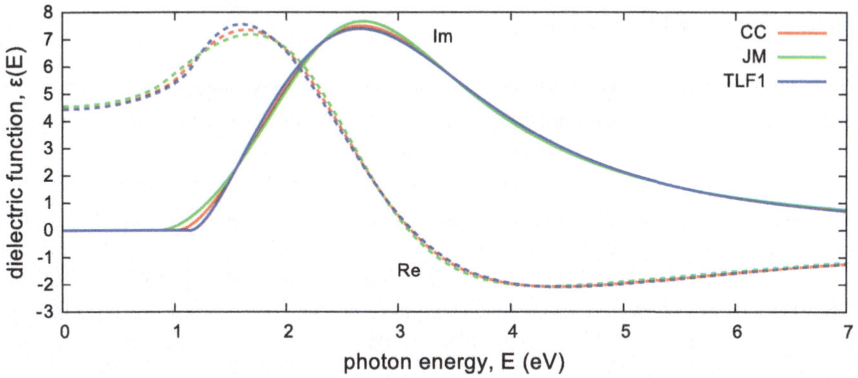

Fig. 3.11 Contributions to the dielectric function of three versions of Tauc–Lorentz models. The dispersion parameters are in Table 3.4

the other versions of the TL model. The dispersion parameters corresponding to the best fits and the parameters $\delta\varepsilon$ representing the rms values of the differences between the models are in Table 3.4. The results for the IBT+HET model (3.134) with one IBT term (3.133) and one HET term (3.129) are introduced for comparison. In Fig. 3.11 the dielectric function of the CC model is compared with the four-parametric JM and TLF1 models. The curves corresponding to the five-parametric ASF and TLF2 models are not displayed since the differences from the CC model are too small to be visible in the figure. A more detailed comparison of these models can be found in [67].

From the physical point of view, there is no reason to prefer one of the TL models over another. The choice of the model giving the best results is specific for individual materials. If we are unsure which model to use, it is possible to model the valence electron excitations with a linear combination of several of the presented models.

3.3.3 Electronic Transitions Involving Localized States

Apart from the extended states forming the valence and conduction bands there are localized occupied and unoccupied states (see Fig. 3.1). In both disordered and crystalline solids there are localized states originating from the broken transition symmetry. In crystalline solids, the broken transition symmetry (disorder) is a consequence of the thermal motion of nuclei forming the crystal while in the disordered solids the disorder is both inherent and caused by the thermal motion. The localized states could originate also from the defects in the materials, such as vacancies, impurities, etc. If the initial and final states are localized states, we talk about localized excitations and if the initial or final state is the extended state, i. e. valence or conduction state, then the excitation is modeled by the subgap absorption tails.

3.3.3.1 Localized Excitations

The localized states are typically observable below the band gap energy as discrete transitions with the photon energies given as the differences between the energy of corresponding occupied and unoccupied localized states. This means that the JDOS cannot be calculated as the correlation between the initial and final states as in the case of IBT (3.131). In most cases, the Gaussian ε-broadened discrete transitions are an adequate model [63, 64], i. e. the same model as for phonons (see Sect. 3.3.1.1). In the case of excitations of π-electrons in diamond-like carbon (DLC) [70–73] or excitations of non-bonding electrons in chalcogenides [74–76] the corresponding absorption structures can be described by the IBT model (3.133). In these cases the degree of localization of the π-electron and non-bonding electron states is relatively low and, therefore, it is possible to represent the density of states as two valence bands and two conduction bands with different bandgap energies.

3.3.3.2 Exponential (Urbach) Tail

The exponential (Urbach) tail is employed to describe the weak absorption below the band gap caused by the transitions of electrons from the localized valence states to the unoccupied extended states and the transitions from the extended valence states to the unoccupied localized states [77] (see Fig. 3.1). The JDOS associated with these transitions are given as the correlation between the density of localized states with the density of states in valence or conduction bands. The density of localized states is given by the narrow peak with typical width in tens of meV which decays exponentially. Since the typical width of extended bands is around 10 eV the correlation gives the JDOS which looks like copies of the extended bands with exponential tails on both of its ends. The exponential tails have limited extent because the DOS of the localized states is bounded by the Fermi energy. The imaginary part of the normalized dielectric function corresponding to localized-to-extended or extended-to-localized transitions can be modeled by [63, 64]

$$
\begin{aligned}
\varepsilon^0_{\mathrm{i,ut}}(E) = \frac{1}{C_{\mathrm{N}} E|E|^{\kappa}} \Bigg\{ &\exp\left(\frac{|E| - E_{\mathrm{g}}}{E_{\mathrm{u}}}\right) \Pi_{E_{\mathrm{g}}/2, E_{\mathrm{g}}}(|E|) \\
&+ \left[1 + \frac{E_{\mathrm{m}} - E_{\mathrm{g}}}{4E_{\mathrm{u}}} - \frac{(2|E| - E_{\mathrm{m}} - E_{\mathrm{g}})^2}{4E_{\mathrm{u}}(E_{\mathrm{m}} - E_{\mathrm{g}})} \right] \Pi_{E_{\mathrm{g}}, E_{\mathrm{m}}}(|E|) \\
&+ \exp\left(\frac{E_{\mathrm{m}} - |E|}{E_{\mathrm{u}}}\right) \Pi_{E_{\mathrm{m}}, E_{\mathrm{m}} + E_{\mathrm{g}}/2}(|E|) - \exp\left(-\frac{E_{\mathrm{g}}}{2E_{\mathrm{u}}}\right) \Pi_{E_{\mathrm{g}}/2, E_{\mathrm{m}} + E_{\mathrm{g}}/2}(|E|) \Bigg\} ,
\end{aligned}
\tag{3.141}
$$

where E_{u} is the parameter called the Urbach energy. It express the slope of the decrease of the exponential tail and the parameter κ modifies the shape of the absorption band. The energy E_{m} is the difference between the energies of tops of the valence and conduction bands if the model describes the localized-to-extended transitions and the difference between the energies of the bottoms of the valence and conduction bands if the model describes extended-to-localized transitions. In principle two

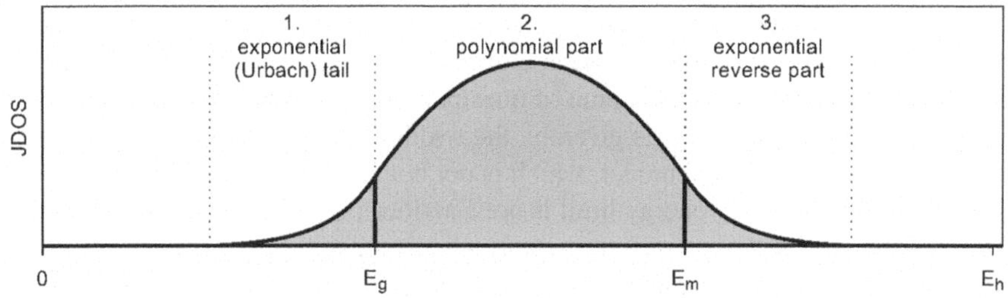

Fig. 3.12 Schematic diagram of JDOS function of exponential (Urbach) tail model ($\kappa = 1$)

models (3.141) with different E_{m} should be used but because the absorption structures corresponding to these transitions are weak they are masked by other absorption structures above the E_{g}. For this reason, it is possible to use one model with the mean value

$$E_{\text{m}} = E_{\text{g}}/2 + E_{\text{h}}/2. \tag{3.142}$$

A schematic diagram of this model constructed from three parts is plotted in Fig. 3.12. In (3.141) the first three terms correspond to the three parts of the model and the fourth constant term ensures the continuous course of the JDOS in energies $E_{\text{g}}/2$ and $E_{\text{m}} + E_{\text{g}}/2$. As was mentioned above the exponential tail has extent limited by the Fermi energy. For this reason the model is constructed such that it is zero below $E_{\text{g}}/2$ and above $E_{\text{m}} + E_{\text{g}}/2$, i. e. it is assumed that the Fermi energy lies in the middle of the bandgap. The analytic expressions for the real part of the normalized dielectric function and the normalization constant C_{N} for $\kappa = 1$ can be found in [64].

When the valence electron excitations are described by the Tauc–Lorentz models, it is possible to use the model of exponential (Urbach) tail, which is not limited by the energy E_{m} but extends to infinity

$$\begin{aligned} \varepsilon^{0}_{\text{i,ut}}(E) = \frac{1}{C_{\text{N}} E |E|^{\kappa}} \Bigg\{ \left[\exp\left(\frac{|E| - E_{\text{g}}}{E_{\text{u}}} \right) - \exp\left(-\frac{E_{\text{g}}}{2E_{\text{u}}} \right) \right] \Pi_{E_{\text{g}}/2, E_{\text{g}}}(|E|) \\ + \frac{A(|E| - E_0)^2}{|E|^{4-\kappa}} \Theta(|E| - E_{\text{g}}) \Bigg\} . \end{aligned} \tag{3.143}$$

The model is parameterized by the energies E_{u}, E_{g} and the integer κ. The values of E_0 and A must be determined so that the imaginary part of the dielectric function and its derivative are continuous at E_{g}.

3.3.4 Free Carrier Contributions

If the valence or conduction band is partially occupied, i.e. the Fermi energy lies inside or sufficiently close to these bands, the indirect intraband transitions contribute to the dielectric response of the material. These transitions are called free carrier contributions (FCC) because they can be interpreted as movement of carriers of charge, i.e. electrons and holes, in the frame of quasiparticle approximation. Although the maximum energy of the FCC is given by the width of the partially filled band, it is possible to use the classical model, which is not bounded by any maximum energy. The absence of the upper energy limit is not a serious problem because it lies in the region of the interband electronic transitions which are, in this region, much stronger than the FCC. Since the dielectric function corresponding to the FCC has a singularity at the zero energy, it is more convenient to describe the free carrier dielectric response using the complex optical conductivity $\hat{\sigma}(\omega) = -\mathrm{i}\omega\epsilon_0[\hat{\varepsilon}(\omega) - 1]$ or equivalently using the transition strength function which is proportional to its real part σ_r. The normalized transition strength function of the Drude model (3.15) is given as follows

$$F^0_\mathrm{fcc}(E) = \frac{2}{\pi}\frac{B}{E^2 + B^2} \,. \tag{3.144}$$

Instead of being bounded by a high energy limit, the Drude model exhibits classical asymptotic behavior. This could be problematic if we intent to model the dielectric response in the high energy X-ray region. In that case, it is better to use the modified version of the Drude model that introduces the high energy limit E_w as follows

$$F^0_\mathrm{fcc}(E) = \frac{(E_\mathrm{w}^2 - E^2)^2\Theta(E_\mathrm{w} - E)}{C_\mathrm{N}(E^2 + B^2)}\,, \quad C_\mathrm{N} = \frac{(E_\mathrm{w}^2 + B^2)^2}{B}\arctan\left(\frac{E_\mathrm{w}}{B}\right) - E_\mathrm{w}B^2 - \frac{5}{3}E_\mathrm{w}^3 \,. \tag{3.145}$$

The dielectric function of this model can be expressed analytically by (3.35) and (3.36). The comparison of this model with the Drude model is in Fig. 3.13.

The temperature dependence of the intraband transitions is a complex phenomenon and it is impossible to describe it by a formula which is universally valid. Roughly speaking, it depends on the position of the Fermi energy within the band

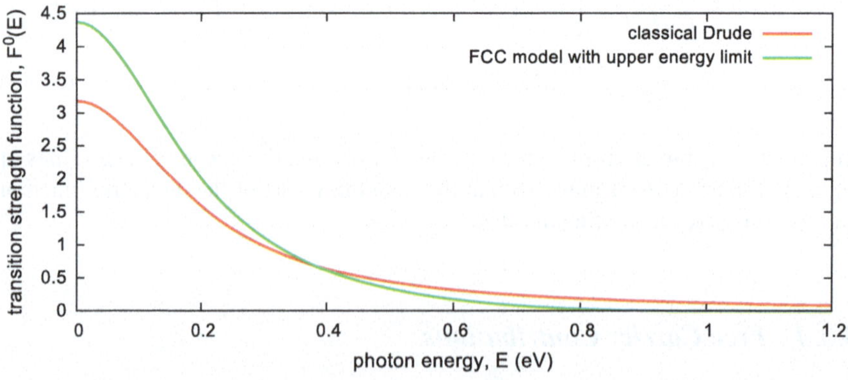

Fig. 3.13 Normalized transition strength functions $F^0(E) = E\varepsilon_\mathrm{i}^0(E)$ for models of free carrier contribution: $B = 0.2\,\mathrm{eV}$, $E_\mathrm{w} = 1\,\mathrm{eV}$

structure. For example, if the Fermi energy lies inside the valence band (like in metals), the electron occupation numbers are redistributed inside the partially filled band and the total transition strength is practically constant. The broadening parameter B increases with the temperature due to the blurring of the Fermi–Dirac statistics which results in the decrease of the static conductivity. If the Fermi energy lies inside the band gap (like in semiconductors) then the conductivity of the material increases with temperature because the density of holes in the valence band and electrons in the conduction band increases. In intrinsic semiconductors the the Fermi energy lies exactly in the middle of the bandgap and the temperature dependence of the transition strength of intraband indirect transitions is given as [78, 79]

$$F^0_{\mathrm{fcc}}(E,T) = \left(\frac{T}{300\,\mathrm{K}}\right)^{3/2} \exp\left[-\frac{E_{\mathrm{g}}}{2k_{\mathrm{B}}}\left(\frac{1}{T} - \frac{1}{300\,\mathrm{K}}\right)\right] F^0_{\mathrm{fcc}}(E)\,, \tag{3.146}$$

where the normalization of $F^0_{\mathrm{fcc}}(E)$ is performed for 300 K.

Although the description of materials in the superconducting state poses considerable theoretical problem, it is quite easy to write the part corresponding to superconductivity (SC) in the dielectric model. The normalized contribution to the transition strength is given as

$$F^0_{\mathrm{sc}}(E) = 2\delta(E) \tag{3.147}$$

and the contribution to the real part of the dielectric function is

$$\varepsilon^0_{\mathrm{r,sc}}(E) = -\frac{2}{\pi E^2}\,. \tag{3.148}$$

3.3.5 Core Electron Excitations

Apart from the valence electrons the core electrons can also be excited to the unoccupied electronic states (see Fig. 3.1). These excitations are manifested by sharp absorption edges in the spectra. The positions of these absorption edges are determined by the energies of the core electron states in the individual atoms forming the material. This means that each element contained in the material is manifested by absorption edges at energies characteristic for this element. The exact configuration of atoms in the material has only a small influence on the positions of these absorption edges.

The excitations of core electrons into conduction band influence the shape of the absorption structures near the absorption edges. These structures are called fine absorption structures and their shape is determined not only by the types and quantities of atoms forming the material but also by their configuration within this material. The excitation of core electrons into higher states contributes to the dielectric response by smooth functions. The strength of these smooth contributions is determined only by the types and quantities of the atoms and it is independent on the exact configuration of these atoms within the material.

The influence of the core electron excitations (CEE) on the dielectric response in the region of valence electron excitations is very small. Therefore, if the experimental data do not extend to very high energies, it is possible to neglect the contribution coming from the CEE or a simple analytical model that is sufficiently accurate at low energies could be used [7, 60, 64, 78]:

$$\hat{\varepsilon}^0_{\mathrm{cee}}(E) = \frac{E_\mathrm{s}}{\pi E^3}\left(\ln\left|\frac{E+E_\mathrm{s}}{E-E_\mathrm{s}}\right| - \frac{2E}{E_\mathrm{s}}\right) + \mathrm{i}\frac{E_\mathrm{s}}{E^3}\Theta(|E|-E_\mathrm{s})\,. \tag{3.149}$$

At low energies this model gives a practically constant contribution to the real part of the dielectric function, which can be expressed as

$$\lim_{E\to 0}\hat{\varepsilon}^0_{\mathrm{cee}}(E) = \frac{2}{3\pi E_\mathrm{s}^2}\,. \tag{3.150}$$

For example, in the case of hydrogenated amorphous silicon these contributions are 1.2×10^{-5} for K and 0.0204 for L core electrons [64].

The model of the CEE (3.149) has classical asymptotic behavior and singularity at energy $E = E_\mathrm{s}$. Similarly as for the HET the correct model of the CEE should fall faster than the classical asymptotic behavior at high energies. The faster decay at high energies could be achieved by placing the term $(|E| + E_\mathrm{a})$ into the denominator of the imaginary part of (3.149). The singularity at $E = E_\mathrm{s}$ is caused by the Heaviside step function and it can be removed by the broadening procedure. Thus the normalized dielectric function of the corrected CEE dispersion model can be written as

$$\hat{\varepsilon}^0_{\mathrm{cee}}(E) = \hat{\beta} * \frac{1}{C_\mathrm{N}E^3(|E|+E_\mathrm{a})}\Theta(|E|-E_\mathrm{s})\,, \tag{3.151}$$

where C_N is the normalization constant given by

$$C_\mathrm{N} = \frac{E_\mathrm{s}[\ln(E_\mathrm{s}) - \ln(E_\mathrm{s}+E_\mathrm{a})] + E_\mathrm{a}}{E_\mathrm{s}E_\mathrm{a}^2}\,. \tag{3.152}$$

If the Lorentzian broadening function (3.55) is used, the result can be expressed analytically as

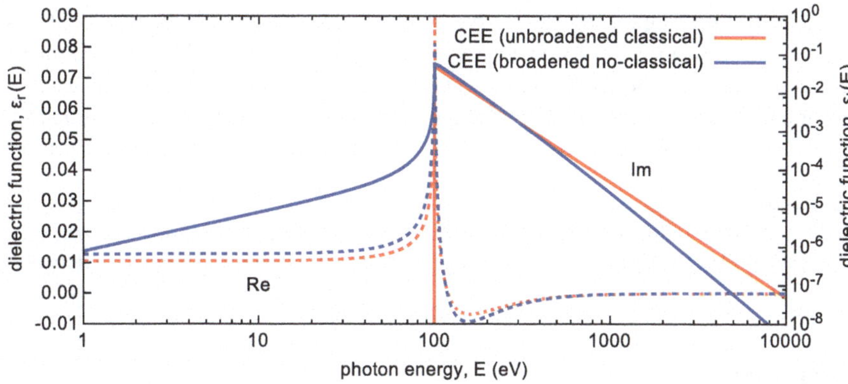

Fig. 3.14 Contributions to the dielectric function corresponding to CEE models (3.149) and (3.153): $N_{\mathrm{cee}} = 500\,\mathrm{eV}^2$, $E_{\mathrm{s}} = 100\,\mathrm{eV}$, $B = 1\,\mathrm{eV}$, $E_{\mathrm{a}} = 500\,\mathrm{eV}$

$$\hat{\varepsilon}^{0}_{\mathrm{cee}}(E) = \frac{1}{\pi C_{\mathrm{N}}}\left\{\frac{2}{E_{\mathrm{a}}^2(\hat{E}^2 - E_{\mathrm{a}}^2)}\ln(E_{\mathrm{a}} + E_{\mathrm{s}}) - \frac{2}{E_{\mathrm{a}}^2\hat{E}^2}\ln(E_{\mathrm{s}}) - \frac{2}{E_{\mathrm{s}}E_{\mathrm{a}}\hat{E}^2}\right.$$
$$+ \frac{1}{\hat{E}^3(E_{\mathrm{a}} - \hat{E})}\left[\frac{1}{2}\ln\!\left((E + E_{\mathrm{s}})^2 + B^2/4\right) - \mathrm{i}\arctan\!\left(\frac{E + E_{\mathrm{s}}}{B/2}\right) + \mathrm{i}\frac{\pi}{2}\right]$$
$$\left. - \frac{1}{\hat{E}^3(E_{\mathrm{a}} + \hat{E})}\left[\frac{1}{2}\ln\!\left((E - E_{\mathrm{s}})^2 + B^2/4\right) - \mathrm{i}\arctan\!\left(\frac{E - E_{\mathrm{s}}}{B/2}\right) - \mathrm{i}\frac{\pi}{2}\right]\right\}. \tag{3.153}$$

In Fig. 3.14, the contributions to the dielectric function calculated by the CEE models are shown in a log-log plot. In this figure, it is observed that the Lorentzian broadening results in nonzero contribution to the imaginary part in the low energy region. This slowly decaying weak absorption tail is unphysical in the region below the bandgap where the exponential (Urbach) tail should be observed. For this reason it is better to use the Gaussian broadening in (3.151) but in this case the broadened dielectric function requires the use of numerical integration.

3.4 Conclusion

The UDM is a collection of models describing various elementary excitations presented in a unified form. The contributions corresponding to these elementary excitations must be combined in order to describe the dielectric response of a specific material. In order to decide which contributions should be included in the model some prior knowledge of the structure of the studied material is needed. Moreover, the exact number and types of the contributions depend on the spectral range in which the model should be valid and the desired precision of the model. The models presented in this chapter are only a basic set of models of elementary excitations and some modifications or extensions of these models are possible. An example of a software that uses the UDM is newAD project [80]. So far the UDM was used to describe the optical response of the following materials: a-Si:H [64, 81], HfO_2 [63, 82], SiO_2 [82, 83], Al_2O_3 [82], Ta_2O_5 [82], TiO_2 [82], MgF_2 [82, 84] and ZnSe [85].

Acknowledgements We are grateful to Dominik Munzar, Adam Dubroka and David Nečas for reading of the manuscript, discussions and useful comments. This research has been supported by the project LO1411 (NPU I) funded by Ministry of Education Youth and Sports of Czech Republic.

References

1. H.A. Kramers, in *Atti Cong. Intern. Fisica, (Transactions of Volta Centenary Congress) Como*, vol. 2 (1927), pp. 545–557
2. R. de Laer Kronig, J. Opt. Soc. Am.; Rev. Sci. Instrum. **12**, 547 (1926)
3. M. Altarelli, D.L. Dexter, H.M. Nussenzveig, D.Y. Smith, Phys. Rev. B **6**, 4502 (1972)
4. E. Shiles, T. Sasaki, M. Inokuti, D.Y. Smith, Phys. Rev. B **22**, 1612 (1980)
5. D.Y. Smith, in *Handbook of Optical Constants of Solids*, vol. 1, ed. by E.D. Palik (Academic Press, Dublin 1985), pp. 35–68
6. V. Lucarini, K.E. Peiponen, J.J. Saarinen, E.M. Vartiainen, *Kramers-Kronig Relations in Optical Materials Research* (Springer, Berlin, 2005)
7. D. Franta, D. Nečas, L. Zajíčková, Thin Solid Films **534**, 432 (2013)
8. Z.C. Yan, G.W.F. Drake, Phys. Rev. A **52**, R4316 (1995)
9. M. Dressel, G. Grüner, *Electrodynamics of Solids: Optical Properties of Electrons in Matter* (University Press, Cambridge, 2002)
10. D. van der Marel, in *Strong interactions in low dimensions* (Kluwer Academic Publishers, Dordrecht, 2004), chap. 8, pp. 237–276
11. B.L. Zhou, J.M. Zhu, Z.C. Yan, Phys. Rev. A **73**, 014501 (2006)
12. J.A.S. Barker, J.J. Hopfield, Phys. Rev. **135**, A1732 (1964)
13. F. Wooten, *Optical Properties of Solids* (Academic Press, New York, 1972)
14. R. Kubo, J. Phys. Soc. Japan **12**, 570 (1957)
15. P.Y. Yu, M. Cardona, *Fundamentals of Semiconductors* (Springer, Berlin, 2001)
16. P.B. Allen, in *Conceptual Foundations of Material Properties:A Standard Model For Calculation Of Ground- and Excited-state Properties*, ed. by M.L. Cohen, S.G. Louie (Elsevier, Amsterdam, 2006), pp. 165–218
17. H.A. Bethe, E.E. Salpeter, *Quantum Mechanics of One- and Two-Electron Atoms* (Springer, Berlin, 1957)
18. L. Van Hove, Phys. Rev. **89**, 1189 (1953)
19. S. Adachi, Phys. Rev. B **35**, 7454 (1987)
20. S. Adachi, Phys. Rev. B **41**, 1003 (1990)
21. C.C. Kim, J.W. Garland, H. Abad, P.M. Raccah, Phys. Rev. B **45**, 11749 (1992)
22. C.C. Kim, J.W. Garland, P.M. Raccah, Phys. Rev. B **47**, 1876 (1993)
23. C. Tanguy, Phys. Rev. Lett. **75**, 4090 (1995)
24. C. Tanguy, Solid State Commun. **98**, 65 (1996)
25. C.M. Herzinger, B.D. Johs, W.A. McGahan, J.A. Woollam, W. Paulson, J. Appl. Phys. **83**, 3323 (1998)
26. B.D. Johs, C.M. Herzinger, J.H. Dinan, A. Cornfeld, J.D. Benson, Thin Solid Films **313–314**, 137 (1998)
27. A.B. Djurišić, E.H. Li, Phys. Status Solidi B **216**, 199 (1999)
28. A.B. Djurišić, Y. Chan, E.H. Li, Semicond. Sci. Technol. **16**, 902 (2001)

29. Y.S. Ihn, T.J. Kim, T.H. Ghong, Y. Kim, D.E. Aspnes, J. Kossut, Thin Solid Films **455–456**, 222 (2004)
30. L.V. Rodríguez-de Marcos, J.I. Larruquert, Opt. Express **24**, 28561 (2016)
31. D. Franta, D. Nečas, L. Zajíčková, I. Ohlídal, Thin Solid Films **571**, 496 (2014)
32. R.N. Bracewell, *The Fourier Transform and Its Applications* (McGraw-Hill, Singapore, 1999)
33. M. Abramowitz, I.A. Stegun (eds.), *Handbook of mathematical functions with Formulas, Graphs, and Mathematical Tables* (National Bureau of Standards, 1964)
34. M.E. Thomas, S.K. Andersson, R.M. Sova, R.I. Joseph, Infrared Phys. Technol. **39**, 235 (1998)
35. D. De Sousa Meneses, M. Malki, P. Echegut, J. Non-Cryst. Solids **352**, 769 (2006)
36. D. De Sousa Meneses, J.F. Brun, B. Rousseau, P. Echegut, J. Phys. Condes. Matter **18**, 5669 (2006)
37. R. Kitamura, L. Pilon, M. Jonasz, Appl. Opt. **46**, 8118 (2007)
38. J. Humlíček, J. Quant. Spectrosc. Radiat. Transf. **27**, 437 (1982)
39. R. Brendel, D. Bormann, J. Appl. Phys. **71**, 1 (1992)
40. D. De Sousa Meneses, G. Gruener, M. Malki, P. Echegut, J. Non-Cryst. Solids **351**, 124 (2005)
41. S.M. Abrarov, B.M. Quine, Appl. Math. Comput. **218**, 1894 (2011)
42. J. Kischkat, S. Peters, B. Gruska, M. Semtsiv, M. Chashnikova, M. Klinkmüller, O. Fedosenko, S. Machulik, A. Aleksandrova, G. Monastyrskyi, Y. Flores, W.T. Masselink, Appl. Opt. **71**, 6789 (2012)
43. J.J. Olivero, R.L. Longbothum, J. Quant. Spectrosc. Radiat. Transf. **17**, 233 (1977)
44. P. Thompson, D.E. Cox, J.B. Hastings, Appl. Crystallogr. **20**, 79 (1987)
45. J. Humlíček, R. Henn, M. Cardona, Phys. Rev. B **61**, 14554 (2000)
46. D.W. Berreman, F.C. Unterwald, Phys. Rev. **174**, 791 (1968)
47. G.S. Jeon, G.D. Mahan, Phys. Rev. B **71**, 184306 (2005)
48. A.B. Kuzmenko, E.B. Kuzmenko, E. van Heumen, F. Carbone, D. van der Marel, Phys. Rev. Lett. **100**, 117401 (2008)
49. J. Humlíček, A. Nebojsa, F. Munz, M. Miric, R. Gajic, Thin Solid Films **519**, 2624 (2011)
50. U. Fano, Phys. Rev. **124**, 1866 (1961)
51. R. Zallen, M.L. Slade, Phys. Rev. B **18**, 5775 (1978)
52. P. Lautenschlager, P.B. Allen, M. Cardona, Phys. Rev. B **31**, 2163 (1985)
53. P. Lautenschlager, M. Garriga, L. Viña, M. Cardona, Phys. Rev. B **36**, 4821 (1987)
54. A. Debernardi, Phys. Rev. B **57**, 12847 (1998)
55. D. Franta, D. Nečas, L. Zajíčková, I. Ohlídal, Opt. Mater. Express **4**, 1641 (2014)
56. G. Harbeke, in *Optical Properties of Solids*, ed. by F. Abelès (North–Holland, Amsterdam, 1972), pp. 21–92
57. F. Bassani, G. Pastori Parravicini, *Electronic States and Optical Transitions in Solids*, *The Science of the Solid State*, vol. 8 (Pergamon Press, USA, 1975)
58. B. Velický, J. Sak, Phys. Status Solidi **16**, 147 (1966)
59. D. Sell, P. Lawaetz, Phys. Rev. Lett. **26**, 311 (1971)
60. D. Franta, A. Dubroka, C. Wang, A. Giglia, J. Vohánka, P. Franta, I. Ohlídal. Appl. Surf. Sci. **421**, 405 (2017)
61. S. Adachi, Phys. Rev. B **38**, 12966 (1988)
62. S. Flügge, H. Marshall, *Rechenmethoden der Quantentheorie*, 2nd edn. (Springer, Berlin, 1952)
63. D. Franta, D. Nečas, I. Ohlídal, Appl. Opt. **54**, 9108 (2015)
64. D. Franta, D. Nečas, L. Zajíčková, I. Ohlídal, J. Stuchlík, D. Chvostová, Thin Solid Films **539**, 233 (2013)
65. D. Campi, C. Coriasso, J. Appl. Phys. **64**, 4128 (1988)
66. D. Campi, C. Coriasso, Mater. Lett. **7**, 134 (1988)
67. D. Franta, M. Černák, J. Vohánka, I. Ohlídal, Thin Solid Films **631**, 12 (2017)
68. G.E. Jellison Jr., F.A. Modine, Appl. Phys. Lett. **69**, 371 (1996)
69. A.S. Ferlauto, G.M. Ferreira, J.M. Pearce, C.R. Wronski, R.W. Collins, X.M. Deng, G. Ganguly, J. Appl. Phys. **92**, 2424 (2002)
70. V. Paret, M.L. Thèye, J. Non-Cryst, Solids **266–269**, 750 (2000)
71. M. Gioti, D. Papadimitriou, S. Logothetidis, Diam. Relat. Mater. **9**, 741 (2000)

72. D. Franta, I. Ohlídal, Acta Phys. Slov. **50**, 411 (2000)
73. D. Franta, L. Zajíčková, I. Ohlídal, J. Janča, Vacuum **60**, 279 (2001)
74. D. Franta, M. Hrdlička, D. Nečas, M. Frumar, I. Ohlídal, M. Pavlišta, Phys. Status Solidi C **5**, 1324 (2008)
75. D. Franta, D. Nečas, M. Frumar, Phys. Status Solidi C **6**, S59 (2009)
76. D. Franta, D. Nečas, I. Ohlídal, M. Hrdlička, M. Pavlišta, M. Frumar, M. Ohlídal, J. Optoelectron. Adv. Matter. **11**, 1891 (2009)
77. J. Tauc, in *Optical Properties of Solids*, ed. by F. Abelès (North–Holland, Amsterdam, 1972), pp. 277–313
78. D. Franta, D. Nečas, L. Zajíčková, I. Ohlídal, Thin Solid Films **571**, 490 (2014)
79. C. Kittel, *Introduction to Solid State Physics*, 5th edn. (Wiley, New York, 1976)
80. D. Franta, D. Nečas, et al. Software for optical characterization, in *newAD2*. http://newad.physics.muni.cz
81. D. Franta, D. Nečas, L. Zajíčková, I. Ohlídal, J. Stuchlík, Thin Solid Films **541**, 12 (2013)
82. D. Franta, D. Nečas, I. Ohlídal, A. Giglia, in *Proceedings of SPIE*. Optical Systems Design 2015: Optical Fabrication, Testing, and Metrology V, vol. 9628 (2015), p. 96281U
83. D. Franta, D. Nečas, I. Ohlídal, A. Giglia, in Photonics Europe 2016: Optical Micro- and Nanometrology VI, Proc. SPIE, vol. 9890 (2016), p. 989014
84. D. Franta, D. Nečas, A. Giglia, P. Franta, I. Ohlídal, Appl. Surf. Sci. **421**, 424 (2017)
85. I. Ohlídal, D. Franta, D. Nečas, Appl. Surf. Sci. **421**, 687 (2017)

Chapter 4
Predicting Optical Properties from Ab Initio Calculations

Pavel Ondračka, David Holec and Lenka Zajíčková

Abstract In this chapter a short overview is given of some of the ab initio methods that can be used to predict optical properties of solids in order to gain insights into the underlying principles and to explain experimentally observed phenomena or predict properties of new materials. Density functional theory is presented as the most popular first principles technique for electronic structure calculations along with a brief description of a more sophisticated many body perturbation theory based on the Green's functions formalism. The Bethe–Salpeter equation is introduced as a mean to calculate optical properties including excitonic effects. Those methods are applied to a model system of crystalline silicon as well as more complicated oxide materials.

4.1 Introduction

Ab initio methods, also called first principles methods, allow calculating certain physical quantities using only fundamental principles without a need for any empirical parameters. A bottom-up approach in materials science builds on such quantum-mechanical calculations and up-scales the obtained properties either for a better understanding of experimental results or for a prediction and planning the experiments. While the ab initio modeling techniques have been around for almost a whole century, it was only in the last two decades that they have seen much broader adoption and applications beyond the simplest crystal systems. This is due to the rapid

P. Ondračka (✉) · L. Zajíčková
Plasma Technologies, CEITEC, Masaryk University, Purkyňova 123,
61200 Brno, Czech Republic
e-mail: pavel.ondracka@gmail.com

P. Ondračka · L. Zajíčková
Faculty of Science, Department of Physical Electronics,
Masaryk University, Kotlářská 2, 61137 Brno, Czech Republic

D. Holec
Department of Physical Metallurgy and Materials Testing,
Montanuniversität Leoben, Franz-Josef-Straße 18, 8700 Leoben, Austria

O. Stenzel and M. Ohlídal (eds.), *Optical Characterization of Thin Solid Films*, Springer Series in Surface Sciences 64,
https://doi.org/10.1007/978-3-319-75325-6_4

increase of the available computing power, advancements in fundamental theories, numerical methods, and increasing availability and user friendliness of the ab initio software. Combination of those factors lowers the entry barrier for researchers interested in the ab initio methods, and allows to apply them to much more sophisticated, hence more realistic, systems than ever before. While the main focus of this book is on the experimental characterization of thin films, this chapter provides a very brief overview of techniques used to predict band structures and optical properties of solids. However, the aim is not a thorough description of the underlying theory and principles because it could not fit in a single chapter.

After reading this chapter, the reader shall have a general idea about possibilities of the standard ab initio techniques, and how they can be used to complement experimental efforts. There are two conceptually different approaches to the utilization of first principles calculations. The first one uses the ab initio calculations for interpreting experimental results, i.e. explaining the observed phenomena and gaining insights into the underlying principles. For example, in the context of optical properties, the information obtained from calculations is not limited only to the final dielectric function but includes also the complete band structure, momentum matrix elements, etc. Such knowledge is essential for understanding and interpreting the measured data. The second field of applications of the ab initio calculations uses their predictive power. Since no empirical input is needed, in principle, the ab initio techniques can be used to model properties of not-yet-known materials. For example, high throughput searches for novel materials via scanning a broad range of possible material candidates and selecting the most promising ones for subsequent in-depth experimental studies, are gaining on importance as they save experimental costs and efforts.

4.2 Band Structure Calculations

4.2.1 Wave Function Quantum Mechanics

The first step and, in theory, the only needed input for any ab initio calculation is a specification of the system. Mathematically, it means to formulate Hamiltonian, $\hat{H}$, a quantum-mechanical operator "setting up the scene". Then, the system is described by a wave function, $\Psi(\boldsymbol{r}, t)$, whose squared norm represents the probability that the system is in a particular configuration $\boldsymbol{r}$ at time t. The Hamiltonian defines the wave function by the Schrödinger equation

$$\hat{H}\Psi(\boldsymbol{r}, t) = i\hbar\frac{\partial\Psi}{\partial t}(\boldsymbol{r}, t) , \tag{4.1}$$

which in the stationary case assumes a form known as the time-independent Schrödinger equation

$$\hat{H}\Psi(\boldsymbol{r}) = E\Psi(\boldsymbol{r}) , \tag{4.2}$$

where E is the total energy of the system. In the context of solid state physics, the system is usually considered to be composed of positively charged ions, each placed at a position $\boldsymbol{R}_j$ and having a charge q_j, and electrons at positions $\boldsymbol{r}_i$ and possessing charge $-e$. Equation (4.2) representing such a many-body problem reads

$$\hat{H}\Psi(\{\boldsymbol{R}_j\}, \{\boldsymbol{r}_i\}) = E\Psi(\{\boldsymbol{R}_j\}, \{\boldsymbol{r}_i\}) . \tag{4.3}$$

While formally simple, the Schrödinger equation is a partial differential equation for the many-body wave function $\Psi(\{\boldsymbol{R}_j\}, \{\boldsymbol{r}_i\})$, and hence its solution is impossible for majority of problems with a practical relevance. The mass of a single proton is $\approx 1836\times$ bigger than the mass of an electron, and therefore the electrons can react to any perturbation much faster. This inspires the Born–Oppenheimer approximation

$$\Psi(\{\boldsymbol{R}_j\}, \{\boldsymbol{r}_i\}) = \psi_N(\{\boldsymbol{R}_j\}) \times \psi(\{\boldsymbol{r}_i\}, \{\boldsymbol{R}_j\}) \tag{4.4}$$

separating the nuclear ($\psi_N(\{\boldsymbol{R}_j\})$) and electronic ($\psi(\{\boldsymbol{r}_i\}, \{\boldsymbol{R}_j\})$) wave functions. It turns out that in most cases the nuclei can be treated using classical mechanics as positively charged particles in a potential landscape given by the electrons. Contrarily, the instantaneous state of electrons is an ab initio solution of an electronic many-body problem in the static configuration of nuclei (e.g., the $\{\boldsymbol{R}_j\}$ now plays only the part of an external parameter). A further simplification is the assumption that the electronic many-body wave function $\psi(\{\boldsymbol{r}_i\})$ takes the form of a product of single electron wave functions

$$\psi(\{\boldsymbol{r}_i\}) = \Pi_i \phi(\boldsymbol{r}_i) , \tag{4.5}$$

known as the Hartree approximation. This form, however, violates the Pauli exclusion principle since the resulting wave function is not anti-symmetrical with respect to the exchange of particles, as required for fermions (particles with half-integer spin, e.g., electrons). This is resolved in the Harftree–Fock (HF) approximation by assuming the electronic wave-function in a form of the Slater's determinant

$$\psi(\{\boldsymbol{r}_i\}) = \frac{1}{\sqrt{N!}} \begin{vmatrix} \phi_1(\boldsymbol{r}_1) & \phi_2(\boldsymbol{r}_1) & \cdots & \phi_N(\boldsymbol{r}_1) \\ \phi_1(\boldsymbol{r}_2) & \phi_2(\boldsymbol{r}_2) & \cdots & \phi_N(\boldsymbol{r}_2) \\ \vdots & \vdots & \ddots & \vdots \\ \phi_1(\boldsymbol{r}_n) & \phi_2(\boldsymbol{r}_n) & \cdots & \phi_N(\boldsymbol{r}_n) \end{vmatrix} . \tag{4.6}$$

Although fulfilling the Pauli exclusion principle (exchange interaction), the Hartree–Fock solution is still based on the single particle orbitals, and hence misses many-body effects except the simple exchange. We note that for (4.6) to be correct, the $\phi_i(\boldsymbol{r}_j)$ must already contain a spin (spin-orbital). i.e. the position vectors $\boldsymbol{r}_j$ already include electron spin-coordinates and their integration includes summation over spin-coordinates.

4.2.2 Density Functional Theory

A new avenue for quantum-mechanical calculations has been opened via density functional theory (DFT) by switching from the many-body wave function to charge density as the main quantity. This fundamentally simplifies the solution, since we are now dealing with a function of only one position, $\boldsymbol{r}$. The ground state electron charge density, $\rho(\boldsymbol{r})$, is related to the many-body wave function as

$$\rho(\boldsymbol{r}) = \sum_i \int \mathrm{d}\boldsymbol{r}_1 \mathrm{d}\boldsymbol{r}_2 \ldots \cancel{\mathrm{d}\boldsymbol{r}_i} \ldots \mathrm{d}\boldsymbol{r}_N \, \psi^*(\boldsymbol{r}_1, \boldsymbol{r}_2, \ldots, \boldsymbol{r}_i \equiv \boldsymbol{r}, \ldots, \boldsymbol{r}_N) \cdot \cdot \psi(\boldsymbol{r}_1, \boldsymbol{r}_2, \ldots, \boldsymbol{r}_i \equiv \boldsymbol{r}, \ldots, \boldsymbol{r}_N) . \quad (4.7)$$

Hohenberg and Kohn [1] proved also a reverse relation, i.e., that any observable (including the total energy) is *uniquely* determined by $\rho(\boldsymbol{r})$ up to an additive constant. Consequently, the many-body problem was reformulated as a variational problem: the ground state charge density minimizes the total energy functional. Finally, Kohn and Sham [2] provided a practical recipe by proving that the ground state charge density, $\rho(\boldsymbol{r})$, of many interacting electrons is identical to that of a ground state of a system of fictitious non-interacting particles with the same elemental charge. The corresponding single-particle wave functions ϕ_i and energies ϵ_i are solutions of the Kohn–Sham (KS) equations [2]

$$\hat{H}_{KS}\phi_i = \epsilon_i \phi_i \, , \quad (4.8)$$

leading to

$$\rho(\boldsymbol{r}) = \sum_{i=1}^{N} \phi_i^*(\boldsymbol{r})\phi_i(\boldsymbol{r}) \, , \quad (4.9)$$

where N is the total number of electrons.

The KS Hamiltonian can be decomposed into four parts

$$\hat{H}_{KS} = -\frac{\hbar^2}{2m_\mathrm{e}}\nabla^2 + \frac{e^2}{4\pi\varepsilon_0}\int d\boldsymbol{r}' \frac{\rho(\boldsymbol{r}')}{|\boldsymbol{r}' - \boldsymbol{r}|} + V_\mathrm{ext} + V_\mathrm{xc} \, , \quad (4.10)$$

where the first term is the kinetic energy of the non-interacting electrons, the second term (called the Hartree potential V_H) is the Coulomb electrostatic potential of the electronic cloud with the density $\rho(\boldsymbol{r})$, the third term is the external potential of atomic nuclei and/or external fields, and the last one is the so-called *exchange-correlation (xc) potential*. The last term represents the quantum-mechanical part of the electron-electron interactions. The external potential is, in contrary to other terms, system specific and is defined by the system geometry and chemistry (e.g., positions and types of atoms).

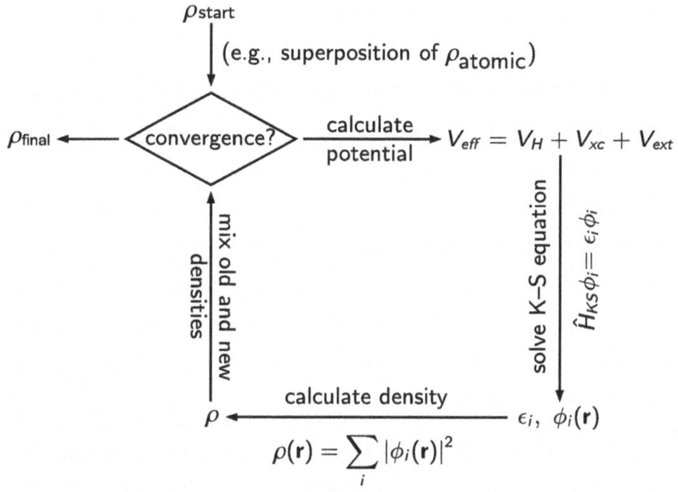

Fig. 4.1 The self-consistent scheme for calculating the electron charge density within the DFT framework

The Kohn–Sham orbitals are solutions of the KS equations (4.8), depending on the charge density, which in turn depends on the KS orbitals themselves. Thus, a self-consistent solution is needed, starting from an initial guess of the electron density (for example a superposition of atomic densities), solving the KS equation, and mixing the old and new electron densities until a converged, self-consistent solution is obtained. This process is schematically depicted in Fig. 4.1. We note that the KS equations can be easily extended to include the spin-polarization [3]. However, in order to give as simple picture as possible only the non-spin-polarized case will be considered throughout this chapter.

It is important to note that while some quantities, such as the charge density and the total energy, have well-defined physical meaning, the KS orbitals are not, in fact, single-electron states in the Hartree–Fock sense. Therefore, interpreting the KS states as the single-particle states does not have a formal justification. Despite that, it was shown that the KS wave functions are very similar to quasiparticle wave functions [4]. Consequently, the interpretation of the KS eigenvalues as quasiparticle energies and their differences as excitation energies, has been successful in some cases.

4.2.3 *Exchange-Correlation Functionals*

Although the KS equations do not contain any other than the Born–Oppenheimer approximation, and are guaranteed to lead to a ground state solution, an exact form of the exchange-correlation potential V_{xc} is not known. Consequently, there exist different levels of approximations [5] for the exchange-correlation energy E_{xc}, from which the exchange-correlation potential can be recovered as

$$V_{\mathrm{xc}}[\rho](\boldsymbol{r}) = \frac{\delta E_{\mathrm{xc}}[\rho]}{\delta \rho(\boldsymbol{r})}. \tag{4.11}$$

- Local Density Approximation (LDA)
  In the LDA scheme, the exchange-correlation energy is modeled by that of a homogeneous electron gas with a density equal to the density of the real system [1]

$$E_{\mathrm{xc}}^{\mathrm{LDA}}[\rho] = \int d^3\boldsymbol{r}\,\rho(\boldsymbol{r}) e_{\mathrm{xc}}^{\mathrm{hom}}(\rho(\boldsymbol{r})), \tag{4.12}$$

  where $e_{\mathrm{xc}}^{\mathrm{hom}}$ is the exchange-correlation energy per particle of the homogeneous electron gas. The exchange energy of homogeneous electron gas is known analytically, the correlation energy has been determined from quantum mechanical Monte Carlo simulations [6] and a simple analytic formula for the correlation part has been also established [7].
  The LDA was quite successful describing the total energies and structural properties. However, it suffers from problems such as overestimated binding (and thus underestimated cell sizes) or a notoriously known underestimation of the band gap [8]. It is worth noting that the failure in a correct determination of the band gap is not entirely caused by the LDA [9]. Moreover, it was shown for some cases that the shifted LDA band structure can be in a good agreement with the quasiparticle band structure calculated with a more sophisticated many-body approach [9].
- Generalized gradient approximation (GGA)
  In the GGA, the first correction to the LDA, the xc energy functional depends not only on the total electron density at a given point $\boldsymbol{r}$, but also on its gradient

$$E_{\mathrm{xc}}^{\mathrm{GGA}}[\rho] = \int d^3\boldsymbol{r}\, f(\rho(\boldsymbol{r}), \nabla\rho(\boldsymbol{r})). \tag{4.13}$$

  There have been developed many parametrization of the GGA functional, some based on a semi-empirical approach including parameters determined from fitting properties of a selected set of materials (e.g. [10]) and others developed fully ab initio, such as the popular PBE functional [11]. While the GGAs are quite successful with correcting some of the LDA problems such as the total energy or the structural parameters [12] (although usually not both at the same time), the improvements over LDA in band energies with respect to the experiment are usually only marginal [13].
- Meta-GGA
  Meta-GGA functionals are another step in ladder of exchange-correlation approximations [14], adding a semi-local kinetic energy density $\tau(\boldsymbol{r})$ as another ingredient

$$E_{\mathrm{xc}}^{\mathrm{meta-GGA}}[\rho] = \int d^3\boldsymbol{r}\, f(\rho(\boldsymbol{r}), \nabla\rho(\boldsymbol{r}), \tau(\boldsymbol{r})), \tag{4.14}$$

$$\tau(\boldsymbol{r}) = -\frac{\hbar^2}{2m_e}\sum_{i=1}^{N} |\nabla\phi_i(\boldsymbol{r})|^2 \ . \tag{4.15}$$

Meta-GGA allows a construction of functional that gives lattice constants, binding and surface energies with a good accuracy [15], in contrast to the GGA functionals that are usually optimized to give only one of those quantities reliably. One particular example worth mentioning due to the present focus on the optical properties, is the modified Becke–Johnson (TB-mBJ) functional, which was developed specifically to provide correct band gaps [16]. It predicts band gaps with the accuracy comparable to computationally extensive GW approaches, which are typically in an excellent agreement with experiments [17] (see Fig. 4.2). However, it was argued that for some materials the good band gap values comes at the cost of worse overall band structure compared to the GGA [17, 18].

- Hybrid functionals
Hybrid functionals combine the DFT with the Hartree–Fock theory by including a fraction of the exact exchange component

$$E_x^{\text{exact}} = -\frac{e^2}{4\pi\varepsilon_0}\sum_{i,j}^{N}\iint d^3\boldsymbol{r}_1 d^3\boldsymbol{r}_2 \frac{\phi_i^*(\boldsymbol{r}_1)\phi_j^*(\boldsymbol{r}_2)\phi_j(\boldsymbol{r}_1)\phi_i(\boldsymbol{r}_2)}{|\boldsymbol{r}_1 - \boldsymbol{r}_2|} \tag{4.16}$$

in combination with fraction of exchange and full correlation from DFT functionals [19]. Due to the mixing of KS local potential with non-local HF potential, a much better description of the band gaps, lattice constants and other properties can be obtained [20]. There are various hybrid functionals available, with the most popular ones being B3LYP [21], PBE0 [22], or HSE [23]. The differences between them boil down mostly to the choice of the exchange-correlation functional used in combination with the exact exchange, the mixing factors of the DFT and Hartree–Fock exchange, and the calculation of screening.

4.2.4 *Beyond the DFT*

4.2.4.1 The *GW* Method

For a rigorous description of the quasiparticle properties, methods beyond the DFT, e.g., many-body perturbation theory using Green's functions, become unavoidable. Here, we highlight only some general concepts, particularly those which are also directly related to the calculation of optical properties. For a thorough review on the GW method, the interested reader is directed, e.g., to [4, 24, 25].

In the many-body formalism of the Green's functions, the quasiparticle energies $\epsilon_{n,k}^{\text{QP}}$ for a general inhomogeneous system can be obtained from the following quasiparticle equation

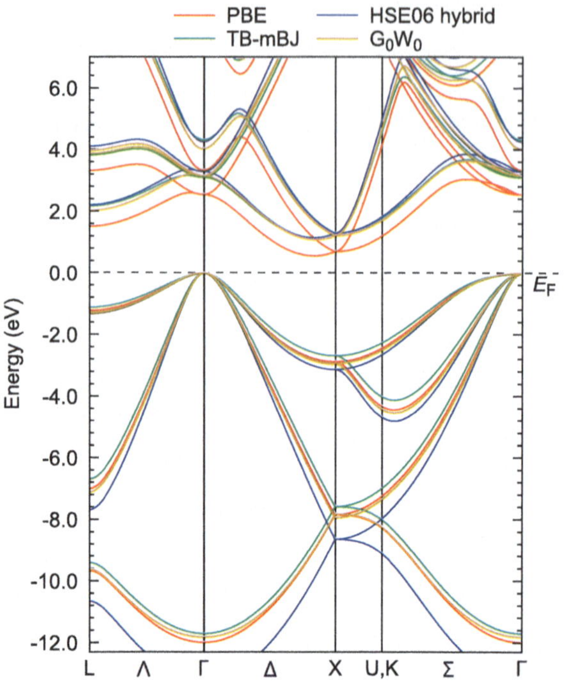

Fig. 4.2 Band structure plot of Si as obtained with different levels of DFT functionals and with the G_0W_0 method on top of the PBE. This highlights the band gap problems of the standard DFT functionals (in this case the PBE). However, it can be seen that even the methods which can reproduce the quasiparticle G_0W_0 band gap such as the hybrid functional (HSE06) or the TB-mBJ functional show some other problems such as a different compression of the valence bands and other minor changes in the band structure

$$(\hat{T} + \hat{V}_{\text{ext}} + \hat{V}_{\text{H}})\psi_{n,\boldsymbol{k}}(\boldsymbol{r}) + \int d^3\boldsymbol{r}'\Sigma(\boldsymbol{r}, \boldsymbol{r}', \epsilon^{\text{QP}}_{n,\boldsymbol{k}})\psi_{n,\boldsymbol{k}}(\boldsymbol{r}') = \epsilon^{\text{QP}}_{n,\boldsymbol{k}}\psi_{n,\boldsymbol{k}}(\boldsymbol{r}) \tag{4.17}$$

with $\hat{T}$ being the kinetic energy operator, $\hat{V}_{\text{ext}}$ and $\hat{V}_{\text{H}}$ are the external and Coulomb (Hartree) potentials, respectively, and $\Sigma(\boldsymbol{r}, \boldsymbol{r}', \epsilon^{\text{QP}}_{n,\boldsymbol{k}})$ is the electron self-energy operator. It represents the potential due to the exchange and correlation effects among electrons. It is in general non-Hermitian operator, thus the $\epsilon^{\text{QP}}_{n,\boldsymbol{k}}$ can be a complex number, with the imaginary part corresponding to a lifetime of the quasiparticle.

The single- and two-particle Green's functions, G and G_2, are defined as

$$G(1,2) = -\frac{\mathrm{i}}{\hbar}\langle N|\mathcal{T}[\psi(2)\psi^{\dagger}(1)]|N\rangle \tag{4.18}$$

and

$$G_2(1,2;1',2') = -\left(\frac{\mathrm{i}}{\hbar}\right)^2 \langle N|\mathcal{T}[\psi(1)\psi(2)\psi^{\dagger}(2')\psi^{\dagger}(1')]|N\rangle\,. \tag{4.19}$$

In this formalism (1) stands for $(\boldsymbol{r}, t)$, the $|N\rangle$ is the ground state configuration of system with N electrons, $\mathcal{T}$ is the time-ordering operator, and $\psi(1)$, $\psi^{\dagger}(1)$ are field operators in Heisenberg picture that annihilate and create electron at (1). The physical meaning of single particle Green's function is the probability amplitude that for $t_2 < t_1$, a hole will propagate from $\boldsymbol{r}_1$ to $\boldsymbol{r}_2$, or for $t_2 > t_1$, an electron will propagate from $\boldsymbol{r}_2$ to $\boldsymbol{r}_1$. When Fourier-transformed to the frequency domain, the poles of the Green's function correspond to the excitation energies.

A set of self-consistent equations, in which the self-energy operator Σ is connected to the single particle Green's function, was formulated by Hedin [26]

$$\Sigma(1,2) = \mathrm{i}\hbar \int d(34) G(1,3^+) W(1,4) \Lambda(3,2,4) \,, \tag{4.20}$$

$$G(1,2) = G_0(1,2) + \int d(34) G_0(1,3) \Sigma(3,4) G(4,2) \,, \tag{4.21}$$

$$\Lambda(1,2,3) = \delta(1,2)\delta(1,3) + \int d(4567) \frac{\delta\Sigma(1,2)}{\delta G(4,5)} G(4,6) G(7,5) \Lambda(6,7,3) \,, \tag{4.22}$$

$$W(1,2) = v(1,2) + \int d(34) v(1,3) P(3,4) W(4,2) \,, \tag{4.23}$$

$$P(1,2) = -\mathrm{i}\hbar \int d(34) G(1,3) \Lambda(3,4,2) G(4,1^+) \,, \tag{4.24}$$

where Λ is the vertex function, P is the polarization function, G_0 is the non-interacting Green's function (corresponding to $\Sigma = 0$), W is the screened Coulomb potential, and 1^+ denotes the state $t \to t + \delta$, with δ being positive infinitesimal. The screened Coulomb potential is connected to the bare Coulomb potential through the inverse microscopic dielectric function ε^{-1} as

$$W(1,2) = \int d(3) \varepsilon^{-1}(1,3) v(3,2) \,. \tag{4.25}$$

In the GW approximation, the second part of the vertex function is neglected, leaving only $\Lambda(1,2,3) = \delta(1,2)\delta(1,3)$ which leads to

$$\Sigma(1,2) = \mathrm{i}\hbar G(1,2) W(1,2) \,, \tag{4.26}$$

$$W(1,2) = v(1,2) + \int d(34) W(1,3) P(3,4) v(4,2) \,, \tag{4.27}$$

and

$$P(1,2) = -\mathrm{i}\hbar G(1,2) G(2,1^+) \,. \tag{4.28}$$

After a Fourier transform to the frequency domain the self-energy takes the following form

$$\Sigma(\boldsymbol{r},\boldsymbol{r}';\omega) = \frac{\mathrm{i}}{2\pi} \int d\omega' G(\boldsymbol{r},\boldsymbol{r}';\omega+\omega') W(\boldsymbol{r},\boldsymbol{r}';\omega') \,. \tag{4.29}$$

To calculate the screened Coulomb potential using (4.25), one needs to calculate the microscopic dielectric function which is connected with the polarizability as

$$\varepsilon(\boldsymbol{r}, \boldsymbol{r}', \omega) = \delta(\boldsymbol{r} - \boldsymbol{r}') - \int d\boldsymbol{r}'' v(\boldsymbol{r} - \boldsymbol{r}'') P(\boldsymbol{r}'', \boldsymbol{r}', \omega) \,. \tag{4.30}$$

A standard approximation for the calculation of the polarization is the random phase approximation (RPA) [27, 28]. It again considers only the first part of the vertex function, leading to (4.28) with G replaced by the non-interacting G_0

$$P^0(1, 2) = -\mathrm{i}\hbar G_0(1, 2) G_0(2, 1^+) \,, \tag{4.31}$$

where P^0 is the independent-particle polarization.

In the most simple case, the density functional theory is used as a starting point for the evaluation of quasiparticle properties using the KS orbitals as a basis. In such case, the non-interacting Green's function is constructed from the KS states as

$$G_0(\boldsymbol{r}, \boldsymbol{r}'; \omega) = \sum_{n\boldsymbol{k}} \frac{\phi_{n\boldsymbol{k}}(\boldsymbol{r}) \phi^*_{n\boldsymbol{k}}(\boldsymbol{r}')}{\hbar\omega - \epsilon_{n\boldsymbol{k}} - \mathrm{sgn}(\epsilon_{\mathrm{F}} - \epsilon_{n\boldsymbol{k}}) \mathrm{i}\delta} \tag{4.32}$$

where ϵ_{F} is the Fermi energy. The independent particle polarization, P^0, is a sum over all independent transitions of non-interacting electrons that respond to total potential (induced screening and external)

$$\begin{aligned} P^0(\boldsymbol{r}, \boldsymbol{r}', \omega) = &\sum_{n\boldsymbol{k}}^{\mathrm{val}} \sum_{n'\boldsymbol{k}'}^{\mathrm{cond}} \phi^*_{n\boldsymbol{k}}(\boldsymbol{r}) \phi_{n'\boldsymbol{k}'}(\boldsymbol{r}) \phi^*_{n'\boldsymbol{k}'}(\boldsymbol{r}') \phi_{n\boldsymbol{k}}(\boldsymbol{r}') \times \\ &\times \left(\frac{1}{\hbar\omega - \epsilon_{\boldsymbol{k}'n'} + \epsilon_{\boldsymbol{k}n} + \mathrm{i}\delta} - \frac{1}{\hbar\omega + \epsilon_{\boldsymbol{k}'n'} - \epsilon_{\boldsymbol{k}n} - \mathrm{i}\delta} \right) . \end{aligned} \tag{4.33}$$

For a crystalline material, (4.30) can be Fourier-transformed to

$$\varepsilon_{\boldsymbol{G},\boldsymbol{G}'}(\boldsymbol{q}, \omega) = \delta_{\boldsymbol{G},\boldsymbol{G}'} - v_{\boldsymbol{G}}(\boldsymbol{q}) P_{\boldsymbol{G},\boldsymbol{G}'}(\boldsymbol{q}, \omega) \tag{4.34}$$

where $\boldsymbol{G}$ and $\boldsymbol{G}'$ are reciprocal lattice vectors, and $\boldsymbol{q}$ is a wave vector from the first Brillouin zone, $v_{\boldsymbol{G}}(\boldsymbol{q}) = 4\pi e^2/|\boldsymbol{q} + \boldsymbol{G}|^2$. The irreducible polarizability in the RPA approximation than takes a form

$$P^0_{G,G'}(q,\omega) = \frac{1}{\Omega}\sum_{n,n',k} \langle n,k|e^{-i(q+G)r}|n',k+q\rangle\langle n',k+q|e^{i(q+G')r'}|n,k\rangle \times$$
$$\times\ [f(\epsilon_{n',k+q}) - f(\epsilon_{n,k})]\left[\frac{1}{\epsilon_{n',k+q} - \epsilon_{n,k} - \hbar\omega - i\delta} + \frac{1}{\epsilon_{n',k+q} - \epsilon_{n,k} + \hbar\omega + i\delta}\right] \quad (4.35)$$

where Ω is the unit cell volume and f is the Fermi distribution function. The first part of the sum, e.g., the momentum matrix elements correspond to the transition probability, while the second part corresponds to the joint density of states (JDOS). After calculating the dielectric tensor ε by (4.30), inverting it to get the ε^{-1}, calculating W (4.25), and using the G_0 (4.32) as approximation of G, we obtain the self-energy defined in (4.26). Finally, we can use this to obtain the quasiparticle energies $\epsilon^{QP}_{n,k}$ in the first-order perturbative approach on top of the KS results

$$\epsilon^{QP}_{n,k} = \epsilon_{n,k} + C_{n,k}\,\mathrm{Re}\langle\phi_{n,k}|\Sigma(\epsilon_{n,k}) - V_{xc}|\phi_{n,k}\rangle\,. \quad (4.36)$$

where $C_{n,k}$ is the quasiparticle renormalization factor

$$C_{n,k} = \left(1 - \mathrm{Re}\langle\phi_{n,k}|\partial\Sigma(\epsilon)/\partial\epsilon|_{\epsilon_{n,k}}|\phi_{n,k}\rangle\right)^{-1}\,. \quad (4.37)$$

This formulation of the perturbation method is usually denoted G_0W_0 to show that both, the Green's function and the screened Coulomb potential are calculated without self-consistency [4]. There are many other schemes (GW_0, GW) with different amounts of self-consistency [29–31].

4.3 Optical Properties

The directly measurable macroscopic dielectric function ε_M is connected to the microscopic dielectric function by the (0,0) element of the inverse dielectric tensor

$$\varepsilon_M(\omega) = \lim_{q\to 0}\frac{1}{[\varepsilon^{-1}(q,\omega)]_{0,0}}\,. \quad (4.38)$$

Here, $q \to 0$ stems from the fact that a wave vector of the light is usually much shorter than a typical wave vector of electrons in the system.

In the case of a microscopic inhomogeneity in the system, all of the elements of the dielectric matrix contribute to the $\varepsilon^{-1}_{0,0}$ component. This is called a local field effects and originates from the fact that in inhomogeneous systems even spatially constant external field induces some field fluctuations on the microscopic scale, and leads to an effective mixing of the transitions even if the polarizability (e.g., in the RPA) contains no explicit electron-hole interaction.

One of the problems associated with (4.38) is that it involves evaluation of $\varepsilon_{\boldsymbol{G},\boldsymbol{G}'}$ for $\boldsymbol{G}$, $\boldsymbol{G}'$ up to a certain artificial limit, and a subsequent inversion to get the $\varepsilon_{0,0}^{-1}$ component. A possible simplification is to skip the inversion of the full dielectric tensor, and to replace the $(0, 0)$ component of the inverse tensor with the inverse of the $(0, 0)$ component. In other words, this neglects the local field effects and simplifies (4.38) to

$$\varepsilon_{\mathrm{M}}^{\mathrm{NLF}}(\omega) = \lim_{q\to 0} \varepsilon_{0,0}(\boldsymbol{q}, \omega) = 1 - \lim_{q\to 0} \left[v(\boldsymbol{q}) P_{0,0}(\boldsymbol{q}, \omega)\right] . \tag{4.39}$$

4.3.1 Bethe–Salpeter Equation

It was shown in the previous section, how to obtain the microscopic dielectric function using the independent transitions RPA (independent particles) approach. This approximation is sufficient for metals and for other materials where the excitonic (electron-hole interaction) effects are negligible due to effective screening. In semiconductors, however, the excitonic effects usually cannot be neglected.

Two most popular frameworks for including excitonic effects into optical response is the Bethe–Salpeter equation (BSE) [32–35] and time-dependent density functional theory [36, 37]. We focus only on the former since it is much better established and more widely used, even though computationally somewhat more demanding.

The Bethe–Salpeter theory is formulated in terms of two-particle propagators, the four-point functions describing the motion of two particles through the system. The BSE for the four-point generalized (reducible) polarization function $\bar{P}$ in the direct space assumes a form [37]

$$\bar{P}(1, 2; 1', 2') = P^0(1, 2; 1', 2') + \int d(3456) P^0(1, 4; 1', 3)\bar{K}(3, 5; 4, 6)\bar{P}(6, 2; 5, 2'), \tag{4.40}$$

where $\bar{P}$ is defined in the terms of one and two particle interacting Green's functions

$$\bar{P}(1, 2; 1', 2') = \mathrm{i}\hbar\left[G_2(1, 2; 1', 2') - G(1, 1')G(2, 2')\right] , \tag{4.41}$$

and K is an interaction kernel which contains both the screened and bare Coulomb interaction

$$\bar{K}(3, 5; 4, 6) = \delta(1, 2)\delta(3^+, 4)\bar{v}(1, 3) - \delta(1, 3)\delta(2, 4)W(1^+, 2) , \tag{4.42}$$

where $\bar{v}$ is the modified Coulomb interaction. $\bar{v}_{\boldsymbol{G}}(\boldsymbol{q})$ is defined as 0 for $\boldsymbol{G} = 0$ and $v_{\boldsymbol{G}}(\boldsymbol{q})$ otherwise, e.g., this is the bare Coulomb potential without the long range term. The $P^0(1, 2; 1', 2')$ is the four-point independent particle polarization

$$P^0(1, 2; 1', 2') = \mathrm{i}\hbar G(1, 2')G(2, 1') . \tag{4.43}$$

It is convenient to deal with the $\bar{P}$ instead of the irreducible polarization P since it can be shown that ε_M can be calculated directly from the polarization function, hence avoiding the costly matrix inversion of the dielectric tensor in (4.38) [37]

$$\varepsilon_\mathrm{M}(\omega) = 1 - \lim_{q\to 0}\left[v_0(\boldsymbol{q})\bar{P}_{0,0}(\boldsymbol{q},\omega)\right] . \tag{4.44}$$

The four-point polarization is related to its two-point analogy used in the calculation of macroscopic dielectric function

$$\bar{P}(1,2) = \bar{P}(1,2;1^+,2^+) . \tag{4.45}$$

Finally, the generalized (reducible) polarizability $\bar{P}$ is connected with the irreducible polarizability by the Dyson-like equation

$$\bar{P}(1,2) = P(1,2) + \int d(34)P(1,3)\bar{v}(3,4)\bar{P}(4,2) . \tag{4.46}$$

When going into the transition space with basis defined by single particle states (usually the KS orbitals), the BSE takes the form of matrix equation and leads to an eigenvalue problem for the effective two particle Hamiltonian $\hat{H}^\mathrm{BSE}$

$$\hat{H}^\mathrm{BSE}_{vc\boldsymbol{k},v'c'\boldsymbol{k}'}\, A^\lambda_{v'c'\boldsymbol{k}'} = E^\lambda\, A^\lambda_{vc\boldsymbol{k}}, \tag{4.47}$$

where v spans over all occupied valence bands, while c goes over empty conduction bands. The Hamiltonian in the Tamm–Dancoff approximation [38] (i.e., neglecting the coupling terms of the Hamiltonian) reads

$$\hat{H}^\mathrm{BSE} = \hat{H}^\mathrm{(diag)} + 2\,\hat{H}^\mathrm{(x)} + \hat{H}^\mathrm{(dir)} . \tag{4.48}$$

The diagonal term of the Hamiltonian, $\hat{H}^\mathrm{(diag)}$, accounts for a contribution of the single-particle transitions

$$\hat{H}^\mathrm{(diag)}_{vc\boldsymbol{k},v'c'\boldsymbol{k}'} = (\epsilon_{c\boldsymbol{k}} - \epsilon_{v\boldsymbol{k}})\,\delta_{vv'}\,\delta_{cc'}\,\delta_{\boldsymbol{k}\boldsymbol{k}'} . \tag{4.49}$$

The electron-hole exchange term, $\hat{H}^\mathrm{(x)}$ is repulsive and is caused by the unscreened Coulomb interaction

$$\hat{H}^\mathrm{(x)}_{vc\boldsymbol{k},v'c'\boldsymbol{k}'} = \int d^3\boldsymbol{r}\int d^3\boldsymbol{r}'\,\phi_{v\boldsymbol{k}}(\boldsymbol{r})\,\phi^*_{c\boldsymbol{k}}(\boldsymbol{r})\,\bar{v}(\boldsymbol{r},\boldsymbol{r}')\,\phi^*_{v'\boldsymbol{k}'}(\boldsymbol{r}')\,\phi_{c'\boldsymbol{k}'}(\boldsymbol{r}') . \tag{4.50}$$

The direct (correlation) term $\hat{H}^\mathrm{(dir)}$ describes the screened attractive electron-hole interaction, due to the screened Coulomb potential

$$\hat{H}^\mathrm{(dir)}_{vc\boldsymbol{k},v'c'\boldsymbol{k}'} = -\int d^3\boldsymbol{r}\int d^3\boldsymbol{r}'\,\phi_{v\boldsymbol{k}}(\boldsymbol{r})\,\phi^*_{c\boldsymbol{k}}(\boldsymbol{r}')\,W(\boldsymbol{r},\boldsymbol{r}')\,\phi^*_{v'\boldsymbol{k}'}(\boldsymbol{r})\,\phi_{c'\boldsymbol{k}'}(\boldsymbol{r}') . \tag{4.51}$$

This decomposition of the Hamiltonian allows to easily turn off and on its different parts, i.e., different interactions. For example, neglecting the $\hat{H}^{(\mathrm{dir})}$ is equivalent to the RPA with local field effects.

Then, the generalized polarizability can be written as [35]

$$\bar{P}_{vc\boldsymbol{k}v'c'\boldsymbol{k}'} = \sum_{\lambda} \frac{A^{\lambda}_{vc\boldsymbol{k}} A^{\lambda *}_{v'c'\boldsymbol{k}'}}{\hbar\omega - E^{\lambda}}, \tag{4.52}$$

and the $\varepsilon_{\alpha\alpha}$ component of the macroscopic dielectric function is subsequently calculated according to the following formula [35]

$$\varepsilon_{\alpha\alpha,\mathrm{M}}(\omega) = 1 + \frac{1}{\Omega}\frac{e^2\hbar^2}{\varepsilon_0 m_0^2} \sum_{\lambda} \left| \sum_{cv\boldsymbol{k}} A^{\lambda}_{vc\boldsymbol{k}} \frac{\langle c\boldsymbol{k}|\hat{p}_{\alpha}|v\boldsymbol{k}\rangle}{\epsilon_{c\boldsymbol{k}} - \epsilon_{v\boldsymbol{k}}} \right|^2 \sum_{\beta=\pm 1} \frac{1}{E_{\lambda} - \beta\hbar(\omega + \mathrm{i}\delta)}. \tag{4.53}$$

where $\hat{p}_{\alpha}$ is the α component of the momentum operator $\hat{\boldsymbol{p}}$.

The BSE calculation is quite time-consuming, especially for large systems, since the BSE Hamiltonian has size $N = N_{\mathrm{v}} N_{\mathrm{c}} N_{\boldsymbol{k}}$ and the diagonalization problem has a time complexity of $O(N^3)$. There are some alternative approaches to speed up the calculations such as the time-evolution algorithm, in which the macroscopic polarizability obtained from the solution of initial-value problem instead of diagonalization of the BSE Hamiltonian, leading to a much better $O(N^2)$ complexity [39].

4.3.2 Usual Workflow

In most cases, a sequence of calculations depicted in Fig. 4.3 is needed in order to get the optical spectra including the excitonic effects. One starts with the KS orbitals and energies, which are used to construct the non-interacting Green's function G_0 and to calculate the screening at the RPA level. Using the single-pass G_0 and W_0, the self-energy is constructed leading to quasiparticle energies which together with the W_0 and KS orbitals are used to construct the BSE two-particle Hamiltonian. A common simplification is to skip the GW step and use the DFT energies (with a possible rigid shift, i.e., the scissor operator). Differences between the dielectric functions calculated with those two schemes, G_0W_0 BSE and PBE BSE, are demonstrated

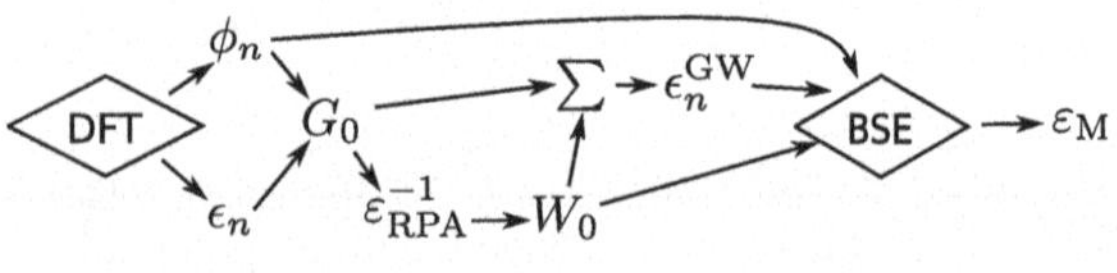

Fig. 4.3 Schematic workflow for the BSE calculation on top of DFT and G_0W_0

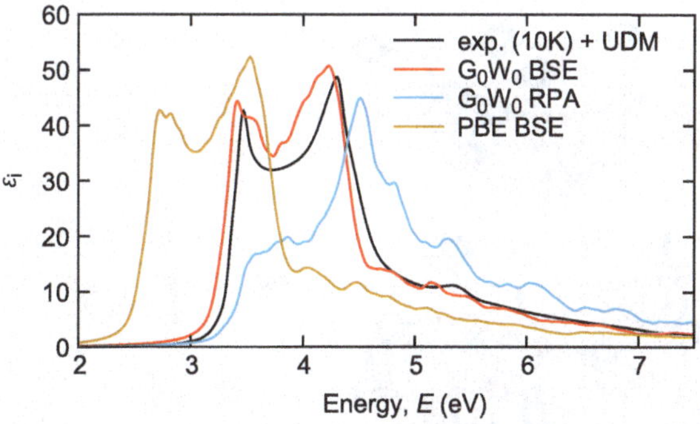

Fig. 4.4 Imaginary part of the dielectric function, ε_i, calculated at different theory levels compared with the experimental ε_i of c-Si determined from ellipsometry at 10 K using UDM described in Chap. 3 [40]. The G_0W_0 calculations were done on top of the PBE-DFT calculations. Notice a significant spectral weight shift between the G_0W_0 BSE and G_0W_0 RPA calculations caused by the inclusion of excitonic effects in the former. In this simple model case, the shape of ε_i calculated with the BSE on top of the PBE-DFT results (PBE BSE) is almost identical to the BSE results on top of the quasiparticle G_0W_0 calculations (G_0W_0 BSE) except for the band gap underestimation of the PBE functional (i.e., rigid energy shift)

in Fig. 4.4 on an example of cubic Si. Figure 4.4 also demonstrates the significant spectral weight shift between the G_0W_0 BSE and G_0W_0 RPA calculations caused by the inclusion of excitonic effects in the former approach.

4.4 Modeling of Complex Systems

As mentioned at the beginning of this chapter, the only needed input for the ab initio calculations is, in the principle, the structure of the system, which defines the external potential V_{ext} related to the atomic cores. This is an easy task for ordered crystalline materials, which structure is unambiguously defined by the primitive cell. However, the situation is more complicated for systems that do not possess any short range order such as solid solutions or amorphous materials. Although an exact description suitable for standard DFT calculations is not readily available in these cases, it is still possible to generate structural models fulfilling the periodic boundary conditions (PBCs) framework, which, in most cases, yield representative results also of such aperiodic structures.

4.4.1 Special Quasi-random Structures

There are several approaches to treat crystalline solid solutions within PBCs. Perhaps the most commonly used one is the so-called supercell approach, in which a

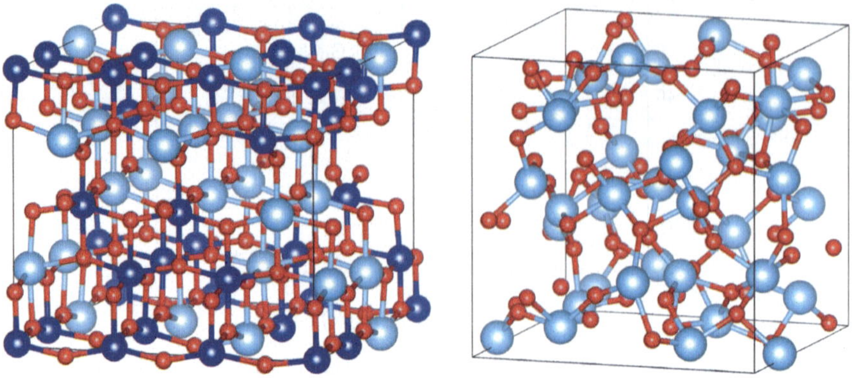

Fig. 4.5 Structural models of: left – $Ti_{0.5}Si_{0.5}O_2$ solid solution in anatase structure created with the SQS method; right – amorphous TiO_2 structure created by simulated annealing

multiple of the unit cell is employed and atoms forming the solid solution are distributed on the corresponding sublattice. For example, in the case of anatase-based $Ti_xSi_{1-x}O_2$, Ti and Si atoms are distributed on the Ti sublattice sites of the underlying anatase structure. A special quasi-random structure (SQS) [41, 42] is a specially designed supercell with atoms closely mimicking short range order parameters of a statistically random solid solution with the same composition. The advantage of this method is that it explicitly includes various local environments, which may lead to non-uniform local relaxations and connected localized features in the electronic structure. The generation of SQS supercells is relatively straightforward and provides a geometrical insight. On the other hand, the accessible compositions are directly limited by the supercell size (e.g., an example 108-atom anatase supercell in Fig. 4.5 contains 72 oxygen atoms and 36 metal atoms, hence the smallest available compositional step in $Ti_xSi_{1-x}O_2$ is $\Delta x \approx 0.03$). Large supercells supposedly treating dilute compositions remain computationally prohibitive. However, this method is an ideal tool for studying trends over a large range of compositions.

4.4.2 Simulated Annealing

A simulated annealing "melt and quench" scheme is a theoretical method to generate models of amorphous structures which follows a physical process of fast quenching and hence freezing a disordered, liquid-like structure with none or only a small degree of order. In the context of DFT, ab initio molecular dynamics are used to perform this process, in which the originally crystalline structure is kept for some time (typically a few ps) at high temperature (e.g., 5000 K) and then over a few ps rapidly cooled down to 0 K. The resulting structure is then relaxed to remove any residual forces on ions. If reliable force fields are available the simulated annealing can be performed also using classical molecular dynamics which significantly shortens the computational time. Since the DFT calculations (either simulated annealing itself, or any subsequent calculation of optical properties using the generated structure) employ PBCs, the aim is to remove crystalline short-range order, while the long range-order cannot be prevented. Therefore, it is crucial to choose a large enough simulation box so that the PBC-induced artificial long-range order does not impact the predicted properties while, at the same time, it is small enough to keep computations affordable.

4.4.3 Example: Refractive Index of $Ti_xSi_{1-x}O_2$

An example of application of solid solution models and amorphous structures is given in Fig. 4.6. There, refractive index, was calculated for different crystalline-based solid solutions well as for an amorphous-like $Ti_xSi_{1-x}O_2$ for several discrete compositions covering the whole quasi-binary SiO_2–TiO_2 tie-line. The selection of crystalline structures was based on the most stable crystal polymorphs of the boundary binary oxides. The predicted values of n are compared with experimentally measured ones using ellipsometry on thin film samples deposited with plasma-enhanced chemical vapor deposition (PECVD) [43]. An excellent agreement between the values predicted using the TB-mBJ potential (see Sect. 4.2.3) for the amorphous structure and experiment is seen. Perhaps the largest discrepancy between theoretical prediction

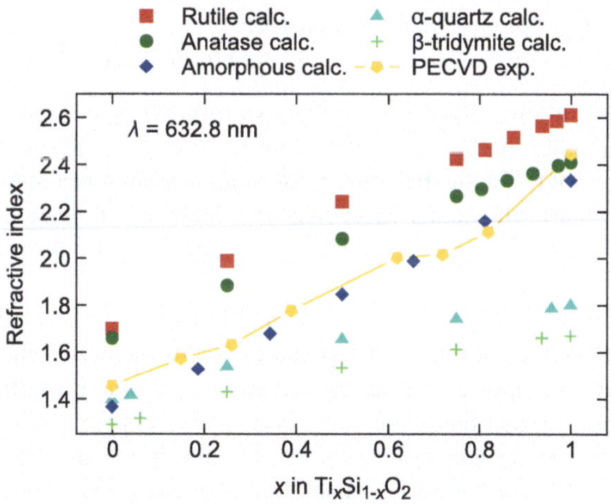

Fig. 4.6 Compositional dependence of calculated refractive index of $Ti_xSi_{1-x}O_2$ mixed oxide at 632.8 nm [43]. The electronic structure was calculated using DFT with the TB-mBJ functional, optical properties were calculated at the RPA level. The crystalline structures based on the anatase, rutile, α-quartz, and β-tridymite were produced by SQS, amorphous structures were constructed by the simulated annealing approach. Experimental refractive index of amorphous $Ti_xSi_{1-x}O_2$ thin films is included for comparison

and measurement is obtained for TiO_2 where experimental values of n agrees better with that of the anatase rather then amorphous structure. Indeed, such structural evolution was confirmed also by X-ray diffraction showing that while the pure TiO_2 sample assumed the anatase structure, all other samples containing Si were X-ray amorphous [43].

4.5 Few Notes on Interpretation of the Results

4.5.1 Predictions Versus Experiment

Despite the great progress on the development of theories and available computational resources the perfect agreement between calculated predictions and experimental data is usually not reached. It is caused by a multiple reasons such as

- approximations in the theory (GW approximation, RPA);
- numerical accuracy – especially for larger systems it is sometimes prohibitively difficult to get well converged calculations with respect to parameters such as number of $\boldsymbol{k}$-points (i.e., discretization of the reciprocal space), number of conduction bands, etc.;
- temperature effects – strictly speaking, ab initio calculations correspond to 0 K (without the zero-point vibrations), and although consideration of electron-phonon interactions is possible [44], it is not yet well established in the ab initio codes, in addition to being computationally expensive;
- models of infinite perfect systems contrast real finite-sized systems with defects.

Sometime extra caution is needed even in cases of good agreements. For example, it was reported that the simple LDA+RPA approach can provide a good estimate for the electronic part of the static dielectric tensor $\varepsilon_{\text{stat}}$ in some cases such as TiO_2 [45]. However, in this case, the actual agreement is a lucky error cancellation since the downshift of the valence band from DFT was compensated by the missing redshift caused by the excitonic effects.

On the other hand, it is fair to conclude that although it is quite challenging to get accurate absolute values of the band gap and optical properties, general trends like relative values and compositional trends are usually qualitatively correct.

Only electronic transitions were discussed in this chapter. Note that a calculation of phonon spectra and evaluation of phonon-photon interactions using the frozen-phonon method or the density functional perturbation theory [46] are relative straightforward. The phonon-assisted indirect electronic transitions can be included as well [47]. Nevertheless, this is not a routine task at present.

4.5.2 *Electronic Versus Optical Band Gap*

The value of the band gap is an important benchmark parameter in the ab initio calculations of electronic and optical properties as well as during a development of new methods and functionals. Since it is just a scalar quantity, it provides a very handy way for comparison and/or validation of the calculations with other reports. Unfortunately, there are some common misconceptions in the band gap comparison. In contrast to the photo-electron spectroscopy (PES) and inverse photo-electron spectroscopy (IPS) needed to get the quasiparticle band structure, it is relatively easy to perform the optical measurements. Because of the availability of experimental data it is common to compare the band gap value obtained from the electronic structure calculations with the experimental optical band gap. As an example let us name titanium dioxide of which two most common crystalline polymorphs, rutile and anatase, have optical band gaps of 3.03 and 3.2 eV, respectively [48]. Numerous papers compare the calculated electronic HOMO-LUMO-like band gap with those values, such as [49, 50]. However, the quasiparticle band gap of rutile was reported to be 3.6 eV [51] by PES and IPS, and the lower optical band gap is caused by below-the-gap excitons.

Another problematic area is the treatment of band gap in amorphous materials. While in crystalline materials the electronic states are delocalized and the absorption onset is sharp, in the amorphous structure the band edges consist of spatially localized electronic states and the absorption onset is slower. This is known as the Urbach tail and it contains the contributions from factors such as a structural disorder and disorder originating from thermal lattice vibrations. Therefore, the HOMO-LUMO gap obtained from the electronic structure calculations can be significantly lower then the experimental one. The commonly used experimental method for extraction of the optical band gap is the so-called Tauc plot fitting [52, 53] justified by the formula

$$\omega\sqrt{\varepsilon_{\mathrm{i}}} \propto (\hbar\omega - E_{\mathrm{g}})\,. \tag{4.54}$$

Within this procedure, the linear part of $\omega\sqrt{\varepsilon}$ near the absorption onset is fitted by a linear function of $\hbar\omega$, and extrapolated to zero. A similar approach can be used also for ab initio data, either for the calculated dielectric function [54] or as a simple fitting of the joint density of states under the assumption of constant matrix elements [55].

4.6 Conclusions

This chapter reviewed ab initio methodology developed for the calculation of electronic structure within the framework of standard DFT, with a particular focus on the evaluation of optical properties. Different levels of accuracy and complexity, from a single-particle DFT employing standard approximations for the quantum-mechanical exchange and correlation electron-electron interactions to a quasiparticle GW approach, were introduced. Prediction of the optical properties using RPA and

Bethe–Salpeter theory were briefly described. Finally, issues related to the modeling of realistic materials, e.g., solid-solutions or amorphous structures, were also mentioned. Selected examples of calculated electronic band structure, dielectric function, and refractive index, suggest that the state-of-the-art ab initio calculations possess the predictive power and qualitative, often also quantitative, accuracy needed for modern, knowledge-based materials science.

Acknowledgements We would like to thank professor Dominik Munzar for reading this chapter, valuable comments and helpful discussion.

References

1. P. Hohenberg, W. Kohn, Phys. Rev. **136**(3B), B864 (1964). https://doi.org/10.1103/PhysRev.136.B864
2. W. Kohn, L.J. Sham, Phys. Rev. **140**(4A), A1133 (1965). https://doi.org/10.1103/PhysRev.140.A1133
3. U. von Barth, L. Hedin, J. Phys. C: Solid State Phys. **5**(13), 1629 (1972). https://doi.org/10.1088/0022-3719/5/13/012
4. M.S. Hybertsen, S.G. Louie, Phys. Rev. B **34**(8), 5390 (1986). https://doi.org/10.1103/PhysRevB.34.5390
5. J.P. Perdew, MRS Bull. **38**(09), 743 (2013). https://doi.org/10.1557/mrs.2013.178
6. D.M. Ceperley, B.J. Alder, Phys. Rev. Lett. **45**(7), 566 (1980). https://doi.org/10.1103/PhysRevLett.45.566
7. J.P. Perdew, Y. Wang, Phys. Rev. B **45**(23), 13244 (1992). https://doi.org/10.1103/PhysRevB.45.13244
8. J.P. Perdew, A. Zunger, Phys. Rev. B **23**(10), 5048 (1981). https://doi.org/10.1103/PhysRevB.23.5048
9. R.W. Godby, M. Schlüter, L.J. Sham, Phys. Rev. B **37**(17), 10159 (1988). https://doi.org/10.1103/PhysRevB.37.10159
10. J.P. Perdew, A. Ruzsinszky, G.I. Csonka, O.A. Vydrov, G.E. Scuseria, L.A. Constantin, X. Zhou, K. Burke, Phys. Rev. Lett. **100**(13), 136406 (2008). https://doi.org/10.1103/PhysRevLett.100.136406
11. J.P. Perdew, K. Burke, M. Ernzerhof, Phys. Rev. Lett. **77**(18), 3865 (1996). https://doi.org/10.1103/PhysRevLett.77.3865
12. J.P. Perdew, A. Ruzsinszky, G.I. Csonka, O.A. Vydrov, G.E. Scuseria, L.A. Constantin, X. Zhou, K. Burke, Phys. Rev. Lett. **100**(13), 136406 (2008). https://doi.org/10.1103/PhysRevLett.100.136406
13. P. Dufek, P. Blaha, V. Sliwko, K. Schwarz, Phys. Rev. B **49**(15), 10170 (1994). https://doi.org/10.1103/PhysRevB.49.10170
14. J.P. Perdew, K. Schmidt, AIP Conf. Proc. **577**(1), 1 (2001). https://doi.org/10.1063/1.1390175
15. J.P. Perdew, A. Ruzsinszky, G.I. Csonka, L.A. Constantin, J. Sun, Phys. Rev. Lett. **103**(2), 026403 (2009). https://doi.org/10.1103/PhysRevLett.103.026403
16. F. Tran, P. Blaha, Phys. Rev. Lett. **102**(22), 226401 (2009). https://doi.org/10.1103/PhysRevLett.102.226401
17. D. Koller, F. Tran, P. Blaha, Phys. Rev. B **85**(15), 155109 (2012). https://doi.org/10.1103/PhysRevB.85.155109
18. D.J. Singh, Phys. Rev. B **82**(20), 205102 (2010). https://doi.org/10.1103/PhysRevB.82.205102
19. A.D. Becke, J. Chem. Phys. **98**(2), 1372 (1993). https://doi.org/10.1063/1.464304
20. J. Heyd, J.E. Peralta, G.E. Scuseria, R.L. Martin, J. Chem. Phys. **123**(17), 174101 (2005). https://doi.org/10.1063/1.2085170

21. K. Kim, K.D. Jordan, J. Phys. Chem. **98**(40), 10089 (1994). https://doi.org/10.1021/j100091a024
22. C. Adamo, V. Barone, J. Chem. Phys. **110**(13), 6158 (1999). https://doi.org/10.1063/1.478522
23. J. Heyd, G.E. Scuseria, M. Ernzerhof, J. Chem. Phys. **118**(18), 8207 (2003). https://doi.org/10.1063/1.1564060
24. F. Aryasetiawan, O. Gunnarsson, Rep. Prog. Phys. **61**(3), 237 (1998), http://stacks.iop.org/0034-4885/61/i=3/a=002
25. L. Hedin, J. Phys.: Condens. Matter **11**(42), R489 (1999). https://doi.org/10.1088/0953-8984/11/42/201
26. L. Hedin, Phys. Rev. **139**(3A), A796 (1965). https://doi.org/10.1103/PhysRev.139.A796
27. J. Lindhard, Kgl. Danske Videnskab. Selskab Mat.-Fys. Medd. **28** (1954)
28. D. Pines, D. Bohm, Phys. Rev. **85**(2), 338 (1952)
29. M. Shishkin, G. Kresse, Phys. Rev. B **75**(23), 235102 (2007). https://doi.org/10.1103/PhysRevB.75.235102
30. B. Holm, U. von Barth, Phys. Rev. B **57**(4), 2108 (1998). https://doi.org/10.1103/PhysRevB.57.2108
31. E.L. Shirley, Phys. Rev. B **54**(11), 7758 (1996). https://doi.org/10.1103/PhysRevB.54.7758
32. E.E. Salpeter, H.A. Bethe, Phys. Rev. **84**(6), 1232 (1951). https://doi.org/10.1103/PhysRev.84.1232
33. G. Strinati, La Rivista del Nuovo Cimento **11**(12), 1 (1988). https://doi.org/10.1007/BF02725962
34. G. Bussi, Physica Scripta **T109**, 141 (2004). https://doi.org/10.1238/Physica.Topical.109a00141
35. C. Rödl, F. Fuchs, J. Furthmüller, F. Bechstedt, Phys. Rev. B **77**(18), 184408 (2008). https://doi.org/10.1103/PhysRevB.77.184408
36. E. Runge, E.K.U. Gross, Phys. Rev. Lett. **52**(12), 997 (1984). https://doi.org/10.1103/PhysRevLett.52.997
37. G. Onida, L. Reining, A. Rubio, Rev. Mod. Phys. **74**(2), 601 (2002). https://doi.org/10.1103/RevModPhys.74.601
38. A.L. Fetter, J.D. Walecka, *Quantum Theory of Many-Particle Systems* (Courier Corporation, 2012)
39. W.G. Schmidt, S. Glutsch, P.H. Hahn, F. Bechstedt, Phys. Rev. B **67**(8), 085307 (2003). https://doi.org/10.1103/PhysRevB.67.085307
40. D. Franta, A. Dubroka, C. Wang, A. Giglia, J. Vohánka, P. Franta, I. Ohlídal, Appl. Surf. Sci. (2017). https://doi.org/10.1016/j.apsusc.2017.02.021
41. S.H. Wei, L.G. Ferreira, J. Bernard, A. Zunger, Phys. Rev. B: Condens. Matter **42**(15), 9622 (1990). https://doi.org/10.1103/PhysRevB.42.9622
42. D. Holec, L. Zhou, R. Rachbauer, P.H. Mayrhofer, in *Density Functional Theory: Principles, Applications and Analysis*, ed. by J.M. Pelletier, J. Morin (Nova Publishers, 2013), pp. 259–284
43. P. Ondračka, D. Holec, D. Nečas, E. Kedroňová, S. Elisabeth, A. Goullet, L. Zajíčková, Phys. Rev. B **95**(19), 195163 (2017). https://doi.org/10.1103/PhysRevB.95.195163
44. F. Giustino, Rev. Mod. Phys. **89**(1), 015003 (2017). https://doi.org/10.1103/RevModPhys.89.015003
45. S.D. Mo, W.Y. Ching, Phys. Rev. B **51**(19), 13023 (1995). https://doi.org/10.1103/PhysRevB.51.13023
46. S. Baroni, S. De Gironcoli, A. Dal Corso, P. Giannozzi, Rev. Mod. Phys. **73**(2), 515 (2001). https://doi.org/10.1103/RevModPhys.73.515
47. J. Noffsinger, E. Kioupakis, C.G. Van de Walle, S.G. Louie, M.L. Cohen, Phys. Rev. Lett. **108**(16), 167402 (2012). https://doi.org/10.1103/PhysRevLett.108.167402
48. H. Tang, H. Berger, P.E. Schmid, F. Lévy, Optical properties of anatase (TiO2). Solid State Commun. **92**(3), 267–271 (1994). https://doi.org/10.1016/0038-1098(94)90889-3
49. R. Asahi, Y. Taga, W. Mannstadt, A. Freeman, Phys. Rev. B **61**(11), 7459 (2000). https://doi.org/10.1103/PhysRevB.61.7459

50. K. Glassford, J. Chelikowsky, Phys. Rev. B **46**(3), 1284 (1992). https://doi.org/10.1103/PhysRevB.46.1284
51. S. Rangan, S. Katalinic, R. Thorpe, R.A. Bartynski, J. Rochford, E. Galoppini, J. Phys. Chem. C **114**(2), 1139 (2010). https://doi.org/10.1021/jp909320f
52. J. Tauc, Mater. Res. Bull. **3**(1), 37 (1968). https://doi.org/10.1016/0025-5408(68)90023-8
53. O. Stenzel, *The Physics of Thin Film Optical Spectra*, vol. 44, Springer Series in Surface Sciences (Springer, Berlin, 2005). https://doi.org/10.1007/3-540-27905-9
54. M. Landmann, T. Köhler, S. Köppen, E. Rauls, T. Frauenheim, W.G. Schmidt, Phys. Rev. B **86**(6), 064201 (2012). https://doi.org/10.1103/PhysRevB.86.064201
55. P. Ondračka, D. Holec, D. Nečas, L. Zajíčková, J. Phys. D: Appl. Phys. **49**(39), 395301 (2016). https://doi.org/10.1088/0022-3727/49/39/395301

Part II
Spectrophotometry and Spectral Ellipsometry

Chapter 5
Optical Characterization of Thin Films by Means of Imaging Spectroscopic Reflectometry

Miloslav Ohlídal, Jiří Vodák and David Nečas

Abstract This chapter focuses on optical characterization of thin films by means of non-microscopic imaging spectroscopic reflectometry. This technique is primarily intended for characterization of thin films with an area non-uniformity in their optical properties. An advantage of the technique is the possibility to measure along a relatively large area of the measured films. The motivation for development and exploitation of this technique is also discussed. Essential features and implementation of the technique are given, as well as the basic experimental set-up of imaging spectroscopic reflectometers and the way the experimental data are obtained. The data processing methods are classified based on the purpose of the thin film measurement. Furthermore, this chapter presents examples of results of imaging spectroscopic reflectometry in the field of thin films. At the end of the chapter, potential applications of imaging spectroscopic reflectometry in other tasks are also briefly mentioned.

M. Ohlídal (✉) · J. Vodák
Institute of Physical Engineering, Brno University of Technology,
Technická 2, 61669 Brno, Czech Republic
e-mail: ohlidal@fme.vutbr.cz

M. Ohlídal
Institute of Physics, Faculty of Mining and Geology,
VŠB-Technical University of Ostrava, Ostrava, Czech Republic

J. Vodák
Department of Physical Electronics, Masaryk University,
Kotlářská 267/2, 61137 Brno, Czech Republic
e-mail: jiri.vodak@yahoo.com

J. Vodák
Czech Metrology Institute, Okružní 31, 63800 Brno, Czech Republic

D. Nečas
Plasma Technologies, CEITEC, Masaryk University, Purkyňova 123,
61200 Brno, Czech Republic
e-mail: yeti@physics.muni.cz

O. Stenzel and M. Ohlídal (eds.), *Optical Characterization of Thin Solid Films*, Springer Series in Surface Sciences 64,
https://doi.org/10.1007/978-3-319-75325-6_5

5.1 Introduction

The production of thin films and thin film systems, which possess novel and sophisticated properties desirable in optical applications, requires increasingly advanced techniques measuring these properties. In the following paragraphs, we will present one of such techniques, the imaging spectroscopic reflectometry (ISR) technique. We will describe the essential features and possibilities of the technique and also its implementation. We will classify the ISR methods and also we will demonstrate selected results achieved by means of them in the field of thin film optical characterization. It should be noted that we will differentiate between the concepts 'technique' and 'method' in the following text. We will use the expression 'technique' when referring to the way of obtaining experimental data. The expression 'method' will be used when describing the determination of thin film optical characteristics from aforementioned experimental data. The ISR technique and the ISR methods are parts of a whole, which we call simply as 'ISR'.

5.2 Motivation for Development and Exploitation of Imaging Spectroscopic Reflectometry

The aim of any manufacturer of thin films for optical applications is to produce ideal thin films fulfilling specific requirements in their optical properties. Unfortunately, thin films produced under real conditions often exhibit various defects influencing those properties. Ignoring the existence of these defects can lead to significantly distorted or even incorrect values of the optical parameters of these films. This issue is addressed in Chap. 10.

One of the defects mentioned above is the area non-uniformity in the optical properties of a thin film. From the point of view of thin film optics we use the term non-uniformity of thin films, if their optical properties, and therefore, their optical parameters (thickness and optical constants), vary along the area of the films. The most frequent type of this defect, which we can encounter in practice, is the non-uniformity in thin film thickness. Even when the thin film is non-uniform only in thickness, well established (non-imaging) optical techniques, such as conventional ellipsometric and conventional spectrophotometric techniques [1–3] can fail as long as the thickness non-uniformity is of a general type.

This is caused by the fact that the diameter of the illuminating light beam in commercial spectrophotometers and ellipsometers used in optical analysis of thin films is relatively large (usually in the order of mm^2 up to tens of mm^2 depending on the angle of light beam incidence). Consequently, local film thickness variations within an illuminated spot on the surface of the film lead to averaging in the parameters of the film. On the other hand, the measurement of optical properties of thin films using conventional techniques is a local measurement. Therefore, the general non-uniformity in the film thickness cannot be assessed in this way. For this purpose, it is necessary to scan the studied area of the film, which is time consuming. Another case, in which it is difficult to expect the correct result using a conventional technique, is the characterization of coating of objects with lateral dimensions smaller than the diameter of the illuminating beam. In the field of spectrophotometry several authors [4–8] solved the problem of thin film thickness non-uniformity with the aim to obtain more correct values of film optical parameters from output experimental data provided by conventional spectrophotometers. They assumed that the film non-uniformity in thickness was of a special shape, namely the shape of a wedge. They also assumed that the sample is illuminated by the light beam of a rectangular cross section two sides of which are oriented parallel with the thickness gradient. Unfortunately, the formulae used by them are not applicable in the case of general thickness non-uniformity.

In [9] the reflectance of thin films which are non-uniform in thickness was expressed by means of an integral over the distribution density of the film thickness. This approach to solving the problem of film non-uniformity in thickness is the most general yet. But also this approach does not provide a distribution (map) of the thin film local thickness.

This map can be acquired by scanning an interesting region of the investigated film by means of an illuminating light beam of a reduced diameter (requested spatial resolution of the map is given by a size of the illuminating spot on the film surface). A studied sample is illuminated by a white light beam, the reflected light is gathered from a small sample area by a fiber and then analyzed by an optical spectrum analyzer [10–13]. The scanning is performed either by a 2D movement of a sample holder relative to the immobile fiber or a 2D movement of the fiber relative to the immobile sample. The technique is very time consuming in the case when the investigated film region is large and/or the requested spatial resolution of the measurement is high. It is the significant drawback of the technique.

Several works also took into account the influence of the thin film thickness non-uniformity during evaluation of measurements of thin film optical parameters by means of conventional ellipsometers [14–16].

Again, the approaches presented there do not provide a map of local thin film thickness. It is, therefore, needed to extend the conventional spectrophotometry and ellipsometry to their imaging versions which add a spatial resolution on a sample to the conventional techniques without a necessity of scanning the sample, and thus open up new ways for characterization the optical properties of thin films that vary along the surface of these films. Such imaging techniques are developed particularly in the last decade.

The often utilized technique in the aforementioned problem has been the imaging ellipsometry. We do not deal with this technique in this book, but the relevant information concerning it can be found elsewhere, e.g. [17–25].

We only mention that this technique has certain drawbacks which cause the fact that imaging ellipsometry to have about the same sensitivity as conventional, i.e. non-imaging ellipsometry technique, but lower trueness of results (when data are not repaired). The imaging ellipsometers utilize an imaging system which creates an image of the studied sample mostly on a chip of a CCD camera. That imaging system is mostly constructed as a microscopic one. This brings a substantial benefit of the spatial resolution on a studied sample, but also leads to certain shortcomings of this imaging technique. The reason of them is that parameters of imaging are not ideal: the angle of incidence is varying in a range given by the numerical aperture of an imaging lens; the numerical aperture must be high to achieve a sufficient spatial resolution; the measurement is averaged over a range of angles circumscribed by this aperture (this problem can be eliminated by a certain correction, which is however not simple); as a result of the possible oblique incidence of the illuminating light beam a sample image is deformed and it is also necessary during the data pre-processing to carry out fusion of images caused by insufficient depth of field of view of the imaging system; the material from which the elements of the imaging system are made defines the spectral range that can be used.

Furthermore, the upgrading the conventional techniques to their imaging versions also increases complexity of the corresponding measurement systems, and, consequently, their price. It is, therefore, desirable to find such an imaging technique that overcomes some of these shortcomings. Such a technique is the non-microscopic imaging spectroscopic reflectometry at normal incidence of light, which utilizes the image of the whole sample. In the following paragraphs we will focus on this technique and we will refer to it as the ISR technique.

Therefore, we will deal neither with the scanning reflectometry techniques [10, 12, 13, 26, 27] nor the microscopic imaging reflectometry [28].

The ISR technique is being developed since the end of 1990s. It has been used and its applicability has been confirmed in many cases [11, 29–41]. It was proven that the ISR technique is a powerful tool for optical characterization of thin films non-uniform in thickness.

5.3 Brief Specification of Non-microscopic Imaging Spectroscopic Reflectometry at Normal Incidence of Light

ISR is specifically intended for characterization of the optical properties of non-uniform thin films. Of course, ISR can also be used to inspect the thin film uniformity. The most general aim of ISR is to obtain maps of local parameters describing the area optical non-uniformity of thin films. However, the most common practical application of ISR is a precise mapping of the thin film thickness and determination of spectral dependence of the optical constants of the film. Regarding its wider aims, ISR can also determine the parameters of the dispersion model used, such as the band gap of a thin film material, the maximum energy limit of the relevant electron transitions, or a parameter proportional to the concentration of electrons participating in the relevant transitions. In some cases, it also can provide maps of the RMS parameter of the upper film boundary roughness.

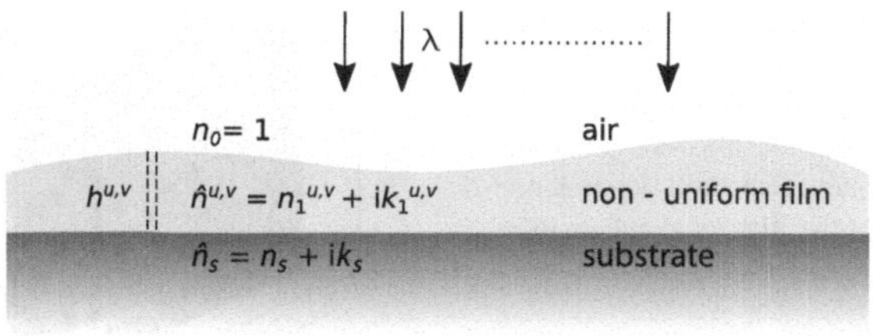

Fig. 5.1 An example of a sample studied by means of ISR

The general situation for a non-uniform thin film is shown in Fig. 5.1. The non-uniform weakly absorbing thin film with thickness and optical constants varying along the area of the film is deposited on an absorbing substrate. A collimated monochromatic light beam impinges perpendicularly on the film from the air. The selected wavelength λ of the beam can vary in a sufficiently wide spectral interval. The response of the whole system on the incident light is given by the local thickness $h^{u,v}$ and the local refractive index $\hat{n}_1^{u,v} = n_1^{u,v} + ik_1^{u,v}$ of the film ($n_1^{u,v}$ is the real local refractive index, $k_1^{u,v}$ is the local extinction coefficient) and the refractive index of the substrate $\hat{n}_s^{u,v} = n_s^{u,v} + ik_s^{u,v}$ (we will assume that the absorption of the substrate is so high and/or the substrate is so thick that its lower boundary does not influence the result). The word 'local' means that these local optical parameters of the film characterize the optical properties of the film in a small part of its area and may differ from optical parameters in other areas. These areas form a continuous matrix covering the whole area of the studied film and are labeled with indices u and v. The measured quantity is the local reflectance of the film in a wide spectral range (NUV, VIS and NIR). In the following text, electromagnetic radiation from this interval will simply be called light. From the perspective of optics, thin films are defined as films in which light is undergoing interference. Since we are dealing with such films, the local reflectance is given by the interference of the light between the film boundaries. This simultaneously implies that the ISR technique can be only applied to non-absorbing or weakly absorbing thin films. When a collimated beam of monochromatic light illuminates a thin film perpendicularly, the interference fringes viewed with an imaging system focused on the film are fringes of equal thickness and they are localized within the film [42].

All instruments applied for the ISR technique utilize an imaging system creating an image of the studied sample which is most often recorded by a CCD camera. This imaging system must be focused on the studied film. It is just the imaging process that assigns each $(u, v)^{\text{th}}$ pixel of the CCD camera to the corresponding $(u, v)^{\text{th}}$ small area of the sample surface. The size of these areas should be small enough so that it is possible to consider that the film is uniform within each of these areas.[1] Then the local reflectance of the small $(u, v)^{\text{th}}$ area of the system in Fig. 5.1 corresponding to the $(u, v)^{\text{th}}$ pixel of the CCD camera is given by the following expression:

[1]When the gradient of thickness non-uniformity is so high (e.g. edges of thin films) that the film cannot be considered uniform within those areas, it is possible to perform correction leading to the correct results.

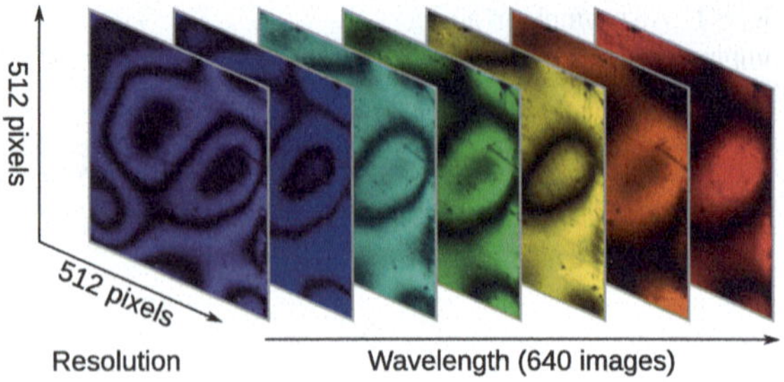

Fig. 5.2 The ensemble of monochromatic images of the studied thin film. This ensemble allows to obtain the map of spectral dependence of the film local reflectance

$$R^{u,v} = \frac{I_r^{u,v}}{I_o^{u,v}} = \left|\hat{r}^{u,v}\right|^2. \tag{5.1}$$

In the equation above, $I_r^{u,v}$ is the intensity of light reflected by the $(u, v)^{\text{th}}$ area of the film, $I_o^{u,v}$ is the intensity of light incident on the $(u, v)^{\text{th}}$ area of the film and $\hat{r}^{u,v}$ is the local reflection coefficient. This coefficient at normal incidence is expressed as follows:

$$\hat{r}^{u,v} = \frac{\hat{r}_1^{u,v} + \hat{r}_2^{u,v} \exp\left(\mathrm{i}\hat{X}^{u,v}\right)}{1 + \hat{r}_1^{u,v}\hat{r}_2^{u,v} \exp\left(\mathrm{i}\hat{X}^{u,v}\right)},$$

where $\hat{r}_1^{u,v}$ and $\hat{r}_2^{u,v}$ are local Fresnel reflection coefficients on the upper and lower boundary, respectively. The symbol $\hat{X}^{u,v}$ denotes the local phase-shift angle.

$$\hat{r}_1^{u,v} = \frac{n_0 - \hat{n}_1^{u,v}}{n_0 + \hat{n}_1^{u,v}}, \; \hat{r}_2^{u,v} = \frac{\hat{n}_1^{u,v} - \hat{n}_s}{\hat{n}_1^{u,v} + \hat{n}_s}, \; \hat{X}^{u,v} = \frac{4\pi}{\lambda}\hat{n}_1^{u,v} h^{u,v}.$$

The ability to measure the local reflectance $R^{u,v}$ in each $(u, v)^{\text{th}}$ area of the film brings a spatial resolution compared with the conventional non-imaging reflectometry, without the need for scanning the sample. Changing the wavelength of incident light between the acquiring the successive monochromatic images we can obtain a relatively large ensemble of these images of the film, and in this manner also the map of spectral dependence of local reflectance (see Fig. 5.2).

5.4 Experimental Set Up of ISR Technique

The experimental set up of the ISR technique is simple. Its principal scheme is shown in Fig. 5.3.

A white light source illuminates the input of a monochromator, the monochromatic light beam with a computer-controlled wavelength emerging from the output of the monochromator is split by a beamsplitter. A part of this beam illuminates the studied sample perpendicularly and, after being reflected from the sample, it goes back through the imaging system which creates an image of the sample on the chip of a CCD camera.

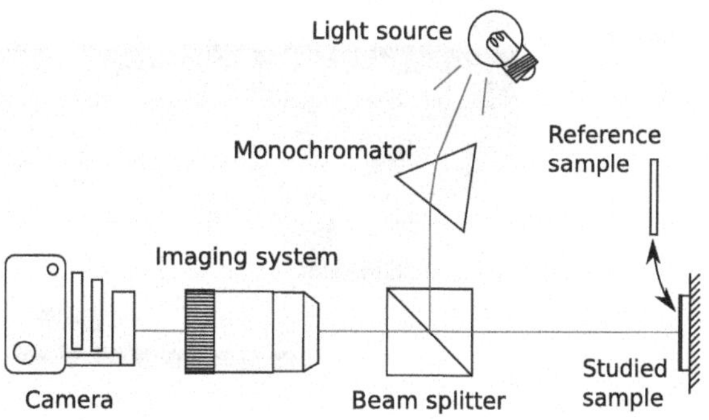

Fig. 5.3 Basic scheme of ISR technique

The normal incidence of a collimated monochromatic light beam on a studied sample brings some benefits. Specifically, we get information from the whole sample surface at once (i.e. without scanning the surface), the interference fringes carrying the necessary information are exactly imaged by means of the imaging system on the CCD camera chip and formulae used for evaluation of required optical parameters are simpler. The disadvantage is that a beamsplitter must be used. The imaging system defines the applicable spectral range. It is better to base this system on reflection optics which works well also in the UV spectral region where the optical properties of thin films are manifested more prominently. The CCD camera must have a good spectral sensitivity within the spectral range, in which the sample is studied. Together with the imaging system, it defines the spatial resolution on a sample. The ISR technique is designed as a relative technique, i.e. the measurement of the studied sample is compared with the measurement of a reference (known) sample under the same conditions. In this way, a possible non-uniformity in the sample illumination is eliminated. Of course, it is necessary to ensure the identical position of the studied and the reference samples. This can be done by means of an appropriate sample holder.

5.5 Imaging Spectroscopic Reflectometers

The concrete implementation of the basic scheme of the ISR technique can be demonstrated by two examples verified in practice.

5.5.1 Imaging Spectroscopic Reflectometer with Wide Spectral Range

The spectral range in which the local reflectance of the studied film is measured must be wide enough, in order to get the information needed for reliable evaluation of the

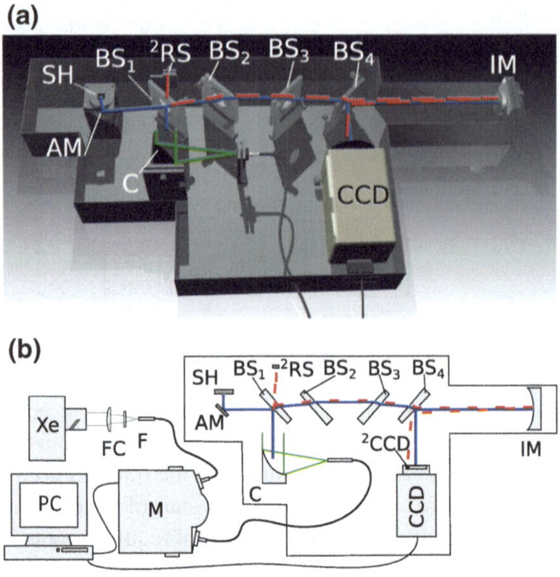

Fig. 5.4 **a** Computer-rendered 3D view of ISRWS (external parts of the whole set up of ISRWS, i.e. the xenon lamp, the monochromator and the control computer are not presented). **b** The three parts of ISRWS experimental set up: The first, illuminating part is a XeUV arc lamp Xe, which is connected to the second part (a monochromator M) by a fiber, a fiber coupler FC and a filter F. The third part (the measuring system) consists of a collimator C, a set of silica wedges BS_{1-4}, an auxiliary mirror AM, a sample holder SH, an imaging mirror IM and a CCD camera CCD. The reference channel of the measuring system consists of a secondary reference channel sample ^{2}RS and a small part of the CCD chip – ^{2}CCD. Everything is controlled by a personal computer – PC

optical parameters. In this context, using imaging systems with refractive optical elements brings issues caused by the dispersion of light in these elements. Furthermore, refractive optical elements manufactured from common optical materials do not work in the UV spectral region in which the optical properties of thin films manifest themselves more prominently. These issues can be effectively resolved by using an imaging system with reflective optical elements. Then, only dispersion of light in a beamsplitter must be tackled. An example of a non-microscopic spectroscopic imaging reflectometer with wide spectral range (ISRWS) which uses such an approach is shown in Fig. 5.4a (computer-rendered 3D view), its scheme is in Fig. 5.4b.

The complete ISRWS system is divided into three distinctive parts connected using optical cables. Although the use of optical cables reduces the overall light throughput on the other hand it allows higher flexibility of the system set up. The three parts of the ISRWS system are: A light source (a UV capable xenon arc lamp Xe), a monochromator M (a commercially available computer controlled Czerny-Turner type monochromator) and an original measuring system itself. The measuring system consists of a collimator (a single off-axis parabolic mirror C), a sample holder SH, a set of fused silica wedges BS_1–BS_4 (some used also as beamsplitters), a spherical imaging mirror IM and a UV-VIS CCD camera. The monochromator with the xenon lamp connected serves as a source of monochromatic light for the measuring system and thanks to the use of fiber optics it can be easily used as a source for other devices as well. In the measuring system the divergent monochromatic light beam is collimated by the collimator and then it is directed at the measured sample using the first fused silica wedge (functioning as a beamsplitter in a way described in Sect. 5.4). In this way a normal incidence of light on the sample can be achieved. Light reflected from the measured sample is then directed through all four silica wedges. The first one serves as the aforementioned beamsplitter, while the others are used to eliminate secondary reflections from the main light path. The fourth wedge is also used as a beamsplitter to allow in-axis imaging by the imaging mirror located behind all the silica wedges. The image created by the optical system is then recorded by the chip of the CCD camera. Light, which initially passes through the BS_1 is not used for imaging of the measured sample. In fact, it contributes to the losses of light intensity. But it can be exploited in a reference channel to measure and subsequently eliminate possible fluctuations of the source light intensity. This is realized using a secondary reference sample which is imaged on the CCD camera chip at the same time as the measured sample (there is a specifically reserved part of the CCD chip for this purpose). The principle of this idea is that the secondary reference sample is never replaced or moved between a series of sample measurements so it is possible to observe intensity changes of the source light. The ISRWS is capable of measuring samples of maximum size about 20 mm × 20 mm while maintaining spatial resolution of 9 lp/mm.

The spectral range spans from 270 to 1000 nm (1.2–4.6 eV). The duration of a measurement of a single sample measurement is typically 30 min (not including the measurement of a reference sample and the background, which need not to be measured every time).

5.5.2 *Imaging Spectroscopic Reflectometer with Enhanced Spatial Resolution*

An example of a very simple instrument allowing an implementation of the ISR technique is the imaging spectroscopic reflectometer with enhanced spatial resolution on a sample (ISRER).

The ISRER was designed as a low cost, simple instrument for optical characterization of thin films with high gradients of non-uniformity. Its computer-rendered 3D view is shown in Fig. 5.5a and its basic scheme is in Fig. 5.5b.

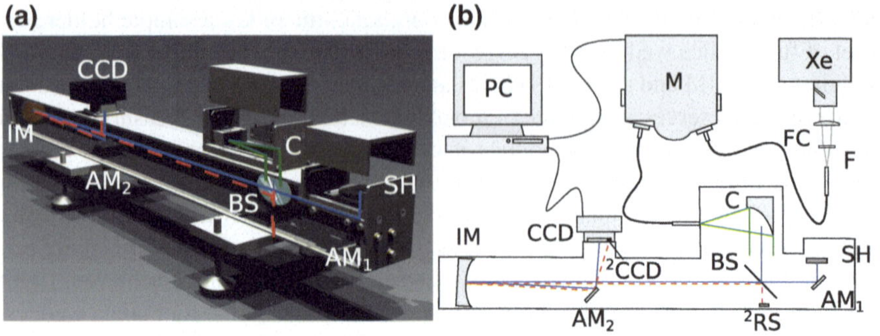

Fig. 5.5 **a** Computer-rendered 3D view of the ISRER (external parts of the whole set up of the ISRER, i.e. a xenon lamp, a monochromator and a control computer are not presented). **b** Basic scheme of the complete ISRER. A xenon UV arc lamp Xe is connected to a monochromator M by a fiber, a fiber coupler FC and a filter F. Measuring system itself consists of a collimator C, a membrane (pellicle) beamsplitter PB, two auxiliary mirrors AM_{1-2}, a sample holder SH, an imaging mirror IM and a CCD camera CCD. The reference channel is realized by a secondary reference sample ^{2}RS and a small part of the CCD chip (^{2}CCD). The whole system is controlled by a personal computer PC

A monochromatic light source (the lamp and the monochromator) for the ISRER is the same as used in the ISRWS, only the measuring part of the ISRER is different. The main difference is the usage of a membrane (pellicle) beamsplitter instead of the four silica wedges. The advantage of the pellicle beamsplitter is the low thickness of the membrane which in a sense eliminates the influence of the secondary reflection (the secondary reflection is so close to the primary reflection that they cannot be distinguished and therefore does not degrade the captured image). Since there is no additional beamsplitter used, the imaging is realized as slightly off-axis imaging (when using an auxiliary mirror AM_2 to reduce the off-axis angle) which brings the benefit that the light intensity hitting the CCD chip does not decrease significantly (use of even an ideal beamsplitter results in 75% loss of intensity). Although the spatial resolution of the ISRER is significantly higher than of the ISRWS, it is still low enough not to be affected by the off-axis imaging setup. The reference channel is realized in a similar way as in the ISRWS by the use of a reference channel sample. The size limit of the measured sample for the ISRER is about 20 mm × 15 mm (a bit less than ISRWS) but the measurement can be done with spatial resolution of 16 µm × 16 µm on a sample. The spectral range from which the wavelengths can be selected spans from 400 to 1000 nm.

The maximum value of thickness gradients is 12.5 µm/mm. The possibilities of the ISRER in the case of thin films with high gradients in thickness are demonstrated in Figs. 5.6 and 5.7.

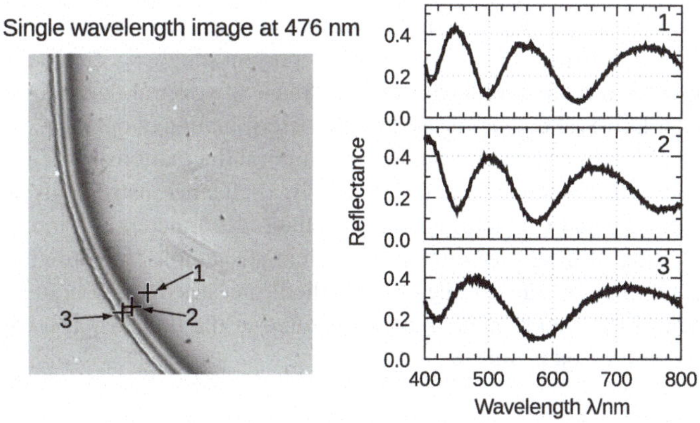

Fig. 5.6 Edge of a HfO_2 film – a map of local reflectance for $\lambda = 400$ nm (raw data multiplied with reflectance of the reference sample) and spectral dependencies of reflectance for three selected CCD pixels

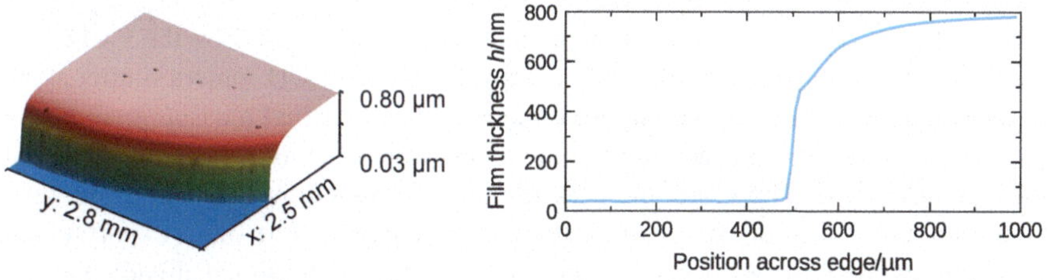

Fig. 5.7 3D representation of the film edge from Fig. 5.6 and profile of the film edge perpendicular to this edge

5.6 Data Acquisition

Both aforementioned imaging spectroscopic reflectometers measure spectral dependencies of local reflectance of a sample studied. From the ensemble of these data the values of spectral dependencies of local relative reflectance $R_r^{u,v}(\lambda_k)$ of the sample are obtained. These values for a given wavelength λ_k are arranged in a matrix. The $(u, v)^{\text{th}}$ element of this matrix corresponds to the $(u, v)^{\text{th}}$ pixel of the CCD camera recording the image of the studied sample at the wavelength λ_k. It means that this matrix element $R_r^{u,v}(\lambda_k)$ corresponds to the $(u, v)^{\text{th}}$ small area on the sample surface which is imagined on the above mentioned $(u, v)^{\text{th}}$ CCD pixel. Indices of these individual small areas imaged on corresponding pixels of the CCD camera take values $u = 1...U; v = 1...V$. U and V are the numbers of pixels of the CCD chip in horizontal and vertical directions and are given by the CCD camera resolution. The matrix as a whole corresponds to the imaged area of the sample. When measuring, the wavelengths λ_k are suitably selected (accordingly to the presumed spectral reflectance of the studied sample) from the usable spectral range of the given imaging spectroscopic reflectometer (ISRM) with a chosen sampling step. In this way an ensemble of matrices is obtained from the set of sample images (see Fig. 5.2). Vectors formed from matrix elements with the same indices u and v represent sought spectral dependencies of the sample local relative reflectance. This 'map' of spectral dependencies of the sample local relative reflectance is used for the determination of optical parameters of a studied thin film. In order to eliminate the temporal fluctuations in the light source intensity, both reflectometers were designed as two-channel instruments. Retrieving the experimental data of a sample using both the reflectometers is a three-step procedure consisting of the measurement of a reference sample (measured at the time t_1), the measurement of the sample to be studied (measured at the time t_2), and the measurement of the background signal (measured at the time t_3). The background signal can be expressed as follows:

$$S^{u,v}(\lambda_k, t_3) = D_f^{u,v}(\lambda_k, t_3) + b^{u,v} I_o(\lambda_k, t_3),$$

where $I_o(\lambda_k, t_3)$ is the intensity of the monochromator output at the wavelength λ_k and at the time t_3, $b^{u,v}$ is the constant of proportionality, $b^{u,v} I_o(\lambda, t_3)$ is the response of the CCD camera to the light scattered inside the reflectometer, $D_f^{u,v}(\lambda_k, t_3)$ is the dark frame (it contains the whole signal which is generated by the CCD chip without being exposed to any light) obtained at the closed CCD camera shutter at the exposure time and the chip temperature identical to the actual measurement of the sample. This dark frame is acquired immediately after obtaining the relevant signal and it is subtracted. Therefore, the dark frame is not mentioned further in the text. After eliminating the dark frame, the whole three-step procedure of the experimental signal processing may be concisely expressed as follows: The signal already without the $D_f^{u,v}$ obtained from a single pixel with coordinates u and v can be written as

$$S_{J,i}^{u,v}(\lambda_k, t_i) = I_o(\lambda_k, t_i)[\eta^{u,v}(\lambda_k) R_i^{u,v}(\lambda_k) + b^{u,v}], \tag{5.2}$$

where index J can take two values: value m, which stands for 'measuring channel' and value s for 'reference channel', index i can be of value 1, 2 or 3 according to the kind of the measurement (1 is for the measurement of the reference sample, 2 is for the measurement the studied sample and 3 is for the measurement of the background i.e. without the sample).

$I_o(\lambda_k, t_i)$ is again the intensity of light on the monochromator output at the wavelength λ_k and at the time t_i, $\eta^{u,v}(\lambda_k)$ describes all the influences of the apparatus, e.g., effects of possible imperfections in optical elements and/or bad pixels of the camera, but also the signal amplification or camera bias, noise etc. $R_i^{u,v}(\lambda_k)$ is the absolute local reflectance of the current sample given by the index. Since the reflectance is equal to 0 for index $i = 3$ (blank measurement without any sample), the first addend in the formula (5.2) is also equal to 0 and only the background $I_o(\lambda_k, t_i) b^{u,v}(\lambda_k, t_3)$ remains. The following formula (5.3) ensures that any temporal instability of the light source of the ISRM is eliminated (i.e. it removes the time dependence of the $I_o(\lambda_k, t)$). It also removes the influence of the background and of any non-uniformity in the illumination of samples:

$$R_r^{u,v}(\lambda_k) = \frac{\frac{S_{m,2}^{u,v}(\lambda_k,t_2)}{S_{s,2}(\lambda_k,t_2)} - \frac{S_{m,3}^{u,v}(\lambda_k,t_3)}{S_{s,3}(\lambda_k,t_3)}}{\frac{S_{m,1}^{u,v}(\lambda_k,t_1)}{S_{s,1}(\lambda_k,t_1)} - \frac{S_{m,3}^{u,v}(\lambda_k,t_3)}{S_{s,3}(\lambda_k,t_3)}} = \frac{R_2^{u,v}(\lambda_k)}{R_1^{u,v}(\lambda_k)}. \tag{5.3}$$

This is the spectral dependence of the local relative reflectance of the studied sample in the spectral range chosen in the measurement. In this way we acquire a map of such local reflectance spectral dependencies (see Fig. 5.6). The duration of a single sample measurement is typically 30 min (without the measurement of a reference sample and the background, which need not to be measured every time).

5.7 Key Features of Imaging Spectroscopic Reflectometry

At the end of our treatise on ISR technique, we will summarize the main features of this technique as follows:

- A CCD camera records *monochromatic images* of a *relatively large area* of a studied film, which are *created by an imaging system within wide span of wavelengths*.
- A small region of the film surface is assigned to a one pixel of the CCD camera by *the process of imaging*.
- These areas are so small that the film can be considered *uniform within individual areas*.
- The ISR technique is *a relative technique*. The spectral dependence of the local reflectance of the studied sample is measured against the spectral dependence of the local reflectance of a reference sample (mostly a silicon single crystal wafer).
- The output experimental data of the ISR technique are the *maps of spectral dependence of the thin film local relative reflectance*.
- *Normal incidence* of the collimated beam of light illuminating the sample *eliminates the necessity of scanning and also image fusion* during postprocessing the output experimental data.

5.8 Methods of Imaging Spectroscopic Reflectometry

As mentioned in the introduction, we consider an ISR method as a way of experimental data processing, through which we determine interesting optical parameters of a thin film. These methods are an integral part of the determination of the optical parameters of thin films. The experimental data obtained by means of the ISR technique are the maps of spectral dependence of the local thin film reflectance. This fact defines the limit of the information content of these data. The different methods of the data processing give a different level of information we can get within this limit. Their specific feature is that they must handle enormous amounts of experimental data (assuming an image 500 × 500 pixels large with a 500-point spectrum in each pixel the number of data points is 1.25×10^8) and also to determine a huge amount of the thin film parameters (the number of parameters searched can be estimated to be about 2.5×10^5). This means that it is not possible to simply use the standard form of Levenberg–Marquardt non-linear least-squares fitting algorithm for the determination of thin film parameters but it is necessary to develop original algorithms for this purpose. On the other hand, this enormous amount of experimental data eliminates random errors of the thin film parameters. Therefore, the determined values of the parameters have only systematic errors. The ISR methods are discussed in detail in Chap. 6, where their mathematical formulation is presented.

It is also important to stress another significant feature of the ISR methods. Most of thin film parameters that are sought are practically always mutually correlated. Then, it is impossible to determine them unambiguously. To overcome this problem, the multi-sample method must be applied to improve the stability of least-squares data fitting (e.g. [43]). The ISR technique, performing independent measurements in individual CCD pixels, inherently provides data for a multi-sample method. Now we will focus on the classification of these methods and, simultaneously, we will present selected demonstrations of individual ISR methods in order to illustrate their possibilities. We will classify the ISR methods from the viewpoint of the way in which the information provided by CCD pixels is used. It should be emphasized that this classification can only be schematically. The reason is that the use of the ISR method depends not only on the task which we solve, but also on our decision what method we want to use. For example, when aiming to determine the local thickness of a thin film with a known spectral dependence of the optical constants, the relevant ISR method can be used as the stand-alone method. But if we aim to characterize a thin film which differs very much from an ideal one (for example a film exhibiting more defects) and/or with a complicated form of spectral dependencies of optical constants, we probably would have to use the method in combination with other methods of film characterization (i.e. conventional ellipsometric or spectrophotometric) and the ISR method should be used as a complementary method. However, sometimes we can also use the relevant ISR method for the latter film as the stand-alone method. The schematic classification is presented in Fig. 5.8.

We will mark an ISR method as the single-pixel method when the spectra of local reflectance measured by individual CCD camera pixels are processed separately. When those spectra are processed simultaneously the method is called the multi-

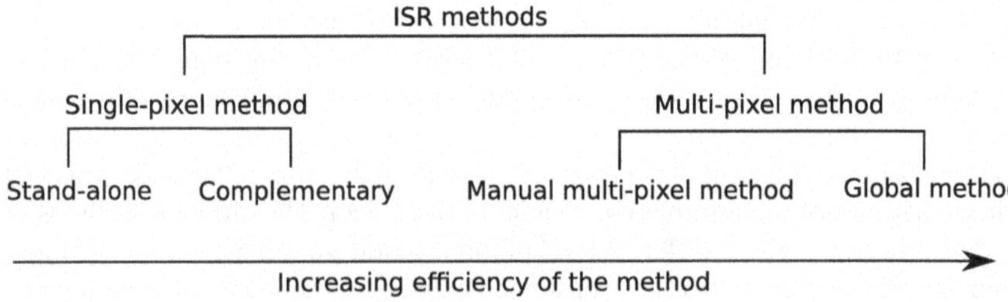

Fig. 5.8 Scheme of ISR methods classification

pixel method. We will start our demonstration of the individual methods with the case where the ISR method can be applied as the stand-alone single-pixel method.

5.8.1 Single-Pixel ISR Method as the Stand-Alone Method

The single-pixel ISR method can be applied as the stand-alone method for optical characterization of a non-uniform thin film, as long as it is possible to suppose that the film does not exhibit another structural defect than non-uniformity in thickness, the film is uniform in optical constants and the spectral dependence of optical constants of this film is relatively simple (or even known) within the interesting spectral range.

Then the number of parameters appearing in the dispersion model describing the spectral dependence of those optical constants is small as opposed to the case when the spectral dependence of the optical constants is complex. In that case all these dispersion parameters and local thickness can be determined independently by a fitting procedure separately in each pixel.

The aforementioned case can be demonstrated using carbon-nitride films, which were deposited onto silicon single crystal wafers by a dielectric barrier discharge with CH_4/N_2 gas mixture (details of the technological procedure used to prepare the films are given in [44]). When treating the experimental data, the dispersion model based on parametrization of the joint density of electronic states (PJDOS) corresponding to amorphous materials [45] was used. It was assumed that the films contain no defects other than the thickness non-uniformity.

It was found that those films can be considered uniform in the optical constants (the determined values of dispersion model parameters were practically the same in all film areas which corresponded to the individual pixels of a CCD camera). Spectral dependence of these optical constants determined from the parameters of the above-mentioned corresponding dispersion model are presented in Fig. 5.9a for a sample selected from a measured file of those films. The 3D map of the local thickness of this carbon-nitride film is shown in Fig. 5.9b.

The maps of the spectral dependence of the local relative reflectance got from ISR measurements exhibit noise. This implies that the maps of the local thickness and the values of the film optical constants determined from individual CCD pixel inevitably exhibit noise as well. The reflectance values $R^{u,v}$ were measured with the statistical relative error about 1% (corresponding to the standard deviation). Using a standard error analysis, it was found that the values of the local thicknesses in the area distributions were determined with the statistical relative error of 1–2% (corresponding to the standard deviation). The same conclusion concerning the accuracy was also found for the optical constants.

The single-pixel ISR method is simple. That is its main advantage. Unfortunately, this method cannot be applied when the characterized non-uniform thin films exhibit a complicated course of spectral dependence of the optical constants requiring the usage of a dispersion model with a larger number of parameters and/or exhibit further defects than non-uniformity in thickness (such as the roughness of boundaries,

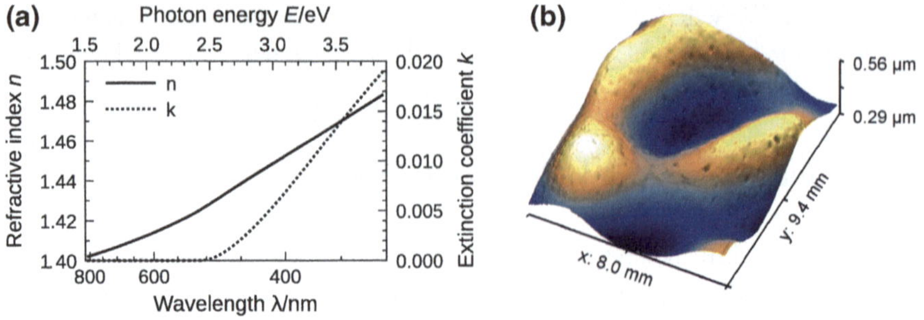

Fig. 5.9 **a** Spectral dependence of the refractive index n and the extinction coefficient k of the selected non-uniform carbon-nitride film. **b** 3D map of the local thickness of this carbon-nitride film determined using the ISR method

very thin overlayers on the upper boundary or very thin transition layers between the substrate and the film). Then it is necessary to complete this ISR method with other methods such as conventional (non-imaging) spectroscopic ellipsometry and conventional (non-imaging) spectrophotometry. The detailed description of this method application to the presented case can be seen in [36]. Other applications can be found in [32, 33].

5.8.2 *Single-Pixel ISR Method as the Complementary Method*

In this case, the single-pixel ISR method is applied in combination with other optical methods (e.g. ellipsometric and/or spectrophotometric). It plays a role of a complementary method to these other methods, i.e. the method which allows us to obtain values of the local thin film parameters by which it is possible to characterize a non-uniformity of the film along its surface (e.g. local thickness or local roughness), while the film optical constants that can be supposed to remain unchanging along the entire surface of the film are found by means of the above mentioned other methods. As the demonstration example of the case where the single-pixel ISR method is used as the complementary method in combination with other optical methods, we will present the optical characterization of a selected sample of considerably non-uniform $SiO_xC_yH_z$ thin films deposited using plasma enhanced chemical vapor deposition onto a silicon single-crystal wafer (the detailed preparation of the film see [11]). Three optical techniques, i.e. conventional variable-angle spectroscopic ellipsometry (VASE), mapping spectroscopic ellipsometry with microspot (μSE), and ISR were used for the film characterization. Both ellipsometric techniques were used to determine spectral dependence of the optical constants of the studied film. Moreover, μSE was used to evaluate uniformity of the film in its optical constants and the type of the film thickness non-uniformity. For this purpose μSE measurement was carried out in 99 sample positions that formed a regular 11×9 grid with 1 mm spacing. Experimental data acquired by the ISR technique were used for the determination of the map of the film local thickness. The experimental data obtained by means of ellipsometric techniques were processed by the Levenberg–Marquardt algorithm using the following structural model of the film: The film is without defects except the thickness non-uniformity, i.e. the material of the film is optically isotropic, it is homogeneous in the direction perpendicular to the sample plane, the film boundaries are sharp and smooth, and the thickness non-uniformity is of the wedge type. Moreover, it is assumed that the film optical constants do not vary within the region corresponding to the microspot used in μSE (the circle 250 μm in diameter for normal incidence). In the case of ISR, the film is assumed as uniform within the region corresponding to a single pixel of the CCD camera. The film complex refractive index was modeled using an expression for SiO_2–like materials based on PJDOS [45]. The single pixel ISR method was utilized as the complementary method to conventional VASE and μSE. Spectral dependencies of the film optical constants were found by fitting VASE data. Subsequently, reflectance spectra in individual pixels obtained by ISR were fitted, utilizing the optical constants obtained by ellipsometry and assuming they were correct. The results obtained are presented in Fig. 5.10.

Since it is not feasible to display the error bars for all the 99 μSE curves in Fig. 5.10, the errors will be summarized numerically. The average three standard deviations error estimate for the refractive index n was about 0.013 in the whole spectral range, whereas for the extinction coefficient k it varied from approximately 0.01 at the UV end of the spectrum to 0.001 at the IR end. It is evident from Fig. 5.10 that the error bars of the conventional VASE and μSE overlap. The results are thus in agreement. No trend was found in the area distribution of refractive index values obtained by means of μSE. Their fluctuations represented random experimental errors and the same can be said about the extinction coefficient. Therefore, no evidence of non-

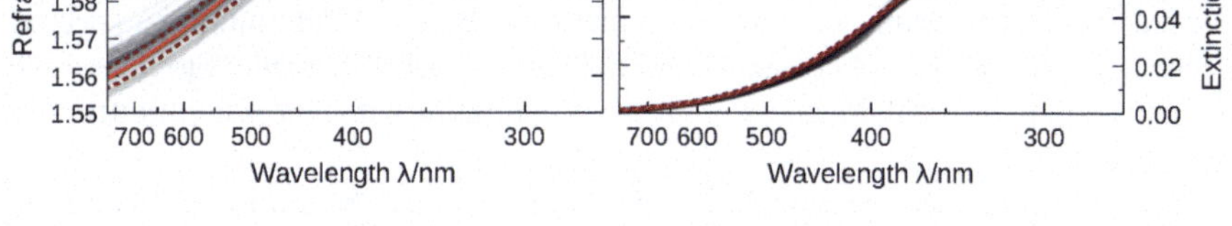

Fig. 5.10 Comparison of spectral dependencies of optical constants, i.e. refractive index and extinction coefficient determined using conventional VASE (solid thick lines), using single-sample μSE in all 99 individual locations (thin shaded lines), and using multi-sample μSE (dotted lines). Error bars corresponding to three standard deviations are plotted for the conventional VASE as dashed lines [11]

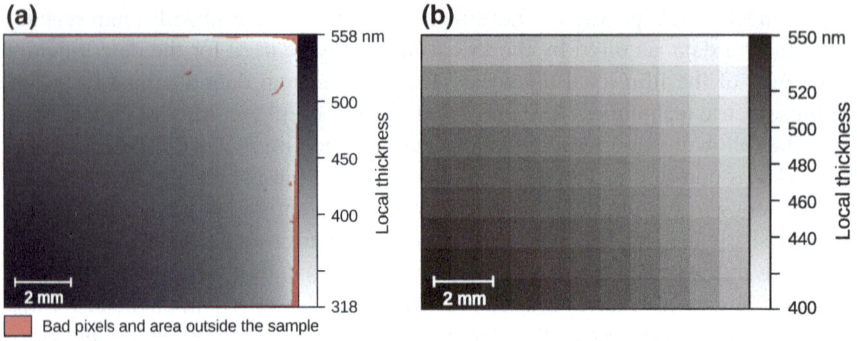

Fig. 5.11 **a** Map of film local thickness obtained by ISR data fitting. Pink regions along the right and top edges correspond to bad pixels and pixels outside of sample. **b** Map of film local thickness determined using μSE [11]

uniformity of optical constants was found within the experimental precision. It is worth pointing out that such conclusion is typical according to our experience. In other words, even if the film is considerably non-uniform in thickness its optical constants can usually be still considered uniform.

Because of the aforementioned consistency of the values of optical constants obtained by conventional VASE and μSE, the spectral dependence of the optical constants determined using conventional VASE was utilized for determining the fine local thickness map by the ISR presented in Fig. 5.11a. The corresponding map found using μSE is shown in Fig. 5.11b.

An exact comparison of the maps of local thickness determined by using ISR and μSE is not possible because it is not possible to ensure exactly the same position of the sample during both the measurements. However, the trend of both the maps allows us to conclude that both the measurements are in agreement. Finally, it can be said that the combination of conventional VASE, μSE and ISR represents a precise tool for optical characterization of thin films non-uniform in thickness. Unfortunately, it is not convenient for routine use because the analysis of the discussed film by means of μSE took approximately five days. Other cases, in which the single-pixel ISR method is applied as the complementary method in combination with other optical methods, are published in [35, 38], where the ISR method is applied in combination with conventional VASE and conventional spectroscopic reflectometry (SR) at near normal incidence. Again, the latter two methods served to determine the spectral dependence of the optical constants and the single-pixel ISR method to determine the fine map of the local thickness of the films exhibiting a thickness non-uniformity only.

In conclusion of this paragraph, it is necessary to make the following note. If the thin film is considerably non-uniform in thickness, it is possible to suspect that the deposition process was not sufficiently uniform and the film material would also be non-uniform along the film, although this material non-uniformity is probably smaller than the thickness non-uniformity. This implies that the film could be non-uniform in the optical constants as well. Unfortunately, the salient feature of reflectance spectra, interference in the film, depends primarily on the product of film thickness and refractive index, called optical thickness. This makes difficult to distinguish non-uniformities in thickness and optical constants. Therefore, in the case where the conventional methods do not determine optical constants correctly the single pixel ISR method utilized together with those conventional methods leads also to incorrect results. This fact limits the applicability of the results of the single-pixel ISR method in the case under consideration.

5.8.3 Manual Multi-pixel ISR Method

When the single-pixel ISR method processes spectral dependencies of the film local reflectance fitting the dispersion model parameters independently in each pixel, the resulting maps exhibit high noise. On the other hand, it is not possible to simply fit the ISR data in all the pixels together using one set of shared dispersion model parameters. The total number of fitting parameters is huge and they are all correlated. Nevertheless, these two problems can be solved, as we will demonstrate in the case of strongly non-uniform thin films deposited from hexamethyldisiloxane on silicon substrates by a single capillary plasma jet at atmospheric pressure. The detailed description of the preparation of these films can be found in [37]. The manual multi-pixel ISR method has been used as the stand-alone method in that case. The procedure had three steps and each step involved only least-squares fitting problems with reasonable numbers of parameters:

First step: Film thickness is fitted independently in each pixel using the model of an ideal thin film and an initial estimate of film optical constants. In this particular case tabulated optical constants of SiO_2 were used. The thickness maps obtained in this way are not yet correct, but this first step is sufficient to distinguish good and bad pixels in the image of the film. The criterion for it is the agreement between experimental local reflectance spectral dependence and its fit in manually selected pixels representatively covering the region of interest. The pixel selection should respect the requirement to cover the full range of values of the sought parameters. The evaluation of this agreement is done subjectively.

Second step: A set of several (e.g. ten) good pixels was selected manually. The experimental data corresponding to those pixels were fitted simultaneously assuming common optical constants. The large variation in thickness within the selected spectrum set helps reducing the correlation between the thickness and dispersion model parameters and improves stability of the fitting procedure (the multi-sample approach). The PJDOS dispersion model for SiO_2 like materials [45] was used for the films, values of Si substrate values were taken from the literature [46].

Third step: Film thickness is again fitted independently in each pixel, but this time utilizing optical constants found in the second step. The spectral dependence of the film real refractive index is presented in Fig. 5.12. Even though the dispersion model

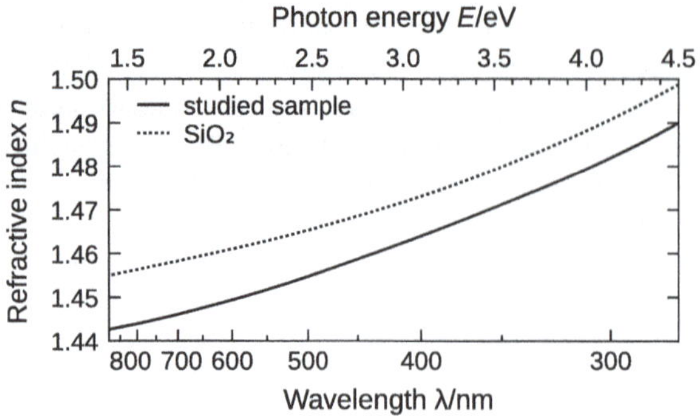

Fig. 5.12 Spectral dependence of the refractive index of the plasma jet thin film compared with the refractive index of thermal SiO_2

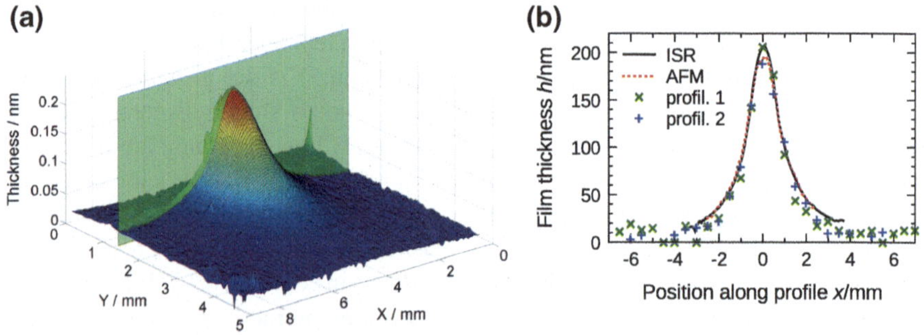

Fig. 5.13 **a** Map of local thickness with selected profiles along the artificial edge of the film. **b** Comparison of thickness profiles along the edge of the film measured by ISR, AFM and the tactile profilometer [37]

permitted absorption, it was found that the film was essentially non-absorbing in the spectral range of the measurement. The extinction coefficient is therefore not shown.

The 3D map of the local thickness of the film obtained as the final result of the procedure described above is presented in Fig. 5.13a. In this figure, the artificial edge created by scratching away a half of the film by means of a scalpel can be seen. The thickness of the film was also measured with a Veeco Dektak profilometer. This measurement was performed with a step of 0.5 mm along the edge (i.e. in the plane in Fig. 5.13a) twice. Finally, the film edge was measured using a Bruker Dimension Icon atomic force microscope in the ScanAsyst mode. Utilizing a motorized table for accurate movement between successive scans, 50 images were acquired and in each the step height was then evaluated.

All measurements were compared using 2D profiles defined by the green plane in Fig. 5.13a. This comparison of the profiles obtained by all three measurement tools is plotted in Fig. 5.13b.

Considering the uncertainty of the individual measuring tools, the agreement of the results obtained can be considered good. Figure 5.13b also shows a strong thickness non-uniformity of the studied thin film. The largest gradient of local thickness in the smooth part of the film (excluding the edges and debris), determined from inter-pixel thickness differences, was approximately 1.6×10^{-4}. The analysis of experimental errors of the discussed ISR method found that the accuracy of local thickness measurements was approximately 2 nm. The limiting factor was the uncertainty of optical constants of the film as they represented the largest uncertainty source. Finally, we can state that the discussed ISR method was successfully applied as the stand-alone method for determining the optical constants and local thickness map of the film strongly non-uniform in thickness.

5.8.4 Global Method

Although the manual multi-pixel method presented in the previous paragraph works well in practice, it has several shortcomings. Firstly, to do the fitting, it is necessary to manually find a small, yet representative, subset of pixels with a good spectral dependence of local reflectance from the region of interest. It is somewhat unsatisfactory that this choice is subjective and is irreconcilable with automation of data analysis. Moreover, not utilizing entire available data means that the contribution of random noise to parameter uncertainties is larger than necessary. If the analysis utilizes all available data, i.e. reflectance curves from all pixels, random errors can become insignificant compared to systematic errors and thus effectively eliminated.

Therefore, on the assumption of a thin film non-uniform in thickness only, an original experimental data processing procedure has been developed, utilizing the specific structure of the least-squares problem related to the main task of ISR. The basic features of this procedure consist in splitting the free parameters into thicknesses (local parameters, possibly different in each pixel) and dispersion model parameters (shared parameters common for all pixels). Subsequently, both kinds of parameters are fitted by turns, utilizing an unmodified Levenberg–Marquardt algorithm. However, this algorithm is used in such a way that the local thicknesses are corrected during the dispersion model parameters fitting step to preserve the effective optical thickness (product of film thickness and refractive index). This brings a substantial improvement in the procedure convergence and permits the analysis of large imaging reflectometry data sets with reasonable computational resources. The reason for using the condition of preservation of the effective optical thickness is that it is the optical thickness what determines the locations of interference minima and maxima in a reflectance spectrum. The minima and maxima thus move in response to changing optical thickness. When the theoretical reflectance curve already corresponds relatively well to the experimental points, the sum of the squared differences will always increase even if the extrema shift only slightly. The fitting algorithm thus becomes unable to progress further by updating thicknesses and dispersion model parameters separately once the tiniest changes are now permitted. This limitation grows more severe with increasing film thickness since the extrema are spaced more closely for thicker films. The precise analysis of the ISR experimental data processing described just above is presented in [47], where this approach was used for the

Table 5.1 Characteristics of the experimental ISR data sets and fitting parameters for the two thin films

Film	$SiO_xC_yH_z$	CN_x:H
Number of fitted spectra M	70,310	85,469
Points per spectrum K	656	628
Data set size [MiB]	176	205
Free shared parameters D	5	5
Shortest fitting time [s]	540	624

first time. It was applied to the case of two thin films of different amorphous materials deposited on silicon substrates, both exhibiting strong thickness non-uniformity.

The first sample was a hydrogenated carbon-nitride film (CN_x:H) prepared in an atmospheric pressure dielectric barrier discharge from CH_4:N_2 =1:10 gas mixture. The second sample was a $SiO_xC_yH_z$ film deposited in a low pressure radio frequency (13.56 MHz) capacitively coupled discharge from the mixture of tetraethoxysilane and methanol (The details of the deposition procedure can be found in [47]). The same structural and dispersion models were used for both films. The films were considered ideal within a single ISR pixel. The complex refractive index was modeled using an expression for SiO_2-like materials based on PJDOS [45]. Straightforward parallelization on data was applied. It was demonstrated that the strategy of preserving quantities corresponding to effective optical thicknesses in individual pixels resulted in the fastest convergence of the least-squares fit. It was also shown that even though a behavior of the algorithm deteriorates above a film thickness of approximately 600 nm, the result was still acceptable. The sizes of the data and fitting parameter sets are summarized in Table 5.1 for both thin films. The table also includes the time duration of the computation running on a reasonably powerful personal computer (six-core AMD Phenom II processor and 16 GiB of RAM). The computation times listed in Table 5.1 evince that the developed fitting procedure made a global ISR data analysis possible, even with relatively modest computational resources.

The reduction of parameter errors and improved reliability of results following from multi-pixel data fitting may be beneficial in the characterization of samples that could be characterized also by other means. However, the key advancement is that a wider range of samples can now be characterized using ISR as a stand-alone method, without resorting to combination with conventional ellipsometry and spectrophotometry. Because this method exploits the experimental data of all pixels in the image of the film in the way that the shared parameters and local parameters are fitted continuously during the fitting procedure, it can be named as the global method.

To show the real power of this method, an application of the method for characterization of a thin film, which is far from an ideal one, is presented here. A ZnSe film prepared by molecular-beam epitaxy (MBE) on a (100) GaAs single crystal substrate can serve as a good example (for detailed deposition parameters see [40]). Epitaxial ZnSe thin films deposited on GaAs substrates by means of MBE have randomly rough

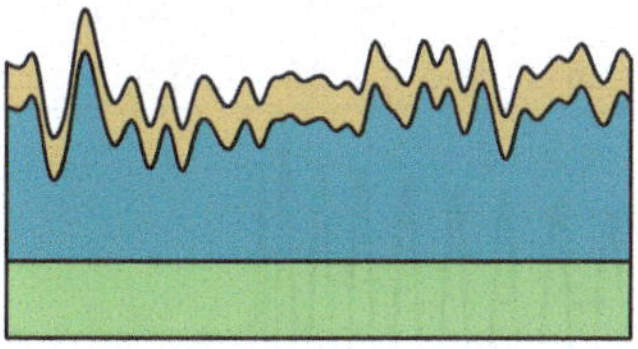

Fig. 5.14 Schematic depiction of the structural model of the film

upper boundaries. This roughness arises from the mosaic (or block) structure of the films [48]. MBE normally produces thin films that are fairly uniform along the substrate plane. However, it is sometimes possible to encounter ZnSe epitaxial thin films whose thickness varies to such extent that they need to be considered non-uniform. Of course, when both imperfections, surface roughness and thickness non-uniformity, occur together, then film characterization is more difficult. The structural model of the film under consideration is shown in Fig. 5.14.

Since the ZnSe thin film with a rough upper boundary was placed in the air, it was covered with a very thin overlayer [49]. This overlayer was modeled as so-called identical rough thin film, i.e. a film with upper and lower boundaries that are exact geometrical copies of each other (see also Fig. 6.4). When lateral correlations play no role, a single number is then sufficient to describe the roughness, the RMS of height irregularities. The substrate–film boundary was assumed to be smooth. Although the film was relatively non-uniform in thickness, within the range of the surface corresponding to a single pixel of a CCD camera, the film was considered uniform in thickness.

The optical constants of the ZnSe film were expressed using a PJDOS model for valence-to-conduction inter-band transitions [50, 51] and fitted. Tabulated values found in earlier studies were used for both the overlayer [49] and the GaAs substrate [52].

Scalar diffraction theory (SDT) was used to model the influence of the upper boundary roughness on reflectance [53]. The expressions resulting from SDT have the form of an infinite series. This series was rewritten into a form suitable for an efficient evaluation by computers. In particular, the computation time was made almost independent on the precision to which the series was evaluated, eliminating the need to make any trade-offs between precision and speed in SDT computations. This is the great advantage in comparison with the Rayleigh–Rice theory (RRT) [53] which has been used for modeling of the upper boundary roughness influence for the optical characterization of the same (discussed above) ZnSe thin film in [38], where the single-pixel ISR method has been used in combination with conventional VASE and SR. The influence of the roughness is described by very complicated formulae within RRT. This is significant in the case of ISR, when large numbers of experimental data must be processed, and would lead to very long data processing times. The fitting algorithm described in this paragraph has led to a satisfactory fit essentially in the whole image of the film. It is illustrated in Fig. 5.15.

Since the optical constants of the ZnSe film were modeled and fitted in the ISR data analysis, it was possible to compare the obtained spectral dependencies with

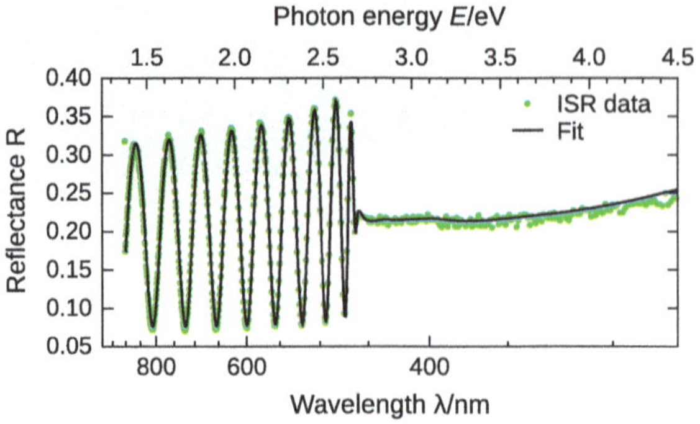

Fig. 5.15 Selected typical ISR spectrum of the ZnSe film on GaAs and its fit by the theoretical model. The spectrum corresponds to a pixel close to the image center [40]

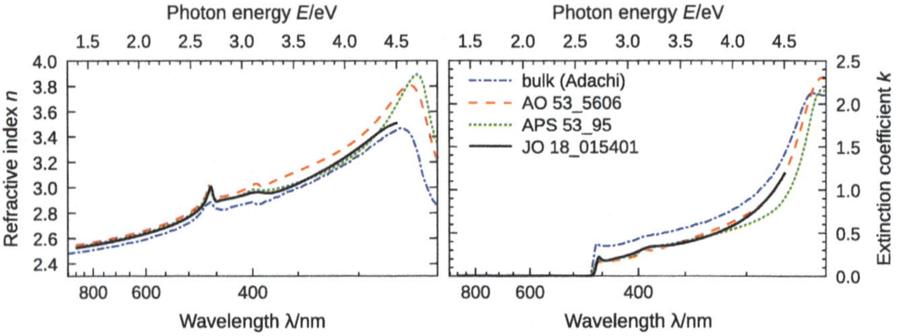

Fig. 5.16 Spectral dependencies of optical constants of the ZnSe film. The curves denoted 'AO 53_5606', 'JO 18_015401', 'APS 53_95' and bulk (Adachi) represent optical constants found for epitaxial ZnSe in [38, 40, 54] and optical constants of bulk ZnSe [55], respectively [40]

those found in other works. This comparison is shown in Fig. 5.16, where also the dependencies for the same film determined in [38] by VASE, for epitaxial ZnSe films studied in [54] and for bulk ZnSe [55] are presented.

The agreement between the results presented in Fig. 5.16 can be considered good. The uniformity of the film in the optical constants was also checked by dividing the ISR data into four quadrants and rerunning the fitting procedure individually for each quadrant. Parameters controlling the overall shape of the complex refractive index curve were fitted, but parameters determining the locations of fine structures in the spectral dependencies were fixed in values obtained from the whole data.

Considering typical experimental errors of the method, all four obtained spectral dependencies were indistinguishable. The initial assumption that the film material could be considered uniform was thus justified.

Maps of local film thickness h, the RMS surface roughness and thickness of the overlayer are presented in Fig. 5.17. The artefacts which can be seen in both the maps of the RMS roughness and overlayer thickness correspond to defects on either the studied sample or the reference sample. The pixels corresponding to these artefacts

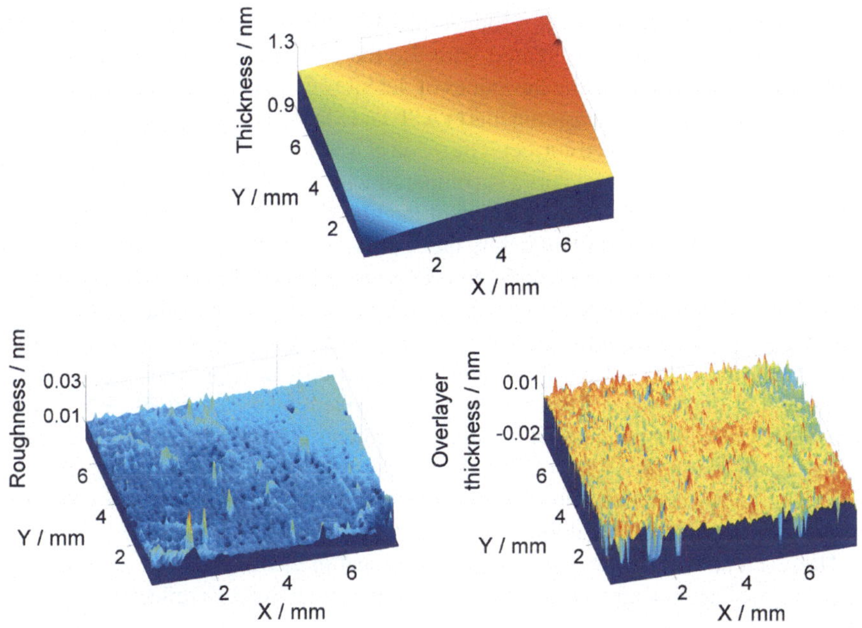

Fig. 5.17 Maps of local thickness of the ZnSe film, RMS roughness of its upper boundary and overlayer thickness

as well as bad pixels (with low quality of reflectance spectra) were removed, and the mean (or typical) values of the RMS roughness and overlayer thickness were determined from the remaining pixels.

The average RMS roughness of 4.7 nm agrees well with other optical [38, 49] and atomic force microscopy [38] studies. The average overlayer thickness value 10.3 nm is somewhat higher than the values found by other methods [38, 49] but still in reasonable agreement.

We can conclude that the ISR technique, when complemented by appropriate data processing approaches, is practical as a stand-alone method of optical characterization of thin films that differ significantly from ideal ones and which, therefore, require complex modeling.

5.9 Precision and Accuracy of ISR

At the beginning of this paragraph, it should be noted that the precision and accuracy of ISR depend, to a great extent, on the problem to be solved. In order to demonstrate the precision and accuracy of the ISR measurements themselves (i.e. the precision and accuracy of reflectance data obtained), a sample of a uniform SiO_2 film of the thickness of 800 nm was deposited on a Si substrate and then measured repeatedly (eight times) by means of the reflectometer ISRER. All the reflectance data obtained in this manner were then compared with the theoretical values calculated by the method described in [56] using the values of the SiO_2 film optical constants obtained from many conventional measurements (VASE and SR, both in many configurations). This curve can be considered correct. The result can be seen in Fig. 5.18. In the top part of this figure, the theoretical curve is compared with the spectral dependence of the local reflectance of a selected small area of the film (imaged onto the relevant single pixel of the CCD camera) obtained by the ISR measurement. In the bottom half of the Fig. 5.18, each point of the graph represents the difference between the calculated theoretical curve at the given wavelength and the relevant mean value of the local reflectance acquired from all CCD pixels corresponding to a region located at the center of the measured sample. This region was selected to be approximately the same as the region utilized (i.e. illuminated) by the conventional techniques (VASE and SR). Different colors indicate different measurements. The relative mean difference was of about 5% in the vicinity of minima of the spectral reflectance dependencies. This value can be considered as the maximum relative difference between the correct and measured values of local thin film reflectance.

However, an ISR user is usually more interested in the reliability of determining sample optical parameter values rather than accuracy of individual reflectance values.

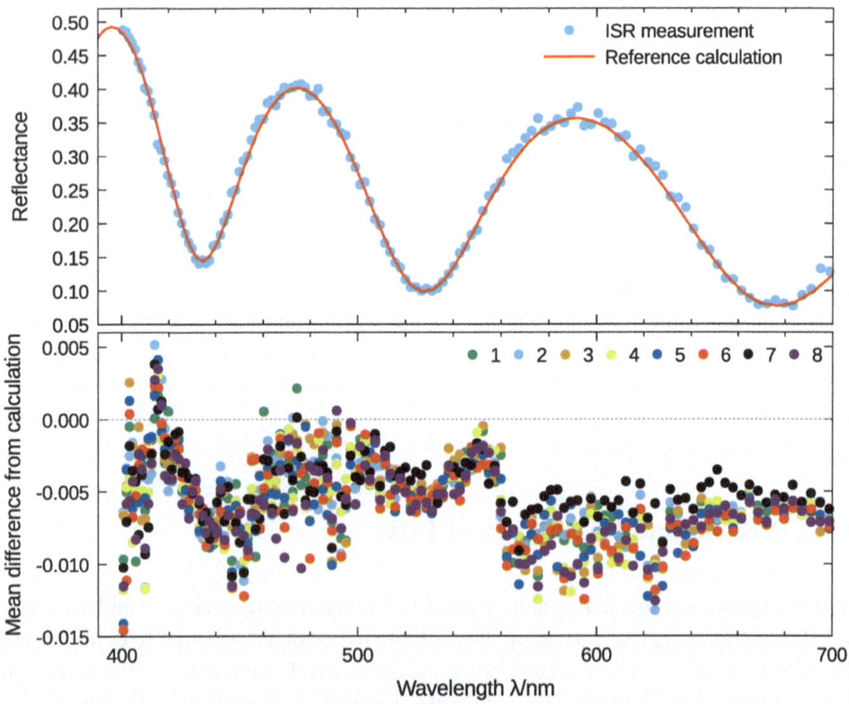

Fig. 5.18 Difference between measured ISRER spectra and theoretical reflectance curves

As it was mentioned in the conclusion of the Sect. 5.8.2, if the main task is to determine the local film thickness, the largest contribution to its error comes from the uncertainty of optical constants of the film material. Whether the optical constants are determined within ISR itself or taken from other measurements (and thus represent an external source of the error), they are seldom known with better accuracy than about 0.01. Consequently, as the technique is sensitive primarily to the optical thickness nh, the absolute thickness values are systematically deviated by a constant multiplicative factor. Depending on a film thickness and other aspects, this systematic deviation can reach up to a few nanometers. This point has to be considered in metrology, but it is moot in characterization of highly non-uniform samples in material research, where the spatial dependence (i.e. a shape) is more important. Therefore, we will further illustrate the precision of ISR results here as more relevant aspect.

Figure 5.19 illustrates the theoretical standard deviation of the thickness of a thin film (on a silicon substrate) determined from a typical ISRWS measurement. It represents the precision limit that cannot be improved without reducing noise or increasing the number of spectral points in a spectral dependence of the local reflectance of a film. The theoretical sensitivity of the measurement is apparently good, with the standard deviation of the fitted film thickness in tens of picometers.

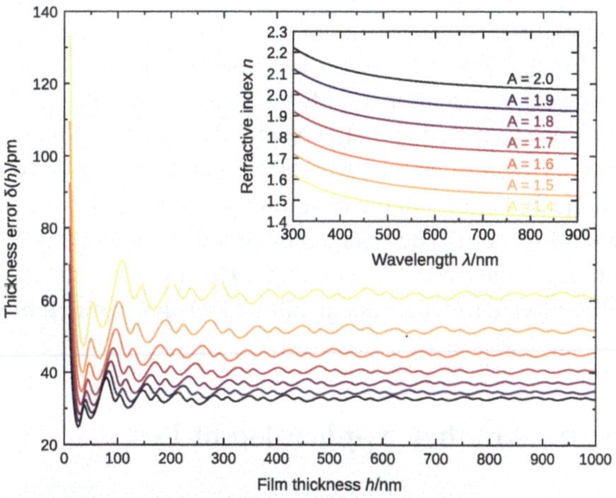

Fig. 5.19 Theoretical standard deviation of the fitted film thickness due to the noise for an ISRWS typical measurement, plotted as a function of film thickness and refractive index. Spectral dependencies of refractive index were modeled using a simple Cauchy formula; values of its constant term A are shown in the plot

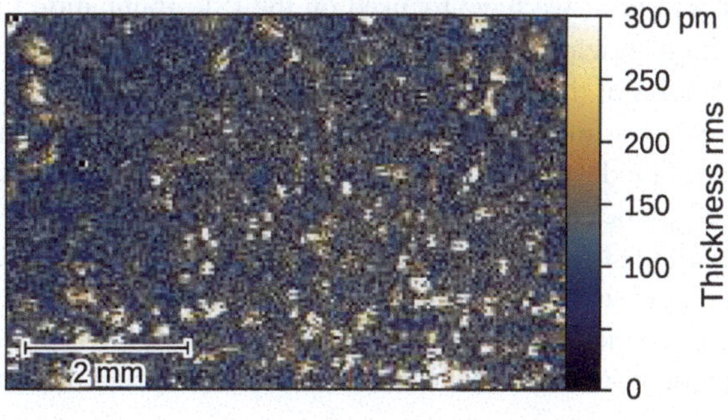

Fig. 5.20 Map of standard deviation of the fitted thickness of an 800 nm thick SiO_2 film obtained from repeated ISRWS measurement

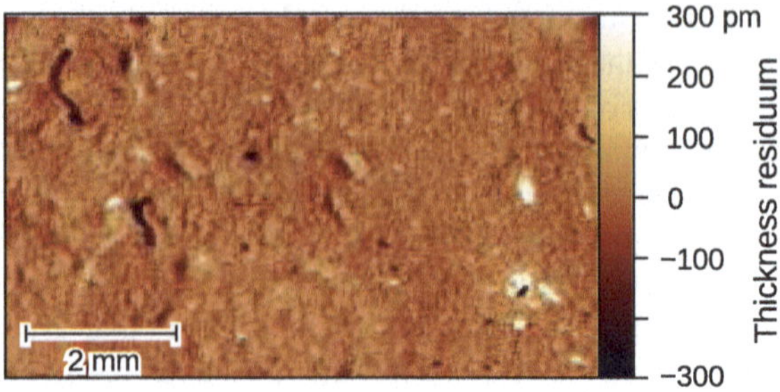

Fig. 5.21 Map of the residuum of fitted thickness of a 300 nm thick TiO_2 thin film (measured by ISRWS). The residuum was obtained by the subtraction of a low-level polynomial fitting the overall spatial dependence of thickness

To answer how does the theoretical estimate corresponds to actual experimental results, the above mentioned measurements of 800 nm thick SiO_2 film uniform in thickness were used. The individual measurements were independently fitted (with fixed optical constants) and the resulting thickness maps statistically analyzed. The map of the local standard deviation of fitted film thickness, which was obtained is shown in Fig. 5.20.

The film is rather uneven and contains spots with relatively large variation in the order of hundreds of picometers on the background of reproducible measurements. The overall mean and median of the map are 128 and 116 pm, respectively, less than twice the theoretical estimates. This verifies the good precision (reproducibility) also in practice.

Finally, Fig. 5.21 shows a complementary demonstration of the consistency of the ISR results. An almost uniform area of a 300 nm thick TiO_2 thin film was measured and the map fitted (with fixed optical constants). Since the uniformity was not perfect, the thickness map was then fitted with a low-order polynomial, and the polynomial subtracted to obtain the residuum plotted in Fig. 5.21. In the ideal case, the residuum would be zero. The map again contains isolated spots where the residuum is of the order of hundreds of picometers. However, the mean square residuum is 52 pm, i.e. about twice the theoretical value (TiO_2 has a much higher refractive index).

5.10 Another Application of ISR

Up to now, we have focused on the ISR application to the basic task of the optical characterization of thin films, i.e. determination of their thickness and spectral dependencies of their optical constants. These films could exhibit some defects like thickness non-uniformity, non-uniformity in upper boundary roughness and their structural model could comprise an overlayer. The ability to measure spatially resolved reflectance in a wide spectral range can be beneficial also in other applications. As an example of such an application can serve the use of ISR for the localization of metallic gold in an organic layer. The metallic gold was reduced from an organo-metallic compound by a localized thermal treatment using a plasma-jet [57]. The plasma treatment of solid surfaces has a lot of interesting and important applications generally. In

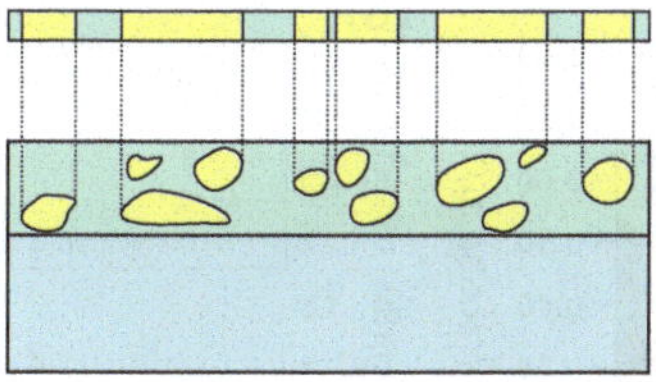

Fig. 5.22 The organogold layer with reduced grains of gold. Grain projection on the film surface is shown

the application mentioned above organogold layers were prepared on a microscopy glass plate by spincoating and then vacuum dried (the detailed description of the layers preparation see [57]).

In these layers, the gold was in the oxidation state +1. By the action of plasma jet, this gold contained in the precursor layers is reduced to metallic gold (i.e. gold in its oxidation state 0) in the form of small grains (see Fig. 5.22).

The projection of these metallic gold grains on the layer surface can be quantitatively evaluated by means of the ISR technique. The models used for the description of light interaction with the studied layers can be various. They have to respect the fact that the layers of interest are approximately 6 μm thick, non-uniform and the refractive index of their organic material is close to that of the glass substrate. Hence, the interference of light is weak in those parts, where the layers are transparent. The parts with metallic gold (the heat treated parts) were not transparent. Therefore, the layers were modeled as thick slabs and the interference was not considered. It was also taken into account that the layers significantly differ from ideal ones, which implies various defects and distortions in the reflectance spectra. Considering also a huge number of ISR experimental data, a simple, robust, model was needed which would agree with experimental data as much as possible. Finally, the layer was modeled as a thick slab formed by separated regions covered by the untreated organic compound or metallic gold. This model works with an area fraction of metallic gold a_f, which is defined as follows:

$$a_f^{u,v} = \frac{A_{gold}^{u,v}}{A_{reg}^{u,v}}.$$

The area of gold covering the region corresponding to a camera pixel (with coordinates u and v) is denoted as $A_{gold}^{u,v}$ and $A_{reg}^{u,v}$ is the area of this whole region. Using $a_f^{u,v}$, the local relative reflectance of the region corresponding to the same CCD camera pixel can be expressed as

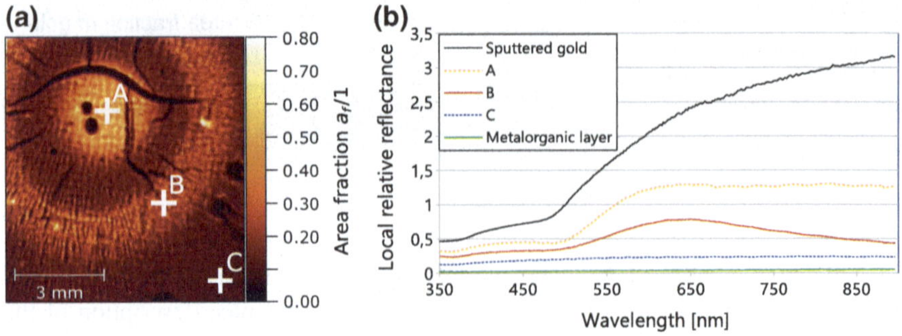

Fig. 5.23 **a** Area metallic gold map obtained by ISR measurements. The scale shows a value of the model parameter a_f–the area fraction of metallic gold in the organogold precursor treated by plasma jet. **b** Local relative reflectance comparison. Measured local relative reflectance of the spots (marked in a) on the measured sample together with measurements of the pure gold layer and the organogold layer without any plasma jet treatment

$$R^{u,v}(\lambda_k) = a_f^{u,v} R_{gold}(\lambda_k) + (1 - a_f^{u,v}) R_{org}(\lambda_k). \quad (5.4)$$

Here, $R_{gold}(\lambda_k)$ is the value of the relative reflectance of gold obtained by measuring the reference sample (a uniform gold layer prepared by magnetron sputtering on a glass sheet) at the wavelength λ_k. $R_{org}(\lambda_k)$ is the value of the relative reflectance of the untreated uniform organogold precursor layer at the same wavelength. The values of $a_f^{u,v}$ were determined by the least square method using the previous equation (5.4). Put together, these values form a map of the area projection of the reduced metallic gold distribution in the studied sample.

The results achieved are shown in Fig. 5.23a. Three spots A, B, C are selected in this figure to illustrate how the corresponding local relative reflectance obtained by the ISR is changing along the surface of the studied film (see Fig. 5.23b). The spot A contains the largest amount of metallic gold (and thus its reflectance is closest to pure metallic gold), the spot B contains less metallic gold and the spot C was not thermally treated and thus it does not contain any metallic gold.

The ISR method was completed by X-ray photoelectron spectroscopy applied at the points A, B, C of the sample surface and confocal microscopy (which provides only qualitative evaluation of the area distribution of metallic gold along the studied sample surface). Results obtained by means of both the additional techniques were consistent with the quantitative results from ISR.

5.11 Conclusion

In the deposition of thin films for optical applications, various factors may cause defects significantly affecting the desired properties of these films. This is valid particularly during the development and tuning of a new deposition technology. Therefore, it is desired to have instruments which can detect the existence of these defects and characterize the influence of them on thin film optical properties. The ISR technique, in conjunction with the adequate data processing methods, is such a suitable tool for this purpose. The advantages of the ISR fully manifest when characterizing thin films with area non-uniformity in their parameters. When such a non-uniformity is of a general type, i.e. it is not possible to describe this non-uniformity analytically, a correct optical characterization of these films by means of the conventional (non-imaging) and frequently used optical methods (e.g. photometric and ellipsometric methods) cannot be performed.

The main aim of ISR in the field of thin film optics is the determination of thin film optical parameters, primarily maps of local thickness, and spectral dependencies of optical constants. With the help of ISR, it is also possible to determine other material parameters appearing in the dispersion models, such as the band gap of a thin film material, the maximum energy limit of the relevant electron transitions, or a quantity proportional to the concentration of electrons participating in the transitions. Eventually, it also allows to determine some structural parameters, such as maps of local RMS roughness of the upper film boundary.

The experimental set up of the ISR technique is simple. The design of imaging spectroscopic reflectometers allows to measure thin film samples up to 20 mm $\times$ 20 mm size, at normal angle of light incidence. The measurements can be done within the spectral range (270–1000) nm, i.e. (1.2–4.6) eV, with the spatial resolution on the sample up to 16 μm $\times$ 16 μm and the maximum value of the local thickness gradient approximately 12.5 μm/mm. The duration of a one sample measurement is typically 30 min (without the measurement of the reference sample and the background, which need not to be measured every time).

The ISR technique provides a tremendous amount of experimental data. This fact implies the necessity of special data processing methods with the aim to determine the sought optical parameters of the film. The use of the appropriate method is determined by a task to be solved. If spectral dependence of optical constants of a thin film is known, the stand-alone single-pixel ISR method can be used for the determination of the map of the film local thickness. If the spectral dependence of optical constants of a thin film is not known, conventional (non-imaging) methods can be used for their determination under the assumption of film uniformity in these constants. The single-pixel ISR method can then be used for the determination of the map of the local thickness of the film as a complementary method. The same task can be solved using the stand-alone manual multi-pixel ISR method.

The multi-pixel approach is equivalent to the multi-sample approach which is inherently present in the ISR technique. This fact further increases the efficiency of this method in solving the basic task of spectroscopic reflectometry. The most powerful ISR method is the global ISR method which enables to address the issue of the characterization of a thin film which is even far from an ideal one. That is, it enables us to determine the unknown spectral dependence of thin film optical constants (assuming the uniformity of the film in optical constants), the map of thin film local thickness and, if necessary, other parameters of the film, such as the RMS roughness of the upper boundary and the mean thickness of the overlayer film. This global method uses an original algorithm for processing the ISR data, which ensures fast convergence of the procedure finding the sought thin film parameters. The above-mentioned ISR technique and the ISR methods form an integral whole – ISR.

It is necessary to mention that all the ISR methods presented here are built on the assumption of thin film uniformity in optical constants. In accordance with our experience, this assumption is fulfilled in the vast majority of real-life cases. In principle, the characterization of thin films exhibiting non-uniformity in optical constants and simultaneously in thickness is also possible using ISR. However, up to date this task still represents a challenge.

The accuracy of optical parameters determined by means of ISR depends on the concrete issue that is being addressed. The optical constants can be seldom determined with a better accuracy than about 0.01. The uncertainty of optical constants causes a systematic deviation of the local thickness of the measured thin film. Depending on the film thickness and other aspects, this systematic deviation can even be a few nanometers. The precision of ISR measurements is good. In the case of the local thickness determination, this precision can be estimated by the value of the RMS deviation in the local thickness which is in the order of 10^1–10^2 picometers.

The applicability range of ISR can be defined as follows: Since the aim of applying the ISR technique is, among other things, to determine a map of the thin film local thickness, it is necessary to exploit the interference pattern which originated in the film. This means there must be interaction between the light beam and the bottom boundary of the film, which affects the local reflection of the film. Thus, only thin films that absorb sufficiently little in the spectral range used can be studied by means of the ISR technique. By other words, the ISR technique can be utilized for dielectric or semiconductor thin films, but not for strongly absorbing thin films (e.g. metal films).

In the previous paragraphs, we dealt with the application of ISR within the field of optical characterization of thin films. We presented not only a solution of the basic task of finding the spectral dependence of optical constants and determination of local thickness maps of thin films, but also pointed out a wider potential of ISR in this field (see Sect. 5.10). Generally speaking, ISR can also be a good choice for the analysis of intentionally modified thin films (like are patterned, locally deposited, locally etched films). In conclusion, we can state that ISR represents the powerful tool for optical characterization of thin films. At the same time, it should be noted that the potential for using ISR outside the thin film optics field is also high. It can be successfully applied wherever it is desirable to know the local reflectance maps along the surface of the studied samples, such as biological, medical objects, etc.

Acknowledgements The work on this chapter was supported by the project LO1411 (NPU I) funded by Ministry of Education Youth and Sports of the Czech Republic. Presented reflectometers were realized using the facilities of Institute of Physical Engineering, Brno University of Technology.

References

1. H.G. Tompkins, W.A. McGahan, *Spectroscopic Ellipsometry and Reflectometry: A User's Guide* (Wiley, 1999). Google-Books-ID: W35tQgAACAAJ
2. H.G. Tompkins, E.A. Irene, *Handbook Of Ellipsometry* (William Andrew Publisher; Springer, Norwich, NY; Heidelberg, Germany, 2005). http://www.books24x7.com/marc.asp?bookid=32277. OCLC: 310147997
3. M. Quinten, *A Practical Guide to Optical Metrology for Thin Films* (Wiley-VCH Verlag GmbH & Co, KGaA, 2012)
4. M.I. Török, Optica Acta: Int. J. Opt. **32**(4), 479 (1985). https://doi.org/10.1080/713821745
5. T. Pisarkiewicz, T. Stapinski, H. Czternastek, P. Rava, J. Non-Cryst. Solids **137**, 619 (1991). https://doi.org/10.1016/S0022-3093(05)80194-2, http://www.sciencedirect.com/science/article/pii/S0022309305801942
6. T. Pisarkiewicz, J. Phys. D: Appl. Phys. **27**(1), 160 (1994). https://doi.org/10.1088/0022-3727/27/1/025, http://stacks.iop.org/0022-3727/27/i=1/a=025
7. E. Márquez, J.M. González-Leal, R. Jiménez-Garay, S.R. Lukic, D.M. Petrovic, J. Phys. D: Appl. Phys. **30**(4), 690 (1997). https://doi.org/10.1088/0022-3727/30/4/026, http://stacks.iop.org/0022-3727/30/i=4/a=026
8. E. Márquez, P. Nagels, J.M. González-Leal, A.M. Bernal-Oliva, E. Sleeckx, R. Callaerts, Vacuum **52**(1–2), 55 (1999). https://doi.org/10.1016/S0042-207X(98)00233-4, http://www.sciencedirect.com/science/article/pii/S0042207X98002334
9. D. Nečas, I. Ohlídal, D. Franta, J. Opt. A: Pure Appl. Opt. **11**(4), 045202 (2009). https://doi.org/10.1088/1464-4258/11/4/045202, http://iopscience.iop.org/1464-4258/11/4/045202
10. L. Abel-Tiberini, F. Lemarquis, M. Lequime, Appl. Opt. **45**(7), 1386 (2006). https://doi.org/10.1364/AO.45.001386
11. D. Nečas, I. Ohlídal, D. Franta, V. Čudek, M. Ohlídal, J. Vodák, L. Sládková, L. Zajíčková, M. Eliáš, F. Vižd'a, Thin Solid Films **571**, Part 3, 573 (2014). https://doi.org/10.1016/j.tsf.2013.12.036
12. J. Spousta, M. Urbánek, R. Chmelík, J. Jiruše, J. Zlámal, K. Navrátil, A. Nebojsa, T. Šikola, Surf. Interface Anal. **34**(1), 664 (2002). https://doi.org/10.1002/sia.1383, http://onlinelibrary.wiley.com/doi/10.1002/sia.1383/abstract
13. M. Urbánek, J. Spousta, K. Navrátil, R. Szotkowski, R. Chmelík, M. Buček, T. Šikola, Surf. Interface Anal. **36**(8), 1102 (2004). https://doi.org/10.1002/sia.1850, http://onlinelibrary.wiley.com/doi/10.1002/sia.1850/abstract
14. I. Ohlídal, D. Nečas, D. Franta, V. Buršíková, Diam. Relat. Mater. **18**(2–3), 364 (2009). https://doi.org/10.1016/j.diamond.2008.09.003, http://www.sciencedirect.com/science/article/pii/S0925963508004342
15. D. Nečas, D. Franta, V. Buršíková, I. Ohlídal, Thin Solid Films **519**(9), 2715 (2011). https://doi.org/10.1016/j.tsf.2010.12.065
16. D. Nečas, I. Ohlídal, D. Franta, J. Opt. **13**(8), 085705 (2011). https://doi.org/10.1088/2040-8978/13/8/085705, http://stacks.iop.org/2040-8986/13/i=8/a=085705
17. L. Stiblert, T. Sandström, Le J. de Phys. Colloq. **44**(C10) (1983). https://doi.org/10.1051/jphyscol:19831016
18. D. Beaglehole, Review of Scientific Instruments **59**(12), 2557 (1989). https://doi.org/10.1063/1.1139897
19. G. Jin, P. Tengvall, I. Lundström, H. Arwin, *Analytical Biochemistry* (1995), p. 69. https://doi.org/10.1006/abio.1995.9959

20. H. Arwin, Thin Solid Films **313–314**, 764 (1998). https://doi.org/10.1016/S0040-6090(97)00993-0, http://www.sciencedirect.com/science/article/pii/S0040609097009930
21. L. Asinovski, D. Beaglehole, M.T. Clarkson, phys. status solidi (a) **205**(4), 764 (2008). https://doi.org/10.1002/pssa.200777855, http://onlinelibrary.wiley.com/doi/10.1002/pssa.200777855/abstract
22. U. Wurstbauer, C. Röling, U. Wurstbauer, W. Wegscheider, M. Vaupel, P.H. Thiesen, D. Weiss, Appl. Phys. Lett. **97**(23), 231901 (2010). https://doi.org/10.1063/1.3524226, http://scitation.aip.org/content/aip/journal/apl/97/23/10.1063/1.3524226
23. M. Fried, G. Juhász, C. Major, P. Petrik, O. Polgár, Z. Horváth, A. Nutsch, Thin Solid Films **519**(9), 2730 (2011). https://doi.org/10.1016/j.tsf.2010.12.067, http://www.sciencedirect.com/science/article/pii/S0040609010016974
24. G. Jin, Y.H. Meng, L. Liu, Y. Niu, S. Chen, Q. Cai, T.J. Jiang, Thin Solid Films **519**(9), 2750 (2011). https://doi.org/10.1016/j.tsf.2010.12.175, http://www.sciencedirect.com/science/article/pii/S0040609010018249
25. P.G. Ellingsen, M.B. Lilledahl, L.M.S. Aas, C.d.L. Davies, M. Kildemo, J. Biomed. Opt. **16**(11), 116002 (2011). https://doi.org/10.1117/1.3643721
26. K. Mahmoud, S. Park, S.N. Park, D.H. Lee, Metrologia **51**(6), S293 (2014). https://doi.org/10.1088/0026-1394/51/6/S293. WOS:000345426000008
27. K. Mahmoud, S. Park, S.N. Park, D.H. Lee, in *Fifth Asia-Pacific Optical Sensors Conference*, vol. 9655, ed. by B. Lee, S.B. Lee, Y. Rao (Spie-Int Soc Optical Engineering, Bellingham, 2015), vol. 9655, p. 965503. WOS:000358602400002
28. K. Kim, S. Kim, S. Kwon, H.J. Pahk, Int. J. Precis. Eng. Manuf. **15**(9), 1817 (2014). https://doi.org/10.1007/s12541-014-0534-3
29. M. Ohlídal, I. Ohlídal, D. Franta, T. Králík, M. Jákl, M. Eliáš, Surf. Interface Anal. **34**(1), 660 (2002). https://doi.org/10.1002/sia.1382, http://onlinelibrary.wiley.com/doi/10.1002/sia.1382/abstract
30. M. Ohlídal, I. Ohlídal, P. Klapetek, M. Jákl, V. Čudek, M. Eliáš, Jpn. J. Appl. Phys. **42**(7S), 4760 (2003). https://doi.org/10.1143/JJAP.42.4760, http://iopscience.iop.org/1347-4065/42/7S/4760
31. I. Ohlídal, M. Ohlídal, P. Klapetek, V. Čudek, M. Jákl, in *Wave-Optical Systems Engineering II* (2004), pp. 260–271. https://doi.org/10.1117/12.509628, http://dx.doi.org/10.1117/12.509628
32. M. Ohlídal, V. Čudek, I. Ohlídal, P. Klapetek, in *Proceedings of SPIE*, ed. by C. Amra, N. Kaiser, H.A. Macleod (2005), vol. 5963, pp. 596, 329–596, 329–9. https://doi.org/10.1117/12.625380, http://proceedings.spiedigitallibrary.org/proceeding.aspx?articleid=876920
33. M. Ohlídal, I. Ohlídal, P. Klapetek, D. Nečas, V. Buršíková, Diam. Relat. Mater. **18**(2–3), 384 (2009). https://doi.org/10.1016/j.diamond.2008.10.012
34. M. Ohlídal, I. Ohlídal, E-J. Surf. Sci. Nanotechnol. **7**, 409 (2009). https://doi.org/10.1380/ejssnt.2009.409
35. I. Ohlídal, M. Ohlídal, D. Nečas, D. Franta, V. Buršíková, Thin Solid Films **519**(9), 2874 (2011). https://doi.org/10.1016/j.tsf.2010.12.069, http://ww.sciencedirect.com/science/article/pii/S0040609010016998
36. M. Ohlídal, I. Ohlídal, P. Klapetek, D. Nečas, A. Majumdar, Meas. Sci. Technol. **22**(8), 085104 (2011). https://doi.org/10.1088/0957-0233/22/8/085104, http://iopscience.iop.org/0957-0233/22/8/085104
37. D. Nečas, V. Čudek, J. Vodák, M. Ohlídal, P. Klapetek, J. Benedikt, K. Rügner, L. Zajíčková, Meas. Sci. Technol. **25**(11), 115201 (2014). https://doi.org/10.1088/0957-0233/25/11/115201
38. D. Nečas, I. Ohlídal, D. Franta, M. Ohlídal, V. Čudek, J. Vodák, Appl. Opt. **53**(25), 5606 (2014). https://doi.org/10.1364/AO.53.005606
39. J. Schaefer, J. Hnilica, J. Sperka, A. Quade, V. Kudrle, R. Foest, J. Vodak, L. Zajickova, Surf. Coat. Technol. **295**, 112 (2016). https://doi.org/10.1016/j.surfcoat.2015.09.047. WOS:000376834700017
40. D. Nečas, I. Ohlídal, D. Franta, M. Ohlídal, J. Vodák, J. Opt. **18**(1), 015401 (2016). https://doi.org/10.1088/2040-8978/18/1/015401, http://stacks.iop.org/2040-8986/18/i=1/a=015401

41. J. Vodák, D. Nečas, M. Ohlídal, I. Ohlídal, Measurement Science and Technology **28**(2), 025205 (2017). https://doi.org/10.1088/1361-6501/aa5534, http://stacks.iop.org/0957-0233/28/i=2/a=025205
42. P. Hariharan, *Optical Interferometry* (Academic Press, 2003). Google-Books-ID: EGdMO3rfVj4C
43. D. Franta, I. Ohlídal, Acta Phys. Slovaca **50**(4), 411 (2000)
44. A. Majumdar, J. Schäfer, P. Mishra, D. Ghose, J. Meichsner, R. Hippler, Surf. Coat. Technol. **201**(14), 6437 (2007). https://doi.org/10.1016/j.surfcoat.2006.12.011, http://www.sciencedirect.com/science/article/pii/S0257897206014800
45. D. Franta, D. Nečas, L. Zajíčková, Opt. Express **15**(24), 16230 (2007). https://doi.org/10.1364/OE.15.016230, https://www.osapublishing.org/oe/abstract.cfm?uri=oe-15-24-16230
46. D. Franta, D. Nečas, L. Zajíčková, I. Ohlídal, Thin Solid Films **571**, 490 (2014). https://doi.org/10.1016/j.tsf.2014.03.059
47. D. Nečas, J. Vodák, I. Ohlídal, M. Ohlídal, A. Majumdar, L. Zajíčková, Appl. Surf. Sci. **350**, 149 (2015). https://doi.org/10.1016/j.apsusc.2015.01.093, http://www.sciencedirect.com/science/article/pii/S016943321500118X
48. V. Holý, K. Wolf, M. Kastner, H. Stanzl, W. Gebhardt, J. Appl. Crystallogr. **27**(4), 551 (1994). https://doi.org/10.1107/S0021889894000208, http://scripts.iucr.org/cgi-bin/paper?wi0141
49. D. Franta, I. Ohlídal, P. Klapetek, A. Montaigne-Ramil, A. Bonanni, D. Stifter, H. Sitter, J. Appl. Phys. **92**(4), 1873 (2002). https://doi.org/10.1063/1.1489068, http://aip.scitation.org/doi/10.1063/1.1489068
50. D. Franta, D. Nečas, L. Zajíčková, Thin Solid Films **534**, 432 (2013). https://doi.org/10.1016/j.tsf.2013.01.081
51. D. Franta, D. Nečas, L. Zajíčková, I. Ohlídal, J. Stuchlík, D. Chvostová, Thin Solid Films **539**, 233 (2013). https://doi.org/10.1016/j.tsf.2013.04.012
52. I. Ohlídal, D. Necas, D. Franta, phys. status solidi (c) **5**(5), 1399 (2008). https://doi.org/10.1002/pssc.200777769, http://onlinelibrary.wiley.com/doi/10.1002/pssc.200777769/abstract
53. J.A. Ogilvy, H.M. Merklinger, J. Acoust. Soc. Am. **90**(6), 3382 (1991). https://doi.org/10.1121/1.401410, http://asa.scitation.org/doi/abs/10.1121/1.401410
54. D. Franta, I. Ohlídal, P. Klapetek, A. Montaigne-Ramil, A. Bonanni, D. Stifter, H. Sitter, Acta Phys. Slovaca **53**(2), 95 (2003)
55. S. Adachi, *Optical properties of crystalline and amorphous semiconductors: materials and fundamental principles* (Boston: Kluwer Academic Publishers, 1999). http://trove.nla.gov.au/version/39699241
56. D. Franta, D. Nečas, I. Ohlídal, A. Giglia, in *Optical Systems Design 2015: Optical Fabrication, Testing, and Metrology V*, vol. 9628 (2015), vol. 9628, pp. 96,281U–96,281U–12. https://doi.org/10.1117/12.2190104
57. J. Vodák, D. Nečas, D. Pavliňák, J.M. Macak, T. Řičica, R. Jambor, M. Ohlídal, Appl. Surf. Sci. **396**, 284 (2017). https://doi.org/10.1016/j.apsusc.2016.10.122

Chapter 6
Data Processing Methods for Imaging Spectrophotometry

David Nečas

Abstract Data processing approaches and algorithms for imaging optical techniques are discussed, with focus on imaging spectrophotometry in the visible and ultraviolet regions. Since the techniques often produce large quantities of data, efficient data fitting methods must be developed. This is covered from the high-level view of attacking the huge least squares problem and splitting it in different ways in order to extract useful information, as well as construction of effective models for individual phenomena and types of samples. Efficient computation of optical quantities is discussed for ideal, non-uniform and rough films together with modelling of optical constants and spectral or angular averaging.

The results shown in Chap. 5 attest the utility of imaging spectrophotometry in thin film characterisation. Here we continue in this topic and describe in more detail the models and data processing methods, focusing again primarily on measurements of thin films in the NIR-VIS-UV spectral region. The salient feature of optical spectra in this region is interference – the pattern of minima and maxima, whose density is given by the film thickness and refractive index. Also other film properties manifest themselves in the spectra in a fairly complex way; rarely it is possible to extract useful information by analysing a short spectrum segment. Data processing, therefore, requires quantitative modelling of entire spectra and searching for the best match between the model and experimental values.

This is in contrast to imaging spectrophotometry techniques for biological and chemical analysis, from IR absorption to fluorescence, where spectrally localised features (peaks) are analysed. Images of interesting quantities are obtained by spectral integration or peak fitting together with various factor and component analyses such as the principal component analysis.

Some of the more general points discussed in Sects. 6.1, 6.3 and 6.4 still apply to spectral imaging in general. Nevertheless, here the concrete models, formulae

D. Nečas (✉)
Plasma Technologies, CEITEC, Masaryk University, Purkyňova 123,
61200 Brno, Czech Republic
e-mail: yeti@physics.muni.cz

O. Stenzel and M. Ohlídal (eds.), *Optical Characterization of Thin Solid Films*, Springer Series in Surface Sciences 64,
https://doi.org/10.1007/978-3-319-75325-6_6

and algorithms are developed for imaging spectroscopic reflectometry (ISR) of thin films and its application to solid matter and material research, because this is how the technique is employed in practice [1–13]. Their adaptation to transmittance mode is straightforward. The basic concepts also carry over to imaging spectroscopic ellipsometry, albeit the particulars have to be adapted.

Finally, we note straight off that imaging brings no fundamental changes to the description of interaction of light with thin film systems and all the theories and approaches developed in Chaps. 2, 3, 7–10 and 14 are in principle usable in ISR. Their practical utility depends on whether they are suitable for the processing of high volumes of optical data that may have somewhat lower accuracy.

6.1 The Challenges

The spectral range and resolution, measurable area and spatial resolution of an imaging instrument depend on the optical system design as well as the pixel resolution of the detector and measurement procedure. What can be, however, be said with certainty is that the pixel resolutions are increasing steadily. Resolution of 512×512 pixels is considered moderate nowadays and instruments can be equipped with 1024×1024, 2048×2048 or even larger detectors. Improved data acquisition rates permit the collection of more points in each spectrum. Consequently, large data sets can be produced by a single measurement, which is both a blessing and a curse. On one hand it enables detailed characterisation of samples and even opens the possibility of completely new approaches, on the other hand the sheer data volume brings new challenges.

Consider, for example, a typical measurement using the imaging spectrophotometer described in Chap. 5, with fine spectral sampling. It takes approximately 40 min and yields 640 images for the individual wavelengths, each 512×512 pixels large. In other words about 1 GiB of experimental data per hour are acquired. Similar data production rates have been noted for other imaging spectrophotometers [4, 10], even though the acquisition of a low-resolution spectra set in a limited range can take as little as one minute.

The data acquisition times are acceptable. However, for an experimental technique to be practical and suited for routine characterisation, it is also necessary that the data analysis time is acceptable. In particular, simple analysis should be sufficiently fast to at least allow interactive experimentation, if instantaneous processing of GiB-large data sets may not be a reasonable demand. And, generally, common data analyses should not take much longer than the acquisition. When this is not satisfied the technique becomes 'specialised' and its applications more limited. A large part of this chapter is, therefore, dedicated to fast data processing, and efficiency will be a key factor in the selection of models and approaches. The second important factor is the number of fitting parameters, which should be low.

The basic task in ISR data processing is the same as in other optical techniques. A model of the measurement is formulated, which includes general sample structure (the number of layers, relations between them, etc.), models of spectral dependencies of optical constants of materials constituting the films (detailed in Chap. 3) and imperfections of both the sample and instrument (Chaps. 9 and 10).

The model determines how the measured reflectance R depends on the independent variables (wavelength λ or position on the sample (x, y)) and on the parameters we are trying to find, such as film thickness, roughness or band gap energy. Frequently it also includes nuisance parameters, for instance related to defects or systematic errors, that may be of no interest to us but are nonetheless necessary because the model has to include all important effects influencing the measured quantities. The model thus defines the function

$$R(\lambda, x, y, \ldots; \boldsymbol{p}) , \tag{6.1}$$

where $\boldsymbol{p} = (p_1, p_2, \ldots, p_P)^{\mathrm{T}}$ is the column vector of parameters and P is the number of parameters. The parameter values are then obtained by fitting this theoretical dependence on the experimental values using the least-squares method (LSM), i.e. by finding parameters $\boldsymbol{p}$ that minimise the sum of squared differences between the model and experimental values:

$$S = \sum_{n=1}^{N} \frac{1}{c_n^2} \left[R(\lambda_n, x_n, y_n, \ldots; \boldsymbol{p}) - R_n^{\mathrm{exp}} \right]^2 = \boldsymbol{d}^{\mathrm{T}} \boldsymbol{d} . \tag{6.2}$$

Index n distinguishes all measured values R_n^{exp} (N in total) and the corresponding values of independent variables (see also Table 6.1). Factor c_n is equal to the standard deviation of the nth measurement and determines its weight, so the vector of weighted differences has components

$$d_n = (R_n - R_n^{\mathrm{exp}})/c_n . \tag{6.3}$$

In the minimum of S it holds

$$\frac{\partial S}{\partial p_\alpha} = 0 . \tag{6.4}$$

In the following we will write $R(\lambda_n, x_n, y_n, \ldots; \boldsymbol{p})$ as $R_n(\boldsymbol{p})$ for brevity.

The dependence of optical quantities on the parameters is highly non-linear, necessitating non-linear least-squares fitting, which is almost universally realised using the Levenberg–Marquardt (LM) method [14–17], although Hanson–Krogh [18] has been also utilised [4], and even a brute-force search of the parameter space [19] (in the case of extremely low number of spectral points).

The LM method is an iterative procedure that we will summarise briefly. In can be written in terms of the Jacobian matrix $\mathbf{J}$

$$J_{n\alpha} = \frac{1}{c_n} \frac{\partial R_n}{\partial p_\alpha} \tag{6.5}$$

Table 6.1 Notation for counts of the various items appearing in this chapter and index variables usually associated with them (the index variables may vary where necessary). The total numbers satisfy the relations $N = K \times M$ and $P = G + M \times L$

Item	Total number	Index	Example
Individual data point	N	n	$R_n^{\exp}$ – a single measured reflectance
Point in a spectrum	K	k	λ_k – kth wavelength
Image pixel	M	m	h_m – film thickness in mth pixel
Fitting parameter	P	α	p_α – αth fitting parameter
Local fitting parameter	L	α	$v_{m,\alpha}$ – αth local parameter in pixel m
Global fitting parameter	G	α	u_α – global αth fitting parameter

with $\alpha = 1, 2, \ldots, P$. Using $\mathbf{J}$ and $\boldsymbol{d}$, we form the approximate Hessian matrix $\mathbf{H}$ (which is positive definite) and gradient vector $\boldsymbol{g}$ as follows

$$\mathbf{H} = \mathbf{J}^{\mathrm{T}}\mathbf{J} \quad \text{and} \quad \boldsymbol{g} = \mathbf{J}^{\mathrm{T}}\boldsymbol{d} \tag{6.6}$$

and solve the system of linear equations for parameter changes $\Delta\boldsymbol{p}$

$$(\mathbf{H} + \mu\mathbf{D})\Delta\boldsymbol{p} + \boldsymbol{g} = \bar{\mathbf{H}}\,\Delta\boldsymbol{p} + \boldsymbol{g} = \mathbf{0}\,. \tag{6.7}$$

The parameter values are then updated

$$\boldsymbol{p} \to \boldsymbol{p} + \Delta\boldsymbol{p}\,. \tag{6.8}$$

This procedure represents one step of the LM method and is repeated until it converges, i.e. the sum of squares S and/or the parameters $\boldsymbol{p}$ cease to change. The resulting set of parameters then satisfies (6.4) (within chosen criteria), although it may not correspond to the global minimum of S, i.e. the best possible agreement.

The damping (also called Marquardt) parameter μ controls the trust region of the method and consequently the behaviour of the parameter update step. Large μ means the LM step is cautious and close to a steepest descend method [17] step for finding the solution of (6.4), while small μ means a confident step close to a step of the Newton method [17]. A successful step is therefore followed by a μ decrease and unsuccessful step by an increase [20].

Matrix $\mathbf{D}$, which augments the Hessian to $\bar{\mathbf{H}}$, is a diagonal matrix that defines how the method scales with parameters. In the simplest case $\mathbf{D} = \mathbf{1}$, corresponding to so called unscaled LM method which converges poorly when parameters are not of comparable magnitude. The convergence can be improved by choosing $\mathbf{D} = \operatorname{diag}\mathbf{H}$, which results in so called scaled LM method – but other choices are possible [16, 17].

In this high-level view the data from all optical techniques are processed in the same manner. The crucial difference between ISR or imaging spectroscopic ellipsometry and single-spot measurements is in the scale of the problem. Continuing the introductory example, assume for instance we would like to fit just the film thickness in each pixel. In addition, five parameters of the dispersion model would be common to a sample region because the film material is uniform there (note that global parameters do not necessarily entail constant sample properties – they can also describe spatial dependencies). For a relatively small region of 200×200 pixels we have 40000 fitting parameters and 25.6×10^6 experimental data points. The Jacobian $\mathbf{J}$ has approximately 10^{12} elements, while the Hessian $\mathbf{H}$ has mere 1.6×10^9 elements. The matrices are in fact relatively sparse so the amounts of non-zero elements are not so huge (we will explore that in more detail in Sect. 6.4). Nevertheless, it is evident that taking a random LM implementation and plugging in the model and an entire ISR data set would be an endeavour doomed to fail.

The core difficulty is the large number of fitting parameters that are, unfortunately, all entangled together. For instance film thickness values in two different pixels may seem independent at the first sight. However, they are tied together via the global parameters of the dispersion model that are strongly correlated with all thicknesses. In the following sections we will therefore break the problem to smaller pieces and start from approaches that are within the capabilities of a stock LM method implementation, gradually progressing to data fitting scenarios involving more and more coupled parameters.

Data abundance is not the only issue in imaging techniques, as they also often suffer from somewhat complementary difficulty, less information in individual spectra. Compared to non-imaging techniques, the spectra are more noisy as the received radiant flux is smaller. Local defects ('bad pixels') are often encountered and also systematic errors are more common. In general, the transition to imaging increases the complexity of the entire measurement system. Many imaging systems are constructed as microscopic, i.e. with magnification greater than unity. This leads to major complications associated with aberrations of the imaging system, depth of focus, numerical aperture (ill-defined angles of incidence and reflection), suitability of materials of the optical components for a wide spectral range including the UV region, etc. [21]. Complex models are used just to obtain somewhat reliable thickness maps [8]. Imaging instruments are therefore often limited compared to their non-imaging counterparts. Complex post-processing, such as image fusion using wavelet reconstruction [22], is required when the angle of incidence is oblique – unless the Scheimpflug principle [23] is satisfied by tilting the detector accordingly. In consequence, the methods are frequently used for semi-quantitative and comparative measurements. Normal-incidence ISR can avoid these problems. On the other hand, the spectra contain less information than, for instance, full three-component ellipsometric spectra at oblique incidence. We will seek to set these two disadvantages – too much and too little information – against each other and turn them to a strength.

6.2 Single-Spectrum Models

A simple and obvious way of breaking the least-squares problem (6.1)–(6.8) to manageable pieces is to avoid global parameters. All parameters are then local, pertaining to particular image pixels and the Hessian $\mathbf{H}$ becomes a block-diagonal matrix, with blocks corresponding to pixels. Symbolically,

$$\mathbf{H} = \begin{pmatrix} \mathbf{H}_1 & & & \\ & \mathbf{H}_2 & & \\ & & \ddots & \\ & & & \mathbf{H}_M \end{pmatrix} . \tag{6.9}$$

Each of the M spectra can be fitted independently, exactly as in single-spot methods. So instead of a single least-squares problem with huge $\mathbf{J}$ and $\mathbf{H}$, we need to solve a large number of small ones.

As each sub-problem has only a few free fitting parameters and K data points, a software implementation can easily employ a standard high-quality LM implementation such as MINPACK [16] or its GSL adaptation [24]. Custom ISR data processing software is common [4, 9, 25], and often a necessity. When ISR data analysis progresses from a proof of concept to routine characterisation and applications, it enables evading overheads invariably incurred by general software frameworks. Parallel computation can obviously shorten the analysis times, and a number of approaches for parallelisation of the LM algorithm were suggested [26]. In practice simple parallelisation on data suffices, taking the fitting of a single spectrum as one basic task and handling these tasks in parallel. Since one spectrum can be fitted quickly (from tens of microseconds for fine spectral resolution down to about a microsecond for coarse spectra – in 2016), it is actually advisable to split the work to larger tasks than individual pixels, for instance image rows, to reduce overheads.

The subproblems can be simplified further using a priori knowledge about the sample. It can come from known tabulated properties of the film and substrate, or from prior characterisation using other methods such as conventional ellipsometry and spectrophotometry. Large non-uniformity (usually among the primary reasons why ISR is employed) can impede the characterisation using conventional optical methods. Nevertheless, with the help of suitable modelling even quite non-uniform films can be characterised successfully [7].

When we know the overall film properties, for instance spectral dependencies of optical constants, mean thickness or roughness values, most fitting parameters can be fixed in the ISR data processing. ISR then usually serves as an auxiliary technique, providing fine maps of selected parameters [1, 2, 4, 9]. An example of this type of analysis is also discussed in Sect. 5.9.

This single-spectrum approach is by far the most common [1–8, 10]. In fact, published works seldom venture beyond single-spectrum data processing. And this should not be seen as a shortcoming, at least not necessarily. Fine mapping of a few selected parameters often brings most of the useful information that can be extracted from ISR data while simultaneously avoiding some of the possible pitfalls.

Single-spectrum data processing and the corresponding models also serve as basic building blocks for more complex schemes described in Sects. 6.3 and 6.4. Therefore, we will start by summarising common film properties and other effects that need to be modelled and by developing corresponding models suitable for ISR from the efficiency and parameter count standpoint.

6.2.1 Thickness

Film thickness is the most useful and common quantity measured in each pixel. Almost all thin films with any substantial spatial variation of properties are non-uniform also in thickness, whereas films non-uniform in thickness frequently do not exhibit measurable variation in other parameters [9, 27]. If the thickness variation is the result of etching (as opposed to non-uniform deposition) there is seldom any reason to even consider local differences in film material. Hence we will start by fitting local thickness in the mth pixel h_m – or simply h, as each spectrum is processed individually and the pixel-distinguishing subscripts can be dropped.

The reflectance is the squared absolute value of the effective Fresnel reflection coefficient r (for normal incidence there is only one)

$$R = |r|^2 = r^* r \,, \tag{6.10}$$

where star* denotes complex conjugation. For a thin film with refractive index $\hat{n}$ and thickness h on opaque substrate we can write

$$r = \frac{r_1 + r_2 U}{1 + r_1 r_2 U} \,, \quad \text{where} \quad U = \exp(\mathrm{i}uh) \quad \text{and} \quad u = \frac{4\pi}{\lambda}\hat{n} \,. \tag{6.11}$$

Here λ is the vacuum wavelength and $r_{1,2}$ are the Fresnel reflection coefficients at the two boundaries (see Chap. 2). The only quantity that depends on the film thickness is U.

In a slightly more general case, the investigated film can be inside a thin film system. Such system is frequently formed by the film and a thin overlayer on its upper boundary and/or a transition layer between the film and substrate. The reflection coefficient is still a simple linear rational expression:

$$r = \frac{r_1 + (t_1 t_1' - r_1 r_1') r_2 U}{1 - r_1' r_2 U} \tag{6.12}$$

where $t_{1,2}$ are the Fresnel transmission coefficients, prime$'$ denotes coefficients for light incoming from the substrate side and all coefficients are effective coefficients

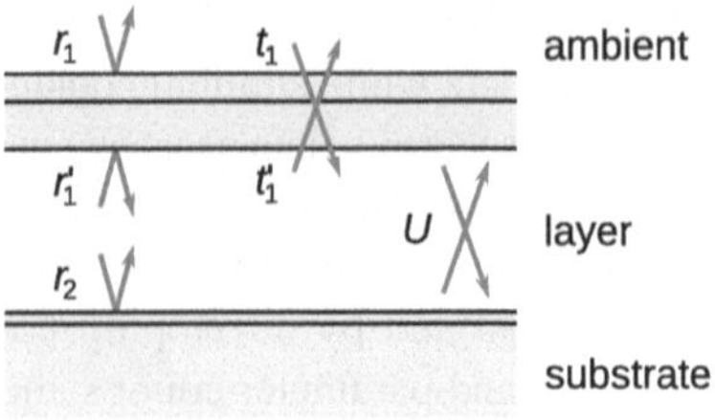

Fig. 6.1 Illustration of effective Fresnel coefficients $r_1, r_1', r_2, t_1, t_1'$ and quantity U occurring in formula (6.12)

for the systems above and below the film (Fig. 6.1). The only quantity depending on the thickness is still U.

All other quantities, optical constants, even Fresnel coefficients, can be precalculated and tabulated for all wavelengths λ_k corresponding to individual spectral points, speeding up the data processing. Nevertheless, the most time consuming operation is the evaluation of complex exponential functions $\exp(\mathrm{i}uh)$ occurring in U (6.11). Since h can vary arbitrarily they cannot be precalculated.

Therefore, it is important to reduce the number of exp evaluations, and this will be a recurring topic in this chapter. Here the reduction is quite straightforward. The derivative of reflectance by h that occurs in $J_{n\alpha}$ can be written

$$\frac{\partial R}{\partial h} = \frac{\partial r}{\partial h} r^* + r \frac{\partial r^*}{\partial h} = 2 \left[\mathrm{Re}\, r \; \mathrm{Re} \frac{\partial r}{\partial h} + \mathrm{Im}\, r \; \mathrm{Im} \frac{\partial r}{\partial h} \right] , \tag{6.13}$$

where for r expressed by (6.11) and (6.12)

$$\frac{\partial r}{\partial h} = \frac{\mathrm{i}uUr_2(1 - r_1^2)}{(1 + r_1 r_2 U)^2} \quad \text{and} \quad \frac{\partial r}{\partial h} = \frac{\mathrm{i}uUr_2 t_1 t_1'}{(1 - r_1' r_2 U)^2} , \tag{6.14}$$

respectively. Therefore, the reflectance and its derivative can be calculated together using just a single exp, a substantial improvement compared to the three that would be required by naïve numerical differentiation.

6.2.2 *Optical Constants*

Spectral dependencies of optical constants can be in principle determined by fitting dispersion model parameters in each pixel. Even the simplest dispersion models have several parameters. The basic PJDOS model for interband transitions [28, 29] has three, the Tauc–Lorentz model [30] has four (or five), and even though the Cauchy formula for refractive index $n = A + B/\lambda^2$ has only two it often needs to be supplemented by an exponential formula for extinction coefficient $k = \alpha \exp(-\beta\lambda)$, bringing again the parameter count to four.

One obvious consequence is the increase of computational demands. The increase is in part unavoidable, but the amount of additional complex calculations compared to fitting thickness alone can be kept small. We first note that formula (6.13) holds for any model parameter; it is not specific to h. If p_α is a parameter of the dispersion model for the complex refractive index of film $\hat{n}_1 = \hat{n}_1(\boldsymbol{p}; \lambda)$ the derivative of the reflection coefficient can be written using the chain rule as

$$\frac{\partial r}{\partial p_\alpha} = \frac{\partial r}{\partial \hat{n}_1} \frac{\partial \hat{n}_1}{\partial p_\alpha} . \tag{6.15}$$

The first factor on the right hand side is not specific to p_α and can be evaluated only once:

$$\frac{\partial r}{\partial \hat{n}_1} = \frac{1}{(1 + r_1 r_2 U)^2} \left[(1 - r_2^2 U^2) \frac{\partial r_1}{\partial \hat{n}_1} + (1 - r_1^2) \left(U \frac{\partial r_2}{\partial \hat{n}_1} + r_2 \frac{\partial U}{\partial \hat{n}_1} \right) \right] . \tag{6.16}$$

Furthermore, when the derivatives are expressed using the refractive indices and reflection coefficients we obtain

$$\frac{\partial r}{\partial \hat{n}_1} = \frac{1}{(1 + r_1 r_2 U)^2} \left[\frac{2 n_0 (1 - r_2^2 U^2)}{(\hat{n}_1 + n_0)^2} + \left(\frac{4\pi i h r_2}{\lambda} - \frac{2 \hat{n}_2}{(\hat{n}_2 + \hat{n}_1)^2} \right) U (1 - r_1^2) \right] , \tag{6.17}$$

where n_0 and $\hat{n}_2$ are the refractive indices of the ambient and substrate, respectively. Although the right hand side of (6.17) is not particularly simple, it is a rational function of quantities that enter the expression for R and have to be calculated anyway. The second factor in (6.15) is specific to the dispersion model and each p_α. For some models, such as the Cauchy formula, the derivatives are very simple functions. In other cases the derivatives are complicated and it may not even be worth calculating them analytically because the evaluation of the derivative has similar computational demands as numerical differentiations.

The idea of fitting the complete dispersion model in each pixel and obtaining two-dimensional maps of all its parameters sounds appealing and was explored in practice [6, 31]. In an attempt to execute we however usually run into the another consequence of increased number of parameters: correlations. One of the main reasons why strong correlations between parameters occur is that the sum (6.2) is very sensitive to positions of interference minima and maxima in the spectrum. And the positions depend only on the optical thickness, i.e. the product of thickness and refractive index, not each separately. In combination with the lower information content of normal reflectance spectra this results in highly correlated parameters, large uncertainties and sensitivity to any imperfections and systematic errors in the experimental spectra.

The parameter correlations are illustrated in Fig. 6.2. We took a simple model of ideal thin film, with optical constants described by the Cauchy formula, on silicon substrate and assumed reflectance measurement in the spectral range from 300 to 800 nm. The correlation coefficient $C_{A,h}$ between parameter A of the Cauchy formula and film thickness h was then calculated according to the LSM and plotted for several A values as a function of film thickness. For most parameter combinations $C_{A,h}$ attains

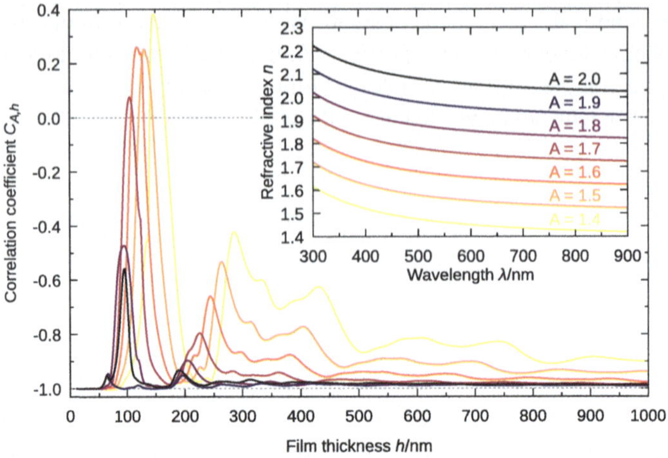

Fig. 6.2 Correlation coefficient between Cauchy formula parameter A and film thickness h as a function of film thickness, plotted for several different values of A (with the corresponding refractive index shown in a subplot). Dashed lines mark correlation coefficient equal to 0 and -1, corresponding to no correlation (the ideal case) and completely negatively correlated A and h (the worst case)

large negative values. In particular when the film is thin ($h \lesssim 50\,\text{nm}$) or the refractive index high the correlation coefficient is very close to -1, indicating that the reliability of fit results will be low. When attempting to fit the complete dispersion model in each pixel it is, therefore, necessary to carefully consider if the resulting parameter maps represent real spatial variation or mere artefacts – and more generally what value they bring.

6.2.3 Non-uniformity

Even though the main application of ISR is the measurement of highly non-uniform films, the areas corresponding to detector pixels are so small that the film often can be assumed uniform within one pixel. When the thickness variation is so large that this assumption is no longer valid we observe a decrease of interference contrast in the spectra. The reflectance then has to be integrated over the thickness distribution within the area corresponding to one pixel (see Sect. 10.4):

$$R(h) = \int \rho(t)\, R(h + t)\, \mathrm{d}t \; , \tag{6.18}$$

where h now stands for the mean film thickness in the pixel (the integration is understood being over the domain of ρ). The thickness density (density of thickness distribution) in the pixel ρ depends on thickness variation $h = h(x, y)$ as well as the

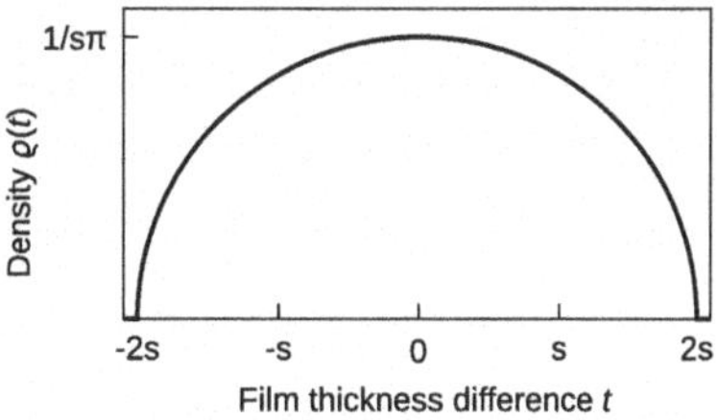

Fig. 6.3 Model film thickness density described by formula (6.20)

pixel shape and have quite a complicated analytical form. Furthermore, integration is a slow operation that impacts the data processing speed.

Fortunately, up to relatively large non-uniformities the precise form of ρ is not critical. The influence on optical quantities can be characterised using a single parameter, the RMS of thickness variation s [7, 32]:

$$s^2 = \int t^2 \rho(t)\, R(h+t)\, \mathrm{d}t\, . \tag{6.19}$$

We can choose $\rho(h)$ in the form corresponding to a simple wedge-shaped film, i.e. constant thickness gradient

$$\rho(t) = \frac{1}{s\pi}\sqrt{1 - \left(\frac{t}{2s}\right)^2} \tag{6.20}$$

that is also reasonable approximation for other film shapes. The density is illustrated in Fig. 6.3.

An efficient method of evaluation of integrals of type (6.18) to low and moderate precision is Gaussian quadrature with ρ as the weight function [33]. An n-point quadrature rule has n abscissae A_j and weights W_j found so that

$$\int \rho(t) R(h+t)\, \mathrm{d}t \approx \sum_{j=1}^{n} W_j R(h + A_j)\, , \tag{6.21}$$

integrates exactly all polynomial functions $R(h)$ up to degree $2n - 1$. Therefore, if $R(h)$ is a smooth function, locally well approximated by polynomials of low order, which almost invariably holds, the recipe (6.21) requires the computation of $R(h)$ for only a few thicknesses to reach sufficient accuracy.

An advantage of the density (6.20) is that it is the same as the weight function for Chebyshev polynomials of the second kind (apart from scaling), for which A_j and W_j have a simple closed form [33, 34] ($j = 1, 2, \ldots, n$):

$$A_j = 2s \cos \frac{\pi j}{n+1} \quad \text{and} \quad W_j = \frac{2}{n+1} \sin^2 \frac{\pi j}{n+1}\, . \tag{6.22}$$

Table 6.2 Abscissae and weights for a few useful low-order Gaussian quadratures. All functions $\rho(t)$ are normalised so that their integrals and second moments (dispersions) are unity, and their first moments (means) are zero

Function $\rho(t)$	Interval	Points	Abscissae	Weights
$\sqrt{1-t^2/4}/\pi$	$[-2, 2]$	3	0	1/2
			$\pm\sqrt{2}$	1/4
		5	0	1/3
			± 1	1/4
			$\pm\sqrt{3}$	1/12
$\exp(-t^2/2)/\sqrt{2\pi}$	$(-\infty, \infty)$	3	0	2/3
			$\pm\sqrt{3}$	1/6
		5	0	8/15
			$\pm[5-\sqrt{10})]^{1/2}$	$3/(140-40\sqrt{10})$
			$\pm[5+\sqrt{10})]^{1/2}$	$3/(140+40\sqrt{10})$
$1/2\sqrt{3}$	$[-\sqrt{3}, \sqrt{3}]$	3	0	4/9
			$\pm 3/\sqrt{5}$	5/18
		5	0	64/225
			$\pm(5/3-2\sqrt{10/63})^{1/2}$	$(322+13\sqrt{70})/1800$
			$\pm(5/3+2\sqrt{10/63})^{1/2}$	$(322-13\sqrt{70})/1800$

We use quadrature rules with low numbers of points because computation speed is important and they are sufficient for ISR. The simplest useful rule implementing (6.18) is the three-point rule (see Table 6.2 for the five-point rule)

$$R(h, s) \approx \frac{1}{2}R(h) + \frac{1}{4}[R(h-\sqrt{2}\,s) + R(h+\sqrt{2}\,s)]\ . \tag{6.23}$$

It integrates exactly $R(h)$ up to polynomials of the fifth degree and requires the evaluation of three exp functions, one in each term. One evaluation can be eliminated by noting that

$$\exp(\mathrm{i}u\sqrt{2}\,s)\,\exp(-\mathrm{i}u\sqrt{2}\,s) = 1\ . \tag{6.24}$$

Therefore,

$$r(h-\sqrt{2}\,s) = \frac{qr_1 + r_2U}{q + r_1r_2U} \quad \text{and} \quad r(h+\sqrt{2}\,s) = \frac{r_1 + r_2qU}{1 + r_1r_2qU}\ , \tag{6.25}$$

where $q = \exp(\mathrm{i}u\sqrt{2}\,s)$. The improvement becomes substantial when we combine evaluation of R with the evaluation of its derivative by h. The factor q is a constant. Hence the derivatives do not introduce any new exp functions:

$$\frac{\partial r(h-\sqrt{2}\,s)}{\partial h} = \frac{\mathrm{i}uqUr_2(1-r_1^2)}{(q+r_1r_2U)^2} \quad \text{and} \quad \frac{\partial r(h+\sqrt{2}\,s)}{\partial h} = \frac{\mathrm{i}uqUr_2(1-r_1^2)}{(1+r_1r_2qU)^2} \,. \tag{6.26}$$

In total only two exp function evaluations are still necessary – compared to nine required by naïve calculation and numerical differentiation.

Comparably low computational demands can be achieved if we instead employ the second-order derivative formula [32]

$$R(h,s) \approx R(h) + \frac{s^2}{2} R''(h) \tag{6.27}$$

to approximate the averaged reflectance (6.18). However, the rule (6.23) remains a good approximation to larger s values than (6.27) and when it starts to fail it does so gracefully, whereas (6.27) is prone to shooting off suddenly to unphysical values.

The film non-uniformity s can be fitted as an independent parameter along with thickness. This requires the evaluation of derivatives of R by s that can again be done using only algebraic manipulation because

$$\frac{\partial r(h \pm \sqrt{2}\,s)}{\partial s} = \pm\sqrt{2}\frac{\partial r(h \pm \sqrt{2}\,s)}{\partial h} \tag{6.28}$$

and the right hand side is given by expressions (6.26). If h and s are the only fitted parameters this can still be a reasonable approach. However, in normal reflectance spectra non-uniformity is often difficult to distinguish from other other effects, in particular absorption [35, 36]. The alternative is two-step processing where in the first step only the film thickness map is determined. The local non-uniformity is then calculated from estimated thickness gradient or spread of thickness value distribution in the neighbourhood of each pixel. The calculated local s values are used in the second step, but as fixed values. Note that it is necessary to use robust estimators that are resistant to local defects [37], e.g. inter-quartile differences for the spread estimation, to suppress the influence of bad pixels and defects on the sample or reflectance normal. The estimated gradient ∇h is related to the pixel's s value

$$2\sqrt{3}\,s = \Lambda \nabla h \,, \tag{6.29}$$

where Λ is the side of sample area corresponding to one pixel. An empirical factor may need to be inserted to (6.29). In particular, CMOS sensors have somewhat smaller fill factors compared to CCD sensors where the light sensitive area is close to 100 % of the pixel. Therefore, the true s is smaller than estimated from the gradient. Another reason for a correction factor may be multiple reflections (or other effects) that cause slight blurring of the images [38].

6.2.4 Spectral and Angular Averaging

If the imaging spectrophotometer has a monochromator and the light beam incident on the sample is collimated, we can usually disregard the finite line width of the monochromator (the exception being films with thickness around 1 μm or more) and essentially always disregard the spread of angles of incidence. Some imaging instruments that can be still considered spectroscopic employ filter carousels or variable band pass filters [2, 8] with large band width; in extreme cases even RGB light sources and detectors were used to obtain somewhat spectrally resolved reflectance [19, 39]. Correct integration over the spectral distribution then becomes crucial. Similarly, integration over the distribution of incidence angle is crucial for instruments with focused light beams and, consequently, large numerical apertures.

Spectral and angular averaging, as well as integration over any other quantity, are similar to thickness averaging (6.18), the only difference being the replacement of thickness h with λ or φ and the replacement of $\rho(h)$ with the corresponding distribution. Gaussian quadratures can be again utilised for efficient integration. For instance a Hermite–Gauss quadrature [33, 34] if the density ρ is Gaussian (abscissae and weights for a few low-order rules are again listed in Table 6.2).

Averaging over multiple quantities should be in principle done using an integral with their multi-dimensional distribution. In practice we replace it with sequential averaging over individual marginal distributions ρ_λ and ρ_φ because the full distribution is unknown:

$$\int \rho_\lambda(t) \left[\int \rho_\varphi(t') R(\lambda + t, \varphi + t') \, \mathrm{d}t' \right] \mathrm{d}t \,, \tag{6.30}$$

As each integration slows down the computation considerably, efficient quadratures become even more important.

For spectral averaging, extra calculations can be avoided by the utilisation of reflectances calculated for neighbour spectral points. For instance a three-point rule for averaged $R^{\mathrm{avg}}(\lambda)$ is implemented

$$R^{\mathrm{avg}}(\lambda_k) \approx W_{-1} R(\lambda_{k-1}) + W_0 R(\lambda_k) + W_1 R(\lambda_{k+1}) \,, \tag{6.31}$$

where the weights W_{-1}, W_0 and W_1 are calculated to approximate the chosen distribution ρ_λ. This is, of course, possible only if the spectral points λ_k are sufficiently close. If they are farther apart it may be still advantageous to calculate the spectrum with a finer spectral step and then average this supersampled spectrum with appropriate weights to obtain $R^{\mathrm{avg}}(\lambda_k)$. In general, when multiple neighbour values are necessary, the rule (6.31) is a discrete correlation

$$R^{\mathrm{avg}}(\lambda_k) \approx \sum_j W_j R(\lambda_{k+j}) \,. \tag{6.32}$$

Discrete correlations can be efficiently evaluated by utilising the correlation theorem and fast Fourier transform [40]. However, when we find this avenue appealing usually it is an indication that the spectral band is simply too broad.

Averaging over the angle of incidence has some specifics. First, for oblique angle of incidence the Fresnel coefficients for p- and s-polarised light differ. The single expression for r in formulae (6.11) or (6.12) must be replaced with separate r_{p} and r_{s} expressions (see Chap. 2) and then

$$R = \frac{1}{2}(|r_{\mathrm{p}}|^2 + |r_{\mathrm{s}}|^2) \tag{6.33}$$

under the assumption the light source, detector and optics are polarisation-unbiased. Notably, u for oblique incidence

$$u = \frac{4\pi}{\lambda}\hat{n}\cos\varphi_1 \ . \tag{6.34}$$

is the same for both polarisations, so U only needs to be computed once for both polarisations. The angles of incidence φ and angle refraction into the film φ_1 only occur as $\cos\varphi$ and $\cos\varphi_1$ in u and the Fresnel coefficients. It is, therefore, useful to work with the cosines instead of the angles themselves. If the spectral dependencies of optical constants are fixed the cosines can and should be precalculated and tabulated. Since $\cos\varphi \approx 1 - \varphi^2/2$ for small φ, the cosines deviate negligibly from unity for normal-incidence collimated beams. This explains why the spread of incidence angles can be disregarded in this case.

Second, the angular distribution is one-sided because $\varphi \geq 0$. If nominally normal incidence of light ($\varphi = 0$) is considered and the light is focused, the mean incidence angle is positive, possibly substantially. Hence, the positions of interference minima and maxima in the spectrum are shifted compared to normal incidence of light. When this is not taken into account, film thickness are systematically underestimated.

The crudest possible approximation that still achieves some correction of the interference extrema positions is based on the calculation of the mean value of $\cos\varphi_1$, which is then taken as the cosine of the effective angle of refraction $\cos\varphi_{1,\mathrm{eff}}$ [8, 41]

$$\cos\varphi_{1,\mathrm{eff}} = \int_0^{\varphi_{\max}} \rho_\varphi(\varphi)\cos\varphi_1(\varphi)\,\mathrm{d}\varphi \ . \tag{6.35}$$

All other angles and Fresnel coefficients are then calculated for this effective angle $\varphi_{1,\mathrm{eff}}$. Although extrema positions are corrected, this ad hoc approximation cannot describe the decreased interference contrast, and since it does not conform to (6.30) it is unclear how to extended or refine it to cover this effect.

Therefore, angular averaging should be preferably based on the integral (6.30). A one-point Gaussian quadrature rule requires the evaluation of R for a single angle of incidence, like the $\cos\varphi_{1,\mathrm{eff}}$ approach, and also leads to a similarly crude approximation. However, it belongs to a framework extendable to multi-point rules that capture the interference contrast reduction and provide better approximations of the averaged reflectance. Unfortunately, ρ_φ is seldom a weight function of a known quadrature rule. Furthermore, it does not even have a fixed form, i.e. functions ρ_φ for two different maximum angles of incidence $\varphi_{\max}$ cannot be obtained one from the another by simple scaling. Therefore, the quadrature rules have to be obtained numerically on case-by-case basis, which is inconvenient.

So, instead, we transform φ to a variable $v = v(\varphi)$ with constant density $\rho_v(v)$ – depending on the light source and optics, it can be $v = \cos\varphi$, $v = 1/\cos^2\varphi$ or a similar quantity. The integral over φ is replaced by

$$\int_{v_{\min}}^{v_{\max}} R(v)\,\mathrm{d}v \tag{6.36}$$

that can be evaluated using the standard Legendre–Gauss quadrature [33, 34] because the weight is now unity (see Table 6.2 for abscissae and weights for selected low-order rules). It is important that the change of variable does not distort the shape of $R(\varphi)$ for small φ in a manner that would prevent the utilisation of Gaussian quadrature – the function remains smooth and approximable by polynomials of low order.

6.2.5 Roughness

Roughness of film boundaries is an ubiquitous phenomenon that influences measured optical quantities and often has to be taken into account. A large body of theoretical approaches have been developed for the description of interaction of light with randomly rough surfaces and media boundaries in general. For specularly reflected light (as opposed to scattered) three are most commonly used, the effective medium approximation (EMA) [42], scalar diffraction theory [43–46] and second order Rayleigh–Rice theory [47–49]. They are described in depth in Sect. 10.3; here we limit the discussion to their suitability for ISR data processing.

The EMA is the simplest of the three and valid only for very fine roughness [50]. Each rough boundary between two media is replaced with a fictitious layer. The layer optical constants are calculated from the optical constants of surrounding media using a mixing formula. Its thickness h_{EMA} is conventionally taken to be proportional to the roughness RMS value σ even though in fact no such relation exists and h_{EMA} is just an effective parameter without definite interpretation [50]. The EMA can be easily adopted in ISR models. Since the optical constants of the fictitious layer are given by the mixing formula only one new parameter appears in the model for each rough boundary, the fictitious thickness h_{EMA}.

Even with EMA the functional dependence of reflectance on model parameters becomes quite complicated, and this holds even more for the derivatives. The calculation of derivatives of R by h and h_{EMA} can still utilise formula (6.14) (the second variant). The derivatives by dispersion model parameters should still be calculated using the chain rule (6.15). However, the derivatives $\mathrm{d}r/\mathrm{d}\hat{n}_1$, that are expressed by

(6.17) for smooth film boundaries, may be as well calculated numerically if the refractive index $\hat{n}_1$ enters the EMA.

The second-order RRT represents the opposite extreme – it has a wide region of applicability [49, 50] but it is very complex. The RRT expressions involve multiple integrals that make it computationally prohibitive for ISR, except perhaps in special applications that justify the computation cost. It should be also noted that in RRT the roughness is described by the power spectral density of spatial frequencies (see Chap. 11 for statistical characterisation of roughness). Even when its functional form is fixed, for instance to Gaussian, it still has to be characterised by at least two parameters that then enter the model as free fitting parameters.

The SDT is moderately complex and can still be computationally accessible when implemented efficiently, as we show below. It is valid only when the rough boundaries are locally smooth, i.e. in the limit of long-wavelength roughness. This makes it in some sense complementary to the EMA and the two approaches can be naturally combined if required.

In SDT the roughness is described by the statistical distribution of z-coordinates of the boundaries. If the thin film system consists of F layers the distribution is $F+1$-dimensional as there are $F+1$ boundaries. The formulae resulting from the SDT express the reflection coefficient of a rough thin film system $\langle r \rangle$ as a statistical average of the reflection coefficient r for the corresponding smooth film system over the $F+1$-dimensional distribution. If the distribution is assumed to be Gaussian the averaged reflection coefficient is equal to the sum of terms involving one to $F+1$ boundaries, i.e. zero to F layers [11, 46]:

$$\langle r \rangle = \sum_{d=0}^{F} \langle r \rangle_d \, . \tag{6.37}$$

The surface roughness (zero-layer) term is simple:

$$\langle r \rangle_0 = r_1 \exp\left(-\frac{1}{2} B_0^2 S_{1,1}\right) \, . \tag{6.38}$$

The others ($d \geq 1$) have the form of a series

$$\langle r \rangle_d = \sum_{p=d}^{\infty} \sum_{\substack{\boldsymbol{m} \\ \sum m_j = p}} \exp(2\mathrm{i}\boldsymbol{m}\hat{\boldsymbol{X}})\, Q_d(\boldsymbol{m})\, H_d(\boldsymbol{m}) \, . \tag{6.39}$$

Elements m_j of vector $\boldsymbol{m} = (m_1, m_2, \ldots, m_F)$ express how many times the light passes through the jth layer. Factors $\exp(2\mathrm{i}\boldsymbol{m}\hat{\boldsymbol{X}})$ describe the phase gained by light by passing through the layers, where $\boldsymbol{X}$ is a vector formed from the phase terms $X_j = 2\pi \hat{n}_j h_j / \lambda$. Factors $Q_d(\boldsymbol{m})$ represent interactions at layer boundaries. They are certain homogeneous polynomial of degree $2p+1$ in Fresnel coefficients for the corresponding system of smooth layers. And finally the factors $H_d(\boldsymbol{m})$ express the influence of roughness. They are multidimensional Gaussians, the arguments of the exponential functions being scalar products of certain vectors with the matrix **S**

$$S_{i,j} = \sigma_i \sigma_j C_{i,j} \,, \tag{6.40}$$

defining the RMS values σ_i of the rough boundaries and correlations $C_{i,j}$ between them.

For a general multilayer with arbitrarily correlated rough boundaries the expressions are still rather complex. They simplify if we consider a single film with rough boundaries, which is the type of sample most likely to be analysed by ISR. The sum (6.37) then consists only of two terms, $\langle r \rangle_0$ and $\langle r \rangle_1$. The latter is given by the classic series [44, 46, 51]

$$\langle r \rangle_1 = \sum_{p=1}^{\infty} \exp(2\mathrm{i}pX_1)\, Q_1(p)\, H_1(p) \,, \tag{6.41}$$

where

$$Q_1(p) = t_1 t_1' (r_1')^{p-1} r_2^p \tag{6.42}$$

and

$$H_1(p) = \exp\left(-\frac{1}{2} D_1^2 S_{1,1} - \frac{1}{2} D_2^2 S_{2,2} - D_1 D_2 S_{1,2} \right) . \tag{6.43}$$

The summation index p enters $H(p)$ via D_j

$$D_j = B_j - B_{j-1} \,, \quad \text{where} \quad B_0 = \frac{4\pi}{\lambda} n_0 \,, B_1 = \frac{4\pi}{\lambda} p\hat{n}_1 \quad \text{and} \quad B_2 = 0 \,. \tag{6.44}$$

The single-layer expressions are reasonably complex. With $Q_1(p)$ in the form (6.42) they do not require any modification when additional very thin layers are considered at the film boundaries, either actual overlayers and transition layers or fictitious layers originating from the EMA. It is sufficient to replace r_1, r_1', t_1 and t_1' in (6.42) and (6.38) with the effective Fresnel coefficients of the corresponding additional layers. More precisely, the replacement is permissible if the additional layers are are so-called identical thin films [52], i.e. films with geometrically identical upper and lower boundaries (Fig. 6.4). Thin overlayers and transition layers essentially always satisfy this condition.

The matrix **S** is equal to

$$\mathbf{S} = \begin{pmatrix} \sigma_1^2 & C\sigma_1\sigma_2 \\ C\sigma_1\sigma_2 & \sigma_2^2 \end{pmatrix} = \begin{pmatrix} \sigma_1 & 0 \\ 0 & \sigma_2 \end{pmatrix} \begin{pmatrix} 1 & C \\ C & 1 \end{pmatrix} \begin{pmatrix} \sigma_1 & 0 \\ 0 & \sigma_2 \end{pmatrix} , \tag{6.45}$$

where $C = C_{12}$ is the correlation coefficient between the two boundaries. It is undefined if one of the boundaries is smooth but we can simply put $C = 0$ in this case.

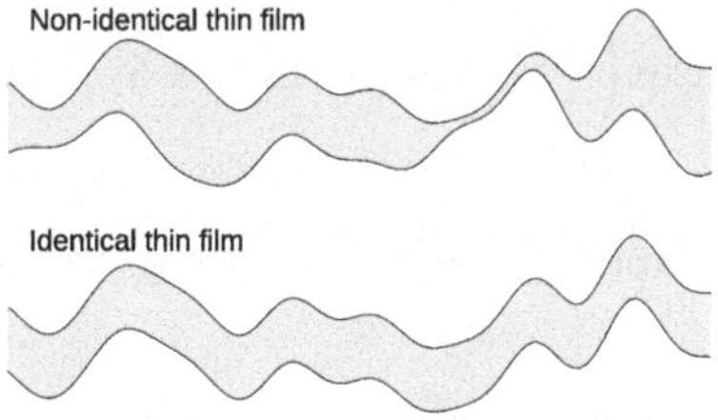

Fig. 6.4 The upper boundary of an identical thin film is the same as the lower boundary, only shifted along the z-axis. A non-identical thin film has geometrically different lower and upper boundaries, even if they may be the same statistically

Alas, formula (6.41) also brought back complex exponential functions. And they arrived in force, each term contains at least one. Combining the exps from (6.42) and (6.43) to a single exp ensures each term contains exactly one, but one per term is still too many.

If the series (6.41) contained only exps of the $\exp(2\mathrm{i}pX_1)$ kind then most of them could be eliminated easily. Writing

$$\exp(2\mathrm{i}pX_1) = \left[\exp(2\mathrm{i}X_1)\right]^p = z^p \tag{6.46}$$

would turn the series into a power series. A simple power series

$$\sum_{p=0}^{\infty} a_p z^p \tag{6.47}$$

can be evaluated using the Horner's algorithm which avoids the computation of powers (or exponentiation) in each term, replacing them by successive multiplications

$$z^p = z^{p-1} z \tag{6.48}$$

that generate the power z^p required in pth term from the power z^{p-1} used in the previous term.

The exponent in SDT factor $H_1(p)$ contains also terms proportional to p^2. Therefore, the simple recurrence relation (6.48) is insufficient. A double recurrence relation, however, makes an efficient summation possible. We start by regrouping the terms according to their dependence on p

$$\langle r \rangle_1 = C \sum_{p=1}^{\infty} B^p A^{p^2} , \tag{6.49}$$

where A, B and C are constants. Specifically,

$$A = \exp\left(-\frac{8\pi^2}{\lambda^2}(S_{1,1} - 2S_{1,2} + S_{2,2})\hat{n}_1^2\right) , \tag{6.50}$$

$$B = r_1' r_2 \exp(2iX_1) \exp\left(\frac{16\pi^2}{\lambda^2}(2S_{1,2} - S_{1,1})\hat{n}_1 n_0\right) \tag{6.51}$$

and

$$C = \frac{t_1 t_1'}{r_1'} \exp\left(-\frac{16\pi^2}{\lambda^2} S_{1,1} n_0^2\right) . \tag{6.52}$$

The evaluation of all three requires two complex exps and one real. Then we define two auxiliary sequences y_p and z_p as follows:

$$y_1 = A , \quad z_1 = ABC , \quad y_{p+1} = A^2 y_p \quad \text{and} \quad z_{p+1} = B y_{p+1} z_p . \tag{6.53}$$

It can be verified that $z_p = CB^p A^{p^2}$ and thus

$$\langle r \rangle_1 = \sum_{p=1}^{\infty} z_p . \tag{6.54}$$

No additional costly operations are necessary; the recipe (6.53) and (6.54) involves only multiplications and additions.

When the series converges fast a few terms can suffice to obtain the sum with a precision needed to model experimental data. Sometimes, therefore, the speed gain may be insubstantial (note that the series converges faster for larger roughness [46]). The reformulated summation scheme however keeps an important advantage. It eliminates the necessity to estimate how many terms are needed and make speed–precision trade-offs [51]. The computational cost is essentially fixed. Once we paid it we can evaluate the sum to a high precision for free. This is particularly useful for numerical differentiation which can lead to nonsensical values if the function is not smooth, and appreciated also in plotting of the theoretical spectra as it prevents spurious jumps in the curves. Numerical differentiation is unfortunately the only sensible option for calculation of derivatives by fitting parameters when SDT is involved.

Frequently only the upper film boundary has to be considered rough; the lower one is smooth. This means $\sigma_2 = 0$ and the expressions (6.50)–(6.52) for A, B and C simplify somewhat. Specifically, $S_{2,2} = S_{1,2} = 0$ and therefore only terms involving $S_{1,1} = \sigma_1^2$ remain in the arguments of the exponential functions. Nevertheless, the overall complexity of the calculation remains the same. This is because the phase gained by light when it travels through the layer still varies due to roughness, even if only one of the boundaries is rough. Only identical thin films (a category including layers with both boundaries smooth as a special case) can be replaced by effective boundaries, allowing to reduce the complexity. The upshot is that the case of films with only one boundary rough is not worth separate implementation.

6.2.6 Thick Layers, Mixing and Incoherent Models

Thin films are not the only type of sample that can be characterised using ISR. In the case of thick layers, for which interference is no longer observed (or becomes negligible), thickness is no longer a meaningful quantity to fit, but it is still possible to observe spatial variation of layer material.

For modelling, we must switch to the incoherent formalism. If the substrate is opaque or back-side reflection can be disregarded, the reflectance of the sample is

$$R = \frac{R_1 + \tilde{R}_2 - 2R_1\tilde{R}_2}{1 - R_1\tilde{R}_2}, \quad \text{where} \quad \tilde{R}_2 = T^2 R_2 . \tag{6.55}$$

In this formula R_1 and R_2 are the reflectances of the lower and upper layer boundary (squared absolute values of reflection coefficients). The expression (6.55) includes the layer transmittance denoted by T for generality. Nevertheless, usually the layer is either transparent ($T = 1$) and $\tilde{R}_2 = R_2$, or it is opaque ($T = 0$) and $R = R_1$. If it is weakly absorbing just so that $T \neq 0, 1$, its thickness appears again in the formulae, or at least the product of thickness and extinction coefficient.

The fitting parameter that characterises the variation of layer material can be chosen in different ways, the basic ones being:

- Explicit formula parametrising the spectral dependencies, for instance using mass density or atomic fraction of an element in a compound. Such formulae can be obtained by parametrised interpolation between experimental curves, if theoretical dependences are difficult to obtain.
- EMA mixing formula in which the parameter is the fraction p of one material. This model is useful if the layer consists of materials mixed at the microscopic level. Depending on the assumptions about the layer structure and constituent materials, different EMA formulae summarised in Sect. 10.3.1 are used to express the dielectric function of the mixture.
- Incoherent (or spatial) averaging, in which we assume a fraction p of the area is covered by one material and the remaining part $1 - p$ by another material. This model is useful when the constituent materials are not mixed at the microscopic level. The reflectance is then given as $R = pR_\text{a} + (1 - p)R_\text{b}$ where R_a and R_b are calculated using (6.55) for the two materials a and b.

An example of the last type of analysis was shown in Sect. 5.13. We remark it is probably the least computationally demanding model discussed here – and arguably the least demanding imaginable. Not only R_a and R_b can be precalculated and tabulated, but the least-squares problem is linear. Therefore, the area fraction is given by the explicit formula

$$p = \frac{1}{C}\left[B + \sum_{k=1}^{K} \frac{1}{c_k^2} R_k^{\text{exp}} (R_{\text{a},k} - R_{\text{b},k})\right], \tag{6.56}$$

where

$$B = -\sum_{k=1}^{K} \frac{1}{c_k^2} R_{b,k}(R_{a,k} - R_{b,k}) \quad \text{and} \quad C = \sum_{k=1}^{K} \frac{1}{c_k^2}(R_{a,k} - R_{b,k})^2 \tag{6.57}$$

are constants. The EMA models are more involved and require non-linear least-squares fitting, but they are still computationally inexpensive.

6.3 Multi-spectra Processing

The limited information in a single normal-incidence spectrum limits the number and reliability of parameters we can hope to obtain in each pixel. Obtaining some sample properties beforehand by independent measurements is helpful but not always possible or practical. When ISR must stand on its own as a thin film characterisation technique, only one option remains how to reduce parameter correlations and improve reliability in general: smarter data processing.

We can take inspiration from multi-sample analysis frequently used with conventional optical techniques. In a multi-sample analysis the experimental data for several related samples are fitted simultaneously using a single consistent model for the entire data set. A related approach is the simultaneous fitting of data from different techniques and instruments. Both are effective at reduction of parameter correlations and improving sensitivity.

In ISR, spectra from individual pixels can play the role of a set of related measurements with some parameters independent for each pixel (usually at least film thickness), others common to all pixels (some or even all dispersion model parameters). However, it is not necessary to advance immediately to the huge general least-squares problem described in Sect. 6.1. We can take more reasonably sized spectra sets and utilise multi-pixel (or multi-spectra) data processing.

6.3.1 Manual Multi-pixel

Optical data fitting software for conventional techniques, such as newAD2 [53] or RefFIT [54], can handle moderate data sets consisting of tens or even a hundred spectra. Such subset must be somehow selected from the full experimental data set which contains from tens of thousands to millions of them. The easiest method is often manual selection based on visual inspection, taking into account:

- quality – exclude spectra with evident defects and distortions,
- spatial coverage – cover the entire region of interest, and
- parameter coverage – attempt to sample the full range of thicknesses and other parameters that vary within the area.

This reduced set of M spectra is then processed, obtaining values of the common parameters. In the subsequent full ISR data fit these parameters are fixed [9].

This procedure does not intrinsically remove the correlation between parameters such as thickness and refractive index for non-absorbing and slightly absorbing films, which remains high (if considerable absorption structures are present then Kramers–Kronig consistency can decrease the correlation when spectra for different film thicknesses are fitted together). Nevertheless, it has several positive effects. By limiting the number of parameters fitted independently in each pixel it often allows to obtain reasonable results even from rather poor spectra. At the same time the poor spectra do not influence the common parameters – and if there are random small distortions and deviations in the chosen spectra, they can cancel out. And finally, the effect of random noise is suppressed by the factor $1/\sqrt{M}$, which can be significant, in particular when individual spectra consist of relatively few points.

It is useful to iterate the procedure twice, because defects and distortions are easier to judge with a reasonable fit of the data at hand. The second iteration can be also performed automatically, based on spatial and parameter coverage and residual sum of squares from the previous iterations. Running the second iteration several times also helps in assessing stability of the results. An example of such approach was shown in Sect. 5.10.

6.3.2 *Zonal Multi-pixel*

In the multi-pixel procedure we assumed some film properties are constant over an entire film area. If they vary, but slowly, or they vary according to an identifiable pattern the procedure can be refined to include the variation. Identifiable pattern may include linear or radial dependence of film properties on coordinates (the latter frequently occurring in the case of local etching and deposition) or distinct areas covered by different materials.

In all these cases we split the sample area into zones and process the data in each zone using the multi-pixel method. The zones are usually defined using the results of a preliminary fit. Distinct areas that require different models can be identified by fitting all spectra with all possible models and then selecting the model that leads to the smallest residual sum of squares (morphological operations such as opening and closing can be used to regularise the zone boundaries if required). Centres of radial patterns can be found from simple fits of local thickness. It is also possible to specify the zones as areas with film thickness within certain ranges, based on the assumption that any other film property would vary with the same geometry as film thickness [9]. Finally, the sample area can be mechanically split into rectangles to obtain a low-resolution map of a slowly varying film property.

Individual zones cover smaller areas and/or ranges of film properties – that was the point of defining them. This also means smaller reduction of parameter correlation and compensation of errors. Uncertainties of the results are thus correspondingly larger. Still, zonal processing represents a reasonable middle ground between single-pixel processing and imposing uniformity in all film parameters.

6.3.3 Timeline Multi-sample

ISR can be used to study kinetics of deposition and etching processes, even in situ [2, 4, 10]. Although the usual approach is to fit each time frame separately and post-process the results, a set of related measurements for the multi-sample method then naturally arises from the sequence of spectra acquired in the same pixel but at different times $t_1, t_2, \ldots, t_T$.

If the film material does not change during the process, an assumption valid for etching and sometimes even for deposition, the multi-sample model is simple. All spectra correspond to the same film, only its thickness (or possibly roughness) is different for each measurement.

If the deposited material slowly changes, the film at time t_j is commonly modelled as the stack of j sublayers. The lower $j - 1$ sublayers correspond to the film at time t_{j-1} and the uppermost one appears newly at time t_j. The thickness of each sublayer is an independent parameter, but most other properties should be still common to all sublayers, only selected parameters, for instance related to film density or composition, should vary. Alternatively, and often preferably, the time evolution is described by a deposition rate model, roughness growth model or film density evolution model, and only parameters of these models are fitted.

6.4 Global Data Processing

In all the data processing approaches discussed above we avoided dealing with the giant least-squares problem with tens of thousands to millions correlated parameters that was formulated in Sect. 6.1. This allowed the utilisation of standard LSM tools. In this section we avoid it no more.

By fitting simultaneously all spectra from a sample area or zone we seek to eliminate the need for manual selections and other ad hoc choices and to achieve the maximum suppression of random noise influence. In fact, when so much data are fitted together the LSM error estimates for global parameters, which decrease as $1/\sqrt{N}$, become so small they are negligible to systematic errors – and thus no longer useful for estimation of uncertainties [25]. Adding more data beyond this point no longer improves the accuracy of global parameters. Hence, it may not be necessary or even useful to fit all spectra together if the data set is very large; we only need to reach the point when random errors are effectively eliminated.

6.4.1 Alternate Global and Local Fitting

Assume for a moment that the local and global parameters are independent. The problem can then be split into two much more approachable parts. The local part with Hessian (6.9) was already handled in Sect. 6.2.

The global part is not demanding either because there are only G global parameters. Therefore, the Hessian is a $G \times G$ matrix and similarly the parameter and gradient vectors have only G elements. The Jacobian matrix (6.5) is still quite large because it has N rows and N is the total number of data points. Even though LM implementations typically take the Jacobian $\mathbf{J}$ and derive other quantities themselves, it is clear from the formulation (6.1)–(6.8) that this is not necessary and the algorithm can be written in terms of $\mathbf{H}$ and $\boldsymbol{g}$. We can avoid constructing $\mathbf{J}$ explicitly if we choose an LM routine which allows supplying directly

$$H_{\alpha\beta} = \sum_{n=1}^{N} \frac{1}{c_n} \frac{\partial R_n}{\partial p_\alpha} \frac{1}{c_n} \frac{\partial R_n}{\partial p_\beta} \quad \text{and} \quad g_\alpha = \sum_{n=1}^{N} \frac{1}{c_n} \frac{\partial R_n}{\partial p_\alpha} \frac{1}{c_n} (R_n - R_n^{\text{exp}}) \tag{6.58}$$

instead of $\mathbf{J}$. The sums (6.58) can be evaluated term by term and no matrices larger than $G \times G$ then ever appear in the calculation. It should be noted that there is, of course, a reason why implementations in terms of $\mathbf{J}$ are more common. Most are geared towards small least-squares problems in which working with $\mathbf{J}$ is preferable because it allows a smarter formulation of the algorithm [16] – but in terms of matrices and operations are better avoided here (such as QR or singular value decomposition of the Jacobian).

Since both parts are easy and can be realised using standard LM routines, a possible data processing procedure seems obvious: fit alternately local and global parameters until both fits converge. Unfortunately, when the LM algorithm is split into separate updates of distinct parameter subsets its convergence tends to be abysmal [17, 25]. Only if the parameter subsets can be fitted almost independently we can hope for a reasonable convergence of the alternate updates.

Are the global and local parameters almost independent in the typical ISR case? No, if film thickness per pixel while the dispersion model is global, the parameter subsets are far from being independent (as illustrated in Fig. 6.2). It is enlightening to look at how exactly the minimisation of S is stymied. The positions of interference minima and maxima in a reflectance spectrum are given by the optical thickness nh, i.e. the product of real refractive index n and film thickness h (disregarding dispersion for now). When the optical thickness changes, the minima and maxima shift. If the theoretical spectrum already corresponds relatively well to the experimental data, the shift will increase S unless the minima and maxima move just slightly. This restricts the magnitude of changes of thicknesses and dispersion model parameters when they are updated separately. The limitation is more severe for thicker films because the minima and maxima are spaced more closely.

Now it is clear that the solution is to choose the fitting parameters more wisely. If we replace the local film thickness with local optical thickness, the dispersion model fitting step can progress without disturbing the positions of interference extrema. The local thickness fitting step then performs some small adjustments of the optical thicknesses, and the entire procedure can converge quickly. At the end, physical film thicknesses are recovered by dividing the optical thicknesses by the refractive index.

Instead of actually changing the model parametrisation we can instead update film thicknesses during the global dispersion model fitting step. In both cases we must define what exactly refractive index means in presence of dispersion. Simple spectral averaging is sufficient:

$$\langle n \rangle = \frac{1}{K} \operatorname{Re} \sum_{k=1}^{K} \hat{n}(\lambda_k) \,. \tag{6.59}$$

The mean index $\langle n \rangle$ is then used for all conversions between physical and optical thicknesses. The procedure is thus as follows. At the beginning of the global fitting step, calculate the mean refractive index and denote it $\langle n \rangle_0$. Whenever the optical constants change, recalculate the mean refractive index $\langle n \rangle$ and use

$$\tilde{h} = h \frac{\langle n \rangle_0}{\langle n \rangle} \tag{6.60}$$

in place of h. Crucially, this must be also done during the computation of derivatives by dispersion model parameters.

Whichever approach is chosen, the reflectance for wavelength λ_k no longer depends only on the refractive index for this wavelength. Formula (6.59) makes it dependent on all $\hat{n}(\lambda_k)$, changing the calculation of derivatives. The chain rule (6.15) becomes

$$\frac{\partial r}{\partial p_\alpha} = \frac{\partial r}{\partial \hat{n}_1} \frac{\partial \hat{n}_1}{\partial p_\alpha} + \frac{\partial r}{\partial \tilde{h}} \frac{\partial \tilde{h}}{\partial p_\alpha} \,. \tag{6.61}$$

The additional term consists of two factors, of which $\partial r / \partial \tilde{h}$ is given by (6.14) and

$$\frac{\partial \tilde{h}}{\partial p_\alpha} = -h \frac{\langle n \rangle_0}{\langle n \rangle^2} \frac{\partial \langle n \rangle}{\partial p_\alpha} = -h \frac{\langle n \rangle_0}{\langle n \rangle^2} \frac{1}{K} \operatorname{Re} \sum_{k=1}^{K} \frac{\partial n(\lambda_k)}{\partial p_\alpha} \,. \tag{6.62}$$

The averaging (6.59) does not need to be performed repeatedly during the computation of the derivatives $\partial r / \partial p_\alpha$ and the derivative $\partial \langle n \rangle / \partial p_\alpha$ appearing in (6.61) can be evaluated only once prior to the computation. Even if the derivatives are all calculated numerically, factoring out and precalculating the derivatives of $\tilde{h}$ by dispersion model parameters is a crucial step that enables an efficient implementation. With this optimisation, global optical constants and local thicknesses can be fitted together within several minutes [25].

When the film is strongly absorbing in a part of the spectral range the mean refractive index formula (6.59) has to be improved. Only the optical constants in the region of transparency are relevant for the preservation of interference minima and maxima. Therefore, we change the mean refractive index definition as follows [11]:

$$\langle n \rangle = \frac{1}{K} \operatorname{Re} \sum_{k=1}^{K} \hat{n}(\lambda_k) \exp\left(-\frac{4\pi}{\lambda_k} h_{\text{mean}} \operatorname{Im} \hat{n}(\lambda_k) \right) . \tag{6.63}$$

Here h_{mean} is a mean film thickness over the analysed area (the procedure is not sensitive to its value). This approach is also illustrated in Sect. 5.11.

6.4.2 *Sparse Levenberg–Marquardt Algorithm*

The ISR least-squares problem has a specific algebraic structure that is well suited for the sparse LM algorithm [17]. The sparse algorithm is a reformulation of the LM algorithm in which the fitting parameters are split into subsets but it remains equivalent to the algorithm for the whole parameter set. In particular, the parameter update step (6.8) does not consist of mere independent updates of the individual subsets; it includes the cross-terms correctly. If **J** and **H** were general dense matrices the splitting would just make the entire procedure more complex. However, as the name suggests, if the matrices have a specific sparse structure the algorithm allows an efficient factorisation of the computation into operations with reasonably-sized matrices and vectors.

Here we formulate the algorithm specifically for ISR, where the two naturally arising two groups of fitting parameters (global and local) correspond to the splitting that is advantageous for sparse LM. With this division, the parameter and gradient vectors have the block structure

$$\boldsymbol{p} = \begin{pmatrix} \boldsymbol{u} \\ \boldsymbol{v} \end{pmatrix} \quad \text{and} \quad \boldsymbol{g} = \begin{pmatrix} \boldsymbol{a} \\ \boldsymbol{b} \end{pmatrix} , \tag{6.64}$$

where $\boldsymbol{u}$ is the vectors of global parameters

$$\boldsymbol{u}^{\mathrm{T}} = (u_1, u_2, \ldots, u_G) \tag{6.65}$$

and $\boldsymbol{v}$ consists of blocks of L local parameters corresponding to individual pixels:

$$\boldsymbol{v}^{\mathrm{T}} = (v_{1,1}, v_{1,2}, \ldots, v_{1,L}, v_{2,1}, v_{2,2}, \ldots, v_{2,L}, \ldots, v_{M,1}, v_{M,2}, \ldots, v_{M,L}) . \tag{6.66}$$

For instance, if film thickness is the only local parameter then $L = 1$ and

$$\boldsymbol{v}^{\mathrm{T}} = (h_1, h_2, \ldots, h_M) . \tag{6.67}$$

The blocks $\boldsymbol{a}$ and $\boldsymbol{b}$ forming the gradient vector $\boldsymbol{g}$ are analogous to $\boldsymbol{u}$ and $\boldsymbol{v}$. The Hessian has the block structure:

$$\mathbf{H} = \begin{pmatrix} \mathbf{U} & \mathbf{W} \\ \mathbf{W}^{\mathrm{T}} & \mathbf{V} \end{pmatrix} , \tag{6.68}$$

where $\mathbf{U}$ and $\mathbf{W}$ are dense $G \times G$ and $G \times ML$ matrices, but $\mathbf{V}$ is a block-diagonal $ML \times ML$ matrix consisting of M blocks of size $L \times L$ (the same structure as (6.9)):

$$\mathbf{V} = \begin{pmatrix} \mathbf{V}_1 & & & \\ & \mathbf{V}_2 & & \\ & & \ddots & \\ & & & \mathbf{V}_M \end{pmatrix} . \tag{6.69}$$

The reason is that derivatives of R by local parameters are non-zero only if the local parameter belongs to the same pixel as the reflectance value:

$$\frac{\partial R_{m,k}}{\partial v_{m',\alpha}} = \frac{\partial R_{m,k}}{\partial v_{m,\alpha}} \delta_{m,m'} , \tag{6.70}$$

where $\delta_{i,j}$ is the Kronecker delta. We write $R_{m,k}$ instead of just R_n to distinguish pixels m and spectral points k as it became important here. Expressing the matrix elements explicitly,

$$U_{\alpha,\beta} = \sum_{m=1}^{M} \sum_{k=1}^{K} \frac{1}{c_{m,k}^2} \frac{\partial R_{m,k}}{\partial u_\alpha} \frac{\partial R_{m,k}}{\partial u_\beta} \tag{6.71}$$

$$V_{\alpha,m,\beta,m'} = \delta_{m,m'} \sum_{k=1}^{K} \frac{1}{c_{m,k}^2} \frac{\partial R_{m,k}}{\partial v_{m,\alpha}} \frac{\partial R_{m,k}}{\partial v_{m,\beta}} \tag{6.72}$$

$$W_{\alpha,m,\beta} = \sum_{k=1}^{K} \frac{1}{c_{m,k}^2} \frac{\partial R_{m,k}}{\partial u_\alpha} \frac{\partial R_{m,k}}{\partial v_{m,\beta}} \tag{6.73}$$

$$a_\alpha = \sum_{m=1}^{M} \sum_{k=1}^{K} \frac{1}{c_{m,k}^2} \frac{\partial R_{m,k}}{\partial u_\alpha} (R_{m,k} - R_{m,k}^{\mathrm{exp}}) \tag{6.74}$$

$$b_{\alpha,m} = \sum_{k=1}^{K} \frac{1}{c_{m,k}^2} \frac{\partial R_{m,k}}{\partial v_{m,\alpha}} (R_{m,k} - R_{m,k}^{\mathrm{exp}}) , \tag{6.75}$$

it can be verified that the number of independent non-zero elements in $\mathbf{H}$ is $G^2/2 + ML^2/2 + GML \approx ML(G + L/2)$. The sums that express all the matrices and vectors consist of approximately $N(G + L + 1)^2/2$ terms. Therefore, they can be calculated in a reasonable time and the required storage is smaller than for the experimental data values (assuming $L(G+L/2) \leq K$, which holds for sensible data analysis). Note that this is not true for the Jacobian, which has $N(G+L)$ non-zero elements and can require order of magnitude larger storage than the experimental data.

In order to perform the LM parameter update step (6.8) we need to solve the system of linear equations (6.7). The augmented Hessian $\bar{\mathbf{H}}$ differs from $\mathbf{H}$ only in the diagonal elements, so augmented $\bar{\mathbf{U}}$ and $\bar{\mathbf{V}}$ can be introduced by modifying their diagonal elements exactly in the same way. A block matrix inversion formula provides the inverse matrix

$$\bar{\mathbf{H}}^{-1} = \begin{pmatrix} \mathbf{X} & -\mathbf{X}\mathbf{Y} \\ -\mathbf{Y}^{\mathrm{T}}\mathbf{X} & \bar{\mathbf{V}}^{-1} + \mathbf{Y}^{\mathrm{T}}\mathbf{X}\mathbf{Y} \end{pmatrix}, \tag{6.76}$$

where

$$\mathbf{Y} = \mathbf{W}\bar{\mathbf{V}}^{-1} \quad \text{and} \quad \mathbf{X} = (\bar{\mathbf{U}} - \mathbf{Y}\mathbf{W}^{\mathrm{T}})^{-1}\,. \tag{6.77}$$

The matrix $\bar{\mathbf{H}}^{-1}$ is huge and dense. However, the evaluation $\bar{\mathbf{H}}^{-1}\boldsymbol{g}$ can be broken down to a sequence operations, listed in Table 6.3, that do not involve huge dense matrices. In total, the evaluation requires $O\big(MG(G+L)^2 + L^3\big)$ operations.

If we choose this approach, the price for generality, flexibility and efficiency is the necessity to reimplement the LM algorithm using the sparse-matrix operations outlined above. While it is not an insurmountable obstacle, it is an obstacle nonetheless.

Table 6.3 Sequence of matrix operations solving the system of linear equations (6.7) in the sparse LM algorithm

Term	Operation	Dimensions	Operations
$\bar{\mathbf{V}}^{-1}$	M inversions	$L \times L$	ML^3
$\mathbf{Y}$	M multiplications	$G \times L$ by $L \times L$	ML^2G
$\mathbf{Y}\mathbf{W}^{\mathrm{T}}$	Multiplication	$G \times ML$ by $ML \times G$	MLG^2
$\mathbf{X}$	Inversion	$G \times G$	G^3
$\mathbf{X}\boldsymbol{a}$	Multiplication	$G \times G$ by $G \times 1$	G^2
$\mathbf{Y}\boldsymbol{b}$	Multiplication	$G \times ML$ by $ML \times 1$	MLG
$\mathbf{X}(\mathbf{Y}\boldsymbol{b})$	Multiplication	$G \times G$ by $G \times 1$	G^2
$\mathbf{X}\boldsymbol{a}$	Multiplication	$G \times G$ by $G \times 1$	G^2
$\mathbf{Y}^{\mathrm{T}}(\mathbf{X}\boldsymbol{a})$	Multiplication	$ML \times G$ by $G \times 1$	MLG
$\mathbf{Y}^{\mathrm{T}}(\mathbf{X}\mathbf{Y}\boldsymbol{b})$	Multiplication	$ML \times G$ by $ML \times 1$	MLG
$\bar{\mathbf{V}}^{-1}\boldsymbol{b}$	M multiplications	$L \times L$ by $L \times 1$	ML^2

6.4.3 Direct Solution

The progress in solvers for large sparse systems of linear equations in recent years has made feasible a direct attack on the problem (6.7) even in the global data processing case. It is possible to write a generic implementation of the LM algorithm using software libraries for linear algebra with sparse matrices, as was demonstrated in the SparseLM software library [55]. The terminology becomes somewhat confusing here because this approach also has the right to be called 'sparse LM' – and often it is called so. To avoid the confusion we will refer to it using the adjective 'direct', with sparsity silently implied.

The direct LM approach is more flexible as it is not limited to cases in which we know how to split the parameter set and factor the problem to employ efficiently the sparse LM algorithm. This is mostly relevant in other fields where large sparse least-squares problems are encountered, such as computer vision, but it could enable new interesting parametrisations also in ISR data processing. The parameter set splitting in sparse LM remains the most efficient option for least-squares problems with the structure (6.68). Nevertheless, direct large-scale LM routines were shown to be relatively competitive, taking only about twice as much time as the sparse LM algorithm for problems of a similar type [55], albeit somewhat smaller.

6.5 Concluding Remarks

Much of this chapter was devoted to speed and reformulation of models and algorithms for fast data processing. Now it is finally time to look at where we stand with respect to the goals stated in Sect. 6.1. An overview of how long it takes to fit an ISR data set for various types of problems is shown in Table 6.4. Even though we did not succeed completely – having to wait several minutes may not particularly invite interactive experimentation and the data processing times for the most complex models exceed acquisition times noticeably – overall, the results are satisfactory.

Table 6.4 ISR data processing times for various types of problems discussed in this chapter, assuming typical data set sizes and personal computer speeds (in 2016)

Data analysis type	Typical computation time
Area fraction	A couple of seconds
Thickness	Tens of seconds
Thickness + manual $\hat{n}(\lambda)$ fit	A minute
Thickness + roughness	Minutes
Thickness + local $\hat{n}_m(\lambda)$	Ten minutes
Thickness + global $\hat{n}(\lambda)$ fit	Tens of minutes
Thickness + roughness + global $\hat{n}(\lambda)$ fit	Hours

Finally, we note that the detailed parameter maps obtained by ISR data fitting are invaluable for overview, illustration and visual inspection. However, they still consist of hundreds of thousands data values. Quantitative analysis thus often requires another data reduction step in which the maps are postprocessed and various overall characteristics (for instance dimensional or statistical) are extracted.

The postprocessing methods and algorithms differ from those used in optical spectra fitting and mostly belong to the category of image or height field processing. Although any image processing software can be utilised in principle, most image analysis programmes are geared towards images in the sense of pictures representing recorded light intensity, not maps of physical quantities in absolute units.

A set of related 'images' of different physical quantities is, however, precisely the type of data that is standard in atomic force microscopy (AFM). The basic quantity imaged in AFM, surface height, is also analogous to the basic quantity imaged in ISR, film thickness. Many tools developed for AFM data analysis are, therefore, useful for ISR data postprocessing. Writing ISR maps in a format understood by AFM software such as Gwyddion [56] enables for instance:

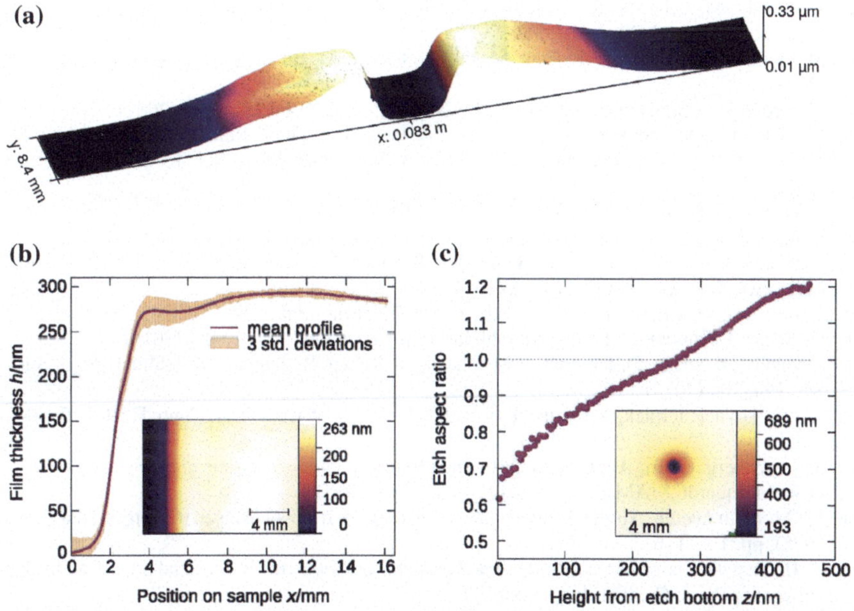

Fig. 6.5 Examples of ISR data postprocessing: **a** large-area map of thickness of a plasma polymer film deposited in dielectric barrier discharge, obtained by merging the results of several measurements; **b** characterisation of uniformity of a similar plasma polymer film near its edge, showing the thickness dependence along the principal axis of the setup and in the perpendicular direction; **c** analysis of the anisotropy of an etch created in an amorphous carbon thin film using an Ar+O_2+H_2O plasma jet. Insets in parts **b** and **c** show the corresponding film thickness maps

- visualisation and data presentation,
- geometrical transforms,
- merging multiple parameter maps together (Fig. 6.5a),
- removal of slowly varying 'background' film thickness for investigation of specific features, i.e. levelling,
- dimensional measurements,
- combination of maps of different quantities using arithmetic operations,
- summarisation and statistical characterisation (Fig. 6.5b [13]),
- detection and correction of outliers, or
- complex measurements based on the detection of 'grains' and other features (Fig. 6.5c).

References

1. M. Ohlídal, I. Ohlídal, D. Franta, T. Králík, M. Jákl, M. Eliáš, Surf. Interface Anal. **34**(1), 660 (2002)
2. J. Spousta, M. Urbánek, R. Chmelík, J. Jiruše, J. Zlámal, K. Navrátil, A. Nebojsa, T. Šikola, Surf. Interface Anal. **33**, 664 (2002)
3. M. Ohlídal, I. Ohlídal, P. Klapetek, M. Jákl, V. Čudek, M. Eliáš, Jap. J. Appl. Phys. **42**, 4760 (2003)
4. M. Urbánek, J. Spousta, T. Běhounek, T. Šikola, Appl. Opt. **46**(25), 6309 (2007)
5. M. Ohlídal, I. Ohlídal, P. Klapetek, D. Nečas, V. Buršíková, Diam. Relat. Mater. **18**, 384 (2009)
6. M. Ohlídal, I. Ohlídal, P. Klapetek, D. Nečas, A. Majumdar, Meas. Sci. Technol. **22**, 085104 (2011)
7. D. Nečas, I. Ohlídal, D. Franta, J. Opt. **13**, 085705 (2011)
8. K. Kim, S. Kim, S. Kwon, H.J. Pahk, Int. J. Precis. Eng. Man. **15**(9), 1817 (2014)
9. D. Nečas, V. Čudek, J. Vodák, M. Ohlídal, P. Klapetek, J. Benedikt, K. Rügner, L. Zajíčková, Meas. Sci. Technol. **25**(11), 115201 (2014)
10. Z. Benzarti, M. Khelifi, I. Halidou, B.E. Jani, J. Electron. Mater. **44**(10), 3243 (2015)
11. D. Nečas, I. Ohlídal, D. Franta, M. Ohlídal, J. Vodák, J. Opt. **18**, 015401 (2016)
12. J. Vodák, D. Nečas, D. Pavliňák, J.M. Macák, T. Řičica, R. Jambor, M. Ohlídal, Appl. Surf. Sci. **396**, 284 (2017)
13. L. Zajíčková, P. Jelínek, A. Obrusník, J. Vodák, D. Nečas, Plasma Phys. Contr. F. **59**(3), 034003 (2017)
14. K. Levenberg, Quart. Appl. Math. **2**(2), 164 (1944)
15. D.W. Marquardt, SIAM, J. Appl. Math. **11**(2), 431 (1963)
16. J.J. More, in *Lecture Notes in Mathematics*, vol. 630, ed. by G.A. Watson (Springer, New York, 1978), pp. 106–116
17. R. Hartley, A. Zisserman, *Multiple View Geometry in Computer Vision*, 2nd edn. (Cambridge University Press, Cambdridge, 2000)
18. R.J. Hanson, F.T. Krogh, ACM T. Math. Softw. **18**(2), 115 (1992)
19. N. Bornemann, E. Dörsam, Opt. Express **21**(19), 21897 (2013)
20. H.B. Nielsen, Damping parameter in Marquardt's method. Technical Report IMM-REP-1999-05, IMM Department of Mathematical Modeling, Technical University of Denmark (1999)
21. L. Asinovski, D. Beaglehole, M.T. Clarkson, Phys. Status Solidi A **205**(4), 764 (2008)
22. B. Forster, D. van de Ville, J. Berent, D. Sage, M. Unser, Microsc. Res. Techniq. **65**(1–2), 33 (2004)
23. T. Scheimpflug, Improved method and apparatus for the systematic alteration or distortion of plane pictures and images by means of lenses and mirrors for photography and for other purposes (1904). GB Patent No. 1196

24. M. Galassi et al., Gnu scientific library reference manual (3rd ed.), http://www.gnu.org/software/gsl/
25. D. Nečas, J. Vodák, I. Ohlídal, M. Ohlídal, A. Majumdard, L. Zajíčková, Appl. Surf. Sci. **350**, 149 (2015)
26. J. Cao, K.A. Novstrup, A. Goyal, S.P. Midkiff, J.M. Caruthers, in *ICS'09 Proceedings of the 23rd international conference on Supercomputing* (2009), pp. 450–459
27. D. Nečas, I. Ohlídal, D. Franta, V. Čudek, M. Ohlídal, J. Vodák, L. Sládková, L. Zajíčková, M. Eliáš, F. Vižd'a, Thin Solid Films **571**, 573 (2014)
28. D. Franta, D. Nečas, L. Zajíčková, Opt. Express **15**, 16230 (2007)
29. D. Franta, D. Nečas, I. Ohlídal, Appl. Opt. **54**, 9108 (2015)
30. G.E. Jellison, F.A. Modine, Appl. Phys. Lett. **69**, 371 (1996)
31. M. Ohlídal, I. Ohlídal, P. Klapetek, M. Jákl, V. Čudek, M. Eliáš, Jpn. J. Appl. Phys. **42**, 4760 (2003)
32. D. Nečas, I. Ohlídal, D. Franta, J. Opt. A-Pure Appl. Opt. **11**, 045202 (2009)
33. D. Zwilinger, *Handbook of Integration* (Jones and Bartlett Publishers, London, 1992)
34. M. Abramowitz, I.A. Stegun, *Handbook of Mathematical Functions with Formulas, Graphs, and Mathematical Tables* (National Bureau of Standards, Washington, 1964)
35. J. Szczyrbowski, J. Phys. D Appl. Phys. **11**, 583 (1978)
36. E. Márquez, J.M. González-Leal, R. Jiménez-Garay, S.R. Lukic, D.M. Petrovic, J. Phys. D Appl. Phys. **30**, 690 (1997)
37. R.A. Maronna, D.R. Martin, V.J. Yohai, *Robust Statistics: Theory and Methods*. Wiley Series in Probability and Statistics (Wiley, New York, 2006)
38. J. Vodák, D. Nečas, M. Ohlídal, I. Ohlídal, Meas. Sci. Technol. **28**, 025205 (2017)
39. V.A. Kotenev, D.N. Tyurin, A.Y. Tsivadze, Inorg. Mater+ **45**(14), 1622 (2009)
40. R.N. Bracewell, *The Fourier Transform & Its Applications* (McGraw-Hill, Singapore, 1999)
41. W.A. Pliskin, R.P. Esch, J. Appl. Phys. **39**, 3274 (1968)
42. D.E. Aspnes, J.B. Theeten, F. Hottier, Phys. Rev. B **20**, 3292 (1979)
43. H.E. Bennett, J. Opt. Soc. Am. **53**, 1389 (1963)
44. I. Ohlídal, K. Navrátil, F. Lukeš, J. Opt. Soc. Am. **61**, 1630 (1971)
45. C.K. Carniglia, Opt. Eng. **18**, 104 (1979)
46. D. Nečas, I. Ohlídal, Opt. Express **22**, 4499 (2014)
47. J.W.S. Rayleigh, *Theory of Sound*, vol. 2 (Macmillan and co., London, 1877)
48. S.O. Rice, Commun. Pure Appl. Math. **4**, 351 (1951)
49. D. Franta, I. Ohlídal, J. Mod. Opt. **45**, 903 (1998)
50. D. Franta, I. Ohlídal, Opt. Commun. **248**, 459 (2005)
51. D. Nečas, I. Ohlídal, D. Franta, M. Ohlídal, V. Čudek, J. Vodák, Appl. Opt. **53**, 5606 (2014)
52. I. Ohlídal, K. Navrátil, F. Lukeš, Opt. Commun. **3**, 40 (1971)
53. D. Franta, D. Nečas, newAD2, http://newad.physics.muni.cz/
54. A. Kuzmenko, RefFIT, https://sites.google.com/site/reffitprogram/home
55. M.I. Lourakis, in *European Conference on Computer Vision*, vol. 2 (2010), pp. 43–56
56. D. Nečas, P. Klapetek, Cent. Eur. J. Phys. **10**(1), 181 (2012)

Chapter 7
In Situ and Ex Situ Spectrophotometric Characterization of Single- and Multilayer-Coatings I: Basics

Olaf Stenzel and Steffen Wilbrandt

Abstract Optical spectrophotometry provides a powerful tool for the characterization of modern coatings, no matter whether they are manufactured for optical or non-optical applications. Spectrophotometry of coatings gives primary access to optical constants and their dispersion as well as to the film thickness. In a second step, the application of sophisticated Kramers–Kronig-consistent dispersion models gives further access to related quantities, including density, porosity, but also charge carrier density, crystalline structure, band structure and possible impurities of the coating. We will present and discuss the state of the art in spectrophotometry of single and multilayer coatings, including their in situ as well as ex situ versions. In situ spectrophotometry allows re-engineering as well as monitoring the deposition process of a growing coating, resulting in excellent specification adherence particularly in the field of optical coatings.

7.1 Introduction

The present chapter deals with the application of spectrophotometry for characterization of thin (solid) films. The general idea of spectrophotometric characterization

---

O. Stenzel (✉) · S. Wilbrandt
Fraunhofer Institute for Applied Optics and Precision Engineering IOF, Albert-Einstein-Str. 7, 07745 Jena, Germany
e-mail: olaf.stenzel@iof.fraunhofer.de; optikbuch@optimon.de

S. Wilbrandt
e-mail: steffen.wilbrandt@iof.fraunhofer.de

O. Stenzel
Abbe School of Photonics, Friedrich-Schiller-University Jena, Albert-Einstein-Straße 6, 07745 Jena, Germany

O. Stenzel and M. Ohlídal (eds.), *Optical Characterization of Thin Solid Films*, Springer Series in Surface Sciences 64,
https://doi.org/10.1007/978-3-319-75325-6_7

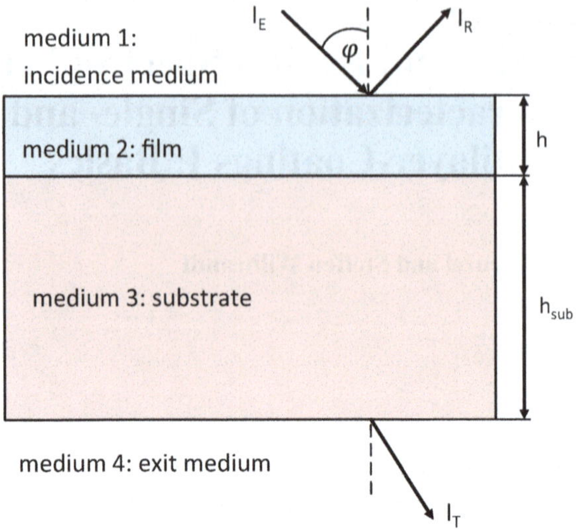

Fig. 7.1 Thin film on a thick substrate, irradiated by light under an incidence angle φ. For details see text

is to bring a thin film sample into interaction with electromagnetic radiation. As the result, certain parameters of the electromagnetic radiation will be modified. In spectrophotometry, the focus is on changes in the *intensity* of the light, which is measured and further used to judge on specific sample properties.

This general situation is visualized in Fig. 7.1.

The intensity I of the light is defined as the amount of light energy penetrating a unit surface area per unit time interval. The *transmittance* T and *reflectance* R of the light are defined through the directed transmitted (I_T) or specularly reflected (I_R) light intensities, divided by the intensity of the incident light (I_E):

$$T \equiv \frac{I_T}{I_E}$$
$$R \equiv \frac{I_R}{I_E} \tag{7.1}$$

As soon as the thin film (system) has been prepared on a transparent substrate, the spectrally resolved measurement of T and R (at any chosen incidence angle φ and any required polarization state of the incident light) appears as a widely used straightforward characterization tool. Alternatively, spectrally resolved ellipsometric measurements become more and more frequently used in coating characterization practice (compare Chap. 9).

The measurement of both T and R under identical conditions provides information on the optical loss L, which is composed from total scatter TS and absorptance A. As a result of energy conservation we have:

$$1 - T - R = L = TS + A \tag{7.2}$$

7.2 Theory

7.2.1 Basics

As it is evident from Fig. 7.1, a light wave which has penetrated a thin film sample, will carry information about the materials which form the sample (i.e. about both film and substrate material constants), as well as about its geometry (here the thicknesses of film h and substrate h_{sub}). Generally, the same will be true for the reflected wave, because all interfaces can principally contribute to the reflectance spectrum. So that we have to expect, that both T and R will be rather complicated functions of all the mentioned construction parameters. Thus, measured T and R spectra can be used to gain information about material properties as well as the sample geometry.

In the model case of optically homogeneous, isotropic, and non-magnetic media, the linear optical *material* properties may be expressed in terms of a scalar frequency-dependent complex dielectric function $\varepsilon(\omega)$ with ω - angular frequency of the electromagnetic radiation [1, 2]. ε is related to the optical constants n and k through the relationship:

$$n(\omega) + ik(\omega)=\sqrt{\varepsilon(\omega)} \equiv \hat{n}(\omega) \tag{7.3}$$

Here $\hat{n}$ is the *complex index of refraction*; its frequency-dependence is called *dispersion*. The absorption coefficient α is defined as:

$$\alpha(\omega) = 2\frac{\omega}{c}k(\omega) \tag{7.4}$$

Let us also mention that a positive imaginary part of the dielectric function results in energy dissipation within a medium. Whenever the dielectric function is purely real, no energy will be dissipated [3].

For characterization purposes, thin films are usually deposited on a much thicker substrate with smooth and parallel surfaces. Therefore, it makes sense to discuss the simplest case of the optical properties of an uncoated substrate first. So we start our discussion from a simplified system like it is shown in Fig. 7.2.

It is rather straightforward to write down the equations for T and R of a bare thick substrate. In accordance to Figs. 7.1 and 7.2, let the incidence medium be numbered as medium 1, while the substrate defines medium 3 ($\hat{n}_3 = \hat{n}_{sub}$ – see Fig. 7.1). Let us further assume, that the incidence (medium 1) and exit media (medium 4) are identical ($n_4 = n_1$). This results in [2]:

Fig. 7.2 Uncoated substrate

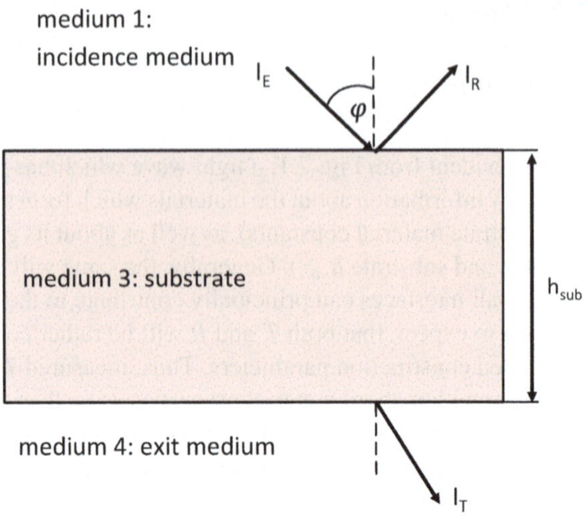

$$
\begin{aligned}
T_{calc} &= \frac{|t_{13}|^2\,|t_{31}|^2\,\mathrm{e}^{-4\pi\nu h_{sub} Im\sqrt{\hat{n}_{sub}^2 - n_1^2 \sin^2\varphi}}}{1-|r_{31}|^4\,\mathrm{e}^{-8\pi\nu h_{sub} Im\sqrt{\hat{n}_{sub}^2 - n_1^2 \sin^2\varphi}}} \\
R_{calc} &= |r_{13}|^2 + \frac{|t_{13}|^2\,|r_{31}|^2\,|t_{31}|^2\,\mathrm{e}^{-8\pi\nu h_{sub} Im\sqrt{\hat{n}_{sub}^2 - n_1^2 \sin^2\varphi}}}{1-|r_{31}|^4\,\mathrm{e}^{-8\pi\nu h_{sub} Im\sqrt{\hat{n}_{sub}^2 - n_1^2 \sin^2\varphi}}} \\
\nu &\equiv \frac{\omega}{2\pi c} = \lambda^{-1}
\end{aligned} \tag{7.5}
$$

Here, symbols of the type t_{ij} and r_{ij} represent nothing else than the Fresnel coefficients for the transmitted and reflected electric field strength at the interface between the ith and jth media, respectively [1, 2]. ω is the angular frequency. Note that at oblique incidence, the Fresnel coefficients are sensitive to the polarization state of the incident light.

Equation (7.5) allow calculating transmittance and reflectance of an uncoated substrate (i.e. a thick slab as shown in Fig. 11.3), taking all internal multiple reflections into account, as well as possible absorption and any effects arising from oblique incidence. Note that (7.5) are obtained when assuming incoherent superposition of all multiple internally reflected wave trains. It cannot be applied to the analysis of thin films, because the latter are usually thin enough to guarantee coherent superposition of internally reflected wave trains.

In the case of normal incidence, (7.5) can be written as:

$$
\begin{aligned}
T_{calc}|_{\varphi=0} &= \frac{(1-R_{13})^2\,\mathrm{e}^{-\alpha_{sub} h_{sub}}}{1-R_{13}^2\mathrm{e}^{-2\alpha_{sub} h_{sub}}} \\
R\ _{calc}|_{\varphi=0} &= \frac{R_{13}\left[1-\mathrm{e}^{-2\alpha_{sub} h_{sub}}\,(2R_{13}-1)\right]}{1-R_{13}^2\mathrm{e}^{-2\alpha_{sub} h_{sub}}}
\end{aligned} \tag{7.6}
$$

Table 7.1 Overview on substrate materials, often used for film characterization purposes

Material	Approximate wavelength range of transparency (nm)	Refractive index
Crystalline Germanium Ge	>2000	$n_{sub} \approx 4.0$ (infrared)
Crystalline Silicon Si	>1000	$n_{sub} \approx 3.45$ (infrared)
BK7, B270	350–4500	$n_{sub} \approx 1.52$ (visible)
Fused silica SiO_2	200–4500	$n_{sub} \approx 1.45$ (visible)
Crystalline Calcium fluoride CaF_2	130–12000	$n_{sub} \approx 1.43$ (visible)
Crystalline Magnesium fluoride MgF_2	115−7500	$n_{sub} \approx 1.38$ (visible)

Here R_{13} denotes the normal incidence intensity reflectance of a single substrate interface:

$$R_{13} = R_{31} = |r_{13}|^2 \tag{7.7}$$

When damping is absent, or even at moderate damping levels, both transmittance and reflectance may be measured and subsequently used for substrate optical characterization. For strong damping, from (7.5) we obtain:

$$\alpha_{sub} h_{sub} \to \infty : \; T_{calc} \to 0; R_{calc} \to R_{13} \tag{7.8}$$

In this case, substrate transmission is completely suppressed, while we still have a reflection signal, originating from the first substrate surface. The latter still contains all information about the substrate optical constants and may therefore be used for substrate optical characterization as well. Nevertheless, in thin film spectrophotometric characterization, it is most convenient to make use of at least semi-transparent substrates, in order to have both transmission and reflection signals available. Often used substrate materials are summarized in Table 7.1.

7.2.2 *Elaboration of Film Thickness and Optical Constants from Single Thin Film Spectra*

7.2.2.1 Basic Equations for Transmittance and Reflectance of a Single Thin Film on a Thick Substrate

It is now rather straightforward to write down the equations for T and R of a single thin film on a thick substrate. Let the film by composed from medium 2, and the substrate from medium 3 ($\hat{n}_3 = \hat{n}_{sub}$ – see Fig. 7.1). Assuming $n_4 = n_1$, in analogy to Sect. 7.2.1, we can write [2]:

$$T_{calc} = \frac{|t_{123}|^2 |t_{31}|^2 e^{-4\pi \nu h_{sub} Im\sqrt{\hat{n}_{sub}^2 - n_1^2 \sin^2\varphi}}}{1 - |r_{321}|^2 |r_{31}|^2 e^{-8\pi \nu h_{sub} Im\sqrt{\hat{n}_{sub}^2 - n_1^2 \sin^2\varphi}}}$$

$$R_{calc} = |r_{123}|^2 + \frac{|t_{123}|^2 |r_{31}|^2 |t_{321}|^2 e^{-8\pi \nu h_{sub} Im\sqrt{\hat{n}_{sub}^2 - n_1^2 \sin^2\varphi}}}{1 - |r_{321}|^2 |r_{31}|^2 e^{-8\pi \nu h_{sub} Im\sqrt{\hat{n}_{sub}^2 - n_1^2 \sin^2\varphi}}} \quad (7.9)$$

Moreover, we have [1, 2]:

$$t_{ijk} = \frac{t_{ij} t_{jk} e^{i\delta}}{1 + r_{ij} r_{jk} e^{2i\delta}}$$

$$r_{ijk} = \frac{r_{ij} + r_{jk} e^{2i\delta}}{1 + r_{ij} r_{jk} e^{2i\delta}} \quad (7.10)$$

Thereby, the superposition of internally reflected light portions within the film is assumed to be completely coherent. The possibly complex phase δ is essential for the description of the thin film interference pattern, it is given by [2]:

$$\delta = \frac{\omega}{c} h \sqrt{\hat{n}_2^2 - n_1^2 \sin^2\varphi} = 2\pi \nu\, h \sqrt{\hat{n}_2^2 - n_1^2 \sin^2\varphi} \quad (7.11)$$

Note that for $h \to 0$, T and R approach the corresponding values of the uncoated substrate, hence the spectrophotometric characterization of ultrathin films ($h << \lambda$) is much more complicated than in the case $h \approx \lambda$ (see next section). In such cases, a spectroellipsometric characterization or even a combination of both approaches may be clearly of use.

7.2.2.2 Information from the Interference Pattern Observed from Dielectric Films

In the case of dielectric or even semiconducting thin films, the couple of (7.9)–(7.11) describes a type of spectra as shown in Fig. 7.3. This figure shows measured spectra of a 211 nm thick zirconium dioxide (zirconia) single film on a fused silica substrate with a thickness of 1 mm. For comparison, the corresponding spectra of the bare (uncoated) substrate are also shown as dashed lines. This is a rather typical thin film spectrum, and it is worth mentioning some of its specific features:

The spectrum may generally be subdivided into several sections, according to the value of the optical loss L as defined in (7.2).

- In the wavenumber region between 10000 cm^{-1} and approximately 35000 cm^{-1}, the spectrum appears almost free of optical losses, because T and R sum up to 1 in terms of the spectrophotometric measurement accuracy (compare later Sect. 8.1). Hence, the dielectric functions of both film and substrate materials are practically real.

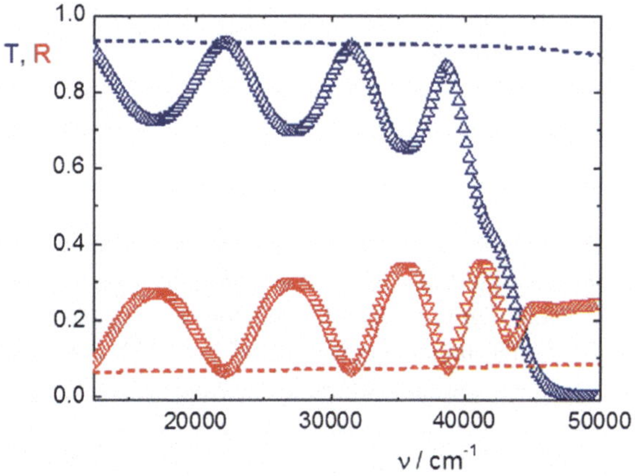

Fig. 7.3 symbols: normal incidence T- and R-spectra of a zirconium oxide thin film on a thick fused silica substrate in the NIR/VIS/UV spectral regions; dashed lines: T and R of the bare substrate

- In such spectral regions, dielectric or semiconductor films of suitable thickness usually show a pronounced interference pattern, which may be identified as a series of subsequent maxima and minima in T and R, observed at discrete wavenumbers ν_j. Certain extrema appear tangential to the bare substrate spectrum; they define what we will call the halfwave (HW) points of the spectrum. The other extrema define quarterwave (QW) points. For normal incidence, the wavenumbers of the extrema are defined by:

$$\text{QW}-\text{points:}\quad n_2h = j\frac{\lambda_j}{4};\, j = 1, 3, 5\ldots$$
$$\text{HW}-\text{points:}\quad n_2h = j\frac{\lambda_j}{4};\, j = 0, 2, 4, \ldots$$
$$\nu_j \equiv \frac{1}{\lambda_j} \tag{7.12}$$

- In the case that the QW transmittance appears to be higher than that of the bare substrate (the QW reflectance lower than that of the substrate), the refractive index of the film will be in-between those of the substrate and the ambient. In the practically relevant case, that the ambient medium is air, and the substrate index $n_{sub} > 1$, we can conclude that the film index is certainly lower than that of the substrate: $n_{sub} > n > 1$ (low index coating).
- In the opposite case (as it is relevant for the example shown in Fig. 7.3), the film refractive index is outside the interval spanned by the substrate and ambient indices. In the practically relevant case, that the ambient medium is air, and the substrate index $n_{sub} > 1$, we can conclude that the film index is certainly higher than that of the substrate (high index coating).
- The dependence of the QW transmittance/reflectance on the refractive indices n and n_{sub} offers the principle possibility for determining the film refractive index by inverting (7.9)–(7.11), independently from knowledge of the film thickness.

On the contrary, when neglecting dispersion, the film thickness may be subsequently estimated from (7.13):

$$h = \frac{1}{4n_2\left(\nu_{j+1} - \nu_j\right)} \tag{7.13}$$

- This type of approach may be extended to the analysis of weakly absorbing films, and is in the basis of the so-called *envelope methods* for film characterization [4, 5]. Note that here knowledge about n_{sub} is usually presumed.
- At oblique incidence, according to (7.11), the interference pattern shifts towards higher wavenumbers (smaller wavelengths). This so-called angular shift offers an alternative method for estimating the film refractive index. Let us assume, that at an angle of incidence φ_a, an interference extremum of arbitrary order j is observed at the wavelength λ_a. At another angle φ_b, the same interference extremum will have shifted to the wavelength λ_b. When neglecting dispersion, from (7.11) we find:

$$n_2 = n_1\sqrt{\frac{\lambda_b^2 \sin^2\varphi_a - \lambda_a^2 \sin^2\varphi_b}{\lambda_b^2 - \lambda_a^2}} \tag{7.14}$$

Note that this approach does not presume knowledge about n_{sub}.

The mentioned spectral characteristics may be used for a first "quick-and-dirty" estimation of refractive index and thickness of the coating in spectral regions with negligible damping. For wavenumbers higher than approximately 35000 cm^{-1}, the optical loss according to the spectra shown in Fig. 7.3 appears to be no more negligible. In such spectral regions, the above type of discussion is no more applicable in the strong sense.

Consequently, in the special case that normal incidence T and R spectra of both the uncoated substrate (Fig. 7.2) and a film-on-substrate system (Fig. 7.1) are available, a simple straightforward optical dielectric thin film characterization may be performed adhering to the following recipe:

(i) First measure T and R of the bare substrate at normal incidence, as well as the substrate thickness h_{sub}. Then, the optical constants of the substrate may be calculated inverting (7.6) [6]. In the case that the substrate is completely intransparent, the substrate optical constants can still be deducted from the reflectance of the substrate surface (see later Sect. 8.2.2).

(ii) Measure T and R of the film-on-substrate system and calculate L according to (11.2). Identify spectral regions where L is negligible (transparency regions).

(iii) In transparency regions, identify HW and QW points in the interference pattern. From QW points, make clear whether you deal with a high index or a low index coating.

(iv) In the case that in the HW points, the T/R-spectra are tangential to those of the substrate, the film may be tackled as a homogeneous film. In this case, calculate

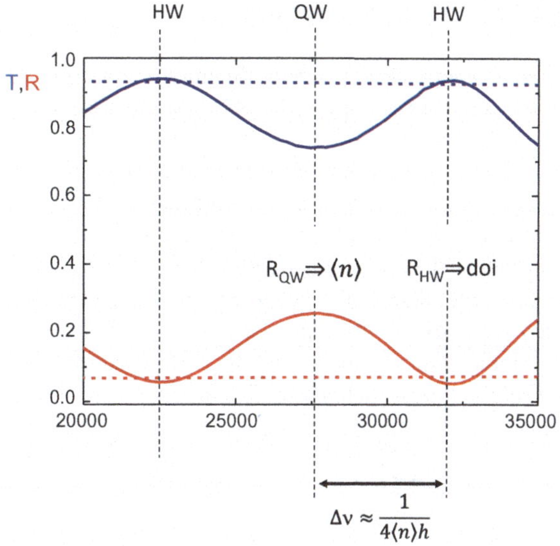

Fig. 7.4 T- and R-spectra of a gradient index film. Full lines show measured spectra of a zirconia sample with a negative refractive index gradient (n_2 decreases with growing distance from the film-substrate interface), dashed lines correspond to the spectra of a bare substrate

the refractive index from the QW points (inverting (7.9)–(7.11)), using substrate data as determined in point ν_j. Note that this procedure is ambiguous for a low index coating, so that one has to identify the physically meaningful solution from side information obtained otherwise. Estimate the film thickness from (7.13).

(v) In the case that $L \approx 0$ and the T/R-spectra are *not* tangential to the substrate spectra in the HW points, the film is expected to show a refractive index gradient (inhomogeneous film) [7]. These effects are no more covered by (7.9)–(7.11). In this case, the measured T and R-values in the HW points embody important information about the so-called *degree of inhomogeneity* (doi), while the corresponding values in the QW points correspond to an average refractive index $\langle n \rangle$, while averaging is performed over the film thickness. So that HW points give information on the doi, while QW points on the average index. This situation is schematically sketched in Fig. 7.4, which shows measured spectra of an inhomogeneous zirconium oxide film. Note that in this particular case, the origin of the refractive index gradient becomes obvious when comparing with the TEM image shown in Fig. 2.2. It stems from a similar sample and confirms a depth-dependent porosity as well as a depth-dependent degree of crystallinity in a real zirconia film, which has a direct impact on the spectrum.

(vi) Having estimated the film thickness as well as the refractive index in the transparency region, the (averaged over the film thickness) extinction coefficient may be estimated from (7.9)–(7.11) in any spectral region.

7.2.2.3 Curve Fit Procedures

A more precise elaboration of film thickness and optical constants is possible in terms of so-called curve fit procedures [8]. In this case, measured (exp) spectral curves are fitted by theoretical (calc) spectra calculated according to (7.9)–(7.11). The sets of optical constants which result in a sufficiently good fit of experimental spectra form a set of possible mathematical solutions to the characterization problem. Once the solution of such fitting attempts is usually ambiguous, side knowledge on the optical constants behavior (compare later Sect. 7.3.1) as well as on the thickness is highly welcome to identify the physically meaningful solution from the set of solutions of the mathematical fitting procedure.

Mathematically, the fit may be performed by minimizing a discrepancy function DF of the type as defined in (7.15):

$$DF = \sqrt{\frac{1}{N}\sum_{l=1}^{N}\left\{\left[T_{exp}(\nu_l) - T_{calc}\left(\nu_l; \hat{n}(\nu_l); h\right)\right]^2 + \left[R_{exp}(\nu_l) - R_{calc}\left(\nu_l; \hat{n}(\nu_l); h\right)\right]^2\right\}} \tag{7.15}$$

Here, the choice of $\{\nu_l\}$ defines a grid of discrete data points which enter into the discrepancy function (7.15) to be minimized (compare later Sect. 8.1.4). Of course, one can include more than two spectra (even including ellipsometric data) into the expression (7.15). The data obtained earlier from the interference pattern (if available) may serve here as a reliable initial approximation for further minimizing (7.15). Examples of curve fits with different degrees of complexity will be presented in Sect. 8.2.3.

7.2.3 *Multilayer Spectra Evaluation*

In the case of multilayer characterization, we have a series of films stacked on the substrate as shown in Fig. 7.5.

Equation (7.9) nevertheless preserve their general structure, although values like t_{123} and r_{123} have to be replaced by the more general expressions t_{stack} and r_{stack}. Symbols characterizing reverse light propagation direction (t_{321} and r_{321}) have to be replaced by new expressions t'_{stack} and r'_{stack}. The new field transmission and reflection coefficients t_{stack}, r_{stack}, t'_{stack} and r'_{stack} may be calculated in terms of the matrix formalism [1, 2]. They carry information about optical constants and thicknesses of all individual layers which compose the film stack shown in Fig. 7.5. Correspondingly, instead of (7.9), we now have the expressions (7.16):

Fig. 7.5 Multilayer system

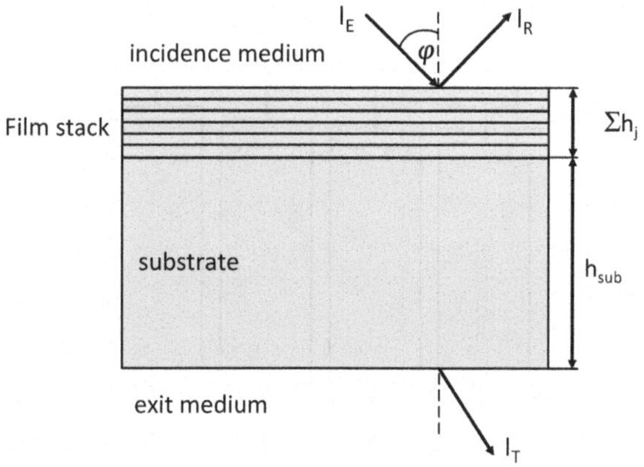

$$T_{calc} = \frac{\left|t_{stack}\left(\left\{\hat{n}_j\right\}, \left\{h_j\right\}\right)\right|^2 |t_{31}|^2 \mathrm{e}^{-4\pi \nu h_{sub} Im\sqrt{\hat{n}_{sub}^2 - n_1^2 \sin^2\varphi}}}{1 - \left|r'_{stack}\left(\left\{\hat{n}_j\right\}, \left\{h_j\right\}\right)\right|^2 |r_{31}|^2 \mathrm{e}^{-8\pi \nu h_{sub} Im\sqrt{\hat{n}_{sub}^2 - n_1^2 \sin^2\varphi}}}$$

$$R_{calc} = \left|r_{stack}\left(\left\{\hat{n}_j\right\}, \left\{h_j\right\}\right)\right|^2 + \frac{\left|t_{stack}\left(\left\{\hat{n}_j\right\}, \left\{h_j\right\}\right)\right|^2 |r_{31}|^2 \left|t'_{stack}\left(\left\{\hat{n}_j\right\}, \left\{h_j\right\}\right)\right|^2 \mathrm{e}^{-8\pi \nu h_{sub} Im\sqrt{\hat{n}_{sub}^2 - n_1^2 \sin^2\varphi}}}{1 - \left|r'_{stack}\left(\left\{\hat{n}_j\right\}, \left\{h_j\right\}\right)\right|^2 |r_{31}|^2 \mathrm{e}^{-8\pi \nu h_{sub} Im\sqrt{\hat{n}_{sub}^2 - n_1^2 \sin^2\varphi}}} \quad (7.16)$$

Here, the individual layers forming the stack are numbered in terms of the subscript j.

In the case that T and R have been measured, a fit of these experimental data may again be performed by minimizing a suitable discrepancy function. Instead of (7.15), one now has to minimize a discrepancy function of the type (7.17):

$$DF = \sqrt{\frac{1}{N}\sum_{l=1}^{N}\left\{\left[T_{exp}(\nu_l) - T_{calc}\left(\nu_l; \left\{\hat{n}_j(\nu_l)\right\}; \left\{h_j\right\}\right)\right]^2 + \left[R_{exp}(\nu_l) - R_{calc}\left(\nu_l; \left\{\hat{n}_j(\nu_l)\right\}; \left\{h_j\right\}\right)\right]^2\right\}} \quad (7.17)$$

When keeping in mind the tremendous number of unknown values $\left\{\hat{n}_j(\nu_l)\right\}$ and $\left\{h_j\right\}$, a reliable re-engineering procedure (i.e. determination of the optical constants and the thicknesses from the measured spectra) appears to be a rather hopeless effort. There are two basic approaches to facilitate the situation:

(a) Reducing the number of parameters to be determined.
(b) Increasing the number of input data, i.e. measured spectra.

(a): In many situations, optical constants of the materials forming the stack are known with sufficient accuracy. In this case, the re-engineering task reduces to the determination of the set of individual layer thicknesses $\left\{h_j\right\}$. This may still be difficult enough, but a further reduction of unknown values may be achieved when making use

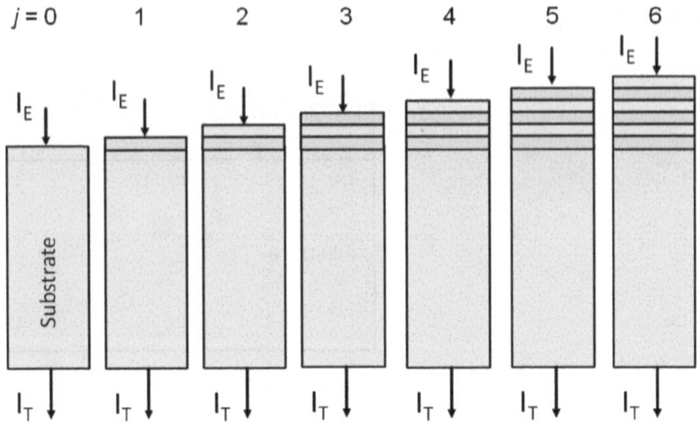

Fig. 7.6 Principle of acquisition of in situ transmission spectra during multilayer coating deposition

of correlations between individual film thicknesses and/or optical constants, defined by systematic deposition errors inherent to the specific thickness monitoring strategy applied during multilayer deposition [9].

(b): On the other hand, the number of measured data entering into (7.17) may be enhanced. T and R measurements may be performed at different angles of incidence, compare later Sect. 3.1. Also, spectroellipsometric data may be included into (7.17) [10].

But the inclusion of more independent ex situ measurement data like oblique incidence spectrophotometry and ellipsometry appears to be both time- and cost consuming because of expansive extra equipment required. An alternative is provided by the use of in situ spectrophotometry (or spectroellipsometry), where numerous experimental data about T and/or R measured on the not yet finished stack are collected directly during the film deposition.

The idea is simple (Fig. 7.6). Let us assume that the film deposition chamber is equipped with a spectrophotometer that allows measuring T and/or R directly during multilayer coating deposition. This will allow, for example, a spectrum recording immediately after each of the individual layers (numbered by j) has been deposited. Typically, a "0th" spectrum is recorded prior to starting deposition, it corresponds to the spectrum of the bare substrate ($j = 0$) and may be used for calibration purposes. Then, the 1st spectrum is recorded after the first layer has been deposited ($j = 1$). It contains information about the optical constants and thickness of the first layer. Then, deposition proceeds with the second layer, again followed by a spectra recording. That second spectrum contains information about optical constants and thicknesses of two layers, and so on. Finally, we will obtain as many spectra as there are individual layers in the stack. This is a tremendous amount of information, and it is obtained from one single spectrometer set-up, which can operate automatically without any additional sample handling. Moreover, when fast spectrometers are used, the amount of extra time necessary for spectra recording during coating deposition is of no relevance.

When fitting in situ spectra one must keep in mind, that optical constants (and even layer thicknesses) in a coating may principally change when the coating is exposed to air and heated up or cooled down to its operation temperature. Therefore, one has to distinguish between in situ and ex situ optical constants. This is particularly relevant for porous coatings, prepared in vacuum conditions by PVD techniques. We will not discuss all the corresponding models here (some simple considerations will be made in Sect. 7.3.3, compare also Sect. 2.3). Instead, we will assume that the in situ relevant optical constants are well-known, and the only task of in situ spectroscopy is the reliable determination of the individual layer thicknesses. This task may be solved by means of a so-called full triangular re-engineering algorithm [11]. In the following we will give a short explanation of this algorithm. The basic principle is visualized in Fig. 7.6. The superscript "^{T}" indicates spectra used in the triangular algorithm.

Let us assume that a first transmission spectrum $T_{meas}^{T\ (1)}(\nu_l)$ is recorded when the first layer has been deposited. This spectrum depends only on the thickness of the first layer h_1. The second spectrum $T_{meas}^{T\ (2)}(\nu_l)$ is recorded after the deposition of the second layer has been completed. It naturally depends on two thickness values h_1 and h_2. This process is repeated until the full multilayer coating deposition has finished. The essence of the full triangular re-engineering algorithm is to determine all thickness values simultaneously by fitting all those transmission spectra by the corresponding theoretical spectra $T_{calc}^{T\ (j)}(\nu_l)$. Thickness calculation is thus achieved by minimizing the *triangular discrepancy function* DF^T [11]:

$$DF^T(h_1, \ldots, h_J) = \left[\frac{1}{JN}\sum_{j=1}^{J}\sum_{l=1}^{N}\left(\frac{T_{meas}^{T\ (j)}(\nu_l) - T_{calc}^{T\ (j)}(\nu_l, h_1, \ldots, h_j)}{\Delta T^T(\nu_l)}\right)^2\right]^{\frac{1}{2}} \tag{7.18}$$

Here J is the number of individual layers deposited at the relevant state of the deposition process. $\Delta T^T(\nu_l)$ is the in situ transmittance measurement error.

So far, in situ spectroscopy combined with the full triangular algorithm is accepted to be the most reliable tool for individual layer thickness determination in complicated optical coatings. For different application examples see references [12–14].

7.3 Further Information Gained from Optical Constants

7.3.1 Basic Classical Dispersion Models and Analytic Properties of the Dielectric Function

As already mentioned, the dielectric function, and consequently the optical constants, appear to be frequency-dependent. This phenomenon is called dispersion. For selected material systems, this frequency dependence may be reliably modelled in

Table 7.2 Optical constants in terms of the oscillator and Drude dispersion models

Bound charge carriers: Lorentzian Single Oscillator model Application: bound electrons in dielectrics and metals	Free charge Carriers: Drude Model Application: Free electrons in metals
$\frac{\varepsilon(\omega)-1}{\varepsilon(\omega)+2} = \frac{\hat{n}^2-1}{\hat{n}^2+2} = \frac{N_{bound}}{3}\frac{q^2}{\varepsilon_0 m}\frac{1}{\omega_0^2-\omega^2-2i\omega\gamma}$	$\varepsilon(\omega) = 1 - \frac{\omega_p^2}{\omega^2+2i\gamma\omega}; \omega_p = \sqrt{\frac{N_{free}q^2}{\varepsilon_0 m}}$
N_{bound}: concentration of bound charge carriers m: mass of bound charge carriers q: their charge ω_0: their resonance frequency γ: damping constant	N_{free}: concentration of free charge carriers m: mass of free charge carriers q: their charge γ: damping constant
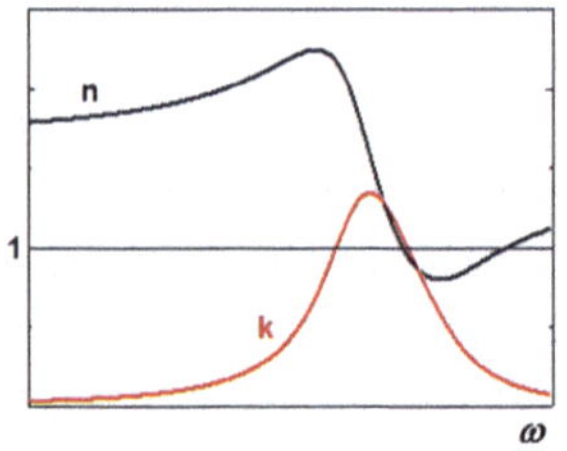	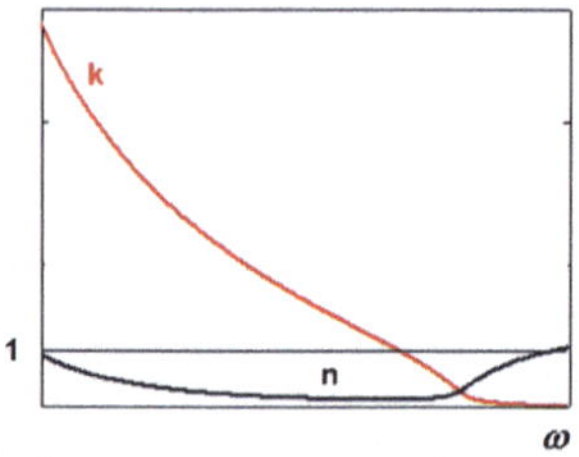

terms of rather compact dispersion models. The Lorentzian oscillator model as well as the Drude model can be regarded as the basic classical dispersion models for the description of optical properties of dielectrics and metals [1, 2, 15]. Their main features are summarized in Table 7.2.

When looking on the picture in the left column of Table 7.2, the most striking feature in the optical response of a system with bound charge carriers is *resonance behavior* of the extinction coefficient at $\omega \to \omega_0$, which results in rather strong damping of the propagating wave. Note that apart from resonance (transparency region), the refractive index increases with increasing frequency, which is called *normal dispersion*. Close to resonance (strong damping), the refractive index decreases with increasing frequency (*anomalous dispersion*).

This type of dispersion is in strong contrast to that described by the Drude model (right column of Table 7.2). Here the refractive index may be significantly smaller than the extinction coefficient as long as the light frequency is well below the plasma frequency. Such a behaviour of the optical constants results in high reflection at the air-material-interface, as it is typically observed at metal surfaces. When describing the optical behavior of metals in terms of the Drude model, parameters like the concentration of free electrons become accessible as well as derived parameters such like dc conductivity σ_{stat} $\left(\sigma_{stat} = \varepsilon_0\omega_p^2/(2\gamma)\right)$ and the damping constants or relaxation times corresponding to the free electrons motion [2].

In rather transparent dielectrics, as a rule, from Table 7.2 we observe that the condition $k \ll n$ is fulfilled. Good metals, i.e. metals where the optical response is dominated by the free electron fraction, show the opposite behavior, namely $k \gg n$.

The obviously correlated behaviour of n and k as visualized in the Table 7.2 can be tackled as the manifestation of a more general relation between the real and imaginary parts of the dielectric function as expressed in terms of the Kramers–Kronig relations [16]:

$$Re\,\varepsilon(\omega) = 1 + \frac{2}{\pi} VP \int_0^{\infty} \frac{Im\,\varepsilon(\xi)\,\xi d\xi}{\xi^2 - \omega^2}$$

$$Im\,\varepsilon(\omega) = -\frac{2\omega}{\pi} VP \int_0^{\infty} \frac{[Re\,\varepsilon(\xi) - 1]}{\xi^2 - \omega^2} d\xi + \frac{\sigma_{stat}}{\varepsilon_0 \omega} \tag{7.19}$$

VP denotes the Cauchy principal value of the integral. (7.19) is in consistency with the relations:

$$Re\,\varepsilon(\omega) = Re\,\varepsilon(-\omega);\, -Im\,\varepsilon(\omega) = Im\,\varepsilon(-\omega) \tag{7.20}$$

It is easily checked, that the dispersion relations given in Table 7.2 are consistent with these fundamental requirements.

Analogous relations my also be formulated for the optical constants [17]. Here we have for insulators as well as conductors:

$$n(\omega) = 1 + \frac{2}{\pi} VP \int_0^{\infty} \frac{k(\xi)\,\xi d\xi}{\xi^2 - \omega^2}$$

$$k(\omega) = -\frac{2\omega}{\pi} VP \int_0^{\infty} \frac{[n(\xi) - 1]}{\xi^2 - \omega^2} d\xi \tag{7.21}$$

Kramers-Kronig-consistency is a strong and useful criterion for the physical relevance of any dispersion law used in coating characterization or design practice.

As a direct consequence from (7.19), the following useful relationships are obtained:

Static dielectric constant of a dielectric:

$$\varepsilon_{stat} = \varepsilon(\omega = 0) = 1 + \frac{2}{\pi} \int_0^{\infty} \frac{Im\,\varepsilon(\omega)}{\omega} d\omega \tag{7.22}$$

The static dielectric constant of a dielectric is thus always larger than 1.

In the *high frequency limit*, contrarily, we find (convergence supposed):

$$Re\,\varepsilon(\omega) = 1 + \frac{2}{\pi} VP \int_0^\infty \frac{Im\,\varepsilon(\xi)\,\xi d\xi}{\xi^2 - \omega^2}\Bigg|_{\omega\to\infty} \to 1 - \frac{2}{\pi\omega^2}\int_0^\infty Im\,\varepsilon(\xi)\,\xi d\xi < 1 \tag{7.23}$$

As a rule, in the extreme ultraviolet (EUV) or soft X-ray spectral ranges, the dielectric function as well as the refractive index are therefore smaller than but close to 1.

An utmost important sum rule interconnects the full concentration of oscillators N (i.e. oscillating electrons on the background of heavy nuclei) with the integral energy dissipation:

$$N = \frac{2\varepsilon_0 m}{\pi q^2}\int_0^\infty Im\,\varepsilon(\omega)\omega d\omega \tag{7.24}$$

Rewriting (7.24) in terms of the optical constants – see (7.3) and (7.4), one immediately obtains:

$$N = \frac{2\varepsilon_0 mc}{\pi q^2}\int_0^\infty n(\omega)\alpha(\omega)\,d\omega \tag{7.25}$$

Here we arrive at the fundament of any quantitative spectrophotometric analysis, where the concentration of any kind of absorption centres (molecules, impurities, and so on) is obtained from the integral over the measured absorbance. Of course, in practice, the integration in (7.25) may be performed only over a finite frequency interval accessible to the measurement.

7.3.2 Often Used Other Dispersion Models

Clearly, the mentioned basic models correspond to rather idealized situations, and in characterization practice, more complicated dispersion models have to be applied. In our treatment, we will restrict on models relevant for description of optical properties from the middle infrared up to the ultraviolet spectral regions. This way we include models relevant for infrared analytics (atomic nuclei vibrations), rather transparent materials for interference coating applications, modelling the absorption edge(s) in (selectively) absorbing materials for light blocking, optoelectronics, and solar energy conversion, as well as metals for reflector optics or light blocking purposes. A schematic overview on optical constants and single film optical behavior of typical dielectric, semiconductor and transparent conductive oxide (TCO) materials from the infrared up to the ultraviolet is presented in Fig. 7.7.

As it is shown in Fig. 7.7, many dielectric or semiconducting materials offer a broad transparency range, which usually extends from the near infrared to the

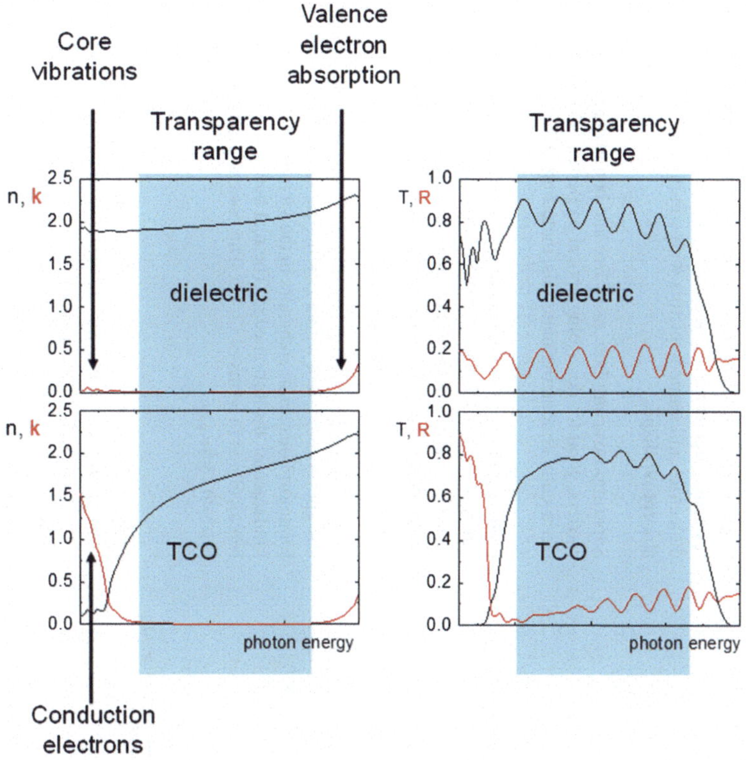

Fig. 7.7 Transparency range in dielectric/undoped semiconducting (on top) or transparent conductive oxide (TCO) materials (on bottom). The photon energy is given by $\hbar\omega$

visible or even the ultraviolet spectral ranges. At the short wavelength (high photon energy) side, transparency is limited by the onset of valence electron excitations, which marks the energy position of the fundamental absorption edge. At the long wavelength (low photon energy) side, it is limited by optical excitation of atomic nuclei vibrations, which occur in the middle infrared spectral range. However, in TCO materials, optical excitation of free electron movement may result in additional transparency range shrinking at low photon energies.

Some often used dispersion models useful for describing properties of dielectrics, metals, and semiconductors from the infrared to the ultraviolet spectral ranges are summarized in Table 7.3.

Table 7.3 Selected dispersion models. Symbols according to [15], f denotes the oscillator strength

Model	Equation	Application/remarks
Lorentzian Multi-oscillator	$\varepsilon(\omega) = 1 + \sum_j \frac{f_j}{\tilde{\omega}_{0j}^2 - \omega^2 - 2i\omega\gamma_j}$	Superposition of Lorentzian oscillators with different resonance frequencies; good for fitting spectra of dielectrics and semiconductors with complicated absorption shapes
Merger of Drude- and multioscillator	$\varepsilon(\omega) = 1 - \frac{\omega_p^2}{\omega^2 + 2i\gamma\omega} + \sum_j \frac{f_j}{\tilde{\omega}_{0j}^2 - \omega^2 - 2i\omega\gamma_j}$	Superimposes contributions of free and bound charge carriers, good for describing optical properties of any electrically conducting materials, including TCOs
Brendel model [18]	$\varepsilon(\omega) = 1 + \frac{1}{\sqrt{2\pi}\sigma} \int_{-\infty}^{\infty} \exp\left[-\frac{(\xi - \bar{\omega}_0)^2}{2\sigma^2}\right] \frac{f}{\xi^2 - \omega^2 - 2i\gamma\omega} d\xi$	Particular case of the multioscillator model; describes inhomogeneously broadened (with respect to a Gaussian statistics) absorption lines. Describes Lorentzian, Gaussian, and Voigt line shapes
Tauc-Lorentz model [19]	$Im\,\varepsilon(\omega) = \frac{const.}{\omega} \frac{\gamma(\omega - \omega_{gap})^2}{(\tilde{\omega}_0^2 - \omega^2)^2 + 4\gamma^2\omega^2}; \ \omega \geq \omega_{gap}$ $Im\,\varepsilon(\omega) = 0; \ \omega < \omega_{gap}$	Merger of the Tauc power law [20] dependence of the absorption coefficient with a Lorentzian. Often used for amorphous semiconductors in the vicinity of the fundamental absorption edge. The real part of ε may be calculated in terms of (7.19). The optical Tauc gap is given by $\hbar\omega_{gap}$
Cody-Lorentz model [21]	$Im\,\varepsilon(\omega) = const. \frac{\omega}{(\omega - \omega_{gap})^2 + \omega_{t2}^2} \frac{\gamma(\omega - \omega_{gap})^2}{(\tilde{\omega}_0^2 - \omega^2)^2 + 4\gamma^2\omega^2}; \ \omega > \omega_{t1}$ $Im\,\varepsilon(\omega) = \frac{\omega_1}{\omega} exp\left[\frac{\omega - \omega_{t1}}{\omega_u}\right]; \quad 0 < \omega \leq \omega_{t1}$	Merger of the Cody power law [21] dependence of the absorption coefficient with a Lorentzian and an Urbach tail [22]. Often used for amorphous semiconductors in the vicinity of the fundamental absorption edge. The real part of ε may be calculated in terms of (7.19). The optical Cody gap is given by $\hbar\omega_{gap}$

7.3.3 Optical Properties of Material Mixtures

Let us now assume a mixture built up from several constituents, numbered by j. Let us further assume, that we know the optical constants (or the dielectric function ε_j) of any of the constituents. Let it be our task to determine the optical constants of the mixture.

We will make the following assumptions:

Let the mixture occupy the full volume V. In the mixture, let us assume that each of the constituents occupies a certain volume fraction V_j. The corresponding volume filling factor p_j of the jth material in the mixture is then defined as:

$$p_j \equiv \frac{V_j}{V} \tag{7.26}$$

Obviously,

$$\sum_j p_j = 1 \tag{7.27}$$

Traditionally, the mixing partners are tackled as small (compared to the wavelength) inclusions numbered by the subscript j, embedded in a certain host medium with a dielectric function ε_h [23]. This assumption leads to the general mixing formula:

$$\frac{\left(\varepsilon_{eff} - \varepsilon_h\right)}{\varepsilon_h + \left(\varepsilon_{eff} - \varepsilon_h\right) L} = \sum_j p_j \frac{\left(\varepsilon_j - \varepsilon_h\right)}{\varepsilon_h + \left(\varepsilon_j - \varepsilon_h\right) L} \tag{7.28}$$

Here L is the so-called depolarization factor, and ε_{eff} is the effective dielectric function of the mixture. Note that $0 \leq L \leq 1$ holds. In the case of spherical inclusions, set $L = 1/3$.

Table 7.4 provides a survey of mixing models that represent special cases of the general formula (7.28).

Knowledge on the optical behavior of mixtures is of extreme practical relevance, because no real material can be regarded as absolutely pure. Contrarily, it may be composed from several crystalline and amorphous phases, it may contain stoichiometric as well as non-stoichiometric fractions as well as several kinds of impurities. Even the zirconia film shown in Fig. 2.2 cannot be regarded as a pure film: It is obviously a mixture of crystalline and amorphous zirconia fractions, and a pore fraction.

In this sense, mixture models even provide a vehicle for understanding the origin of the difference between in situ and ex situ optical constants as mentioned in Sect. 7.2.3. Indeed, when a coating is prepared in vacuum conditions, the pores are empty, and so the pore fraction is characterized by a "pore refractive index" that is equal to 1. At atmosphere, however, water may penetrate into the pores, changing the pore's refractive index to a value of approximately 1.33. The resulting effects in the optical

Table 7.4 Survey of mixing models [2]

Model	Equation	Application/remarks
Parallel nanolaminate	$\varepsilon_{eff} = \sum_j p_j \varepsilon_j$	Nanolaminate oriented parallel to the orientation of the electric field vector. Obtained from (7.28) when setting $L = 0$
Vertical nanolaminate	$\varepsilon_{eff}^{-1} = \sum_j p_j \varepsilon_j^{-1}$	Nanolaminate oriented vertical to the orientation of the electric field vector. Obtained from (7.28) when setting $L = 1$
Maxwell Garnett	$\frac{(\varepsilon_{eff}-\varepsilon_l)}{\varepsilon_l+(\varepsilon_{eff}-\varepsilon_l)L} = \sum_{j\neq l} p_j \frac{(\varepsilon_j-\varepsilon_l)}{\varepsilon_l+(\varepsilon_j-\varepsilon_l)L}$	Guest-host system with the lth mixing partner acting as the host. Obtained from (7.28) when setting $\varepsilon_h = \varepsilon_l$
Lorentz-Lorenz	$\frac{(\varepsilon_{eff}-1)}{1+(\varepsilon_{eff}-1)L} = \sum_j p_j \frac{(\varepsilon_j-1)}{1+(\varepsilon_j-1)L}$	Guest-host system with vacuum acting as the host. Obtained from (7.28) when setting $\varepsilon_h = 1$
Bruggeman	$0 = \sum_j p_j \frac{(\varepsilon_j-\varepsilon_{eff})}{\varepsilon_{eff}+(\varepsilon_j-\varepsilon_{eff})L}$	Molecular mixtures, obtained from (7.28) when setting $\varepsilon_h = \varepsilon_{eff}$. Also known as effective medium approximation (EMT or EMA)

properties of the coating may be calculated in terms of suitable mixing models and are known as the vacuum-to-air shift [15] (compare also Chap. 2, Sect. 2.3).

7.3.4 An Empirical Extension of the Multi-oscillator Model: The Beta Distributed Oscillator (β_do) Model

As it is indicated in Fig. 7.7, in a limited spectral range, the merger of the Drude and Lorentzian multioscillator model is well suited for describing the dielectric function of a large variety of materials. When the spectral range includes the fundamental absorption edge, the required number of Lorentzian oscillators for accurate modelling the dielectric function increases and the resulting large number of parameters often results in numerical instabilities in the fitting process. A reduction of the required number of parameters could be achieved when a suitable distribution function for the oscillator's strength is used. A prominent example is the Brendel model (Table 7.3), which makes use of a Gaussian distribution of resonance frequencies. Another promising approach is given by the Beta distribution:

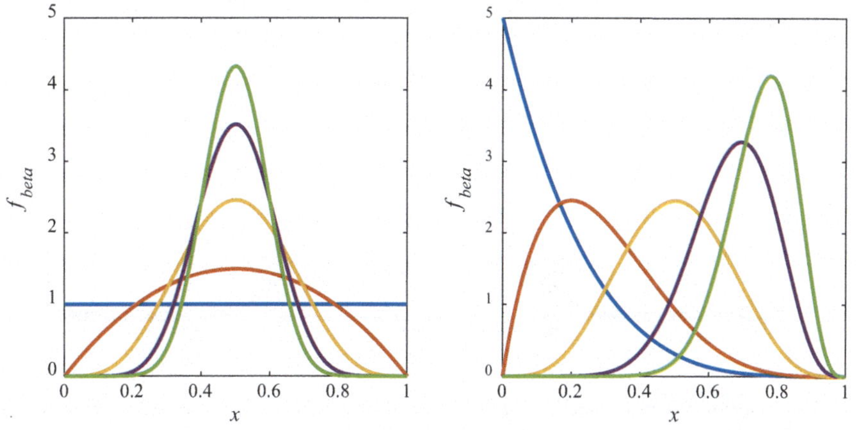

Fig. 7.8 Probability density functions of the beta distribution (left: $\alpha = \beta$, right: $\beta = 5$, blue: $\alpha = 1$, red: $\alpha = 2$, yellow: $\alpha = 5$, violet: $\alpha = 10$, green: $\alpha = 15$)

$$\begin{aligned} f_{beta}(x,\alpha,\beta) &= \frac{\Gamma(\alpha+\beta)}{\Gamma(\alpha)\Gamma(\beta)}x^{\alpha-1}(1-x)^{\beta-1} = \frac{x^{\alpha-1}(1-x)^{\beta-1}}{B(\alpha,\beta)} \quad \text{for } 0 \le x \le 1 \\ f_{beta}(x,\alpha,\beta) &= 0 \quad \text{for } x < 0 \text{ or } x > 1 \end{aligned} \tag{7.29}$$

where $\Gamma(z)$ is the gamma function and $B(\alpha, \beta)$ is the beta function, defined by:

$$B(\alpha,\beta) = \int_0^1 x^{\alpha-1}(1-x)^{\beta-1}dx \tag{7.30}$$

The beta function can be easily generalized to cover an arbitrary interval $[\nu_{min}, \nu_{max}]$:

$$f_{beta}(\xi,\alpha,\beta,\nu_{min},\nu_{max}) = \frac{(\xi-\nu_{min})^{\alpha-1}(\nu_{max}-\xi)^{\beta-1}}{B(\alpha,\beta)(\nu_{max}-\nu_{min})^{\alpha+\beta-1}} \tag{7.31}$$

When $\alpha = \beta$ the distribution will be symmetric (Fig. 7.8 left). For $\alpha = \beta = 1$ a uniform [0,1] distribution and for $\alpha = \beta \to \infty$ a delta function at the midpoint can be generated. In the case $\alpha \neq \beta$ the distribution function will be skewed (Fig. 7.8 right).

For practical application of the beta distribution for modelling the oscillator strength distribution in optical materials a further property seems prospective. Even the case of normally distributed oscillator strength (Brendel model) can be approximated quite well by a symmetric beta distribution. In Fig. 7.9 the (truncated) normal distribution with mean value 0.5 and standard deviation 0.1 is shown (circle). The shape is very close to a beta distribution with $\alpha = \beta = 13$ (solid line).

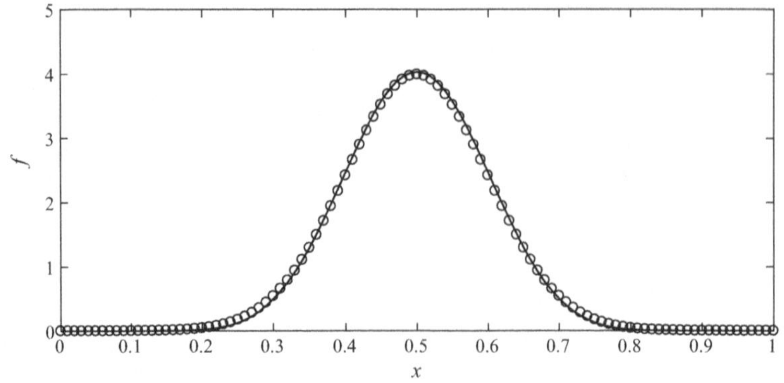

Fig. 7.9 Probability density functions of the normal distribution (circle, mean: 0.5, standard deviation: 0.1) and beta distribution (solid line, $\alpha = \beta = 13$)

To apply the beta distribution to the Lorentzian multi oscillator model the dependence of the susceptibility $\hat{\chi}_j(\nu)$ of a *j*th single oscillator from wavenumber ν can be used (compare Table 7.3):

$$\hat{\chi}_j(\nu) = \frac{f_j}{\tilde{\nu}_{0j}^2 - \nu^2 - 2i\nu\Gamma_j} \tag{7.32}$$

For compatibility reasons to the current implementation of the oscillator model used in LCalc software [24], a slightly different formulation will be used:

$$\hat{\chi}_j(\nu) = \frac{J_j}{\pi}\left(\frac{1}{\nu_{0j} - \nu - i\Gamma_j} + \frac{1}{\nu_{0j} + \nu + i\Gamma_j}\right) \tag{7.33}$$

Let us introduce the complex function X(ξ,ν). It is defined according to:

$$X(\xi, \nu) = \frac{J_{beta,j}}{\pi}\left(\frac{1}{\xi - \nu - i\Gamma_{beta,j}} + \frac{1}{\xi + \nu + i\Gamma_{beta,j}}\right) \tag{7.34}$$

To replace the *j*th single oscillator by a set of beta distributed oscillators located in the interval $[\nu_{min,j}, \nu_{max,j}]$, we write:

$$\hat{\chi}_{beta,j}(\nu) = \int_{\nu_{min,j}}^{\nu_{max,j}} f_{beta}\left(\xi, \alpha_j, \beta_j, \nu_{min,j}, \nu_{max,j}\right) X(\xi, \nu)\, d\xi \tag{7.35}$$

Instead of the single Lorentzian line, as defined by the imaginary part of (7.32), expression (7.35) describes an inhomogeneously broadened absorption structure, which might be used for modelling the absorption edge shape in thin solid films.

Next, the integral function will be replaced by a finite sum. Thereby, an equidistant grid of *N* wavenumbers can be used

$$\nu_{s,j} = \nu_{min,j} + s \cdot \Delta\xi \text{ with } s \in [1, N] \text{ and } \Delta\xi = \frac{\nu_{max,j} - \nu_{min,j}}{N+1} \tag{7.36}$$

Then, the susceptibility of the set of beta distributed oscillators ("β_do") can be calculated by

$$\begin{aligned}\hat{\chi}_{beta,j}(\nu) &= \sum_{s=1}^{N} f_{beta}\left(\nu_{s,j}, \alpha, \beta, \nu_{min,j}, \nu_{max,j}\right) X\left(\nu_{s,j}, \nu\right)\Delta\xi \\ &= \frac{1}{(N+1)\,\mathrm{B}(\alpha,\beta)} \sum_{s=1}^{N} \frac{\left(\nu_{s,j} - \nu_{min,j}\right)^{\alpha-1}\left(\nu_{max,j} - \nu_{s,j}\right)^{\beta-1}}{\left(\nu_{max,j} - \nu_{min,j}\right)^{\alpha+\beta-2}} X\left(\nu_{s,j}, \nu\right)\end{aligned} \tag{7.37}$$

Additionally, it is convenient to replace the beta function also by a sum (compare (7.30)):

$$B(\alpha, \beta) = \frac{1}{N+1} \sum_{s=1}^{N} \frac{\left(\nu_{s,j} - \nu_{min,j}\right)^{\alpha-1}\left(\nu_{max,j} - \nu_{s,j}\right)^{\beta-1}}{\left(\nu_{max,j} - \nu_{min,j}\right)^{\alpha+\beta-2}} \tag{7.38}$$

From (7.31), (7.33)–(7.35), the susceptibility can be calculated by:

$$\hat{\chi}_{beta,j}(\nu) = \frac{\sum_{s=1}^{N} w_{s,j} \frac{J_{beta,j}}{\pi}\left(\frac{1}{\nu_{s,j} - \nu - \mathrm{i}\Gamma_{beta,j}} + \frac{1}{\nu_{s,j} + \nu + \mathrm{i}\Gamma_{beta,j}}\right)}{\sum_{s=1}^{N} w_{s,j}} \tag{7.39}$$

with weight factors

$$w_{s,j} = \frac{\left(\nu_{s,j} - \nu_{min,j}\right)^{\alpha_j-1}\left(\nu_{max,j} - \nu_{s,j}\right)^{\beta_j-1}}{\left(\nu_{max,j} - \nu_{min,j}\right)^{\alpha_j+\beta_j-2}} = \left(\frac{s}{N+1}\right)^{\alpha_j-1}\left(\frac{N+1-s}{N+1}\right)^{\beta_j-1} \tag{7.40}$$

Equations (7.39) and (7.40) essentially define what we will further call the beta distributed oscillator model (β_do model). In Fig. 7.10 the individual contributions from a beta distributed oscillator set to real and imaginary part of the susceptibility are shown.

The impact of the line width to real and imaginary part of susceptibility for a beta distributed set of oscillators is shown in Fig. 7.11. In trend, the width of the imaginary part of the susceptibility decreases with the line width of the underlying individual oscillators. When the line width becomes small in comparison to the width of the beta distribution, the resulting shape becomes dominated by the beta distribution.

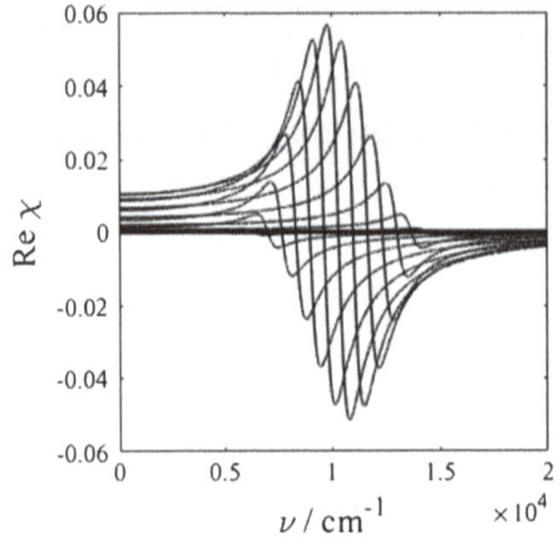

Fig. 7.10 Real (left) and imaginary part (right) of susceptibility of individual equally spaced Lorentzian oscillators defined by the β_do model ($N = 15$, $\nu_{min} = 5000\,\mathrm{cm}^{-1}$, $\nu_{max} = 15000\,\mathrm{cm}^{-1}$, $J_{beta} = 1000\,\mathrm{cm}^{-1}$, $\Gamma_{beta} = 500\,\mathrm{cm}^{-1}$, $\alpha = \beta = 5$)

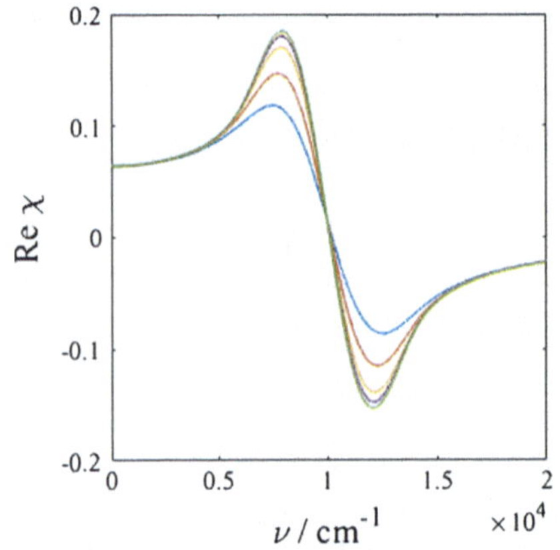

Fig. 7.11 Real (left) and imaginary part (right) of the susceptibility in the β_do model ($N = 1000$, $\nu_{min} = 5000\,\mathrm{cm}^{-1}$, $\nu_{max} = 15000\,\mathrm{cm}^{-1}$, $J_{beta} = 1000\,\mathrm{cm}^{-1}$, blue: $\Gamma_{beta} = 1000\,\mathrm{cm}^{-1}$, red: $\Gamma_{beta} = 500\,\mathrm{cm}^{-1}$, yellow: $\Gamma_{beta} = 200\,\mathrm{cm}^{-1}$, violet: $\Gamma_{beta} = 100\,\mathrm{cm}^{-1}$, green: $\Gamma_{beta} = 50\,\mathrm{cm}^{-1}$)

7.4 Conclusions

In this chapter, basic theoretical concepts and equations necessary for spectrophotometric characterization of thin films and film systems have been introduced. Particularly the β_do model turns out to be extremely useful in coating characterization practice, including typical inorganic dielectric coatings, but also metal coatings, semiconductor films, and even organic molecular films. In Chap. 8, corresponding examples will be presented and discussed.

References

1. M. Born, E. Wolf, *Principles of Optics* (Pergamon Press, Oxford, 1968)
2. O. Stenzel, *The Physics of Thin Film Optical Spectra: An Introduction*. Springer Series in Surface Sciences, vol. 44, 2nd edn. (Springer, Berlin, 2015)
3. R. Gross, A. Marx, Festkörperphysik (Walter de Gruyter GmbH, Berlin, 2014)
4. J.C. Manifacier, J. Gasiot, J.P. Fillard, A simple method for the determination of the optical constants n, k and the thickness of a weakly absorbing thin film. J. Phys. E: Sci. Instrum. **9**, 1002–1004 (1976)
5. I. Ohlidal, K. Navrátil, Simple method of spectroscopic reflectometry for the complete optical analysis of weakly absorbing thin films: application to silicon films. Thin Solid Films **156**, 181–190 (1988)
6. E. Nichelatti, Complex refractive index of a slab from reflectance and transmittance: analytical solution. J. Opt. A: Pure Appl. Opt. **4**, 400–403 (2002)
7. A.V. Tikhonravov, M.K. Trubetskov, B.T. Sullivan, J.A. Dobrowolski, Influence of small inhomogeneities on the spectral characteristics of single thin films. Appl. Opt. **36**, 7188–7198 (1997)
8. J.H. Dobrowolski, F.C. Ho, A. Waldorf, Determination of optical constants of thin film coating materials based on inverse synthesis. Appl. Opt. **22**, 3191–3200 (1983)
9. T.V. Amotchkina, M.K. Trubetskov, V. Pervak, B. Romanov, A.V. Tikhonravov, On the reliability of reverse engineering results. Appl. Opt. **51**, 5543–5551 (2012)
10. V. Janicki, J. Sancho-Parramon, O. Stenzel, M. Lappschies, B. Görtz, C. Rickers, C. Polenzky, U. Richter, Optical characterization of hybrid antireflective coatings using spectrophotometric and ellipsometric measurements. Appl. Opt. **46**, 6084–6091 (2007)
11. A.V. Tikhonravov, M.K. Trubetskov, On-line characterization and reoptimization of optical coatings. Proc. SPIE **5250**, 406–413 (2004)
12. T.V. Amotchkina, M.K. Trubetskov, V. Pervak, S. Schlichting, H. Ehlers, D. Ristau, A.V. Tikhonravov, Comparison of algorithms used for optical characterization of multilayer optical coatings. Appl. Opt. **50**, 3389–3395 (2011)
13. S. Wilbrandt, O. Stenzel, N. Kaiser, Experimental determination of the refractive index profile of rugate filters based on in situ measurements of transmission spectra. J. Phys. D **40**, 1435–1441 (2007)
14. S. Wilbrandt, O. Stenzel, M. Bischoff, N. Kaiser, Combined in situ and ex situ optical data analysis of magnesium fluoride coatings deposited by plasma ion assisted deposition. Appl. Opt. **50**, C5–C10 (2011)
15. O. Stenzel, *Optical Coatings: "Material Apects in Theory and Practice"* (Springer, Berlin, 2014)
16. L.D. Landau, E.M. Lifschitz, Lehrbuch der theoretischen Physik, Bd. VIII: Elektrodynamik der Kontinua [engl.: Textbook of Theoretical Physics, Vol. VIII: Electrodynamics of Continuous Media] (Akademie, Berlin, 1985)
17. V. Lucarini, J.J. Saarinen, K.E. Peiponen, E.M. Vartiainen, *Kramers-Kronig relations in Optical Materials Research* (Springer, Berlin, 2005)
18. R. Brendel, D. Bormann, An infrared dielectric function model for amorphous solids. J. Appl. Phys. **71**, 1–6 (1992)
19. G.E. Jellison, Spectroscopic ellipsometry data analysis: measured versus calculated quantities. Thin Solid Films **313**(314), 33–39 (1998)

20. J. Tauc, R. Grigorovic, A. Vancu, Optical properties and electronic structure of amorphous germanium. Phys. Stat. Sol. **15**, 627–637 (1966)
21. A.S. Ferlauto, G.M. Ferreira, J.M. Pearce, C.R. Wronski, R.W. Collins, X. Deng, G. Ganguly, Analytical model for the optical functions of amorphous semiconductors from the near-infrared to ultraviolet: applications in thin film photovoltaics. J. Appl. Phys. **92**, 2424–2436 (2002)
22. F. Urbach, The long-wavelength edge of photographic sensitivity and of the electronic absorption of solids. Phys. Rev. **92**, 1324 (1953)
23. D.E. Aspnes, J.B. Theeten, F. Hottier, Investigation of effective-medium models of microscopic surface roughness by spectroscopic ellipsometry. Phys. Rev. B **20**, 3292–3302 (1979)
24. O. Stenzel, S. Wilbrandt, K. Friedrich, N. Kaiser, Realistische Modellierung der NIR/VIS/UV-optischen Konstanten dünner optischer Schichten im Rahmen des Oszillatormodells. Vak. Forsch. Prax. **21**(5), 15–23 (2009)

Chapter 8
In Situ and Ex Situ Spectrophotometric Characterization of Single- and Multilayer-Coatings II: Experimental Technique and Application Examples

Steffen Wilbrandt and Olaf Stenzel

Abstract In the previous chapter, the theoretical background for characterization of single layer and multilayer coatings has been outlined. In this chapter, important aspects for the underlying experimental techniques will presented. Furthermore, we demonstrate the application of different dispersion models for characterization of uncoated substrates, single layer coatings of dielectrics, semiconductors, metals and organic coatings. Thereby, the focus has been set to the β_do model. Finally, the interplay of in situ and ex situ spectroscopy will be demonstrated for a multilayer antireflection coating (V-coating).

8.1 Experimental Techniques in Spectrophotometry

Let us for a moment return to Fig. 1.1. Imagine the very simplest case—a monochromatic plane light wave that is incident onto the sample with a light intensity I_E. In a complex notation, the electric field E of that light wave depends on the time t and the coordinates r according to (2.1).

S. Wilbrandt (✉)
Fraunhofer Institute for Applied Optics and Precision Engineering IOF, Albert-Einstein-Str. 7, 07745 Jena, Germany
e-mail: steffen.wilbrandt@iof.fraunhofer.de

O. Stenzel
Fraunhofer Institute for Applied Optics and Precision Engineering IOF, Albert-Einstein-Str. 7, 07745 Jena, Germany
e-mail: olaf.stenzel@iof.fraunhofer.de; optikbuch@optimon.de

O. Stenzel
Abbe School of Photonics, Friedrich-Schiller-University Jena, Albert-Einstein-Straße 6, 07745 Jena, Germany

O. Stenzel and M. Ohlídal (eds.), *Optical Characterization of Thin Solid Films*, Springer Series in Surface Sciences 64,
https://doi.org/10.1007/978-3-319-75325-6_8

Once our focus is on spectrophotometry, it is the intensity of the light that has to be discussed in more detail. It is given by (2.2). In ex situ coating characterization practice, the experimental determination of T_{exp} and R_{exp} as defined in (2.3) is usually performed by means of commercial spectrophotometers, which may be roughly divided into dispersive and Fourier transform spectrophotometers [1]. Let us mention in this context, that for coating characterization purposes, highest absolute accuracy in intensity measurements is utmost important for getting reliable results from minimizing discrepancy functions like (7.15) while highest spectral resolution is usually not of use. Thereby, T_{calc} and R_{calc} are theoretical spectra, which are calculated within a certain layer model.

8.1.1 *Spectral Resolution*

A usual and convenient assumption in the applied layer model is, that the film is thin enough for observing interference phenomena that arise from multiple internal reflections in the film. On the other hand, the substrate should be thick enough so that multiple internal reflections within the substrate superimpose incoherently, e.g. without observable interference. This defines a constraint to the allowed spectral resolution in the corresponding measurement: A too high resolution would resolve interference effects in the substrate, which is no more consistent with the assumed incoherent superposition of internally reflected light trains within the thick substrate. As a rough estimate, the spectral resolution in the T- and R-measurements at near normal incidence should therefore be restricted so that (8.1) is fulfilled [2]:

$$\Delta\nu > \frac{1}{2\pi n_{sub} h_{sub}} \text{ or } \Delta\lambda > \frac{\lambda^2}{2\pi n_{sub} h_{sub}} \tag{8.1}$$

Here, $\Delta\nu$ or $\Delta\lambda$ denote the spectral bandwidth of the incident light (the absolutely monochromatic wave as assumed in (2.2) is only a convenient model assumption, which is never observed in reality in a strong sense). Condition (8.1) is easily fulfilled in characterization practice, because commercial spectrophotometers usually allow setting the spectral bandwidth to a sufficiently large value; or a suitably thick substrate has to be chosen.

On the other hand, highest accuracy in intensity measurements is not so easily achieved. Fortunately, stochastic measurement errors in T or R are not so crucial when performing curve fit s by minimizing (7.15) [3]. The bad news is that systematic measurement errors are highly disturbing [3].

A common source for systematic measurement errors is the limited spectral resolution of the device caused, for example, by the finite width of entrance or exit monochromator slits in a dispersive spectrophotometer. These slits represent rectangular apertures through with light enters into and exits from the monochromator. In an ideal spectrometer, the effect of the finite width of the slit can be modelled in terms of a point spread function with a triangular shape (Fig. 8.1). The obvious result

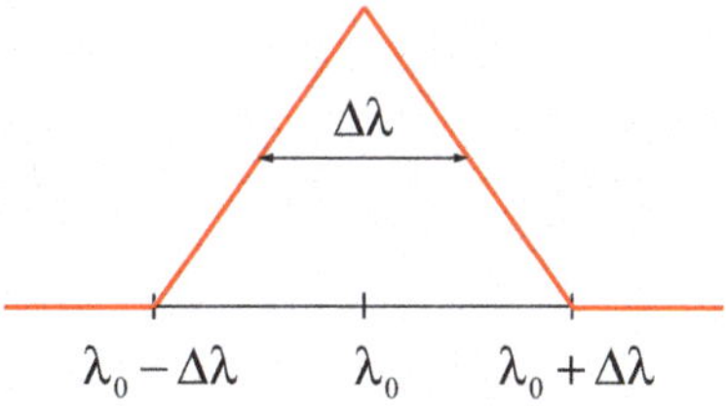

Fig. 8.1 Point spread function of a slit in an ideal spectrometer

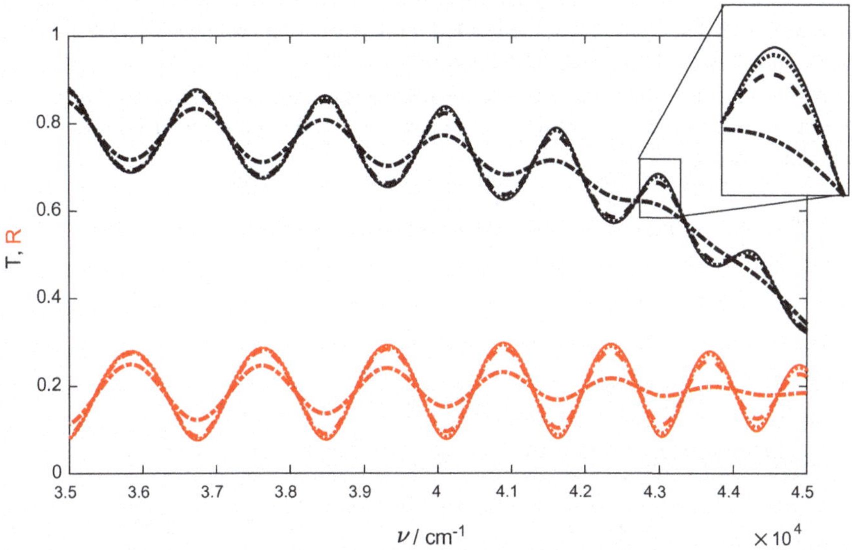

Fig. 8.2 Theoretical transmittance (black) and reflectance (red) assuming a monochromatic wave (solid line) and with effects of a finite spectral bandwidth assuming a triangular point spread function (dotted line: 1 nm, dashed line: 2 nm and dash-dot line: 5 nm slit width)

is a systematic measurement error in transmittance and reflectance, which is largest at the extrema positions of the spectra (Fig. 8.2). The limited spectral resolution of a spectrometer will decrease the measured photometric values at the maxima positions and increase it at the minima position.

In real devices, a more complicated point spread function will be obtained. To take this effect into account, either the point spread function may be included in the calculation of theoretical spectra or a sufficiently high spectral resolution has to be selected during measurement. For an estimation of the required spectral resolution, the impact of the point spread function to a single layer coating can be investigated analytically.

For the special case of vanishing damping, the transmittance at normal incidence of a single layer (refractive index n) on a semi-infinite substrate (refractive index n_{sub}) can be calculated by [2]:

$$T\left(n, n_{sub}, \delta\right) = \frac{n_{sub}\left(\frac{2}{1+n}\right)^2\left(\frac{2n}{n+n_{sub}}\right)^2}{1+\left(\frac{1-n}{1+n}\right)^2\left(\frac{n-n_{sub}}{n+n_{sub}}\right)^2+2\left(\frac{1-n}{1+n}\right)\left(\frac{n-n_{sub}}{n+n_{sub}}\right)cos2\delta} \tag{8.2}$$

with

$$2\delta = 4\pi\nu nh \tag{8.3}$$

It is obvious, that the extrema in transmittance correspond to multiples of π for the 2δ term (weak dispersion presumed). Odd multiples are quarter-wave (QW) points, even multiples are half-wave (HW) points.

When applying the triangular point spread function to the transmittance, the expected transmittance $\tilde{T}$ obtained by means of a real spectrometer could be estimated according to:

$$\tilde{T} \approx \frac{1}{4}T\left(n, n_{sub}, \delta - \Delta\delta\right) + \frac{1}{2}T\left(n, n_{sub}, \delta\right) + \frac{1}{4}T\left(n, n_{sub}, \delta + \Delta\delta\right) \tag{8.4}$$

with

$$\Delta\delta = \frac{\pi nh}{\lambda_0^2}\Delta\lambda \tag{8.5}$$

For the measurement errors at the QW- and HW-points, we can deduct:

$$\Delta T_{QW} = T - \tilde{T} \approx -\frac{2n^2 n_{sub}\left(n^2 - n_{sub}^2\right)\left(n^2 - 1\right)}{\left(n^2 + n_{sub}\right)^4}\left(\Delta\delta\right)^2 \tag{8.6}$$

$$\Delta T_{HW} = T - \tilde{T} \approx \frac{2n^2 n_{sub}\left(n^2 - n_{sub}^2\right)\left(n^2 - 1\right)}{n^4\left(1 + n_{sub}\right)^4}\left(\Delta\delta\right)^2 \tag{8.7}$$

From here we see, that the following relations are valid:

$$\left|\Delta T_{QW}\right| < \left|\Delta T_{HW}\right| \text{ for } n > n_{sub} \tag{8.8}$$

$$\left|\Delta T_{QW}\right| > \left|\Delta T_{HW}\right| \text{ for } n < n_{sub} \tag{8.9}$$

In both cases, the largest effect will be observed at the transmittance maxima. For a given accepted tolerance ΔT, the required spectral resolution can be now estimated. For $n > n_{sub}$ we get:

$$\Delta\lambda < \frac{\lambda_0^2\left(1 + n_{sub}\right)^2}{\pi h}\sqrt{\frac{\Delta T}{2n_{sub}\left(n^2 - n_{sub}^2\right)\left(n^2 - 1\right)}} \tag{8.10}$$

and for $n < n_{sub}$:

$$\Delta\lambda < \frac{\lambda_0^2 \left(n^2 + n_{sub}\right)^2}{\pi n h} \sqrt{\frac{\Delta T}{2n^2 n_{sub}\left(n^2 - n_{sub}^2\right)\left(n^2 - 1\right)}} \tag{8.11}$$

Together with (8.1), this will confine the spectral resolution suitable for spectrophotometric measurements of the film-on substrate system.

8.1.2 Sample Illumination

Basically, possible illumination configurations can be classified by the relation between the incident light spot and the field of view of the detector (Fig. 8.3 on left). When both areas are identical in shape and size, the reversibility of light enables the use of identical optical configurations for illumination and detection in the case of fiber optic based approaches. This will reduce costs for development, but make this approach sensitive to alignment errors. Even a small misalignment between illumination and detector optics will change the throughput of light and may therefore lead to measurement errors. To overcome this problem, spot sizes for illuminated and collimated light should be different. In principle, either a small spot for the illuminated light and a large spot for collimated light (Fig. 8.3 in center) or vice versa (Fig. 8.3 on right) may be selected. In the case of a small illumination spot, spatial homogeneity of the light source is not required, but is essential for the collimation optic of the detector. Furthermore, in this set-up, any additional light arriving from other sources (ambient light caused by electron beam gun or plasma/ion source) may be problematic. In the opposite case, spatial homogeneity of the light source is crucial, but not required for the collimation optic in front of the detector. In general, spatial homogeneity can be optimized for both cases using diffuser plates, Ulbricht sphere s, light mixing rods, or micro optics arrays.

Diffusing plates and Ulbricht sphere result in significant reduction of the light throughput of the system, while micro optics arrays are expansive. Bearing in mind that unwanted deposition on optics can result in measurement errors, Ulbricht spheres will be advantageous here, because the small port size in relation to the inner surface make it nearly insensitive to unwanted deposition. If the Ulbricht sphere is build up from ceramics, it can withstand high temperatures, and any contaminations can be removed by sandblasting. When the Ulbricht sphere is used as in chamber housing for the light source, it may additionally shield the detector from ambient light.

8.1.3 Transmission and Reflection Measurements

Many commercial spectrophotometers are multiple purpose devices, primarily optimized for performing absorbance measurements in gas or liquid cell geometries. Thin film sample measurements often require the application of optional measure-

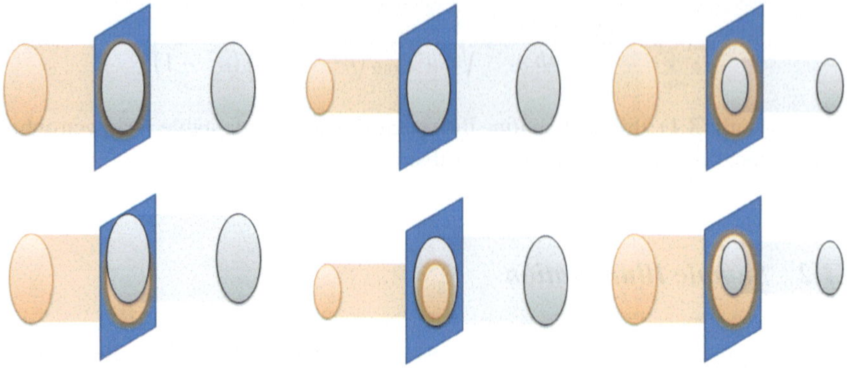

Fig. 8.3 Possible illumination configurations for optimal aligned (top) and slightly misaligned set-up (bottom); On left: Identical incident light spot and field of view of the detector in shape and size in center: small spot for the illuminated light and a large spot for collimated light; On right: Large spot for the illuminated light and a small spot for collimated light

ment accessories, which have to be mounted into the sample compartment and are usually offered for performing reflection measurements at different angles of incidence. The quantification of their systematic measurement errors requires severe own efforts, while corresponding information as included into the manuals—if ever—is usually not very helpful.

In this connection, it is worth mentioning that in many spectrophotometers, I_E, I_T or I_R cannot be measured as directly as it is indicated in Fig. 7.1. Instead, after having interacted with the sample, the light has to pass a certain sequence of optical components before reaching the intensity detector. Therefore, in measurement practice, T and R are accessible from the following set of standard measurements:

- Measurement of an intensity I_{100}, corresponding to an empty sample compartment, i.e. with no sample in the light path (Baseline or Auto Zero measurement)
- Measurement of I_T or I_R with the sample in the light path (sample measurement)
- Measurement of I_0 with the light path blocked (dark signal measurement).

From these intensity data, T and R are obtained in terms of (8.12) [4]:

$$T = \frac{I_T - I_0}{I_{100} - I_0}; \; R = \frac{I_R - I_0}{I_{100} - I_0} \tag{8.12}$$

In the case of transmittance measurement at (near) normal incidence, the implementation is straight forward (Fig. 8.4).

In the case that the reflected light has been measured relative to a reference mirror with the reflectance R_{ref}, instead of (8.12), we have:

$$T = \frac{I_T - I_0}{I_{100} - I_0}; \; R = R_{ref} \frac{I_R - I_0}{I_{100} - I_0} \tag{8.13}$$

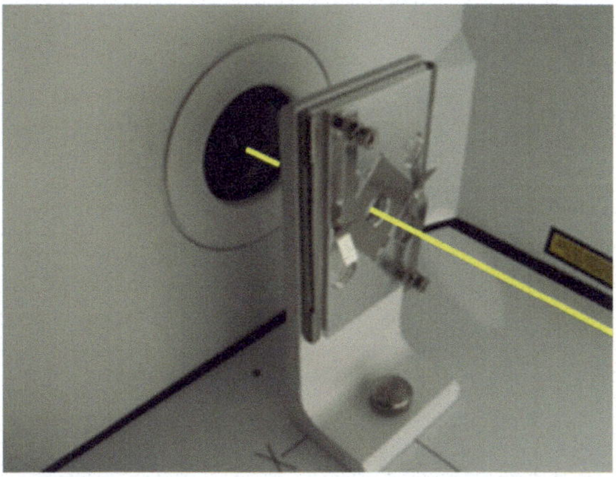

Fig. 8.4 Light path for transmission measurement used in a Perkin Elmer Frontier Optica FTIR

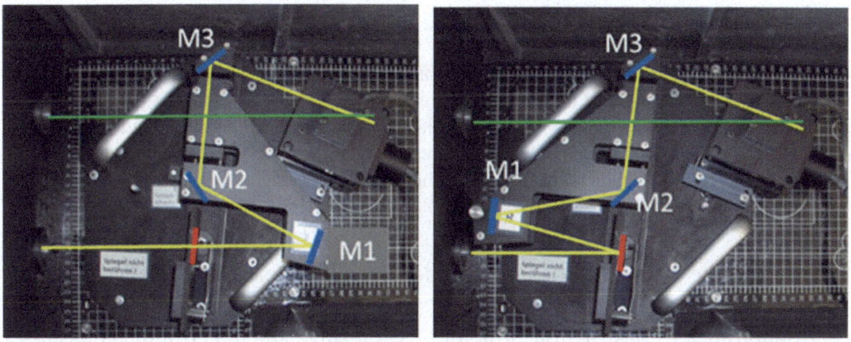

Fig. 8.5 Light path for transmission (left) and reflection measurement (right) in a VN accessory for $\psi = 6°$

Thereby, normal incidence cannot be realized for geometrical reasons. Reference-free (or absolute) -measurements may be performed by means of special accessories exploiting the so-called VN—measurement principle. The corresponding light path for near normal angle of incidence is shown in Fig. 8.5. It is obvious, that the underlying principle can be easily adapted to oblique incidence. In this case, light polarization as well as effects like beam splitting and displacement must be considered [4].

Basically, two movable mirrors are required to direct the transmitted (Fig. 8.5 left) and reflected light (Fig. 8.5 right). Geometrical constraints may require an additional mirror to direct the light from the sample toward the detector.

A more detailed description is provided in [4] and references cited therein. We also mention here that the VW- and IV-measurement principle s [4] give direct access to R^2 instead of R. They are therefore not suitable for the measurement of very low reflectance values, but strong in the measurement of high reflectances.

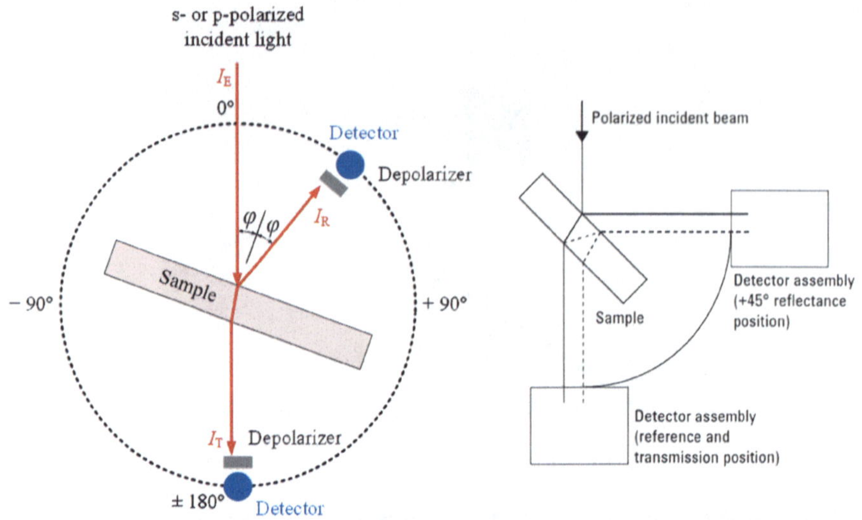

Fig. 8.6 On left: Schematic of the Cary 7000 UMS. Light incident onto the sample can be s- or p-polarized [2]. Absolute specular reflection or transmission can be measured. The detector module allows mounting of an optional depolarizer immediately in front of the detector; On right: Schematic of the Agilent UMA, an absolute variable angle reflectance and transmission accessory, in 45° measurement geometry. Pictures are adapted from [5, 6], and are printed with kind permission by Agilent Technologies Deutschland GmbH

Recently, the Agilent company has developed a measurement system which is adapted to the direct measurement of I_E, I_T or I_R by combining the typical spectrophotometer construction principle with a fully automated mini-goniometer set-up (Cary 7000UMS—compare Fig. 8.6 [5, 6]). It is mounted into an extra sample compartment (the Agilent UMA—Universal Measurement Accessory). The movable detector allows performing direct (reference-free) I_T or I_R measurements at practically any reasonable angle of incidence (Fig. 8.6, on left—compare with Fig. 7.1). The broad detector area even allows collecting the multiply internally reflected light trains which broaden the light beam in oblique incidence conditions (Fig. 8.6, on right, compare also [4]). The recently published TRACK-method for optical thin film characterization [7] is based on the measurement possibilities offered by this innovative spectrometer construction principle.

8.1.4 Pre-processing of Spectra

In principle, measured spectra could be directly used for characterization and any pre-processing of them is not necessarily required. Nevertheless, it may be useful to eliminate superfluous data from measurements.

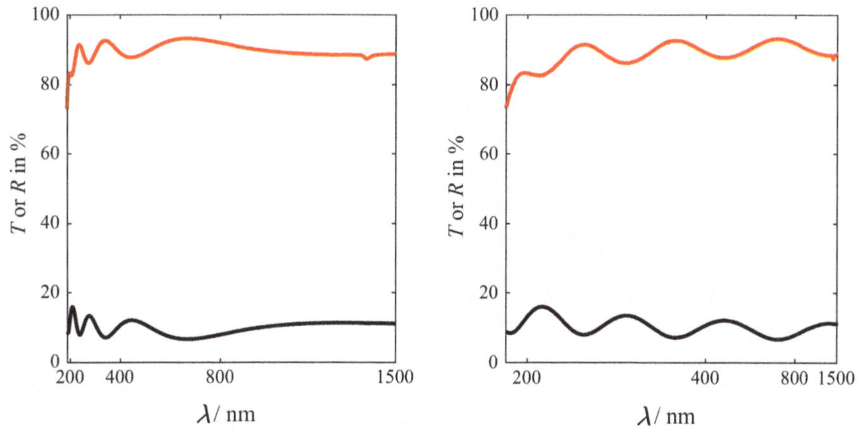

Fig. 8.7 Transmittance and reflectance of an alumina single layer on a fused silica substrate using a wavelength (left) and wavenumber grid (right)

In general, experimental spectra will contain a certain level of random noise. It could be either reduced by averaging multiple measurements on the same sample or by applying some filter to the measured data. It is obvious, that the first approach does not require any a priori knowledge on the spectral characteristic. In the case of normally distributed noise, the noise level for N repeated measurements $\Delta y\,(N)$, each individual measurement with a noise level $\Delta y\,(1)$, can be estimated by

$$\Delta y\,(N) = \frac{\Delta y\,(1)}{\sqrt{N}} \tag{8.14}$$

Therefore, a substantial noise reduction by averaging will commonly result in a significant increase of measurement duration and applying a filter to the measurement data could be prospective. Clearly, the underlying parameters and the algorithm must be carefully selected to minimize resulting systematic errors. Thereby, any a priori knowledge of the spectral characteristic could be very helpful. Common filters used in spectroscopy are Fourier filters and Savitzky-Golay filters [8]. Both filter types can preserve the major features in the spectra and do not affect the grid of the measurement data.

The spatial response of gratings used in dispersive spectrophotometers favors an equidistant wavelength grid for measurements. In the case of single layer coatings on a substrate, this grid will destroy the quasi-periodicity of the interference pattern as shown in Fig. 8.7 on left. Therefore, using an equidistant wavelength grid does not seem to be an efficient choice for coating characterization.

This is not astonishing, because (8.3) will favor a reciprocal stretching of the axes. This could be achieved, when spectra plotted in an equidistant wavenumber grid instead. In this case, the interference pattern appears nearly periodic (Fig. 8.7

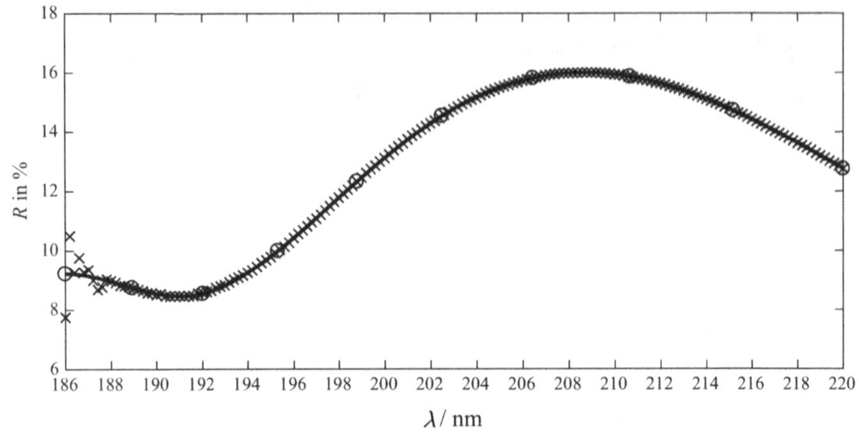

Fig. 8.8 Measured (cross)and pre-processed transmittance spectra (solid line, circles indicate grid) of an alumina single layer on fused silica substrate measured with the 6° VN accessory in the DUV

right). For this reason, an equidistant wavenumber grid appears as the better choice for coating characterization.

Furthermore, a data grid different to the measurement grid may also be useful to eliminate redundant data and to accelerate the characterization process. For this reason, an optional adaption of the data grid during filtering should be considered.

In Fig. 8.8 measured and pre-processed transmittance data of an alumina single layer deposited on a fused silica substrate in the deep ultraviolet (DUV) spectral range is shown. Here, a cubic spline interpolation was used for pre-processing transmittance data. Instead of the huge number of measurement data (crosses), much less data (circles) are used for characterization purposes. Nevertheless, the few wavenumber points (circles) considered for spectra fitting contain all relevant information about the interference pattern.

8.1.5 Specifics of In Situ Spectrophotometry

Photometric measurements in the deposition chamber have already been reported in the previous millennium [9–14]. A overview of the state of the art on optical monitoring techniques can be found at [15]. In this section, we will address only some selected aspects on optical monitoring. Thereby, the broadband monitoring system (OptiMon) developed at the Fraunhofer IOF will be used as an example for a possible implementation. For optical coatings, it clearly will be an advantage to have these measurements available for process control. In contrast to conventional thickness monitoring techniques (e.g. quartz crystal), which only control non-optical properties (e.g. mass), photometric measurements grant access to optical properties of the layer, and particularly to their optical thickness. Significant progress in sensitivity of detectors, miniaturization and degree of integration of electronic circuits used for sensors, as well as decreasing costs of components and increasing processing speed of computers have led to the result that nowadays photometric measurements are widely used in commercial deposition plants for deposition process control. A large variety of monitoring systems are known and can be classified with respect to spectral range, measurement object, termination criteria and error compensation strategy.

8.1.5.1 Classification of In Situ Monitoring Systems by Spectral Range

The accessible spectral range of a monitoring system is limited by the light source and the spectrometer. Commonly, the complete visible spectral range and parts of the ultraviolet and near infrared spectral regions are covered. *In situ* monitoring systems with broader spectral range are not very common and significantly more expensive.

In general, available solutions can be subdivided into single wavelength and broadband monitoring systems. In single wavelength monitoring systems, transmittance and/or reflectance are either measured at a fixed or variable single wavelength (the latter version is also called *monochromatic* monitoring). Single or monochromatic monitoring systems are known to be very sensitive to random measurement errors [3, 16, 17]. To improve the signal-to-noise ratio, commonly log-in amplifiers are used. In the case of a single fixed wavelength, commonly a monochromatic light source (e.g. laser) is used for illumination. Alternatively, a broadband light source and a monochromator are applied. The monochromator can be either located between light source and sample or between sample and detector. In practice, the location of monochromator in front of the detector (as part of the detector) is preferred, because in this case light from other sources (e.g. plasma/ion source, electron beam gun) is damped by the monochromator, so that the resulting signal-to-noise ratio is better.

In a broadband monitoring system, a broadband light source and a polychromator are used. Depending on the selected type of polychromator, the number of parallel measured wavelength could reach a few thousand. Nowadays, image sensors with up to 250 million pixels for photographic applications are in development [18]. In practice, the useful number of pixels is limited by the optical resolution of the spectrometer, and linear arrays with 2048 pixels are sufficient for most applications. The impact of random noise in a broadband monitoring system is reverse proportional to the square root of (independent) pixels [16]. Therefore, broadband monitoring systems are significantly less sensitive to random noise compared to single wavelength or monochromatic monitoring. Furthermore, broadband monitoring can give access to dispersion of the refractive index. The outlined advantages of broadband monitoring often make it to the preferred approach for new deposition plants.

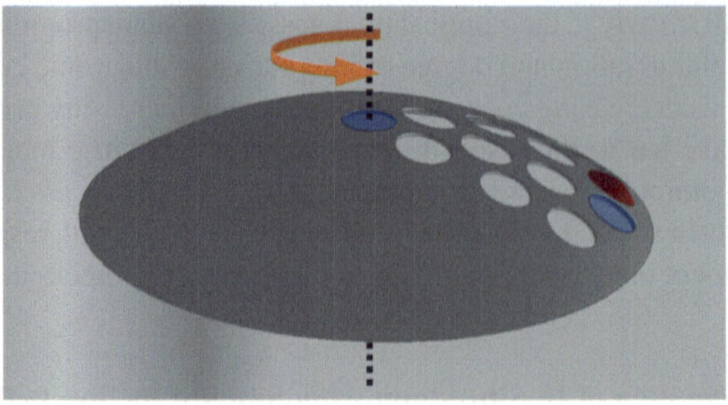

Fig. 8.9 Classification of in situ monitoring systems by measurement object direct monitoring: photometric measurement is performed directly on a relevant sample (red), semi-direct monitoring: photometric measurement is performed on a plane extra substrate (witness glass, blue) located spatially close to the sample, identical performance of the coating on the witness sample and the optical parts is reasonably expected indirect monitoring: photometric measurement is performed on a witness glass (blue) usually mounted at a fixed position in the deposition chamber (e.g. center of the rotating substrate holder), identical performances of the coatings on the witness sample and the optical parts cannot be presumed

8.1.5.2 Classification of In Situ Monitoring Systems by Measurement Object

Depending on the selected measurement object, in situ monitoring approaches are subdivided into *direct*, *semi-direct* and *indirect monitoring* methods [19]—see Fig. 8.9. In the case of *direct monitoring*, the photometric measurement is performed directly on a relevant sample. This approach is commonly preferred for samples with a simple geometry (e.g. plane substrates). In practice, optical parts may have a more complicate geometry (e.g. lenses, prisms) and are not suited for measurements in a generalized measurement configuration. Therefore, a plane extra substrate (the so-called witness sample) is often used for monitoring purposes. In the case of *semi-direct monitoring*, the witness sample is located spatially close to the relevant optical parts, so that an identical performance of the coating on the witness sample and the optical parts is reasonably expected. Direct as well as semi-direct monitoring approaches commonly require a precise synchronization between the movement of the samples and the measurement. The required information could be deducted from different types of sensors. Often, a rotary encoder could be mounted on the driving axes outside the deposition chamber. Alternatively, inductive, capacitive, optical or magnetic sensors inside the plant are used. Moreover, in direct broadband monitoring, the rotating sample holder can be used as a "natural" chopper wheel.

In the case of *indirect monitoring*, the witness glass is usually mounted at a fixed position in the deposition chamber, for example in the center of the rotating substrate holder. Therefore, identical performances of the coatings on the witness sample and the optical parts cannot be expected, so that differences in optical constants, layer thicknesses and further properties must be known. This will cause a serious disadvantage of *indirect monitoring*, because any drift in the assumed behavior will result in systematic deposition errors and finally limits the achievable accuracy. On the other hand, the use of a fixed witness sample simplifies the design of the measurement system, because synchronization between measurement and movement of the optical parts is no longer required. This will also result in more flexibility on selecting integration time and measurement rate. Additionally, in the case of indirect monitoring, different witness samples can be used during one deposition run, when a witness sample changer is used. Nevertheless, the current trend is towards the application of direct and semi-direct monitoring approaches, because of the superior optical performance.

Further classification of monitoring systems can be performed with respect to deposition termination criteria and deposition error compensation strategies. However, a discussion of these topics will lead us into the field of deposition process optimization and thus beyond the narrower field of optical coating characterization, so that it shall not be performed here. Interested readers are referred to [20].

8.1.5.3 Process Photometer OptiMon

The process photometer OptiMon developed at Fraunhofer IOF (Fig. 8.10) is a broadband monitoring system for industrial deposition plants (e.g. OptoTech OAC-90F, Bühler Syrus pro LCIII). It can be used for direct or semi-direct monitoring and is commonly used to terminate the deposition of homogeneous layers. The halogen light source is located in an MACOR-Ulbricht sphere integrated into the evaporation stop blend and is used to generate a large illuminated spot on the sample, while the collimating optic collects light only from a much smaller sample area, corresponding to the arrangement shown in Fig. 8.3 on right.

This approach provides sufficient tolerance to measure transmittance and relative reflectance of the sample [21], but limits the spectral range to approximately 360—2500 nm. Depending on the used spectrometer, the usable spectral range may

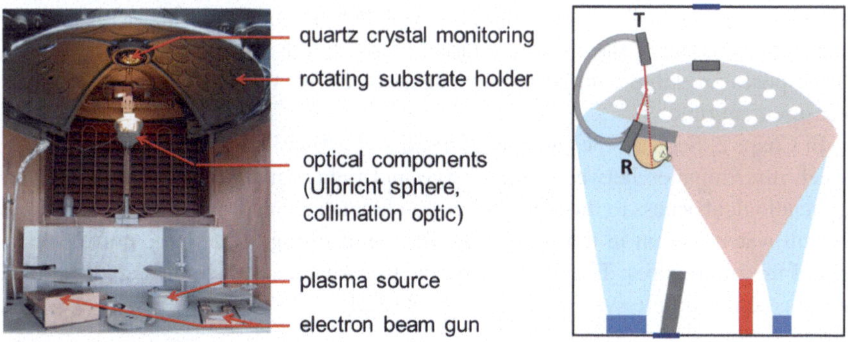

Fig. 8.10 Simultaneous in situ measurement of transmittance and reflectance with process photometer OptiMon in a deposition plant

be further reduced. Currently, two different spectrometers provided by Jeti GmbH [22] are supported:

- PS2000 for wavelength up to 1000 nm
- PS2000 NIR for wavelength in the range 900–1650 nm.

Both devices have built-in hardware for synchronization between sample movement and measurement and internally calculate photometric values from intensity measurements. For determination of layer thicknesses from in situ spectra of homogeneous multilayer systems, a re-engineering software package developed by Alexander Tikhonravov and Michael Trubetskov specifically adapted to the OptiMon system is used. Thereby, the layer thickness during deposition is determined by minimizing (7.17), while breaks between the layer deposition are used to adapt all layer thickness according to (7.18).

Robust re-engineering algorithms use in situ spectra for determining the film thickness assuming that the optical constants are known. They may have been obtained earlier from ex situ measurements performed with suitable single film samples. However, in ex situ conditions, optical constants may be different from those relevant in the vacuum chamber because of atmospheric water which has penetrated into pores in the film. Shift measurements provide a convenient tool to judge differences between ex situ and in situ optical constants. Fortunately, they may be performed by means of the same in situ spectrophotometers.

8.1.6 *Shift Measurement*

As a further application of in situ spectrophotometry, let us mention measurements of the air-to-vacuum shift of real coatings. In Chap. 2, we have already been in touch with such kind of measurement, when characterizing PIAD zirconia coatings with respect to their porosity. Figure 2.6 shows the change in transmittance of a zirconia film when it is brought from air into vacuum. Once the effective refractive index of a porous coating is dependent on whether the pores are filled with water or not, measurements of the shift give direct access to the porosity of a film, as long as the pores are large enough to exchange water with the surrounding within the measurement time.

In Chap. 2, Sect. 2.3.3e) we have formulated a simple model of a porous layer, which discriminates between rather large and rather small pores, and gives phenomenological access to the water migration kinetics in a porous film. Let y express the full water content in the film, y_l the degree of filling of the large pores, and y_s that of the small pores. The filling or evacuation kinetics of the pores are described in terms of the simple system of differential (2.21):

$$\begin{aligned} p_{\text{"large pores"}} \frac{dy_l}{dt} &= p_{\text{"large pores"}} \kappa_{\perp}(y_0 - y_l) - p_{\text{"small pores"}} \frac{dy_s}{dt} \\ \frac{dy_s}{dt} &= p_{\text{"large pores"}} \kappa_{\text{II}}(y_l - y_s). \end{aligned} \tag{8.15}$$

All symbols have the same meaning as introduced in Sect. 2.3.3e. System (8.15) may be used for calculating the evacuation kinetics as well as the filling kinetics of the pores, dependent on initial conditions and ambient parameter setting.

In order to calculate evacuation kinetics, we shall assume that a sample was held at atmosphere for a time long enough so that all pores are essentially filled with water. Then, at $t = 0$, we assume that it is suddenly brought into vacuum (no water in the ambient). In this case, the following conditions hold:

$$y_0 = 0; \;\; y_l(t = 0) = 1; \;\; y_s(t = 0) = 1 \tag{8.16}$$

The full amount of water in the film y is obtained by solving (8.15) and (8.16) according to:

$$\begin{aligned}
y(t) &= \frac{p_{\text{"large pores"}} f_1 + p_{\text{"small pores"}} f_4}{q} e^{-f_3 t} - \frac{p_{\text{"large pores"}} f_2 + p_{\text{"small pores"}} f_3}{q} e^{-f_4 t} \\
f_1 &= \frac{\kappa_{II} p - \kappa_{\perp} + q}{2}; \; f_2 = \frac{\kappa_{II} p - \kappa_{\perp} - q}{2}; \; f_3 = \frac{\kappa_{II} p + \kappa_{\perp} - q}{2}; \; f_4 = \frac{\kappa_{II} p + \kappa_{\perp} + q}{2} \\
q &= \sqrt{\kappa_{\perp}^2 + 2\kappa_{\perp}\kappa_{II}(p_{\text{"small pores"}} - p_{\text{"large pores"}}) + \kappa_{II}^2 p^2} \\
p &\equiv p_{\text{"small pores"}} + p_{\text{"large pores"}}
\end{aligned} \tag{8.17}$$

The system of (8.15) also allows calculating the kinetics of water penetration into the pores after film preparation in vacuum conditions. We shall assume now, that at $t = 0$, the pores are initially empty. At $t = 0$, the system is suddenly exposed to (humid) air, so that the ambient parameter y_0 is set equal to 1. Then, instead of (8.16), we now have the conditions (8.18):

$$y_0 = 1; \;\; y_l(t = 0) = 0; \;\; y_s(t = 0) = 0 \tag{8.18}$$

The corresponding solution is:

$$\begin{aligned}
y(t) &= f_4(p_{\text{"small pores"}}\kappa_{II} - f_5)\frac{(1 - e^{-f_3 t})}{q\kappa_{II}} - f_3(p_{\text{"small pores"}}\kappa_{II} - f_6)\frac{(1 - e^{-f_4 t})}{q\kappa_{II}} \\
f_5 &= \frac{\kappa_{II}(p_{\text{"small pores"}} - p_{\text{"large pores"}}) + \kappa_{\perp} - q}{2}; \; f_6 = \frac{\kappa_{II}(p_{\text{"small pores"}} - p_{\text{"large pores"}}) + \kappa_{\perp} + q}{2}
\end{aligned} \tag{8.19}$$

This model calculation results in some important practical conclusions. According to the definition of the shift as given by (2.12), evacuation or venting processes will result in a continuous change in the optical film thickness with time, which is easily accessible by means of in situ spectroscopic tools. Thereby, as it follows from (8.17) and (8.19), that shift may be analytically described as the sum of two different exponential functions with damping constants f_3 and f_4. In practice, their determination may be a straightforward procedure, but their interpretation is not: According to (8.17), f_3 and f_4 are involved functions of the porosity and the exchange rates κ. Nevertheless, in special cases (see Table 8.1), simplified and physically transparent expressions for the dependence of water content on time may be derived:

Table 8.1 Water migration kinetics: Special cases

Condition	Application in practice	Process	Water content in the film
$\kappa_{\perp} >> \kappa_{II}$	Moderately porous layer	Venting	$y(t) \approx p_{''\text{small pores}''}(1-e^{-p_{''\text{large pores}''}\kappa_{II}t})+p_{''\text{large pores}''}(1-e^{-\kappa_{\perp}t})$
		evacuation	$y(t) \approx p_{''\text{small pores}''}e^{-p_{''\text{large pores}''}\kappa_{II}t}+p_{''\text{large pores}''}e^{-\kappa_{\perp}t}$
$\kappa_{\perp} = \kappa_{II} \equiv \kappa$ $p << 1$	Almost dense layer	Venting	$y(t) \approx p_{''\text{small pores}''}(1-e^{-p_{''\text{large pores}''}\kappa t})+p_{''\text{large pores}''}(1-e^{-\kappa t})$
		evacuation	$y(t) \approx p_{''\text{small pores}''}e^{-p_{''\text{large pores}''}\kappa t}+p_{''\text{large pores}''}e^{-\kappa t}$
$p_{''\text{small pores}''} << p$	Strongly porous layer	Venting	$y(t) \approx p_{''\text{large pores}''}(1-e^{-\kappa_{\perp}t})$
		evacuation	$y(t) \approx p_{''\text{large pores}''}e^{-\kappa_{\perp}t}$

Thus, the *moderately porous layers* correspond to the situation sketched earlier in Fig. 2.12. The large pores are in direct correspondence with the ambient, their filling or evacuation kinetics are defined by a time constant dominated by the value of $\kappa_{\perp}^{-1}$. Small pores, however, have been postulated to exchange water only with the fraction of the large pores. Their filling or evacuation kinetics are therefore dominated by a time constant given by $(p_{''\text{large pores}''}\kappa_{II})^{-1}$. Both time constants are accessible from measurements of the time evolution of the optical film thickness.

In *strongly porous layers* as introduced in Fig. 8.11 the effects caused by filling or evacuation of the large pores are expected to be clearly dominant. The kinetics are practically defined by a simple single exponential function with the time constant $\kappa_{\perp}^{-1}$.

In *almost dense layers*, however, large open pores are no more expected to be relevant (Fig. 8.11). The introduction of two exchange rates $\kappa_{\perp}$ and κ_{II} does no more make sense, and the distinction between large and small pores now lacks its formerly obvious geometrical interpretation. According to (8.15), "large pores" merely have to be interpreted as pores which are able to exchange water with other pores *and* the ambient, while "closed pores" do *only* interact with other pores. Within this interpretation, the equations provided in Table 8.1 describe a minor and slow change of the optical thickness with time. In a typical shift measurement, those layers appear to be stable. At longer time scales, as they are typical for storage or aging effects, small gradual changes in the optical behavior may be recorded.

It turns out that shift measurements may give access to qualitative features even of the pore size distribution, although pore diameters are not explicitly present in equations like (2.21) or (8.15). Nevertheless, pore diameters have an implicit impact on the postulated values of the exchange rates as well as the volume fractions of small and large pores. We strongly believe that spectrophotometric shift measure-

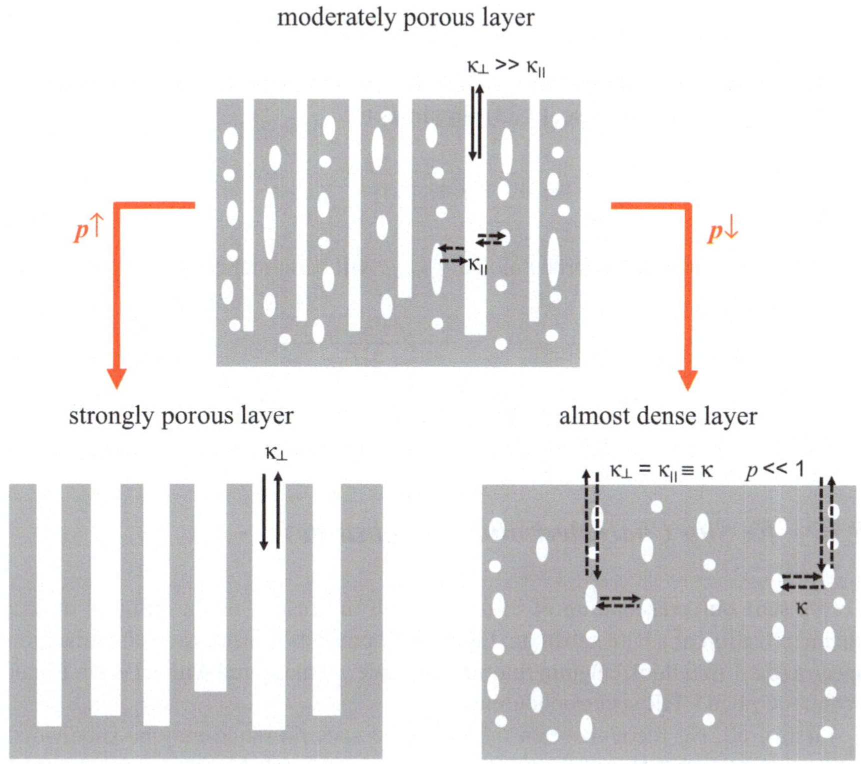

Fig. 8.11 Geometrical visualization of the porosity regimes introduced in Table 8.1. p is here the full porosity

ments do have the potential for determining the pore size distribution in a similar way as it may be done today be means of spectroellipsometric porosimetry [23, 24]. For another alternative approach, see also Sect. 15.2.3 of this book (*Effect of hydrocarbons absorption in thin films at 193 nm*) in this regard.

8.2 Examples

8.2.1 Basics

Classical and often used dispersion models as well as the β_do model [25] have been outlined in Sect. 7.3. For characterization, we will use the wavenumber grid (compare Sect. 8.1.4). In general, a merge of different dispersion models will be used for the following examples.

$$\varepsilon(\nu) = \varepsilon_{\infty}(\nu) + \chi_{Drude}(\nu) + \chi_{\beta\_do}(\nu) \tag{8.20}$$

Thereby, ε_{∞} is the contribution of a single Lorentzian oscillator with assumed negligible line width. In this case, (7.33) simplifies to

$$\varepsilon_{\infty}(\nu) = 1 + \frac{2J\nu_0}{\pi\left(\nu_0^2 - \nu^2\right)} \tag{8.21}$$

The contribution from the Drude model χ_{Drude} will be modelled by

$$\chi_{Drude}(\nu) = -\frac{\nu_{Drude}^2}{\nu^2 + 2i\Gamma_{Drude}\nu} \tag{8.22}$$

For the β_do model contribution χ_{β_do} will be calculated using (7.39) and (7.40).

8.2.2 Ex Situ Characterization of Substrates

Let us start our presentation of examples with the results of the infrared optical characterization of a bare substrate. Figure 8.12 presents transmission and reflection spectra of a 1 mm thick calcium fluoride substrate, as measured with a Perkin Elmer Frontier Optica FTIR spectrophotometer.

First of all, we recognize that the measured spectral range may be subdivided into two sections: a transparency region, where a remarkable transmission signal may be recorded. This transparency range corresponds to wavenumbers higher than approximately 800 cm^{-1}.

In this transparency range, measured transmission and reflection data are available for characterization. Fortunately, for normal incidence, according to Nichelatti [26] equations for transmittance and reflectance of the uncoated substrate may be

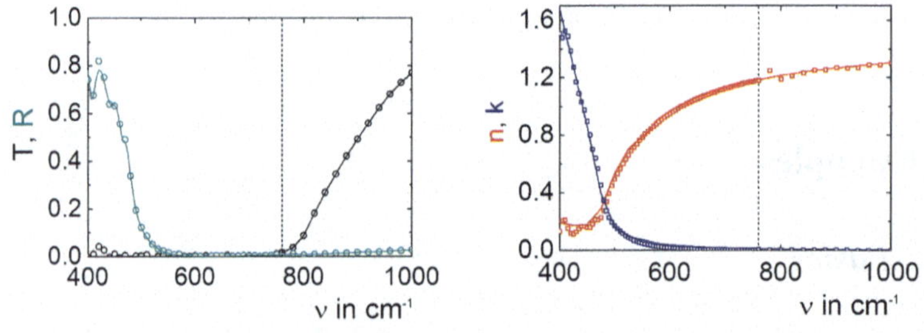

Fig. 8.12 Left: Measure (circle) and modelled (line) transmittance (black) and reflectance (dark cyan) of an uncoated CaF_2 substrate; Right: Calculated (squares) and literature data [27] (line) of the refractive index (red) and extinction coefficient (blue) dashed line marks wavenumber limit used for explicit solution (right) and fit (left)

inverted analytically, so that the calculation of n_{sub} and k_{sub} appears to be a rather straightforward task.

At lower wavenumbers, the transmission is suppressed, so that it is the reflectance only that gives us access to the optical constants. Here, we can make use of a Lorentzian multioscillator model (7.33) to fit the measured reflectance by minimizing a discrepancy function corresponding to the second term in (7.15).

The thus obtained optical constants are presented in Fig. 8.12 on the right together with literature data [27]. The excellent agreement with the literature data confirms us about the consistency of the presented approach.

8.2.3 Ex Situ Characterization of Single Layer Coatings

8.2.3.1 Dielectric Coatings

Here, we present characterization examples for single layer coatings built from hafnia and zirconia. In the transparency region of the corresponding coating, the application of the Lorentzian multi-oscillator model is known to work fine [28]. The task becomes a little bit more challenging, when the fundamental absorption edge is included into the characterization. Commonly, the number of Lorentzian oscillators must be increased. In the case of the hafnia layer, a set of at least 10 Lorentzian oscillators would be required. Thereby, two oscillators have a zero linewidth and only affect the refractive index. Nevertheless, already 28 parameters are used for modelling the dispersion of the optical constants. Therefore, the application of the β_do model (Sect. 7.3.4) seems promising. In fact, a merger of the β_do model (5 parameters with $\alpha = \beta$, $N = 1000$) and (8.21) is required to achieve a practically identical result (Fig. 8.13). This results in a total of only 7 fitting parameters. Thereby, the calculated optical constants show a similar spectral dependence as probably higher densified hafnia layers characterized by the universal dispersion model (Chap. 3 and [29], asterisks). Results from [30] (circles) look also similar, but seem to underestimate refractive index dispersion in the ultraviolet spectral range.

Next, the β_do model is applied to a zirconia single layer coating deposited on fused silica. Here, the coating is opaque for wavenumbers above approximately 47000 cm^{-1} (Fig. 8.14). Nevertheless, spectra could be fitted again using a merger of the β_do model and (8.21). The determined optical constants are in good agreement with data available from [30] (Table 8.2).

8.2.3.2 Semiconductor Coating

The characterization of amorphous germanium (a-Ge) in the UV/VIS/NIR/MIR spectral range using a merger of (8.21), (8.22) and (7.33) appears to be quite challenging a task [31]. Here, considering measurements from the same coating at different φ, on different substrates and coatings with different layer thickness has been required to

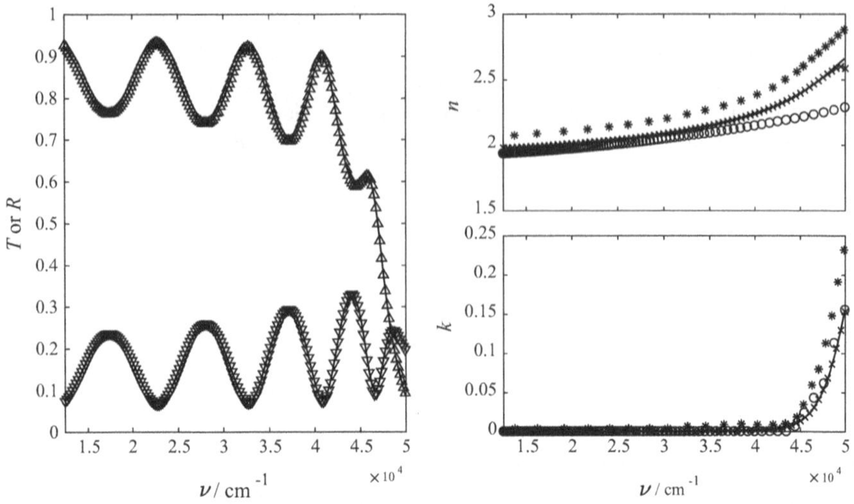

Fig. 8.13 Left: Measured transmittance (upward triangle) and reflectance (downward triangle) and corresponding modelled spectra (solid line) of a hafnia single layer coating on fused silica substrate; Right, Top: Refractive index of hafnia calculated with the β_do model (solid line) and multi-oscillator model (cross) in comparison with published data ([29] asterisks, [30] circles); Right, Bottom: Extinction coefficient calculated with the new model (solid line) and multi-oscillator model (cross)

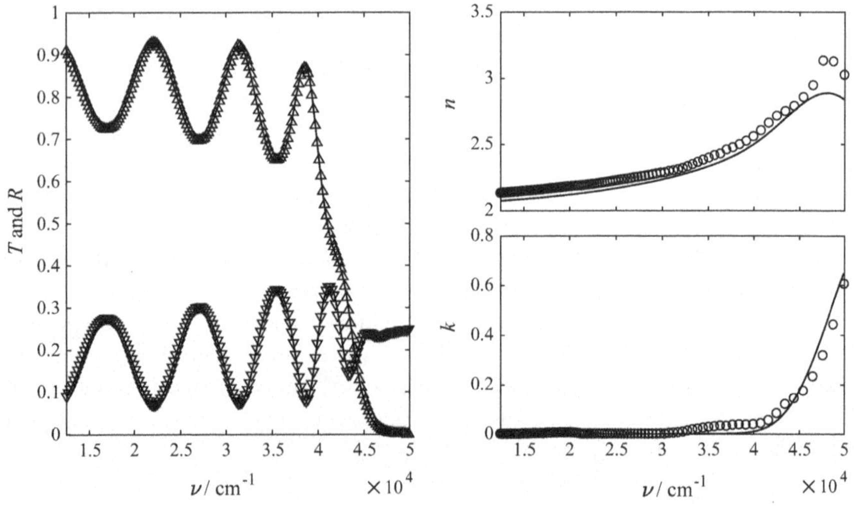

Fig. 8.14 Left: Measured transmittance (downward triangle) and reflectance (upward triangle) and corresponding modelled spectra (solid line) of a zirconia single layer coating on fused silica substrate; Right: Calculated refractive index (solid line, left axes) and extinction coefficient (dotted line, right axes) of zirconia

extract approximately 300 parameters for the multi-oscillator model. It is obvious, that application of the β_do model instead of (7.33) could significantly reduce the number of required parameters and improve the stability of the fitting process. Here, only experimental data from an approximately 100 nm single layer coating on a CaF_2 substrate have been included into the discrepancy function. Transmittance and reflectance spectra measured with the Perkin Elmer Frontier Optica FTIR at near normal incidence and with the 6° and 60° VN-accessory for the Perkin Elmer Lambda 900 are used (Fig. 8.15) for characterization. The calculated optical constants (underlying model parameters are summarized in Table 8.3) are quite smooth and in good agreement with previously published data obtained from the multi-oscillator model [31]. The calculated layer thickness is 102.1 nm and close to expected value. A total of 9 fitting parameters summarized in Table 8.3 is used.

Table 8.2 Model parameters for optical constants of hafnia and zirconia

	$\chi_{\beta\_do}(\nu)$					$\varepsilon_\infty(\nu)$	
	$\nu_{min,1}$ in cm^{-1}	$\nu_{max,1}$ in cm^{-1}	$J_{beta,1}$ in cm^{-1}	$\Gamma_{beta,1}$ in cm^{-1}	$\alpha = \beta$	ν_0 in cm^{-1}	J in cm^{-1}
hafnia	19170	98695	60095	96.3	35.9	87857	300119
zirconia	30253	74582	47423	97.2	13.7	74826	307380

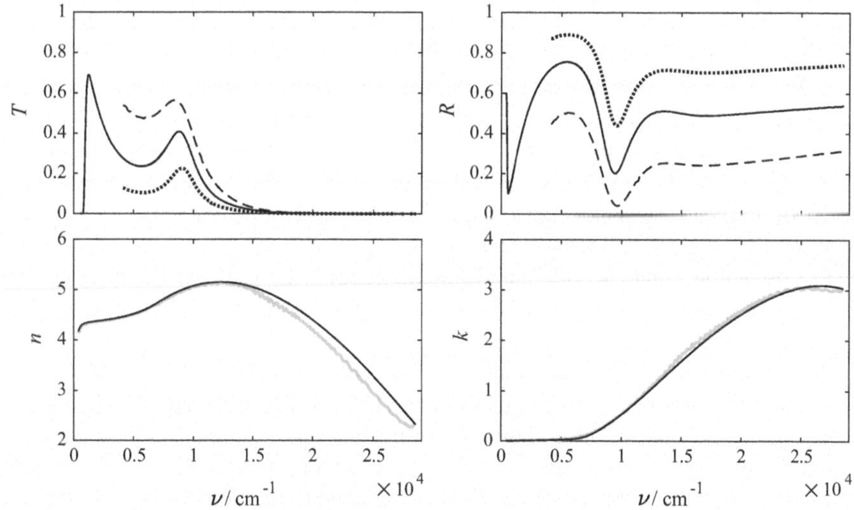

Fig. 8.15 Top: measured transmittance (left) and reflectance (right) of an approximately 100 nm thick a-Ge layer on a CaF_2 substrate measured at near normal incidence in a Frontier Optica FTIR and Lambda 900 equipped with a 6° VN accessory (solid line) and at 60° for s-(dotted line) and p-polarization (dashed) measured in a Lambda 900 equipped with a 60° VN accessory Bottom: Calculated refractive index (left) and extinction coefficient (right) using a merger of a single Lorentzian oscillator with zero linewidth, Drude model and multi-oscillator model (gray line) and β_do model (black line)

Table 8.3 Model parameters for optical constants of a-Ge

$\chi_{Drude}(\nu)$		$\chi_{\beta\_do}(\nu)$					$\varepsilon_{\infty}(\nu)$	
ν_{Drude} in cm^{-1}	Γ_{Drude} in cm^{-1}	$\nu_{min,1}$ in cm^{-1}	$\nu_{max,1}$ in cm^{-1}	$J_{beta,1}$ in cm^{-1}	$\Gamma_{beta,1}$ in cm^{-1}	$\alpha = \beta$	ν_0 in cm^{-1}	J in cm^{-1}
511.5	5.15	6329.7	36176	388443	376.1	2.79	63703	532173

Table 8.4 Model parameters for optical constants of Cu single layer

$\chi_{Drude}(\nu)$		$\chi_{\beta\_do}(\nu)$					$\varepsilon_{\infty}(\nu)$	
ν_{Drude} in cm^{-1}	Γ_{Drude} in cm^{-1}	$\nu_{min,1}$ in cm^{-1}	$\nu_{max,1}$ in cm^{-1}	$J_{beta,1}$ in cm^{-1}	$\Gamma_{beta,1}$ in cm^{-1}	$\alpha - 1$	ν_0 in cm^{-1}	J in cm^{-1}
69368	307.6	17711	41413	127101	530.0	2.1e-07	38777	68360

8.2.3.3 Metal Coating

The fit of metal thin film spectra is another difficult task, because the transmittance spectra are suppressed in broad spectral regions, and no interference pattern is observed that could give us valuable a priori information according to what has been discussed in Sect. 7.2.2.2. Nevertheless, reliable spectra fits are possible in terms of a merger (8.21), (8.22) and (7.33), as earlier demonstrated in [2, 32]. Again, we now replace the multioscillator model (7.33) by the β_do model. Underlying model parameters are summarized in Table 8.4. Here, the parameters $\alpha = \beta$ are close to 1 and therefore, the set of oscillators is nearly uniformly distributed.

In Fig. 8.16 we see the spectra fits of an approximately 120 nm thick copper film on fused silica (top). The corresponding optical constants resemble what has been presented in Table 7.2, and we recognize the expected high extinction coefficients ($k \gg n$) in broad spectral regions. The drop in the reflectance around a wavenumber of 20000 cm^{-1} (corresponding to a wavelength of 500 nm) is responsible for the typical color of clean copper surfaces.

Let us denote, that the calculated thickness could be estimated only from the weak transmittance signal around the wavenumber 20000 cm^{-1}. Nevertheless, the calculated layer thickness of 129 nm is close to the expected value.

The underlying model parameters are summarized in Table 8.4. According to the theoretical considerations discussed in Sect. 7.3.1, the parameters of the Drude function shall give information about the plasma frequency ω_p and damping constant of the corresponding metal. In order to provide an impression, Table 8.5 presents corresponding values obtained from the fit compared to literature data.

Note that the thin film relaxation times are generally lower than the reported bulk values; this is a physically consistent result, because real films produced by technologically relevant deposition techniques contain plenty of defects, which give rise to the lower relaxation times for free electrons motion.

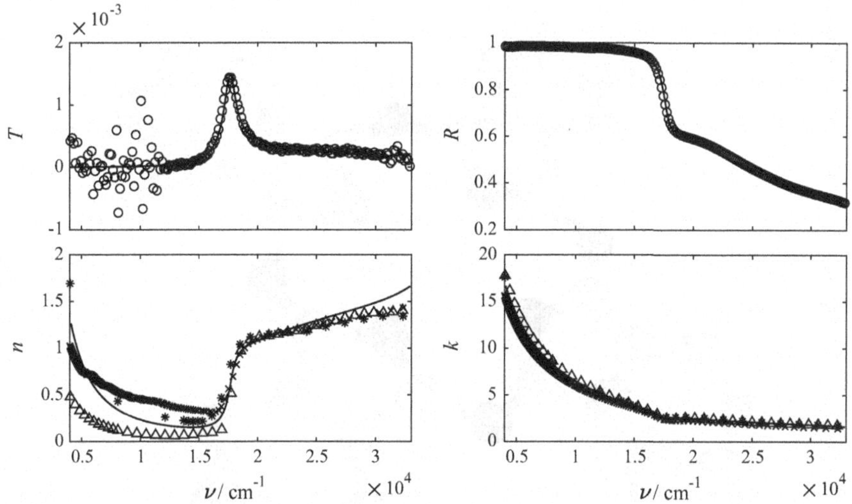

Fig. 8.16 Left: Measured (circle) and modelled (solid line) transmittance and reflectance of an approximately 130 nm thick Cu layer on a fused silica substrate; Right: Modelled refractive index and extinction coefficient (solid line) and literature data (asterisk: [33], triangle [34], cross [35])

Table 8.5 Drude function parameters as obtained from the fit of the Cu film spectra

Model	Estimated plasmon energy		Estimated relaxation time	
	Our fit (single film)	Literature data	Our fit (single film)	Literature data
β_do	8.6 eV	9.3 eV (bulk) [36]	8.6 fs	16-35 fs (bulk) [36]
Lorentzian	9.1 eV		9.7 fs	

8.2.3.4 Organic Dye Coating

Finally, we want to apply the new model to so-called Q absorption band of an approximately 20 nm thick free base phthalocyanine (H_2Pc, Fig. 8.17 [37]) layer deposited on a fused silica substrate. Corresponding transmission and reflection spectra are shown in (Fig. 8.17 left). For this material, application of the multioscillator model to this spectral range is known to be problematic because inhomogeneous broadening of the lines should be considered [38] so that the Brendel model (Table 7.3) may be used instead [39, 40].

When using the β_do model, we have to consider contributions to the optical constants which arise from absorptions outside of the Q-band. To do this, an extended version of (8.21) is used:

$$\varepsilon_\infty(\nu) = 1 + \frac{J_3}{\pi}\left(\frac{1}{\nu_{03} - \nu - \mathrm{i}\Gamma_3} + \frac{1}{\nu_{03} + \nu + \mathrm{i}\Gamma_3}\right) + \frac{2J_4\nu_{04}}{\pi\left(\nu_{04}^2 - \nu^2\right)} \tag{8.23}$$

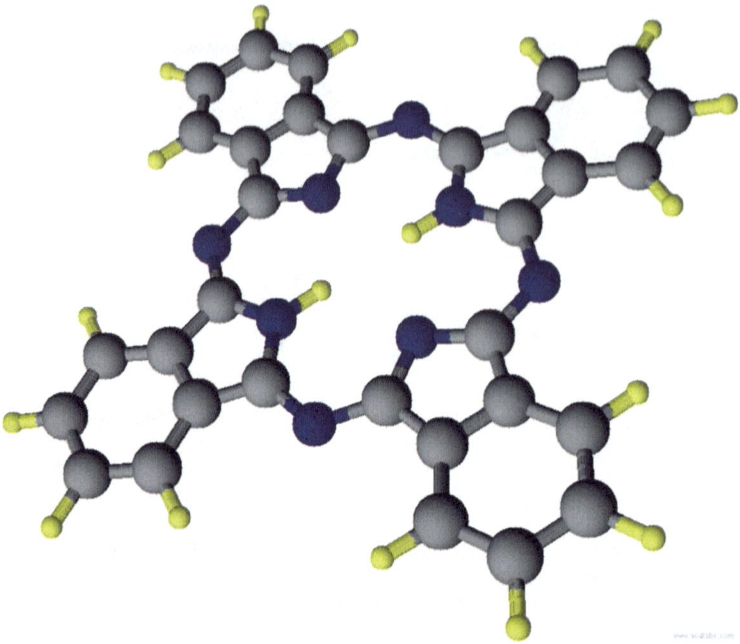

Fig. 8.17 Structure of the H_2Pc molecule calculated with [37]. Hydrogen atoms are shown in yellow, carbon in grey, and nitrogen in navy. Printed with permission of Advanced Chemistry Development, Inc

Table 8.6 Model parameters for optical constants of the H_2Pc single layer

	β_do model				
j	$\nu_{min,j}$ in cm^{-1}	$\nu_{max,j}$ in cm^{-1}	$J_{beta,j}$ in cm^{-1}	$\Gamma_{beta,j}$ in cm^{-1}	$\alpha_j = \beta_j$
1	12907	15110	474.1	2.75	3.60
2	1003.9	30056	7189	369.3	85.5
	Lorentzian oscillators				
	ν_{0j} in cm^{-1}		J_j in cm^{-1}	Γ_j in cm^{-1}	
3	24031		1245.8	1473.0	
4	42754		93302	0	

The calculated model parameters are summarized in Table 8.6 (Fig. 8.18).

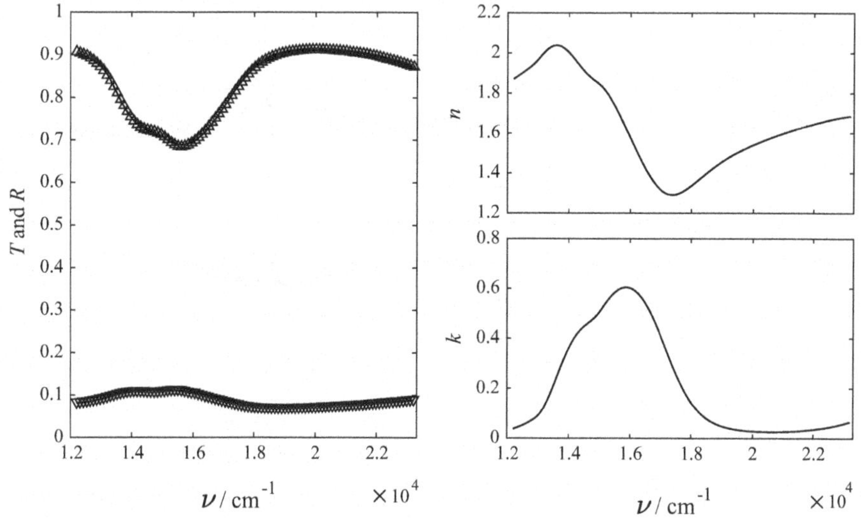

Fig. 8.18 Left: Modelled (line) and measured transmittance (up triangle) and reflectance (down triangle) of an approximately 20 nm thick H_2PC layer on a fused silica substrate; Right: Modelled refractive index (top) and extinction coefficient (bottom)

8.2.4 Interplay of Ex Situ and In Situ Spectroscopy: Preparation and Characterization of a V-Coating

The increased number of parameters and the multiplicity of mathematical solutions make the characterization of multilayer coatings to a quite challenging task. The latter may be addressed by including additional measurement data into the characterization process. Thereby, including of recorded in situ measurement data seems prospective, but may result in further complications when optical constants depends on environment conditions (compare Sect. 8.1.6). Here, some basic concepts will be applied to one of the simplest multilayer coating: a two-layer antireflection coating for 1030 nm at 31° angle of incidence ("V-coating") with a high laser induced damage threshold (LIDT) in the femtosecond regime. In this case, high band gap materials are prospective [41] so that alumina has been selected as high index material and aluminum fluoride as low index material. It is well known, that fluoride coatings are not well-suited for preparation under conditions of ion assistance and therefore, electron beam evaporation without assistance was used for this layer. The resulting porous structure of the coating results in a significant air-to-vacuum shift (Fig. 8.19), which has to be taken into account. The corresponding refractive indices of the AlF_3 film as modelled in terms of (8.21) are shown in Fig. 8.20.

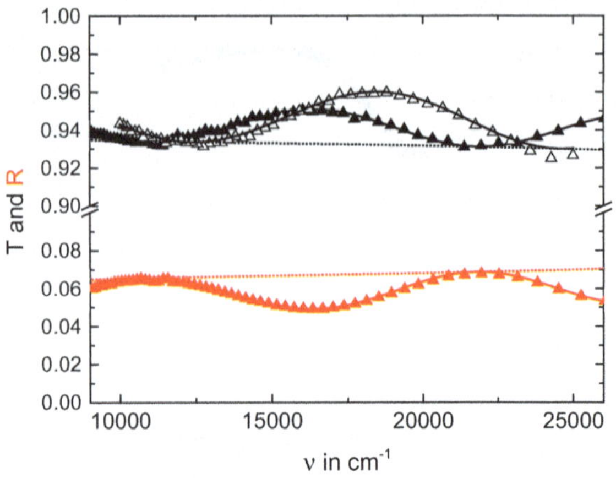

Fig. 8.19 Modelled (solid line) and measured in situ (hollow triangle) and ex situ (filled triangle) transmittance (black) and reflectance (red) of an aluminum fluoride single layer deposited on fused silica (transmittance and reflectance of the uncoated substrate: dotted line)

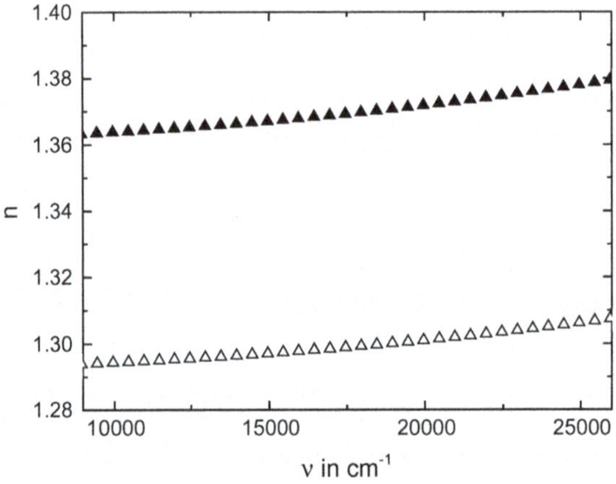

Fig. 8.20 Modelled in situ (hollow triangle) and ex situ (filled triangle) refractive index of an approximately 334 nm thick AlF_3 layer deposited on fused silica

In the case of alumina deposition, two different approaches have been considered:

1. weak assistance and moderate heating during deposition (used later in the design AR1)
2. neither assistance nor heating (used later in the design AR2).

In both cases, ex situ and in situ optical constants have been determined. The corresponding designs AR1 and AR2 for the V-coating (Fig. 8.21) are nearly identical.

For both coatings, the in situ measured transmittance has been in a good agreement with the theoretical performance. In contrast, the ex situ reflectance of AR2 shows significant deviations from the theoretical performance (Fig. 8.22 on right).

Obviously, the assumed optical constants of the AlF_3 layer are not correct when it is deposited on non-assisted alumina (AR2). This may be explained by a different

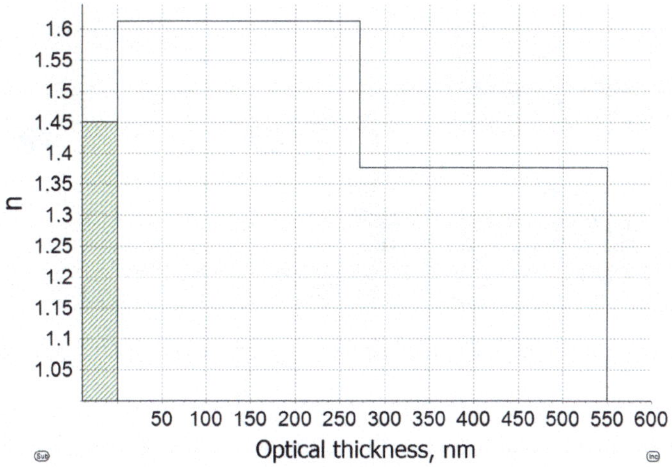

Fig. 8.21 Refractive index profile of the V-coating with weak assistance (AR1: 167.88 nm Al_2O_3, 201.66 nm AlF_3) and no assistance (AR2: 167.55 nm Al_2O_3, 201.70 nm AlF_3)

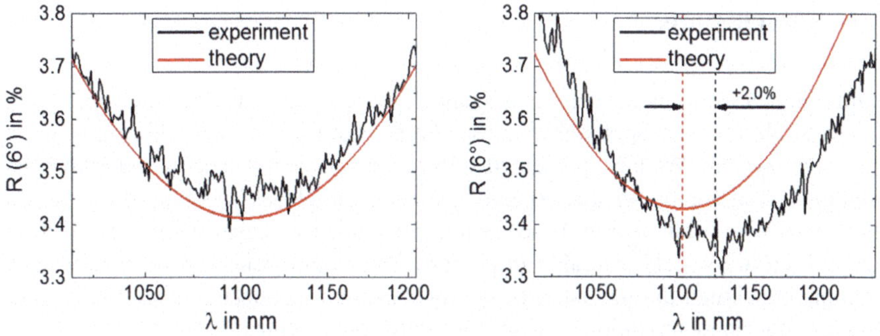

Fig. 8.22 Left: Measured and calculated reflectance of the V coating AR1; Right: Measured and calculated reflectance of the V coating AR2

fluoride growth on a slightly porous alumina layer when being compared with the growth on a fused silica substrate. On porous alumina, the porosity of the resulting fluoride layer also seems to be increased which results in a slightly decreased in situ refractive index, which in turn leads to an increased geometrical thickness when the layer growth is monitored by optical means. That increased thickness explains the observed difference in the minima positions of ex situ measured and calculated reflectance (Fig. 8.22 on right).

Furthermore, the measured LIDT for both the coatings obtained from a series of 429 fs pulses at 1030 nm depends on deposition conditions (Fig. 8.23). The LIDT is 1.89 J/cm^2 for AR1 and 1.32 J/cm^2 for AR2 [42]. Therefore, the weak assistance applied for alumina preparation in AR1 does not only result in a better agreement

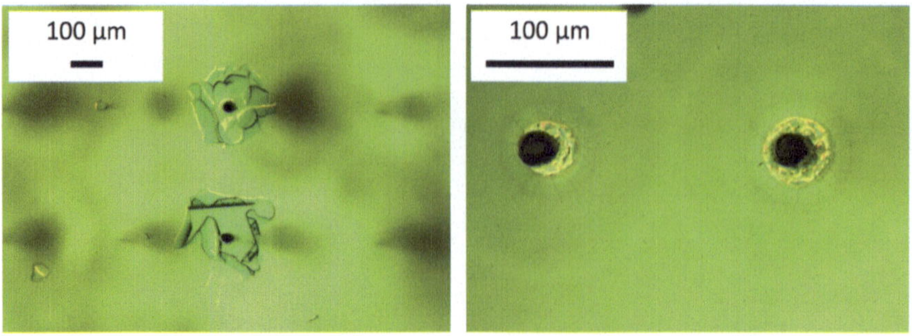

Fig. 8.23 Damage of V-coating AR1 (left) and AR2 (right) after a series of 429 fs pulses at 1030 nm

between experimental and theoretical performances, but also in an improved sub-picosecond LIDT of the coating.

8.3 Conclusions

In this chapter, combined with Chaps. 2 and 7, we have demonstrated the application of spectrophotometric approaches to the characterization of single thin films and a multilayer system. We have presented selected experimental aspects and numerous examples from coating characterization practice. Emphasis was placed on ex situ characterization of single films based on normal incidence transmission and reflection data, which are widely available in practice. For completeness, a more advanced example concerned the inclusion of in situ transmission spectra as well as ex situ data into the characterization strategy for multilayer coatings.

We would like to summarize our analysis in terms of the following theses:

- Spectrophotometry allows determination of the optical constants of thick substrates as well as thin films. This has been demonstrated for selected dielectric, metal and semiconductor films, including an organic dye layer.
- Additionally, spectrophotometry gives access to geometrical construction parameters like the film thickness.
- In complicated samples (for example multilayer coatings), the inclusion of oblique incidence spectra and/or in situ spectra obtained during film deposition may be helpful for enhancing the reliability of the characterization results.
- The obtained optical constants give further access to related quantities, including density, porosity, but also charge carrier density, band structure and possible impurities of the coating. The latter may again be related to results of non-optical characterization techniques like electron microscopy, X-ray-reflection, stoichiometry investigations and the like, and thus contribute to the completion of a physical picture on the nature of the samples investigated.

References

1. H. Kuzmany, *Festkörperspektroskopie. Eine Einführung* (Springer, Berlin Heidelberg New York, 1989)
2. O. Stenzel, *The Physics of Thin Film Optical Spectra: An Introduction*. Springer Series in Surface Sciences, vol. 44, 2nd edn. (Springer, Berlin Heidelberg, 2015)
3. A.V. Tikhonravov, M.K. Trubetskov, M.A. Kokarev, T.V. Amotchkina, A. Duparre, E. Quesnel, D. Ristau, S. Günster, Effect of systematic errors in spectral photometric data on the accuracy of determination of optical parameters of dielectric thin films. Appl. Opt. **41**, 2555–2560 (2002)
4. O. Stenzel, *Optical Coatings: "Material Aspects in Theory and Practice"* (Springer, Berlin Heidelberg, 2014)
5. T. Burt, H. ChuanXu, A. Jiang, Performance of compact visual displays—measuring angular reflectance of optically active materials using the Agilent Cary 7000 Universal Measurement Spectrophotometer (UMS), Agilent Application Note 5991-2508EN, 2013
6. R. Francis, T. Burt, Optical characterization of thin films using a new Universal Measurement Accessory for the Agilent Cary UV-Vis-NIR spectrophotometers, Agilent Application Note 5991-1356EN, 2013
7. R. Vernhes, L. Martinu, TRACK—A new method for the evaluation of low-level extinction coefficient in optical films. Opt. Expr. **23**, 28501–28521 (2015)
8. A. Savitzky, M.J.E. Golay, Smoothing and differentiation of data by simplified least squares procedures. Anal. Chem. **36**, 1627–1639 (1964)
9. B. Vidal, A. Fornier, E. Pelletier, Optical monitoring of nonquarterwave multilayer filters. Appl. Opt. **17**, 1038–1047 (1978)
10. B. Vidal, A. Fornier, E. Pelletier, Wideband optical monitoring of nonquarterwave multilayer filters. Appl. Opt. **18**, 3851–3856 (1979)
11. P.J. Bruce, I.J. Hodgkinson, A stable optical photometer for vacuum coating. J. Vac. Sci. Technol., A **3**, 436–437 (1985)
12. F.J. van Milligen, H.A. Macleod, Development of an automated scanning monochromator for monitoring thin films. Appl. Opt. **24**, 1799–1802 (1985)
13. I. Powell, J.C.M. Zwinkels, Development of optical monitor for control of thin film deposition. Appl. Opt. **25**, 3645–3652 (1986)
14. R.P. Netterfield, P.J. Martin, K.H. Müller, In-situ optical monitoring of thin film deposition. SPIE **1988**, 10–15 (1012)
15. B.T. Sullivan, G. Carlow, An overview of optical monitoring techniques, in *Optical Interference Coatings, OSA Technical Digest* (Optical Society of America, 2010, paper TuC1
16. A.V. Tikhonravov, M.K. Trubetskov, T.V. Amotchkina, Investigation of the effect of accumulation of thickness errors in optical coating production by broadband optical monitoring. Appl. Opt. **45**, 7026–7034 (2006)
17. A.V. Tikhonravov, M.K. Trubetskov, Elimination of cumulative effect of thickness errors in monochromatic monitoring of optical coating production: theory. Appl. Opt. **46**, 2084–2090 (2007)
18. http://www.canon.com/news/2015/sep07e.html
19. R.R. Willey, *Practical Design and Production of Optical Thin Films*, 2nd edn. (CRC Press, 2002). ISBN 9780824708498
20. A.V. Tikhonravov, M.K. Trubetskov, T.V. Amotchkina, Optical monitoring strategies, in ed. by Piegari, Flory, *Optical Thin Films and Coatings* (Woodhead Publishing Limited, 2013)
21. J. Gäbler, O. Stenzel, S. Wilbrandt, N. Kaiser, Optische in-situ Prozessverfolgung und -steuerung des Aufdampfens optischer Beschichtungen durch gleichzeitige Messungen des Transmissions- und Reflexionsvermögens der wachsenden Schicht. Vak. Forsch. Prax. **25**(6), 22–28 (2013)
22. http://www.jeti.com
23. M.R. Baklanov, K.P. Mogilnikov, V.G. Polovinkin, F.N. Dultsev, Determination of pore size distribution in thin films by ellipsometric porosimetry. J. Vac. Sci. Technol. B, 1385–1391 (2000)

24. C. Murray, C. Flannery, I. Streiter, S.E. Schulz, M.R. Baklanov, K.P. Mogilnikov, C. Himcinschi, M. Friedrich, D.R.T. Zahn, T. Gessner, Comparison of techniques to characterise the density, porosity and elastic modulus of porous low-k SiO xerogel films
25. S. Wilbrandt, O. Stenzel, Empirical extension to the multioscillator model: the beta-distributed oscillator model. Appl. Opt. **56**, 9892-9899 (2017)
26. E. Nichelatti, Complex refractive index of a slab from reflectance and transmittance: analytical solution. J. Opt. A: Pure Appl. Opt. **4**, 400–403 (2002)
27. E.D. Palik (ed.), *Handbook of Optical Constants of Solids*, vol. I, II (Academic Press, Orlando, 1998), pp. 753–763, 766–775, 830–835
28. K. Friedrich, S. Wilbrandt, O. Stenzel, N. Kaiser, K.H. Hoffmann, Computational manufacturing of optical interference coatings: method, simulation results, and comparison with experiment. Appl. Opt. **49**, 3150–3162 (2010)
29. D. Franta, D. Nečas, I. Ohlídal, Universal dispersion model for characterization of optical thin films over a wide spectral range: application to hafnia. Appl. Opt. **54**, 9108–9119 (2015)
30. Center for Nanolithography Research, Rochester Institute of Technology, http://www.rit.edu/~w-lith/thinfilms/thinfilms/thinfilms.html
31. S. Wilbrandt, C. Franke, V. Todorova, O. Stenzel, N. Kaiser, Infrared optical constants determination by advanced FTIR techniques, in *Optical Interference Coatings 2016, OSA Technical Digest* (online) (Optical Society of America, 2016), paper MC.10
32. O. Stenzel, S. Wilbrandt, N. Kaiser, Spektralphotometrische Messungen zur Charakterisierung und Qualitätskontrolle von Oberflächen und Schichten, Colloquium optische Spektrometrie (COSP), 17./18.03.2014
33. H.J. Hagemann, W. Gudat, C. Kunz, Optical constants from the far infrared to the x-ray region: Mg, Al, Cu, Ag, Au, Bi, C, and Al2O3. J. Opt. Soc. Am. **65**, 742–744 (1975)
34. S. Babar, J.H. Weaver, Optical constants of Cu, Ag, and Au revisited. Appl. Opt. **54**, 477–481 (2015)
35. M.R. Querry, Optical constants, Contractor Report CRDC-CR-85034 (1985)
36. H. Ehrenreich, H.R. Philipp, Optical properties of Ag and Cu. Phys. Rev. **128**, 1622–1629 (1962)
37. ACD/ChemSketch (Freeware), version 2016.2, Advanced Chemistry Development, Inc., Toronto, ON, Canada, www.acdlabs.com, 2016
38. The linear optical constants of thin phthalocyanine and fullerite films from the near infrared up to the UV spectral regions: Estimation of electronic oscillator strength values
39. R. Brendel, D. Bormann, An infrared dielectric function model for amorphous solids. J. Appl. Phys. **71**, 1–6 (1992)
40. A. Franke, A. Stendal, O. Stenzel, C. von Borczyskowski, Gaussian quadrature approach to the calculation of the optical constants in the vicinity of inhomogeneously broadened absorption lines. Pure Appl. Opt. **5**, 845–853 (1996)
41. M. Jupe et al., Mixed oxide coatings for advanced fs-laser applications. Proc. SPIE **6720**, 67200U (2007)
42. Final report, Optische Komponenten und Baugruppen mit hohen Lebensdauern für Ultrakurzpuls-Laser und Systeme (Ultra-LIFE), FKZ: 13N11555, Technische Informationsbibliothek

Chapter 9
Ellipsometry of Layered Systems

Ivan Ohlídal, Jiří Vohánka, Martin Čermák and Daniel Franta

Abstract In this chapter the theoretical aspects of ellipsometry and their applications in optics of layered systems are presented. The basic formulae of the theory of ellipsometric measurements are introduced. For this purpose the Jones and Stokes–Mueller matrix formalisms are used. By using these formalisms the individual types of ellipsometry and the most utilized ellipsometric techniques are briefly described. Furthermore, the matrix formalisms enabling us to derive the formulae for the optical quantities of optically isotropic and anisotropic layered systems are described as well. Applications of the matrix formalisms in practice are illustrated by means of three examples.

9.1 Introduction

Ellipsometry is a very useful experimental tool allowing to perform optical characterization of thin film systems efficiently. Therefore, a considerable attention has been devoted to developing both the theoretical and experimental aspects of ellipsometry. During the last two decades, enormous progress has been achieved in this field. In this chapter we will deal with the theoretical aspects of ellipsometry and their applications in optics of thin films. The results will be formulated with the help of matrix formalisms that allow us to derive formulae for the ellipsometric quantities of thin films in a systematic way. Specifically, we will focus on the Jones matrix formalism, Stokes–Mueller matrix formalism, matrix formalism for optically isotropic layered systems and matrix formalism for optically anisotropic layered systems (Yeh matrix formalism).

Applications of both the Jones and Stokes–Mueller formalisms are used to describe the theory of ellipsometric measurements within conventional, generalized and Mueller-matrix ellipsometries together with ellipsometric techniques utilized for

---

I. Ohlídal (✉) · J. Vohánka · M. Čermák · D. Franta
Faculty of Science, Department of Physical Electronics, Masaryk University,
Kotlářská 2, 611 37 Brno, Czech Republic
e-mail: ohlidal@physics.muni.cz

O. Stenzel and M. Ohlídal (eds.), *Optical Characterization of Thin Solid Films*, Springer Series in Surface Sciences 64,
https://doi.org/10.1007/978-3-319-75325-6_9

practical measurements. Furthermore, we present three examples demonstrating the use of the matrix formalisms. The applications of the matrix formalisms to reflection of light from an isotropic inhomogeneous layer, uniaxial anisotropic layer and the reflection and transmission of light by a transparent slab covered with layered systems are performed.

Only ellipsometric formalisms concerning specular reflection and transmission corresponding to Snell's law will be presented. Ellipsometry corresponding to the scattered light will not be considered.

9.2 Matrix Formalisms

9.2.1 Jones Formalism

Let us assume that a polarized monochromatic plane wave is incident on a thin film system. After the interaction between the incident wave and the system, an outgoing monochromatic plane wave emerges from this system. A schematic diagram of this situation is presented in Fig. 9.1. The Cartesian coordinate systems (x, y, z) and (x', y', z') are connected with the incident and outgoing plane waves propagating along z and z', respectively. The wave vectors $\mathbf{k}_\mathrm{i}$ and $\mathbf{k}_\mathrm{o}$ need not be mutually parallel. It will be assumed that x and x' and/or y and y' coordinate axes are parallel and/or perpendicular to the plane of incidence given by the wave vector of the incident wave and the normals to the boundaries of the system. Both the plane waves mentioned above are described by the Jones vectors $\hat{\mathbf{E}}_\mathrm{i}$ (incident) and $\hat{\mathbf{E}}_\mathrm{o}$ (outgoing) [1–4]. The Jones vectors are the complex two-dimensional vectors given as

$$\hat{\mathbf{E}}_\mathrm{i} = \begin{pmatrix} \hat{E}_{\mathrm{i}x} \\ \hat{E}_{\mathrm{i}y} \end{pmatrix} = \begin{pmatrix} \hat{E}_\mathrm{ip} \\ \hat{E}_\mathrm{is} \end{pmatrix} \quad \text{and} \quad \hat{\mathbf{E}}_\mathrm{o} = \begin{pmatrix} \hat{E}_{\mathrm{o}x} \\ \hat{E}_{\mathrm{o}y} \end{pmatrix} = \begin{pmatrix} \hat{E}_\mathrm{op} \\ \hat{E}_\mathrm{os} \end{pmatrix}, \tag{9.1}$$

where $\hat{E}_{\mathrm{i}x}$ and $\hat{E}_{\mathrm{i}y}$ and/or $\hat{E}_{\mathrm{o}x}$ and $\hat{E}_{\mathrm{o}y}$ denote the complex amplitudes of the electric fields of the p- and s-polarized incident and/or outgoing monochromatic plane waves, respectively. The Jones vector fully describes the polarization state (polarization) of the polarized monochromatic plane wave. Moreover, it will be assumed that the thin

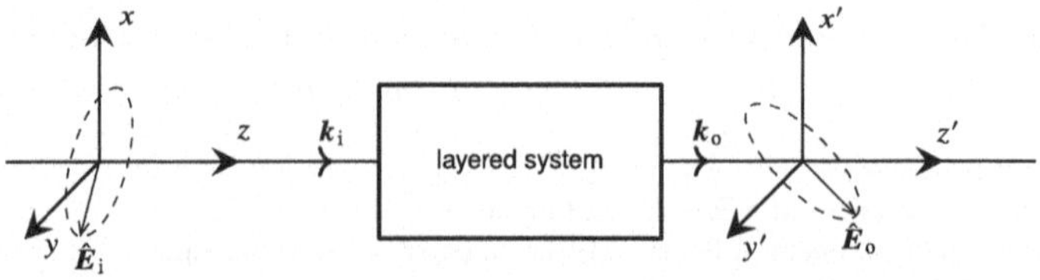

Fig. 9.1 Schematic diagram of the incident wave on the system and the emergent wave from this system

film system is linear, i.e. nonlinear and depolarization effects are absent in the system. It is then possible to write the following matrix equation:

$$\begin{pmatrix} \hat{E}_{\mathrm{op}} \\ \hat{E}_{\mathrm{os}} \end{pmatrix} = \begin{pmatrix} \hat{J}_{\mathrm{pp}} & \hat{J}_{\mathrm{ps}} \\ \hat{J}_{\mathrm{sp}} & \hat{J}_{\mathrm{ss}} \end{pmatrix} \begin{pmatrix} \hat{E}_{\mathrm{ip}} \\ \hat{E}_{\mathrm{is}} \end{pmatrix}, \tag{9.2}$$

or if the 2×2 matrix, called the Jones matrix, is denoted by $\hat{\mathbf{J}}$ then we can write this as:

$$\hat{\boldsymbol{E}}_{\mathrm{o}} = \hat{\mathbf{J}}\hat{\boldsymbol{E}}_{\mathrm{i}}. \tag{9.3}$$

Note that in general, the elements $\hat{J}_{ij}$ of the Jones matrix are complex numbers. The Jones matrices can be also written as

$$\begin{aligned} \hat{\mathbf{J}} &= \begin{pmatrix} \hat{J}_{\mathrm{pp}} & \hat{J}_{\mathrm{ps}} \\ \hat{J}_{\mathrm{sp}} & \hat{J}_{\mathrm{ss}} \end{pmatrix} = \begin{pmatrix} \hat{r}_{\mathrm{pp}} & \hat{r}_{\mathrm{ps}} \\ \hat{r}_{\mathrm{sp}} & \hat{r}_{\mathrm{ss}} \end{pmatrix} \quad \text{(reflection mode)}, \\ \hat{\mathbf{J}} &= \begin{pmatrix} \hat{J}_{\mathrm{pp}} & \hat{J}_{\mathrm{ps}} \\ \hat{J}_{\mathrm{sp}} & \hat{J}_{\mathrm{ss}} \end{pmatrix} = \begin{pmatrix} \hat{t}_{\mathrm{pp}} & \hat{t}_{\mathrm{ps}} \\ \hat{t}_{\mathrm{sp}} & \hat{t}_{\mathrm{ss}} \end{pmatrix} \quad \text{(transmission mode)}, \end{aligned} \tag{9.4}$$

where $\hat{r}_{\mathrm{pp}}, \hat{r}_{\mathrm{ps}}, \hat{r}_{\mathrm{sp}}, \hat{r}_{\mathrm{ss}}$ are the Fresnel reflection coefficients and $\hat{t}_{\mathrm{pp}}, \hat{t}_{\mathrm{ps}}, \hat{t}_{\mathrm{sp}}, \hat{t}_{\mathrm{ss}}$ are the Fresnel transmission coefficients.

In ellipsometry, it is convenient to introduce the normalized Jones matrix as

$$\hat{\mathbf{J}}_{\mathrm{n}} = \begin{pmatrix} \hat{\rho}_1 & \hat{\rho}_2 \\ \hat{\rho}_3 & 1 \end{pmatrix} = \begin{pmatrix} \hat{J}_{\mathrm{pp}}/\hat{J}_{\mathrm{ss}} & \hat{J}_{\mathrm{ps}}/\hat{J}_{\mathrm{ss}} \\ \hat{J}_{\mathrm{sp}}/\hat{J}_{\mathrm{ss}} & 1 \end{pmatrix}. \tag{9.5}$$

When the ellipsometric measurements are performed in reflected light (reflection mode), the elements of the normalized Jones matrix are expressed as follows

$$\hat{\rho}_1 = \frac{\hat{r}_{\mathrm{pp}}}{\hat{r}_{\mathrm{ss}}}, \qquad \hat{\rho}_2 = \frac{\hat{r}_{\mathrm{ps}}}{\hat{r}_{\mathrm{ss}}}, \qquad \hat{\rho}_3 = \frac{\hat{r}_{\mathrm{sp}}}{\hat{r}_{\mathrm{ss}}}. \tag{9.6}$$

In the case of ellipsometric measurements carried out in the transmitted light (transmission mode), they are expressed as follows:

$$\hat{\rho}_1 = \frac{\hat{t}_{\mathrm{pp}}}{\hat{t}_{\mathrm{ss}}}, \qquad \hat{\rho}_2 = \frac{\hat{t}_{\mathrm{ps}}}{\hat{t}_{\mathrm{ss}}}, \qquad \hat{\rho}_3 = \frac{\hat{t}_{\mathrm{sp}}}{\hat{t}_{\mathrm{ss}}}. \tag{9.7}$$

It is practical to describe the polarization states of light waves using the relative amplitudes and differences of the phases belonging to their p- and s-polarizations, i.e. it is advantageous to define the following complex quantities

$$\hat{\chi}_{\mathrm{i}} = \frac{\hat{E}_{\mathrm{ip}}}{\hat{E}_{\mathrm{is}}}, \qquad \hat{\chi}_{\mathrm{o}} = \frac{\hat{E}_{\mathrm{op}}}{\hat{E}_{\mathrm{os}}}. \tag{9.8}$$

The dependence of $\hat{\chi}_o$ on $\hat{\chi}_i$ is called the polarization transfer function and it is given by

$$\hat{\chi}_o = \frac{\hat{\rho}_1 \hat{\chi}_i + \hat{\rho}_2}{\hat{\rho}_3 \hat{\chi}_i + 1}. \tag{9.9}$$

9.2.2 Stokes–Mueller Formalism

The Jones vectors can be used to describe fully polarized monochromatic plane waves. In order to describe the partially polarized states of light, it is necessary to use the Stokes–Mueller formalism. The Stokes–Mueller formalism can, of course, also be employed for strictly defined polarization states, which makes it more general than the Jones matrix formalism. The Stokes vector is defined as [5, 6]

$$\boldsymbol{S} = \begin{pmatrix} S_0 \\ S_1 \\ S_2 \\ S_3 \end{pmatrix} = \begin{pmatrix} I_0 \\ I_{\uparrow} - I_{\leftarrow} \\ I_{\nearrow} - I_{\nwarrow} \\ I_{\circlearrowright} - I_{\circlearrowleft} \end{pmatrix}, \tag{9.10}$$

where I_0 denotes the total intensity of the wave and $I_{\uparrow}$, $I_{\leftarrow}$, $I_{\nwarrow}$, $I_{\nearrow}$, $I_{\circlearrowright}$ and $I_{\circlearrowleft}$ represent intensities transmitted by ideal polarizers transmitting the linearly polarized light along the axis tilted with respect to the plane of incidence by $0, \pi/2, -\pi/4, \pi/4$ and left and right circularly polarized light, respectively. The Stokes vectors fully describe the polarization states of light waves.

While the Jones formalism works with the electric field amplitudes which can not be measured directly, the Stokes formalism works with the intensities of light which can be measured experimentally.

The Stokes vector $\boldsymbol{S}(\hat{\boldsymbol{E}})$ corresponding to a fully polarized monochromatic wave described by the Jones vector $\hat{\boldsymbol{E}}$ can be calculated using the following formula:

$$\boldsymbol{S}(\hat{\boldsymbol{E}}) = \begin{pmatrix} \hat{E}_p \hat{E}_p^* + \hat{E}_s \hat{E}_s^* \\ \hat{E}_p \hat{E}_p^* - \hat{E}_s \hat{E}_s^* \\ \hat{E}_p \hat{E}_s^* + \hat{E}_s \hat{E}_p^* \\ \mathrm{i}\hat{E}_p \hat{E}_s^* - \mathrm{i}\hat{E}_s \hat{E}_p^* \end{pmatrix}. \tag{9.11}$$

From the above formula it is evident that the Stokes vector is independent on the total phase of the Jones vector. Of course, this total phase is not measurable in ellipsometric and reflectometric measurements. The components of the Stokes vector in (9.11) are given as a linear combinations of terms in the form $\hat{E}_j^* \hat{E}_k$ with indices j, k taking values p, s. It is trivial to show that the relation between the components of the Stokes vector and these terms can be inverted, i.e. the terms $\hat{E}_j^* \hat{E}_k$ can be determined if the Stokes vector is known.

In Sect. 9.2.1, it was shown that for non-depolarizing sample, the Jones vector of the outgoing wave is related to the Jones vector of the incident wave by (9.3). Therefore, it is possible to write the following relation:

$$\hat{E}^*_{\mathrm{o}j}\hat{E}_{\mathrm{o}k} = \sum_{l=\mathrm{p,s}}\sum_{m=\mathrm{p,s}} \hat{J}^*_{jl}\hat{J}_{km}\hat{E}^*_{\mathrm{i}l}\hat{E}_{\mathrm{i}m}, \tag{9.12}$$

where $\hat{E}_{\mathrm{i}j}$ and $\hat{E}_{\mathrm{o}j}$ are the components of the Jones vectors corresponding to the incident and outgoing waves, respectively. Since the quadratic terms $\hat{E}^*_j\hat{E}_k$ determine the Stokes vectors and vice versa, the above equation can be also understood as a linear relation between the Stokes vector $\boldsymbol{S}(\hat{\boldsymbol{E}}_\mathrm{i})$ of the incident wave and the Stokes vector $\boldsymbol{S}(\hat{\boldsymbol{E}}_\mathrm{o})$ of the outgoing wave as follows:

$$\boldsymbol{S}(\hat{\boldsymbol{E}}_\mathrm{o}) = \mathbf{M}(\hat{\mathbf{J}})\boldsymbol{S}(\hat{\boldsymbol{E}}_\mathrm{i}), \qquad \text{where} \qquad \mathbf{M}(\hat{\mathbf{J}}) = \begin{pmatrix} M_{00} & M_{01} & M_{02} & M_{03} \\ M_{10} & M_{11} & M_{12} & M_{13} \\ M_{20} & M_{21} & M_{22} & M_{23} \\ M_{30} & M_{31} & M_{32} & M_{33} \end{pmatrix}. \tag{9.13}$$

The 4×4 real matrix $\mathbf{M}(\hat{\mathbf{J}})$ is called the Mueller matrix of the layered system [6–8]. The elements of this Mueller matrix can be calculated as

$$\begin{aligned}
M_{00} &= \frac{1}{2}\left(|\hat{J}_{\mathrm{pp}}|^2 + |\hat{J}_{\mathrm{ss}}|^2 + |\hat{J}_{\mathrm{sp}}|^2 + |\hat{J}_{\mathrm{ps}}|^2\right), \\
M_{01} &= \frac{1}{2}\left(|\hat{J}_{\mathrm{pp}}|^2 - |\hat{J}_{\mathrm{ss}}|^2 + |\hat{J}_{\mathrm{sp}}|^2 - |\hat{J}_{\mathrm{ps}}|^2\right), \\
M_{02} &= \mathrm{Re}\left(\hat{J}_{\mathrm{ps}}\hat{J}^*_{\mathrm{pp}} + \hat{J}_{\mathrm{ss}}\hat{J}^*_{\mathrm{sp}}\right), \quad M_{03} = -\mathrm{Im}\left(\hat{J}_{\mathrm{ps}}\hat{J}^*_{\mathrm{pp}} + \hat{J}_{\mathrm{ss}}\hat{J}^*_{\mathrm{sp}}\right), \\
M_{10} &= \frac{1}{2}\left(|\hat{J}_{\mathrm{pp}}|^2 - |\hat{J}_{\mathrm{ss}}|^2 - |\hat{J}_{\mathrm{sp}}|^2 + |\hat{J}_{\mathrm{ps}}|^2\right), \\
M_{11} &= \frac{1}{2}\left(|\hat{J}_{\mathrm{pp}}|^2 + |\hat{J}_{\mathrm{ss}}|^2 - |\hat{J}_{\mathrm{sp}}|^2 - |\hat{J}_{\mathrm{ps}}|^2\right), \\
M_{12} &= \mathrm{Re}\left(\hat{J}_{\mathrm{ps}}\hat{J}^*_{\mathrm{pp}} - \hat{J}_{\mathrm{ss}}\hat{J}^*_{\mathrm{sp}}\right), \quad M_{13} = \mathrm{Im}\left(-\hat{J}_{\mathrm{ps}}\hat{J}^*_{\mathrm{pp}} + \hat{J}_{\mathrm{ss}}\hat{J}^*_{\mathrm{sp}}\right), \\
M_{20} &= \mathrm{Re}\left(\hat{J}_{\mathrm{sp}}\hat{J}^*_{\mathrm{pp}} + \hat{J}_{\mathrm{ss}}\hat{J}^*_{\mathrm{ps}}\right), \quad M_{21} = \mathrm{Re}\left(\hat{J}_{\mathrm{sp}}\hat{J}^*_{\mathrm{pp}} - \hat{J}_{\mathrm{ss}}\hat{J}^*_{\mathrm{ps}}\right), \\
M_{22} &= \mathrm{Re}\left(\hat{J}_{\mathrm{ss}}\hat{J}^*_{\mathrm{pp}} + \hat{J}_{\mathrm{sp}}\hat{J}^*_{\mathrm{ps}}\right), \quad M_{23} = \mathrm{Im}\left(-\hat{J}_{\mathrm{ss}}\hat{J}^*_{\mathrm{pp}} + \hat{J}_{\mathrm{sp}}\hat{J}^*_{\mathrm{ps}}\right), \\
M_{30} &= \mathrm{Im}\left(\hat{J}_{\mathrm{sp}}\hat{J}^*_{\mathrm{pp}} + \hat{J}_{\mathrm{ss}}\hat{J}^*_{\mathrm{ps}}\right), \quad M_{31} = \mathrm{Im}\left(\hat{J}_{\mathrm{sp}}\hat{J}^*_{\mathrm{pp}} - \hat{J}_{\mathrm{ss}}\hat{J}^*_{\mathrm{ps}}\right), \\
M_{32} &= \mathrm{Im}\left(\hat{J}_{\mathrm{ss}}\hat{J}^*_{\mathrm{pp}} + \hat{J}_{\mathrm{sp}}\hat{J}^*_{\mathrm{ps}}\right), \quad M_{33} = \mathrm{Re}\left(\hat{J}_{\mathrm{ss}}\hat{J}^*_{\mathrm{pp}} - \hat{J}_{\mathrm{sp}}\hat{J}^*_{\mathrm{ps}}\right).
\end{aligned} \tag{9.14}$$

So far we have considered only fully polarized monochromatic waves. In order to describe partially polarized waves, it is useful to represent these partially polarized waves by incoherent superpositions of fully polarized waves with different polarization states, i.e. different Jones vectors. If the sample is non-depolarizing and the incident wave is monochromatic but only partially polarized, then the formula (9.13) can be used to express the Stokes vector of the outgoing wave by means of the statistical ensemble corresponding to the incident monochromatic wave with frequency ω. Thus, the Stokes vector $\boldsymbol{S}_\mathrm{o}$ of the outgoing wave is given as

$$\boldsymbol{S}_\mathrm{o} = \left\langle \mathbf{S}(\hat{\boldsymbol{E}}_\mathrm{o}) \right\rangle = \left\langle \mathbf{M}(\hat{\mathbf{J}})\mathbf{S}(\hat{\boldsymbol{E}}_\mathrm{i}) \right\rangle = \mathbf{M}(\hat{\mathbf{J}})\boldsymbol{S}_\mathrm{i}, \tag{9.15}$$

where $\mathbf{M}(\hat{\mathbf{J}})$ is calculated from the Jones matrix $\hat{\mathbf{J}}$ using (9.14) and $\boldsymbol{S}_\mathrm{i} = \langle \boldsymbol{S}(\hat{\boldsymbol{E}}_\mathrm{i}) \rangle$ is the mean value of the Stokes vector of the incident wave calculated using the statistical ensemble consisting of the Jones vectors $\hat{\boldsymbol{E}}_\mathrm{i}$ corresponding to fully polarized waves (see (9.11)).

If we want to take into account the influence of the temporal coherence, it is necessary to consider a wave which consists not only of waves with differently polarized states but also with different frequencies. This corresponds to polychromatic light described by the spectral density of the intensity (see e.g. [5]). In this case the incident and outgoing waves are characterized by the spectral densities of the Stokes vectors denoted $\boldsymbol{G}_\mathrm{i}(\omega)$ and $\boldsymbol{G}_\mathrm{o}(\omega)$, respectively. The zeroth components of these vectors are the spectral densities of the intensity while the other components correspond to spectral densities of intensities defined in (9.10). It is possible to write

$$\boldsymbol{G}_\mathrm{o}(\omega) = \mathbf{M}(\hat{\mathbf{J}}(\omega))\boldsymbol{G}_\mathrm{i}(\omega), \tag{9.16}$$

where $\mathbf{M}(\hat{\mathbf{J}}(\omega))$ is calculated using (9.14) and $\hat{\mathbf{J}}(\omega)$ is the Jones matrix dependent on the frequency. If the detection of the polarization states of the outgoing wave is independent on frequency, then one can write the following formula:

$$\boldsymbol{S}'_\mathrm{o} = \int \boldsymbol{G}_\mathrm{o}(\omega)\mathrm{d}\omega = \int \mathbf{M}(\hat{\mathbf{J}}(\omega))\boldsymbol{G}_\mathrm{i}(\omega)\mathrm{d}\omega, \tag{9.17}$$

where $\boldsymbol{S}'_\mathrm{o}$ is the Stokes vector measured by the detector. If the polarization states of the incident wave do not depend on frequency, i.e. the direction of $\boldsymbol{G}_\mathrm{i}(\omega)$ does not change with frequency, then the spectral density can be expressed as $\boldsymbol{G}_\mathrm{i}(\omega) = \boldsymbol{S}'_\mathrm{i} w_\mathrm{spectral}(\omega)$, where $w_\mathrm{spectral}(\omega)$ denotes the normalized spectral distribution of the intensity of the incident wave and the symbol $\boldsymbol{S}'_\mathrm{i}$ denotes the frequency independent Stokes vector. Note that the Stokes vectors $\boldsymbol{S}'_\mathrm{i}$ and $\boldsymbol{S}'_\mathrm{o}$ are the mean values calculated for the incident and outgoing waves, respectively, over the spectral interval of polychromatic light. Under these assumptions it is possible to write the following formula:

$$\boldsymbol{S}'_\mathrm{o} = \mathbf{M}\boldsymbol{S}'_\mathrm{i}, \tag{9.18}$$

where

$$\mathbf{M} = \int \mathbf{M}(\hat{\mathbf{J}}(\omega)) w_\mathrm{spectral}(\omega)\mathrm{d}\omega. \tag{9.19}$$

If the sample is depolarizing, then the outgoing wave can be only partially polarized even when the incident wave is monochromatic and completely polarized. The depolarizing character of the sample can be taken into account if the sample is not described by one Jones matrix but it is described by the statistical ensemble of Jones matrices. If it is assumed that the mean values over statistical ensemble of Jones matrices can be expressed using the probability density $w_{\text{sample}}(\xi)$ of one variable ξ, then one can express the Stokes vector of the outgoing wave as

$$\boldsymbol{S}'_{\text{o}} = \langle \mathbf{M} \rangle \boldsymbol{S}'_{\text{i}}, \tag{9.20}$$

where

$$\langle \mathbf{M} \rangle = \iint \mathbf{M}(\hat{\mathbf{J}}(\omega, \xi)) w_{\text{spectral}}(\omega) w_{\text{sample}}(\xi) \mathrm{d}\omega \mathrm{d}\xi. \tag{9.21}$$

This formula takes into account not only the depolarization caused by the sample but also the influence of the temporal coherence. The results expressed by (9.19) and (9.21) are important from the practical point of view because the elements of the Mueller matrix can be determined experimentally.

For a given Stokes vector the degree of polarization P is defined as

$$P = \frac{1}{S_0} \sqrt{S_1^2 + S_2^2 + S_3^2}. \tag{9.22}$$

The parameter varies from the value of 1 for the fully polarized states to value 0 for completely unpolarized states. The inequalities $S_0 \geq 0$ and $0 \leq P \leq 1$ must be understood as conditions that restrict the physically realizable polarization states. The fact that the components of the Stokes vector must fulfill the mentioned inequality contrasts with the case of Jones vectors, where any two-dimensional complex vector represents a valid polarization state. It is trivial to show that the Stokes vectors corresponding to fully polarized waves, i.e. calculated by means of the formula (9.11), always give $P = 1$. It is also possible to prove that if $P = 1$, then there is a Jones vector $\hat{\boldsymbol{E}}$ (unique up to a total phase factor) such that the relation (9.11) is valid.

The following special form of the Mueller matrix is often encountered in practice

$$\mathbf{M} = \begin{pmatrix} M_{00} & M_{01} & 0 & 0 \\ M_{01} & M_{00} & 0 & 0 \\ 0 & 0 & M_{22} & M_{23} \\ 0 & 0 & -M_{23} & M_{22} \end{pmatrix}. \tag{9.23}$$

This Mueller matrix occurs, for example, for optically isotropic layered systems, layered systems formed by anisotropic materials with each principal axis either lying in the plane of incidence or perpendicular to the plane of incidence (see e.g. Sect. 9.4.2). The same form of the Mueller matrix is encountered even if these systems exhibit depolarization caused by a finite spectral line width, thickness non-uniformity, area non-uniformity in optical constants, or as a consequence of back-side reflections in non-absorbing substrates. In this special case, the elements of the normalized Mueller matrix $\mathbf{M}_n$ defined as $\mathbf{M} = M_{00}\mathbf{M}_n$ determine the associated ellipsometric parameters I_s, I_c, I_n as follows: $M_{n01} = -I_n$, $M_{n22} = I_c$, $M_{n23} = I_s$. The values of these parameters can be measured by phase modulated ellipsometry (see Sect. 9.3.4.3). For ellipsometry in reflection mode the quantities appearing in the Mueller matrix (9.23) can be calculated from (9.14) if we set $\hat{J}_{pp} = \hat{r}_p$, $\hat{J}_{ss} = \hat{r}_s$, $\hat{J}_{ps} = \hat{J}_{sp} = 0$. The resulting expressions are

$$M_{00} = R = \frac{\langle|\hat{r}_p|^2\rangle + \langle|\hat{r}_s|^2\rangle}{2}, \qquad M_{01} = \frac{\langle|\hat{r}_p|^2\rangle - \langle|\hat{r}_s|^2\rangle}{2},$$
$$M_{22} = \frac{\langle\hat{r}_p\hat{r}_s^*\rangle + \langle\hat{r}_p^*\hat{r}_s\rangle}{2}, \qquad M_{23} = -\mathrm{i}\frac{\langle\hat{r}_p\hat{r}_s^*\rangle - \langle\hat{r}_p^*\hat{r}_s\rangle}{2}, \tag{9.24}$$

where the angled brackets denote the mean values calculated as indicated in (9.21), i.e. averaging over the statistical ensemble of Jones matrices and over spectral distribution of incident light. If the reflection coefficients $\hat{r}_p$ and $\hat{r}_s$ are replaced by the Fresnel transmission coefficients $\hat{t}_p$, $\hat{t}_s$, analogous expressions are obtained for the transmission mode. In general these quantities fulfill the inequality $I_s^2 + I_c^2 + I_n^2 \leq 1$, with the equality occurring in the case without depolarization [9].

Although only the influence of the temporal coherence was included in the considerations presented above, it is possible to modify the results to include also the influence of the spatial coherence of the light beam. The generalization to probability density function $w_{\text{sample}}(\xi_1, \xi_2, \ldots)$ of several variables is straightforward. An example of the depolarization by an area non-uniformity of the sample is presented in detail in Chap. 10. The ideas presented here can be used also in the context of quantum mechanics where the states of the light are described by the density operator [10].

9.2.3 Matrix Formalism for Isotropic Layered Systems

The reflection and transmission coefficients of layered systems formed by L isotropic thin films (see Fig. 9.2) can be calculated using a simple matrix formalism. This system will be called the isotropic layered system. In isotropic layered systems with

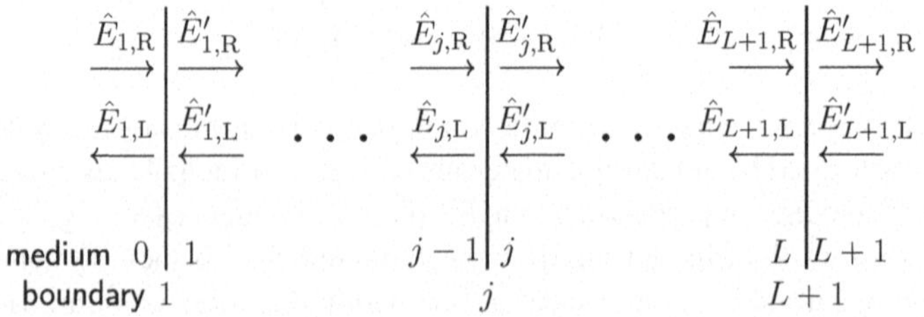

Fig. 9.2 Schematic diagram of isotropic layered system

parallel boundaries, it is possible to investigate the propagation of s- and p-polarized waves separately. In this formalism the s- or p-polarized waves are described by means of complex vectors

$$\begin{pmatrix} \hat{E}_{j,\mathrm{R}} \\ \hat{E}_{j,\mathrm{L}} \end{pmatrix} \quad \text{and} \quad \begin{pmatrix} \hat{E}'_{j,\mathrm{R}} \\ \hat{E}'_{j,\mathrm{L}} \end{pmatrix} \tag{9.25}$$

that determine the amplitudes of the electric intensity components tangent to the boundary. The amplitudes with primes correspond to the right side of the jth boundary while the amplitudes without primes correspond to the left side of the jth boundary. The subscripts L and R distinguish the amplitudes of the left-going waves from the amplitudes of the right-going waves. The boundary conditions imply that the tangential components of the electric and magnetic fields must be preserved at the jth boundary. The amplitudes $\hat{E}_{j,\mathrm{S}}$ and $\hat{H}_{j,\mathrm{S}}$ will denote the amplitudes corresponding to the total electric and magnetic fields at the jth boundary. These total tangential fields are expressed as

$$\hat{E}_{j,\mathrm{S}} = \hat{E}_{j,\mathrm{R}} + \hat{E}_{j,\mathrm{L}} = \hat{E}'_{j,\mathrm{R}} + \hat{E}'_{j,\mathrm{L}}, \tag{9.26}$$

$$\hat{H}_{j,\mathrm{S}} = \hat{Y}_{j-1}\hat{E}_{j,\mathrm{R}} - \hat{Y}_{j-1}\hat{E}_{j,\mathrm{L}} = \hat{Y}_j\hat{E}'_{j,\mathrm{R}} - \hat{Y}_j\hat{E}'_{j,\mathrm{L}}, \tag{9.27}$$

where the symbol $\hat{Y}_j$ denotes the optical admittance of the medium in which the waves propagate. These admittances are different for the s- and p-polarizations:

$$\hat{Y}_j = \hat{n}_j \cos\hat{\psi}_j \quad \text{for the s-polarization,} \quad \hat{Y}_j = \frac{\hat{n}_j}{\cos\hat{\psi}_j} \quad \text{for the p-polarization,} \tag{9.28}$$

where $\hat{n}_j$ denotes the complex refractive index of the jth medium ($\hat{n}_j = n_j + \mathrm{i}k_j$, where n_j and k_j are the real refractive index and extinction coefficient, respectively) and $\hat{\psi}_j$ denotes the refraction angle of the wave propagating in the jth medium (in the ambient this angle corresponds to the incidence angle $\hat{\psi}_0 = \varphi$).

The boundary conditions can be expressed in the so called admittance notation [11]:

$$\begin{pmatrix} \hat{E}_{j,\mathrm{S}} \\ \hat{H}_{j,\mathrm{S}} \end{pmatrix} = \hat{\mathbf{V}}_{j-1} \begin{pmatrix} \hat{E}_{j,\mathrm{R}} \\ \hat{E}_{j,\mathrm{L}} \end{pmatrix} = \hat{\mathbf{V}}_j \begin{pmatrix} \hat{E}'_{j,\mathrm{R}} \\ \hat{E}'_{j,\mathrm{L}} \end{pmatrix}, \tag{9.29}$$

where the admittances matrices $\hat{\mathbf{V}}_j$ and their inverses $\hat{\mathbf{V}}_j^{-1}$ are defined as:

$$\hat{\mathbf{V}}_j = \begin{pmatrix} 1 & 1 \\ \hat{Y}_j & -\hat{Y}_j \end{pmatrix}, \qquad \hat{\mathbf{V}}_j^{-1} = \frac{1}{2}\begin{pmatrix} 1 & \hat{Y}_j^{-1} \\ 1 & -\hat{Y}_j^{-1} \end{pmatrix}. \tag{9.30}$$

The equation (9.29) can be rewritten as

$$\begin{pmatrix} \hat{E}_{j,\mathrm{R}} \\ \hat{E}_{j,\mathrm{L}} \end{pmatrix} = \hat{\mathbf{W}}_j \begin{pmatrix} \hat{E}'_{j,\mathrm{R}} \\ \hat{E}'_{j,\mathrm{L}} \end{pmatrix}, \qquad \text{where} \qquad \hat{\mathbf{W}}_j = \hat{\mathbf{V}}_{j-1}^{-1}\hat{\mathbf{V}}_j. \tag{9.31}$$

The matrix $\hat{\mathbf{W}}_j$ is called the refraction matrix and it can be expressed as

$$\hat{\mathbf{W}}_j = \frac{1}{2}\begin{pmatrix} 1 + \hat{Y}_j/\hat{Y}_{j-1} & 1 - \hat{Y}_j/\hat{Y}_{j-1} \\ 1 - \hat{Y}_j/\hat{Y}_{j-1} & 1 + \hat{Y}_j/\hat{Y}_{j-1} \end{pmatrix} = \frac{\hat{c}_{j-1}}{\hat{c}_j}\frac{1}{\hat{t}_j}\begin{pmatrix} 1 & \hat{r}_j \\ \hat{r}_j & 1 \end{pmatrix}, \tag{9.32}$$

where $\hat{r}_j$ and $\hat{t}_j$ are the Fresnel reflection and transmission coefficients for the jth boundary, respectively, and the quantities $\hat{c}_j$ are given as

$$\hat{c}_j = 1 \quad \text{for s-polarization,} \qquad \hat{c}_j = \cos\hat{\psi}_j \quad \text{for p-polarization.} \tag{9.33}$$

In order to prove the second equality in (9.32) the following expression for the Fresnel coefficients must be used

$$\hat{r}_j = \frac{\hat{Y}_{j-1} - \hat{Y}_j}{\hat{Y}_{j-1} + \hat{Y}_j}, \qquad \hat{t}_j = \frac{\hat{c}_{j-1}}{\hat{c}_j}\frac{2\hat{Y}_{j-1}}{\hat{Y}_{j-1} + \hat{Y}_j}. \tag{9.34}$$

The tangential components of the electric fields on the left and right side of the jth layer are connected by the following equation:

$$\begin{pmatrix} \hat{E}'_{j,\mathrm{R}} \\ \hat{E}'_{j,\mathrm{L}} \end{pmatrix} = \hat{\mathbf{U}}_j \begin{pmatrix} \hat{E}_{j+1,\mathrm{R}} \\ \hat{E}_{j+1,\mathrm{L}} \end{pmatrix}, \quad \text{where} \quad \hat{\mathbf{U}}_j = \begin{pmatrix} \mathrm{e}^{-\mathrm{i}\hat{\phi}_j} & 0 \\ 0 & \mathrm{e}^{\mathrm{i}\hat{\phi}_j} \end{pmatrix}, \tag{9.35}$$

where $\hat{\phi}_j = (2\pi/\lambda)h_j\hat{n}_j\cos\hat{\psi}_j$ with λ denoting the wavelength of the light and h_j denoting the thickness of the jth layer. The matrix $\hat{\mathbf{U}}$ introduced in the above equation is called the phase matrix.

For the system containing L layers, the following relation between the amplitudes of the electric fields on the left and right sides of this system can be expressed as

$$\begin{pmatrix} \hat{E}_{1,\mathrm{R}} \\ \hat{E}_{1,\mathrm{L}} \end{pmatrix} = \hat{\mathbf{P}} \begin{pmatrix} \hat{E}'_{L+1,\mathrm{R}} \\ \hat{E}'_{L+1,\mathrm{L}} \end{pmatrix} \quad \text{where} \quad \hat{\mathbf{P}} = \hat{\mathbf{W}}_1\hat{\mathbf{U}}_1\hat{\mathbf{W}}_2\cdots\hat{\mathbf{U}}_L\hat{\mathbf{W}}_{L+1}. \tag{9.36}$$

The matrix $\hat{\mathbf{P}}$ is called the system transfer matrix. By using the elements of the system transfer matrix $\hat{P}_{ij}$ the Fresnel reflection and transmission coefficients can be expressed as

$$\hat{r} = \left(\frac{\hat{E}_{1,\mathrm{L}}}{\hat{E}_{1,\mathrm{R}}}\right)_{\hat{E}'_{L+1,\mathrm{L}}=0} = \frac{\hat{P}_{21}}{\hat{P}_{11}}, \qquad \hat{t} = \frac{\hat{c}_0}{\hat{c}_{L+1}}\left(\frac{\hat{E}'_{L+1,\mathrm{R}}}{\hat{E}_{1,\mathrm{R}}}\right)_{\hat{E}'_{L+1,\mathrm{L}}=0} = \frac{\hat{c}_0}{\hat{c}_{L+1}}\frac{1}{\hat{P}_{11}}. \tag{9.37}$$

$$\hat{r}' = \left(\frac{\hat{E}'_{L+1,\mathrm{R}}}{\hat{E}'_{L+1,\mathrm{L}}}\right)_{\hat{E}_{1,\mathrm{R}}=0} = -\frac{\hat{P}_{12}}{\hat{P}_{11}}, \quad \hat{t}' = \frac{\hat{c}_{L+1}}{\hat{c}_0}\left(\frac{\hat{E}_{1,\mathrm{L}}}{\hat{E}'_{L+1,\mathrm{L}}}\right)_{\hat{E}_{1,\mathrm{R}}=0} = \frac{\hat{c}_{L+1}}{\hat{c}_0}\frac{\det\hat{\mathbf{P}}}{\hat{P}_{11}}. \tag{9.38}$$

The symbols $\hat{r}$ and $\hat{r}'$ denote the Fresnel reflection coefficients for the incident wave falling onto the system from the left and right, respectively. The same is valid for the Fresnel transmission coefficients $\hat{t}$ and $\hat{t}'$.

The sequence of matrices $\hat{\mathbf{W}}_j$ and $\hat{\mathbf{U}}_j$ in (9.36) implies introducing new matrices $\hat{\mathbf{I}}_j$ by using the association rule as follows:

$$\hat{\mathbf{I}}_j = \hat{\mathbf{V}}_j\hat{\mathbf{U}}_j\hat{\mathbf{V}}_j^{-1} = \begin{pmatrix} \cos\hat{\phi}_j & \mathrm{i}\hat{Y}_j^{-1}\sin\hat{\phi}_j \\ \mathrm{i}\hat{Y}_j\sin\hat{\phi}_j & \cos\hat{\phi}_j \end{pmatrix} \tag{9.39}$$

The matrix $\hat{\mathbf{I}}_j$ is called the interference matrix of the jth layer (some researchers call it the characteristic matrix). From the foregoing it is clear that the system transfer matrix can be written as

$$\hat{\mathbf{P}} = \hat{\mathbf{V}}_0^{-1}\hat{\mathbf{I}}\hat{\mathbf{V}}_{L+1}, \quad \text{where} \quad \hat{\mathbf{I}} = \hat{\mathbf{I}}_1\hat{\mathbf{I}}_2\cdots\hat{\mathbf{I}}_L. \tag{9.40}$$

The matrix $\hat{\mathbf{I}}$ is called the interference matrix of the whole layered system. The meaning of the interference matrix $\hat{\mathbf{I}}$ is given by the following matrix equation

$$\begin{pmatrix}\hat{E}_{1,\mathrm{S}}\\ \hat{H}_{1,\mathrm{S}}\end{pmatrix} = \hat{\mathbf{V}}_0\begin{pmatrix}\hat{E}_{1,\mathrm{R}}\\ \hat{E}_{1,\mathrm{L}}\end{pmatrix} = \hat{\mathbf{I}}\hat{\mathbf{V}}_{L+1}\begin{pmatrix}\hat{E}'_{L+1,\mathrm{R}}\\ \hat{E}'_{L+1,\mathrm{L}}\end{pmatrix} = \hat{\mathbf{I}}\begin{pmatrix}\hat{E}_{L+1,\mathrm{S}}\\ \hat{H}_{L+1,\mathrm{S}}\end{pmatrix}. \tag{9.41}$$

From (9.40), it is evident that the elements of the transfer matrix $\hat{\mathbf{P}}$ can be expressed using the elements of the interference matrix $\hat{\mathbf{I}}$. Thus, on the basis of (9.37) and (9.38) it is possible to write the Fresnel coefficients of the layered system with the help of the elements of the interference matrix $\hat{\mathbf{I}}$ of this system. For example, the Fresnel reflection and transmission coefficients of the system corresponding to the incidence of light from the left side are given as:

$$\hat{r} = \frac{\hat{I}_{11} - \hat{Y}_0^{-1}\hat{Y}_{L+1}\hat{I}_{22} + \hat{Y}_{L+1}\hat{I}_{12} - \hat{Y}_0^{-1}\hat{I}_{21}}{\hat{I}_{11} + \hat{Y}_0^{-1}\hat{Y}_{L+1}\hat{I}_{22} + \hat{Y}_{L+1}\hat{I}_{12} + \hat{Y}_0^{-1}\hat{I}_{21}}, \tag{9.42}$$

$$\hat{t} = \frac{\hat{c}_0}{\hat{c}_{L+1}}\frac{2}{\hat{I}_{11} + \hat{Y}_0^{-1}\hat{Y}_{L+1}\hat{I}_{22} + \hat{Y}_{L+1}\hat{I}_{12} + \hat{Y}_0^{-1}\hat{I}_{21}}, \tag{9.43}$$

where $\hat{I}_{ij}$ are elements of the matrix $\hat{\mathbf{I}}$. The symbols $\hat{Y}_0$ and $\hat{Y}_{L+1}$ denote the admittances of the ambient and substrate, respectively.

9.2.4 Matrix Formalism for Anisotropic Layered Systems

This matrix formalism, known as the Yeh formalism, concerns the propagation of monochromatic plane waves in layered systems formed by homogeneous anisotropic media. A schematic diagram of the anisotropic layered system consisting of L anisotropic layers is shown in Fig. 9.3. The media are numbered $0, \ldots, L+1$ with media 0 representing the ambient and media $L+1$ representing the substrate. The boundaries are numbered $1, \ldots, L+1$ with jth boundary separating the $(j-1)$th media from the jth media. The Cartesian coordinate system (x, y, z) is chosen such that the z axis is normal to the boundaries of the layered system.

In the jth medium, the electric field corresponding to a propagation of a monochromatic plane wave can be expressed using the complex wavevector $\hat{\boldsymbol{k}}_j$ and amplitude $\hat{\boldsymbol{E}}_j$ as

$$\boldsymbol{E}_j(\boldsymbol{r}, t) = \mathrm{Re}\left(\hat{\boldsymbol{E}}_j \mathrm{e}^{\mathrm{i}\hat{\boldsymbol{k}}_j \boldsymbol{r} - \mathrm{i}\omega t}\right). \tag{9.44}$$

In anisotropic media the characteristics of light propagation depend on the direction of propagation. The Maxwell equations imply the following wave equations [2, 5, 12]

$$\hat{\boldsymbol{k}}_j \times (\hat{\boldsymbol{k}}_j \times \hat{\boldsymbol{E}}_j) + k_0^2 \hat{\boldsymbol{\varepsilon}}_j \hat{\boldsymbol{E}}_j = 0, \tag{9.45}$$

where $k_0 = \omega/c = 2\pi/\lambda$. The symbol $\hat{\boldsymbol{\varepsilon}}_j$ denotes the tensor (matrix) of the dielectric functions describing the optical response of the jth medium. In general this tensor is complex and does not posses any symmetry. In the case of optically non-absorbing anisotropic materials without optical activity, the dielectric function tensor is symmetric [5, 12] and it is, therefore, possible to find such a coordinate system in which it is diagonal. This result is valid also for the absorbing anisotropic materials with relatively high crystallographic symmetry [5].

The continuity of the electric and magnetic fields on boundaries implies that the components of the wavevector $\hat{\boldsymbol{k}}_j$ tangent to the boundaries must be the same on the both sides of the boundary. Therefore, in the layered system with mutually parallel boundaries that are perpendicular to the z axis, it is possible

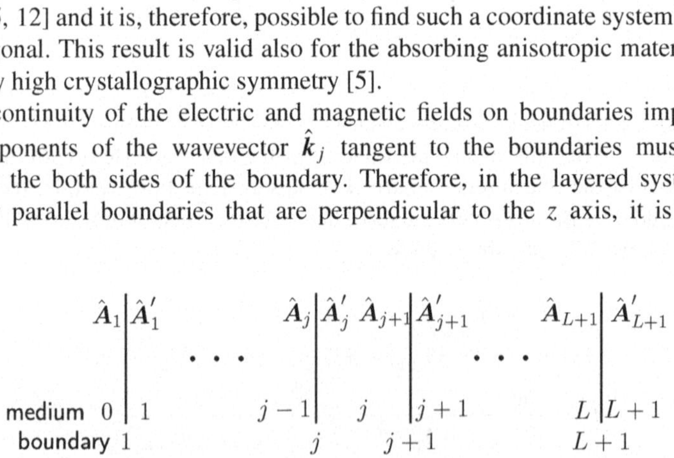

Fig. 9.3 Schematic diagram of anisotropic layered system

to write $\hat{\boldsymbol{k}}_j = (\hat{k}_x, \hat{k}_y, \hat{k}_{j,z})$. The wave equation (9.45) can be rewritten into the following matrix equation

$$\begin{pmatrix} k_0^2\hat{\varepsilon}_{j,xx} - \hat{k}_y^2 - \hat{k}_{j,z}^2 & k_0^2\hat{\varepsilon}_{j,xy} + \hat{k}_x\hat{k}_y & k_0^2\hat{\varepsilon}_{j,xz} + \hat{k}_x\hat{k}_{j,z} \\ k_0^2\hat{\varepsilon}_{j,yx} + \hat{k}_y\hat{k}_x & k_0^2\hat{\varepsilon}_{j,yy} - \hat{k}_x^2 - \hat{k}_{j,z}^2 & k_0^2\hat{\varepsilon}_{j,yz} + \hat{k}_y\hat{k}_{j,z} \\ k_0^2\hat{\varepsilon}_{j,zx} + \hat{k}_{j,z}\hat{k}_x & k_0^2\hat{\varepsilon}_{j,zy} + \hat{k}_{j,z}\hat{k}_y & k_0^2\hat{\varepsilon}_{j,zz} - \hat{k}_x^2 - \hat{k}_y^2 \end{pmatrix} \begin{pmatrix} \hat{E}_{j,x} \\ \hat{E}_{j,y} \\ \hat{E}_{j,z} \end{pmatrix} = 0. \tag{9.46}$$

If the matrix in front of $\hat{\boldsymbol{E}}_j$, which can be considered a function of $\hat{k}_{j,z}$, is denoted by $\hat{\mathbf{N}}(\hat{k}_{j,z})$, then the equation takes the form:

$$\hat{\mathbf{N}}(\hat{k}_{j,z})\hat{\boldsymbol{E}}_j = 0. \tag{9.47}$$

This matrix equation has a nontrivial solution only if the determinant of the matrix in front of $\hat{\boldsymbol{E}}_j$ vanishes, i.e. $\det \hat{\mathbf{N}}(\hat{k}_{j,z}) = 0$. This leads to quartic equations for $\hat{k}_{j,z}$ with four roots $\hat{k}_{j\alpha,z}$ distinguished by the index $\alpha = 1, 2, 3, 4$. The corresponding wavevectors will be denoted by $\hat{\boldsymbol{k}}_{j\alpha} = (\hat{k}_x, \hat{k}_y, \hat{k}_{j\alpha,z})$ and the corresponding polarization vectors will be denoted by $\hat{\boldsymbol{p}}_{j\alpha}$. The polarization vector $\hat{\boldsymbol{p}}_{j\alpha}$ is the solution of (9.47) for a given root $\hat{k}_{j\alpha,z}$, i.e. $\hat{\mathbf{N}}(\hat{k}_{j\alpha,z})\hat{\boldsymbol{p}}_{j\alpha} = 0$. Moreover, we will assume that it fulfills the normalization condition $\hat{\boldsymbol{p}}_{j\alpha} \cdot \hat{\boldsymbol{p}}_{j\alpha} = 1$. The electric field in the jth medium can be expressed using the polarization vectors as

$$\boldsymbol{E}_j(\boldsymbol{r}, t) = \mathrm{Re}\left(\sum_{\alpha=1}^{4} \hat{A}_{j\alpha}^0 \hat{\boldsymbol{p}}_{j\alpha} \mathrm{e}^{\mathrm{i}\hat{k}_x x + \mathrm{i}\hat{k}_y y + \mathrm{i}\hat{k}_{j\alpha,z} z - \mathrm{i}\omega t}\right), \tag{9.48}$$

where $\hat{A}_{j\alpha}^0$ denotes the amplitudes corresponding to different polarizations of light waves. An analogous expression can be written also for the magnetic field

$$\boldsymbol{H}_j(\boldsymbol{r}, t) = \mathrm{Re}\left(\frac{1}{\mu_0\omega}\sum_{\alpha=1}^{4} \hat{A}_{j\alpha}^0 \hat{\boldsymbol{q}}_{j\alpha} \mathrm{e}^{\mathrm{i}\hat{k}_x x + \mathrm{i}\hat{k}_y y + \mathrm{i}\hat{k}_{j\alpha,z} z - \mathrm{i}\omega t}\right), \tag{9.49}$$

where the polarization vectors for the magnetic field are related to polarization vectors $\hat{\boldsymbol{p}}_{j\alpha}$ as

$$\hat{\boldsymbol{q}}_{j\alpha} = \hat{\boldsymbol{k}}_{j\alpha} \times \hat{\boldsymbol{p}}_{j\alpha}. \tag{9.50}$$

In order to write the conditions for continuity of $\boldsymbol{E}$ and $\boldsymbol{H}$ at the boundaries, it is convenient to introduce the amplitudes $\hat{A}_{j\alpha}$ and $\hat{A}'_{j\alpha}$ that correspond to the amplitudes occurring on the left and right side of the jth boundary

$$\hat{A}_{j\alpha} = \hat{A}_{j-1\alpha}^0 \mathrm{e}^{\mathrm{i}\hat{k}_{j-1\alpha,z} z_j}, \qquad \hat{A}'_{j\alpha} = \hat{A}_{j\alpha}^0 \mathrm{e}^{\mathrm{i}\hat{k}_{j\alpha,z} z_j}, \tag{9.51}$$

where z_j denotes the z coordinate of the jth boundary. It is evident that the amplitudes $\hat{A}'_{j\alpha}$ and $\hat{A}_{j+1\alpha}$ are related as

$$\hat{A}'_{j\alpha} = \hat{A}_{j+1\alpha} e^{i\hat{k}_{j\alpha,z}(z_j - z_{j+1})} = \hat{A}_{j+1\alpha} e^{-i\hat{k}_{j\alpha,z} h_j}, \tag{9.52}$$

where the symbol $h_j = z_{j+1} - z_j$ denotes the thickness of the jth layer. If we arrange the amplitudes $\hat{A}_{j\alpha}$ and $\hat{A}'_{j\alpha}$ into column vectors $\hat{\boldsymbol{A}}_j$ and $\hat{\boldsymbol{A}}'_j$ then the formula above could be written in the matrix form as

$$\hat{\boldsymbol{A}}'_j = \hat{\mathbf{T}}_j \hat{\boldsymbol{A}}_{j+1}, \tag{9.53}$$

with the matrix $\hat{\mathbf{T}}_j$ defined as

$$\hat{\mathbf{T}}_j = \begin{pmatrix} e^{-i\hat{k}_{j1,z}h_j} & 0 & 0 & 0 \\ 0 & e^{-i\hat{k}_{j2,z}h_j} & 0 & 0 \\ 0 & 0 & e^{-i\hat{k}_{j3,z}h_j} & 0 \\ 0 & 0 & 0 & e^{-i\hat{k}_{j4,z}h_j} \end{pmatrix}. \tag{9.54}$$

The conditions for the continuity of $\boldsymbol{E}$ and $\boldsymbol{H}$ on the jth boundary now take a very simple form

$$\sum_{\alpha=1}^{4} \boldsymbol{e}_x \hat{\boldsymbol{p}}_{j-1\alpha} \hat{A}_{j\alpha} = \sum_{\alpha=1}^{4} \boldsymbol{e}_x \hat{\boldsymbol{p}}_{j\alpha} \hat{A}'_{j\alpha}, \quad \sum_{\alpha=1}^{4} \boldsymbol{e}_y \hat{\boldsymbol{p}}_{j-1\alpha} \hat{A}_{j\alpha} = \sum_{\alpha=1}^{4} \boldsymbol{e}_y \hat{\boldsymbol{p}}_{j\alpha} \hat{A}'_{j\alpha}, \tag{9.55}$$

$$\sum_{\alpha=1}^{4} \boldsymbol{e}_x \hat{\boldsymbol{q}}_{j-1\alpha} \hat{A}_{j\alpha} = \sum_{\alpha=1}^{4} \boldsymbol{e}_x \hat{\boldsymbol{q}}_{j\alpha} \hat{A}'_{j\alpha}, \quad \sum_{\alpha=1}^{4} \boldsymbol{e}_y \hat{\boldsymbol{q}}_{j-1\alpha} \hat{A}_{j\alpha} = \sum_{\alpha=1}^{4} \boldsymbol{e}_y \hat{\boldsymbol{q}}_{j\alpha} \hat{A}'_{j\alpha}, \tag{9.56}$$

In the matrix form these equations are written as

$$\hat{\mathbf{D}}_{j-1} \hat{\boldsymbol{A}}_j = \hat{\mathbf{D}}_j \hat{\boldsymbol{A}}'_j, \tag{9.57}$$

where the matrix $\hat{\mathbf{D}}_j$ is defined as

$$\hat{\mathbf{D}}_j = \begin{pmatrix} \boldsymbol{e}_x \hat{\boldsymbol{p}}_{j1} & \boldsymbol{e}_x \hat{\boldsymbol{p}}_{j2} & \boldsymbol{e}_x \hat{\boldsymbol{p}}_{j3} & \boldsymbol{e}_x \hat{\boldsymbol{p}}_{j4} \\ \boldsymbol{e}_y \hat{\boldsymbol{q}}_{j1} & \boldsymbol{e}_y \hat{\boldsymbol{q}}_{j2} & \boldsymbol{e}_y \hat{\boldsymbol{q}}_{j3} & \boldsymbol{e}_y \hat{\boldsymbol{q}}_{j4} \\ \boldsymbol{e}_y \hat{\boldsymbol{p}}_{j1} & \boldsymbol{e}_y \hat{\boldsymbol{p}}_{j2} & \boldsymbol{e}_y \hat{\boldsymbol{p}}_{j3} & \boldsymbol{e}_y \hat{\boldsymbol{p}}_{j4} \\ \boldsymbol{e}_x \hat{\boldsymbol{q}}_{j1} & \boldsymbol{e}_x \hat{\boldsymbol{q}}_{j2} & \boldsymbol{e}_x \hat{\boldsymbol{q}}_{j3} & \boldsymbol{e}_x \hat{\boldsymbol{q}}_{j4} \end{pmatrix}. \tag{9.58}$$

The relation between the amplitudes on the left and right sides of the jth boundary can be expressed as

$$\hat{\boldsymbol{A}}_j = \hat{\mathbf{B}}_j \hat{\boldsymbol{A}}'_j \quad \text{where} \quad \hat{\mathbf{B}}_j = \hat{\mathbf{D}}_{j-1}^{-1} \hat{\mathbf{D}}_j. \tag{9.59}$$

The amplitudes on the left side of the first boundary (i.e. on the left side of the layered system) and the amplitudes on the right side of the last boundary (i.e. on the right side of the layered system) are related as

$$\hat{\mathbf{A}}_1 = \hat{\mathbf{X}}\hat{\mathbf{A}}'_{L+1} \qquad \text{where} \qquad \hat{\mathbf{X}} = \hat{\mathbf{B}}_1\hat{\mathbf{T}}_1\hat{\mathbf{B}}_2\hat{\mathbf{T}}_2\hat{\mathbf{B}}_3 \cdots \hat{\mathbf{T}}_L\hat{\mathbf{B}}_{L+1}. \tag{9.60}$$

It is assumed that the anisotropic layered system is placed between two isotropic media corresponding to ambient and substrate. In isotropic media, it is possible to choose the polarization vectors $\hat{\boldsymbol{p}}_{j\alpha}$ in such a way, that the vectors of amplitudes take the following form

$$\hat{\boldsymbol{A}}_j = \begin{pmatrix} \hat{A}_{j\text{p,R}} \\ \hat{A}_{j\text{p,L}} \\ \hat{A}_{j\text{s,R}} \\ \hat{A}_{j\text{s,L}} \end{pmatrix}, \qquad \hat{\boldsymbol{A}}'_j = \begin{pmatrix} \hat{A}'_{j\text{p,R}} \\ \hat{A}'_{j\text{p,L}} \\ \hat{A}'_{j\text{s,R}} \\ \hat{A}'_{j\text{s,L}} \end{pmatrix}, \tag{9.61}$$

where the subscripts p and s distinguish the p- and s-polarizations and the subscripts L and R distinguish left- and right-going waves.

Then the following matrix equation can be written for the anisotropic layered system if the wave is incident from the left side

$$\begin{pmatrix} \hat{A}_{1\text{p,R}} \\ \hat{A}_{1\text{p,L}} \\ \hat{A}_{1\text{s,R}} \\ \hat{A}_{1\text{s,L}} \end{pmatrix} = \begin{pmatrix} \hat{X}_{11} & \hat{X}_{12} & \hat{X}_{13} & \hat{X}_{14} \\ \hat{X}_{21} & \hat{X}_{22} & \hat{X}_{23} & \hat{X}_{24} \\ \hat{X}_{31} & \hat{X}_{32} & \hat{X}_{33} & \hat{X}_{34} \\ \hat{X}_{41} & \hat{X}_{42} & \hat{X}_{43} & \hat{X}_{44} \end{pmatrix} \begin{pmatrix} \hat{A}'_{L+1\text{p,R}} \\ 0 \\ \hat{A}'_{L+1\text{s,R}} \\ 0 \end{pmatrix}, \tag{9.62}$$

where $\hat{A}_{1\text{p,R}}$, $\hat{A}_{1\text{s,R}}$ are the amplitudes of the wave incident on the first boundary, $\hat{A}_{1\text{p,L}}$, $\hat{A}_{1\text{s,L}}$ are the amplitudes of the reflected wave and $\hat{A}'_{L+1\text{p,R}}$, $\hat{A}'_{L+1\text{s,R}}$ are the amplitudes of the transmitted wave. The Fresnel reflection and transmission coefficients of the anisotropic multilayer system can then be expressed as

$$\hat{r}_{\text{pp}} = \left(\frac{\hat{A}_{1\text{p,L}}}{\hat{A}_{1\text{p,R}}}\right)_{\hat{A}_{1\text{s,R}}=0}, \qquad \hat{r}_{\text{ps}} = \left(\frac{\hat{A}_{1\text{p,L}}}{\hat{A}_{1\text{s,R}}}\right)_{\hat{A}_{1\text{p,R}}=0},$$

$$\hat{r}_{\text{sp}} = \left(\frac{\hat{A}_{1\text{s,L}}}{\hat{A}_{1\text{p,R}}}\right)_{\hat{A}_{1\text{s,R}}=0}, \qquad \hat{r}_{\text{ss}} = \left(\frac{\hat{A}_{1\text{s,L}}}{\hat{A}_{1\text{s,R}}}\right)_{\hat{A}_{1\text{p,R}}=0},$$

$$\hat{t}_{\text{pp}} = \left(\frac{\hat{A}'_{L+1\text{p,R}}}{\hat{A}_{1\text{p,R}}}\right)_{\hat{A}_{1\text{s,R}}=0}, \qquad \hat{t}_{\text{ps}} = \left(\frac{\hat{A}'_{L+1\text{p,R}}}{\hat{A}_{1\text{s,R}}}\right)_{\hat{A}_{1\text{p,R}}=0},$$

$$\hat{t}_{\text{sp}} = \left(\frac{\hat{A}'_{L+1\text{s,R}}}{\hat{A}_{1\text{p,R}}}\right)_{\hat{A}_{1\text{s,R}}=0}, \qquad \hat{t}_{\text{ss}} = \left(\frac{\hat{A}'_{L+1\text{s,R}}}{\hat{A}_{1\text{s,R}}}\right)_{\hat{A}_{1\text{p,R}}=0}. \tag{9.63}$$

The following formulae are obtained for these Fresnel coefficients from (9.62):

$$\hat{r}_{\rm pp} = \frac{\hat{X}_{21}\hat{X}_{33} - \hat{X}_{23}\hat{X}_{31}}{\hat{X}_{11}\hat{X}_{33} - \hat{X}_{13}\hat{X}_{31}}, \qquad \hat{r}_{\rm ps} = \frac{\hat{X}_{11}\hat{X}_{23} - \hat{X}_{21}\hat{X}_{13}}{\hat{X}_{11}\hat{X}_{33} - \hat{X}_{13}\hat{X}_{31}},$$
$$\hat{r}_{\rm sp} = \frac{\hat{X}_{41}\hat{X}_{33} - \hat{X}_{43}\hat{X}_{31}}{\hat{X}_{11}\hat{X}_{33} - \hat{X}_{13}\hat{X}_{31}}, \qquad \hat{r}_{\rm ss} = \frac{\hat{X}_{11}\hat{X}_{43} - \hat{X}_{41}\hat{X}_{13}}{\hat{X}_{11}\hat{X}_{33} - \hat{X}_{13}\hat{X}_{31}},$$
$$\hat{t}_{\rm pp} = \frac{\hat{X}_{33}}{\hat{X}_{11}\hat{X}_{33} - \hat{X}_{13}\hat{X}_{31}}, \qquad \hat{t}_{\rm ps} = \frac{-\hat{X}_{13}}{\hat{X}_{11}\hat{X}_{33} - \hat{X}_{13}\hat{X}_{31}},$$
$$\hat{t}_{\rm sp} = \frac{-\hat{X}_{31}}{\hat{X}_{11}\hat{X}_{33} - \hat{X}_{13}\hat{X}_{31}}, \qquad \hat{t}_{\rm ss} = \frac{\hat{X}_{11}}{\hat{X}_{11}\hat{X}_{33} - \hat{X}_{13}\hat{X}_{31}}. \tag{9.64}$$

Note that the Yeh matrix formalism describes the change of polarization states of the wave outgoing from the system with regard to the wave incident on this system.

The Yeh matrix formalism is quite different from the Jones and Stokes–Mueller matrix formalisms. The Jones and Stokes–Mueller formalisms are well-suited for describing the relation between the polarization states of the incident wave and the wave reflected or transmitted by the sample. The Yeh formalism, on the other hand, is not used to describe the same relation. It serves as an efficient tool for calculating the mentioned changes of polarization states caused by reflection or transmission by the anisotropic multilayer system.

If the jth layer consists of an optically isotropic material then the polarization vectors could be chosen as described in (9.61). Assuming that the x–z plane is the plane of incidence, the wavevectors and polarization vectors are given as

$$\begin{aligned}
\hat{\boldsymbol{k}}_1 &= (\hat{k}_x, 0, \hat{k}_{j,z}), & \hat{\boldsymbol{p}}_1 &= (\hat{k}_{j,z}/\hat{k}_j, 0, -\hat{k}_x/\hat{k}_j), & \hat{\boldsymbol{q}}_1 &= (0, \hat{k}_j, 0),\\
\hat{\boldsymbol{k}}_2 &= (\hat{k}_x, 0, -\hat{k}_{j,z}), & \hat{\boldsymbol{p}}_2 &= (\hat{k}_{j,z}/\hat{k}_j, 0, \hat{k}_x/\hat{k}_j), & \hat{\boldsymbol{q}}_2 &= (0, -\hat{k}_j, 0),\\
\hat{\boldsymbol{k}}_3 &= (\hat{k}_x, 0, \hat{k}_{j,z}), & \hat{\boldsymbol{p}}_3 &= (0, 1, 0), & \hat{\boldsymbol{q}}_3 &= (-\hat{k}_{j,z}, 0, \hat{k}_x),\\
\hat{\boldsymbol{k}}_4 &= (\hat{k}_x, 0, -\hat{k}_{j,z}), & \hat{\boldsymbol{p}}_4 &= (0, 1, 0), & \hat{\boldsymbol{q}}_4 &= (\hat{k}_{j,z}, 0, \hat{k}_x),
\end{aligned} \tag{9.65}$$

where $\hat{k}_x = n_0 \sin\varphi$ and

$$\hat{k}_j = k_0\sqrt{\hat{\varepsilon}_j}, \qquad \hat{k}_{j,z} = \sqrt{k_0^2\hat{\varepsilon}_j - \hat{k}_x^2}. \tag{9.66}$$

The matrices $\hat{\mathbf{D}}_j$ (9.58) and $\hat{\mathbf{T}}_j$ (9.54) then take the following forms:

$$\hat{\mathbf{D}}_j = \begin{pmatrix} \hat{k}_{j,z}/\hat{k}_j & \hat{k}_{j,z}/\hat{k}_j & 0 & 0\\ \hat{k}_j & -\hat{k}_j & 0 & 0\\ 0 & 0 & 1 & 1\\ 0 & 0 & -\hat{k}_{j,z} & \hat{k}_{j,z} \end{pmatrix}, \quad \hat{\mathbf{T}}_j = \begin{pmatrix} \mathrm{e}^{-\mathrm{i}\hat{\phi}_j} & 0 & 0 & 0\\ 0 & \mathrm{e}^{\mathrm{i}\hat{\phi}_j} & 0 & 0\\ 0 & 0 & \mathrm{e}^{-\mathrm{i}\hat{\phi}_j} & 0\\ 0 & 0 & 0 & \mathrm{e}^{\mathrm{i}\hat{\phi}_j} \end{pmatrix}, \tag{9.67}$$

where $\hat{\phi}_j = \hat{k}_{j,z}h_j$.

We should also mention the Berreman matrix formalism [13]. In this formalism the propagation of a monochromatic plane wave is described by means of a system of four ordinary differential equations (ODE) which give the dependence on the z coordinate. The differential equations are formulated using 4×4 matrices and four-dimensional vectors whose components are (E_x, E_y, H_x, H_y), i.e. this vector specifies dependence on the z coordinate of the electric and magnetic field amplitudes that are perpendicular to this axis. In this way the Berreman formalism is very similar to the Yeh matrix formalism (see the definition of the matrix (9.58)). In Berreman formalism it is assumed that the optical properties of the materials can depend on the z coordinate in an arbitrary way. The Berreman formalism is very general because it assumes that both the dielectric and permeability tensors can exhibit anisotropy and that optical rotation tensors describing the optical activity can be non-vanishing. All of these tensors are assumed to be dependent of the z coordinate in the Berreman formalism.

The application of the Berreman formalism to layered systems was performed in Schubert's paper [14]. In this paper it was shown how to derive systematically optical quantities usable in ellipsometry for layered systems consisting of homogeneous anisotropic nonmagnetic media without optical activity. Moreover, a model of continuously twisted anisotropic material was investigated in that paper.

The Berreman formalism was used, for example, to study surfaces of uniaxial TiO_2 within generalized ellipsometry in [15].

9.3 Theory of Ellipsometric Measurements

Ellipsometry allows to measure the changes of the polarization states of light waves interacting with the studied sample. Therefore, ellipsometry offers the possibility to perform the optical characterization of the layered systems. Ellipsometric methods can be classified into the following three groups: conventional ellipsometry, generalized ellipsometry and Mueller-matrix ellipsometry.

9.3.1 Conventional Ellipsometry

The methods of conventional ellipsometry are applied to samples characterized by diagonal Jones matrices ($\hat{J}_{ps} = \hat{J}_{sp} = 0$). The normalized Jones matrix (9.5) is then given as

$$\hat{\mathbf{J}}_n = \begin{pmatrix} \hat{\rho} & 0 \\ 0 & 1 \end{pmatrix} \quad \text{where} \quad \hat{\rho} = \frac{\hat{r}_p}{\hat{r}_s} \quad \text{or} \quad \hat{\rho} = \frac{\hat{t}_p}{\hat{t}_s}, \tag{9.68}$$

where the symbols $\hat{r}_p = \hat{r}_{pp}$ and $\hat{r}_s = \hat{r}_{ss}$ denote the Fresnel reflection coefficients of the p- and s-polarized waves and the symbols $\hat{t}_p = \hat{t}_{pp}$ and $\hat{t}_s = \hat{t}_{ss}$ denote the Fresnel transmission coefficients of the p- and s-polarized waves. From the ellipsometric point of view, the samples are then unambiguously described by one complex quantity $\hat{\rho}$ called the ellipsometric ratio. This ratio $\hat{\rho}$ can be written as

$$\hat{\rho} = \tan\psi\, \mathrm{e}^{\mathrm{i}\Delta}, \tag{9.69}$$

where the angles ψ and Δ are called the azimuth and phase change. These angles represent the ellipsometric parameters of the system in the reflection or transmission modes. The polarization transfer function (9.9) takes a very simple form in conventional ellipsometry

$$\hat{\chi}_{\mathrm{o}} = \hat{\rho}\hat{\chi}_{\mathrm{i}}. \tag{9.70}$$

9.3.2 Generalized Ellipsometry

The methods of generalized ellipsometry are utilized for characterizing samples described by non-diagonal Jones matrices. Within this group of ellipsometric methods, the samples are unambiguously described by three complex quantities $\hat{\rho}_1, \hat{\rho}_2, \hat{\rho}_3$ (see (9.5)–(9.7)). Then the polarization transfer function expressed in (9.9) is given by Möbius transformation. The methods of generalized ellipsometry are employed for optical characterization of layered systems with films and substrates formed by anisotropic materials whose principal axes are situated in general positions with respect to the plane of incidence (generalized ellipsometry can also be utilized for systems containing materials with optical activity).

Within generalized ellipsometry, one must know the response of the layered system for three independent polarization states of the incident wave, at least. Therefore, it is necessary to know three output polarization states $\hat{\chi}_{\mathrm{o1}}, \hat{\chi}_{\mathrm{o2}}, \hat{\chi}_{\mathrm{o3}}$ corresponding to three polarization states $\hat{\chi}_{\mathrm{i1}}, \hat{\chi}_{\mathrm{i2}}, \hat{\chi}_{\mathrm{i3}}$ of the incident waves. In this case, the components of the normalized Jones matrix (9.5) can be calculated using the following equations [2]

$$\hat{\rho}_1 = \frac{\hat{\chi}_{\mathrm{o1}}\hat{\chi}_{\mathrm{o2}}(\hat{\chi}_{\mathrm{i1}} - \hat{\chi}_{\mathrm{i2}}) + \hat{\chi}_{\mathrm{o3}}\hat{\chi}_{\mathrm{o1}}(\hat{\chi}_{\mathrm{i3}} - \hat{\chi}_{\mathrm{i1}}) + \hat{\chi}_{\mathrm{o2}}\hat{\chi}_{\mathrm{o3}}(\hat{\chi}_{\mathrm{i2}} - \hat{\chi}_{\mathrm{i3}})}{\hat{D}}, \tag{9.71}$$

$$\hat{\rho}_2 = -\frac{\hat{\chi}_{\mathrm{o1}}\hat{\chi}_{\mathrm{o2}}(\hat{\chi}_{\mathrm{i1}} - \hat{\chi}_{\mathrm{i2}})\hat{\chi}_{\mathrm{i3}} + \hat{\chi}_{\mathrm{o3}}\hat{\chi}_{\mathrm{o1}}(\hat{\chi}_{\mathrm{i3}} - \hat{\chi}_{\mathrm{i1}})\hat{\chi}_{\mathrm{i2}} + \hat{\chi}_{\mathrm{o2}}\hat{\chi}_{\mathrm{o3}}(\hat{\chi}_{\mathrm{i2}} - \hat{\chi}_{\mathrm{i3}})\hat{\chi}_{\mathrm{i1}}}{\hat{D}},$$

$$\hat{\rho}_3 = -\frac{\hat{\chi}_{\mathrm{o3}}(\hat{\chi}_{\mathrm{i1}} - \hat{\chi}_{\mathrm{i2}}) + \hat{\chi}_{\mathrm{o2}}(\hat{\chi}_{\mathrm{i3}} - \hat{\chi}_{\mathrm{i1}}) + \hat{\chi}_{\mathrm{o1}}(\hat{\chi}_{\mathrm{i2}} - \hat{\chi}_{\mathrm{i3}})}{\hat{D}},$$

where

$$\hat{D} = \hat{\chi}_{\mathrm{o3}}\hat{\chi}_{\mathrm{i3}}(\hat{\chi}_{\mathrm{i1}} - \hat{\chi}_{\mathrm{i2}}) + \hat{\chi}_{\mathrm{o2}}\hat{\chi}_{\mathrm{i2}}(\hat{\chi}_{\mathrm{i3}} - \hat{\chi}_{\mathrm{i1}}) + \hat{\chi}_{\mathrm{o1}}\hat{\chi}_{\mathrm{i1}}(\hat{\chi}_{\mathrm{i2}} - \hat{\chi}_{\mathrm{i3}}). \tag{9.72}$$

The Yeh matrix formalism is closely connected with generalized ellipsometry, because if the Yeh matrix $\hat{\mathbf{X}}$ of the system of anisotropic layers is known, then the polarization transfer function (9.9) for reflected light can be calculated as follows [12]

$$\hat{\chi}_o = \frac{\hat{X}_{23}(\hat{X}_{11} - \hat{\chi}_i\hat{X}_{31}) + \hat{X}_{21}(\hat{\chi}_i\hat{X}_{33} - \hat{X}_{13})}{\hat{X}_{43}(\hat{X}_{11} - \hat{\chi}_i\hat{X}_{31}) + \hat{X}_{41}(\hat{\chi}_i\hat{X}_{33} - \hat{X}_{13})}. \tag{9.73}$$

A detailed description of generalized ellipsometry and its practical applications are presented, for example, in [1, 16].

9.3.3 Mueller-Matrix Ellipsometry

Within the Mueller-matrix ellipsometry, the values of the general Mueller matrix (9.13) are measured. The simplest instrument that allows us to measure all the elements of the Mueller matrix is the single-channel Mueller-matrix ellipsometer. This ellipsometer consists of the light source, polarizing optics, sample, analyzing optics and detector. The light intensity I detected by the detector can be expressed as

$$I = I_0 \mathbf{A}\mathbf{M}\boldsymbol{S}_\mathrm{P}, \tag{9.74}$$

where $\boldsymbol{S}_\mathrm{P}$ is the Stokes vector of light leaving the polarizing optics and incident on the sample, $\boldsymbol{A}$ is the first row of the Mueller matrix of the analyzing optics, I_0 denotes the detector responsivity and $\mathbf{M}$ is the Mueller matrix describing the sample. This equation is derived under the assumption that the detector measures only the total intensity of light leaving the analyzing optics (this intensity corresponds to the first component of the Stokes vector of light falling onto the detector, which is calculated using only the first row of the matrix $\mathbf{A}$). From (9.74), it is evident that the values of the sixteen elements of the Mueller matrix can be determined if at least sixteen independent discrete settings of the polarizing and analyzing optics are used (i.e. sixteen different combinations of $\boldsymbol{S}_\mathrm{P}$ and $\boldsymbol{A}$). The sixteen linear algebraic equations generated by (9.74) for sixteen configurations of $\boldsymbol{A}$ and $\boldsymbol{S}_\mathrm{P}$ can be solved in order to obtain all the Mueller matrix elements. If less than sixteen independent settings of $\boldsymbol{A}$ and $\boldsymbol{S}_\mathrm{P}$ are used, then the Mueller matrix can not be fully determined. The number of the Mueller matrix elements that can possibly be determined is dependent on the type of the ellipsometer employed.

The discussion above took into account the case of the general Mueller matrix with sixteen independent elements. In many cases, there are dependencies between these elements or some of these elements may vanish. For example, the Mueller matrix can take the special form (9.23). If the specific form of the Mueller matrix is taken into account, it is possible to fully determine the Mueller matrix with smaller number of independent configurations of $\boldsymbol{A}$ and $\boldsymbol{S}_\mathrm{P}$. If these configurations are chosen appropriately, it is sufficient to use the same number of configurations as the number of independent Mueller matrix elements.

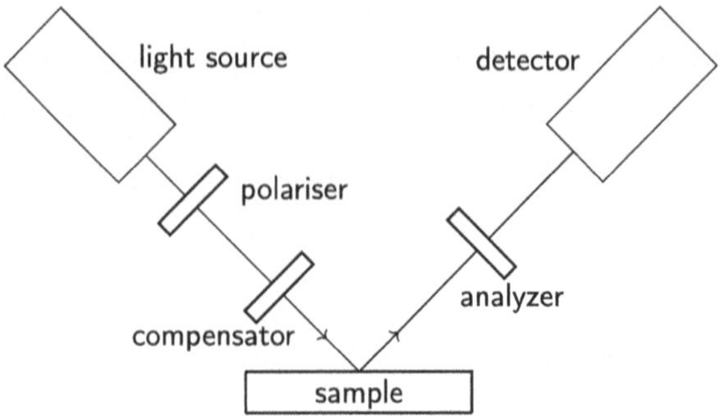

Fig. 9.4 Schematic diagram of PCSA ellipsometer working in reflection mode

9.3.4 Techniques for Conventional and Generalized Ellipsometry

9.3.4.1 Null Ellipsometry

Within null ellipsometry, the polarizer–compensator–sample–analyzer (PCSA) ellipsometer is utilized in most cases (see Fig. 9.4). An equivalent ellipsometer to the PCSA type is the PSCA type.

The mathematical formulae usable for conventional ellipsometry will be derived below. Assuming that the polarizer, analyzer and compensator are optically ideal elements, the Jones vector of the wave incident on the detector is given as

$$\hat{\boldsymbol{E}}_{\mathrm{D}} \propto \hat{\mathbf{J}}_{\mathrm{A}}\hat{\mathbf{R}}(\phi_{\mathrm{A}})\hat{\mathbf{J}}_{\mathrm{n}}\hat{\mathbf{R}}(-\phi_{\mathrm{C}})\hat{\mathbf{J}}_{\mathrm{C}}\hat{\mathbf{R}}(\phi_{\mathrm{C}}-\phi_{\mathrm{P}})\hat{\boldsymbol{E}}_{\mathrm{P}}, \tag{9.75}$$

where the normalized Jones matrix $\hat{\mathbf{J}}_{\mathrm{n}}$ is given by (9.68) and

$$\hat{\mathbf{R}}(\alpha) = \begin{pmatrix} \cos\alpha & \sin\alpha \\ -\sin\alpha & \cos\alpha \end{pmatrix}, \quad \hat{\mathbf{J}}_{\mathrm{C}} = \begin{pmatrix} 1 & 0 \\ 0 & -\mathrm{i} \end{pmatrix}, \quad \hat{\mathbf{J}}_{\mathrm{A}} = \begin{pmatrix} 1 & 0 \\ 0 & 0 \end{pmatrix}, \quad \hat{\boldsymbol{E}}_{\mathrm{P}} = \begin{pmatrix} 1 \\ 0 \end{pmatrix}. \tag{9.76}$$

The matrices $\hat{\mathbf{R}}(\alpha)$, $\hat{\mathbf{J}}_{\mathrm{C}}$ and $\hat{\mathbf{J}}_{\mathrm{A}}$ represent, in this order, the transformation of the Jones vector under the effect of a coordinate rotation, the Jones matrix of the quarter-wave compensator and the Jones matrix of analyzer transmitting p-polarized waves. The vector $\hat{\boldsymbol{E}}_{\mathrm{P}}$ represents the p-polarized wave from the light source passing through the polarizer. The symbols ϕ_{P}, ϕ_{C} and ϕ_{A} denote the azimuth angles of the polarizer, compensator and analyzer, respectively. The s-component of the vector $\hat{\boldsymbol{E}}_{\mathrm{D}}$ is always equal to zero and the p-component is given as:

$$\begin{aligned} \hat{E}_{\mathrm{D,p}} \propto\ & \hat{\rho}\cos(\phi_{\mathrm{C}}-\phi_{\mathrm{P}})\cos\phi_{\mathrm{A}}\cos\phi_{\mathrm{C}} - \mathrm{i}\hat{\rho}\sin(\phi_{\mathrm{C}}-\phi_{\mathrm{P}})\cos\phi_{\mathrm{A}}\sin\phi_{\mathrm{C}} \\ & + \cos(\phi_{\mathrm{C}}-\phi_{\mathrm{P}})\sin\phi_{\mathrm{A}}\sin\phi_{\mathrm{C}} + \mathrm{i}\sin(\phi_{\mathrm{C}}-\phi_{\mathrm{P}})\sin\phi_{\mathrm{A}}\cos\phi_{\mathrm{C}}. \end{aligned} \tag{9.77}$$

The null ellipsometry is based on the fact that the azimuth angles of the polarizer, compensator and analyzer can be configured so that no light intensity is recorded by the detector. In practice, the azimuth angle of the compensator is fixed in selected value and only the angles of the polarizer and analyzer are changed. The condition of vanishing light intensity at the detector leads to the following formula for the ellipsometric ratio:

$$\hat{\rho} = -\tan\phi_{\rm A}\frac{\tan\phi_{\rm C} + {\rm i}\tan(\phi_{\rm C} - \phi_{\rm P})}{1 - {\rm i}\tan(\phi_{\rm C} - \phi_{\rm P})\tan\phi_{\rm C}}. \tag{9.78}$$

The technique of null ellipsometry is mainly utilized in the methods of conventional ellipsometry. It is especially useful for monochromatic methods of ellipsometry because it allows to achieve high precision, when a practically parallel beam, produced by a laser source, is used. In principle, this ellipsometry is also usable for generalized ellipsometry but it is rarely used in this role because of long measurement times.

9.3.4.2 Rotating-Analyzer Ellipsometry

Rotating-analyzer ellipsometry [1, 3] is mostly applied with polarizer–sample–analyzer ellipsometers. The polarizer is fixed in a certain position of the azimuth angle $\phi_{\rm P}$ ($\phi_{\rm P} \neq 0, \pi/2, \pi, 3\pi/2$) and the analyzer is rotating, i.e. the azimuth angle $\phi_{\rm A}$ is a function of time. In conventional ellipsometry the Jones vector of the light wave incident onto the detector is expressed using the following equation:

$$\hat{\boldsymbol{E}}_{\rm D} \propto \hat{\mathbf{J}}_{\rm A}\hat{\mathbf{R}}(\phi_{\rm A}(t))\hat{\mathbf{J}}_{\rm n}\hat{\mathbf{R}}(-\phi_{\rm P})\hat{\boldsymbol{E}}_{\rm P}. \tag{9.79}$$

The light flux recorded by the detector obeys the following equation

$$I(t) = |\hat{\boldsymbol{E}}_{\rm D}(t)|^2 \propto 1 + \gamma_{\rm s}\sin(2\phi_{\rm A}(t)) + \gamma_{\rm c}\cos(2\phi_{\rm A}(t)). \tag{9.80}$$

If $\phi_{\rm A}(t)$ is a linear function of time, then the coefficients $\gamma_{\rm s}$, $\gamma_{\rm c}$ are determined as the Fourier components of the harmonics of the intensity $I(t)$. For the conventional ellipsometry, the following equation can be derived (see e.g. [1, 3])

$$\tan\psi = \tan\phi_{\rm P}\sqrt{\frac{1+\gamma_{\rm c}}{1-\gamma_{\rm c}}}, \qquad \cos\Delta = \frac{\gamma_{\rm s}}{\sqrt{1-\gamma_{\rm c}^2}}. \tag{9.81}$$

There is also a modified version of the rotating-analyzer ellipsometry, in which the analyzer is fixed and polarizer is rotated. These techniques of rotating-analyzer (rotating-polarizer) ellipsometry can also be employed for measurements in generalized ellipsometry. For generalized ellipsometry, the equations for $\gamma_{\rm s}$ and $\gamma_{\rm c}$ are different from those for conventional ellipsometry. These equations are presented, for example, in [1, 17, 18].

9.3.4.3 Phase-Modulated Ellipsometry

In phase-modulated ellipsometry, PCSA or PSCA ellipsometers are employed (see e.g. [1, 3, 19]). The polarizer, compensator and analyzer are fixed in selected positions. The phase retardation of the compensator is a function of time $\delta(t)$. In the conventional ellipsometry the Jones vector of the light wave incident onto the detector of the PCSA ellipsometer is expressed as follows:

$$\hat{\boldsymbol{E}}_{\rm D} \propto \hat{\mathbf{J}}_{\rm A}\hat{\mathbf{R}}(\phi_{\rm A})\hat{\mathbf{J}}_{\rm n}\hat{\mathbf{R}}(-\phi_{\rm C})\hat{\mathbf{J}}_{\rm C}(t)\hat{\mathbf{R}}(\phi_{\rm C}-\phi_{\rm P})\hat{\boldsymbol{E}}_{\rm P}, \quad \text{where} \quad \hat{\mathbf{J}}_{\rm C}(t) = \begin{pmatrix} 1 & 0 \\ 0 & {\rm e}^{{\rm i}\delta(t)} \end{pmatrix}. \tag{9.82}$$

The symbol $\hat{\mathbf{J}}_{\rm C}(t)$ represents the Jones matrix of the compensator. From (9.82) it is implied that

$$\begin{aligned}\hat{E}_{\rm D,p}(t) \propto &(\sin\phi_{\rm A}\sin\phi_{\rm C} + \hat{\rho}\cos\phi_{\rm A}\cos\phi_{\rm C})\cos(\phi_{\rm P}-\phi_{\rm C}) \\ &+ {\rm e}^{{\rm i}\delta(t)}(\sin\phi_{\rm A}\cos\phi_{\rm C} - \hat{\rho}\cos\phi_{\rm A}\sin\phi_{\rm C})\sin(\phi_{\rm P}-\phi_{\rm C}).\end{aligned} \tag{9.83}$$

If it is assumed that $\delta(t) = \mathcal{A}\sin(\Omega t)$, with $\mathcal{A}$ being the amplitude and Ω being the frequency of the periodic signal of the modulation of the phase retardation, then the periodic signal registered by the detector can be processed by the Fourier analysis. In this way, it is possible to obtain the values of the associated ellipsometric parameters $I_{\rm s}$, $I_{\rm c}$, $I_{\rm n}$ (see e.g. [2, 3, 19]) related to the azimuth ψ and phase change Δ as

$$I_{\rm s} = \sin 2\psi \sin\Delta\,, \qquad I_{\rm c} = \sin 2\psi \cos\Delta\,, \qquad I_{\rm n} = \cos 2\psi. \tag{9.84}$$

The techniques of phase-modulated ellipsometry can be employed also for measurements in generalized ellipsometry (see e.g. [20]).

9.3.5 Techniques for Mueller-Matrix Ellipsometry

The techniques and ellipsometers for the Mueller-matrix ellipsometry are more complicated than those for conventional and generalized ellipsometries. Not all types of Mueller-matrix ellipsometers allow to determine all the Mueller matrix elements. The overview of the Mueller-matrix ellipsometric techniques together with schematic diagrams showing the elements of the Mueller matrix which can be determined is presented in Figs. 9.5 and 9.6 [21]. In the dual-rotating-polarizer configuration [6, 21–23] (see Fig. 9.5), the optical elements are rotated with different angular frequencies and the Mueller matrix elements are determined by the Fourier analysis of a single signal recorded by the detector. In the simplest configuration with the rotating polarizer and analyzer shown in Fig. 9.5a, only nine elements of the Mueller matrix are determined. The versions in Fig. 9.5b, c have a rotating compensator placed in

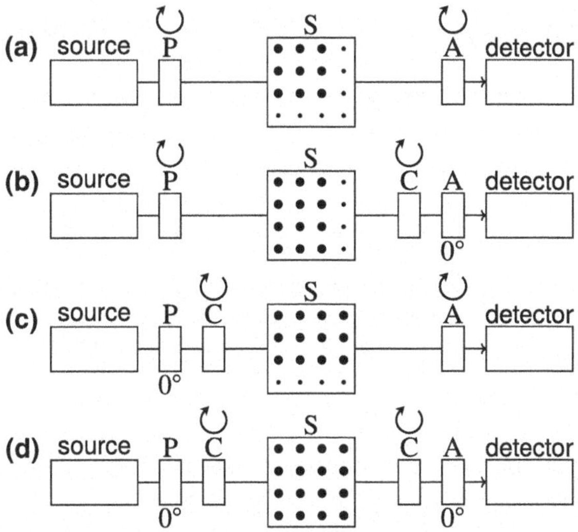

Fig. 9.5 Mueller-matrix ellipsometers with rotating elements: P, C, S and A denote the polarizers, compensators, samples and analyzers, respectively. The arrows indicate rotated elements and the angles specify the fixed positions of elements owing to the plane of incidence

front of the sample or behind the sample. These configurations allow to determine twelve elements of the Mueller matrix. In the most complicated version shown in Fig. 9.5d the polarizer and analyzer are fixed but rotating compensators are placed in front of the sample and behind the sample. In this configuration all 16 elements of the Mueller matrix can be determined. The two techniques with the phase modulation presented in Fig. 9.6 have all the optical elements in fixed positions but the modulation of the phase retardation is used for all the compensators [6, 21, 24–26]. The periodic signal recorded by the detector is processed by the Fourier analysis in order to determine the values of the Mueller matrix elements. From Fig. 9.6 it is evident that dual phase modulation on both sides of the sample must be used in order to determine the values of all sixteen Mueller matrix elements.

Mueller matrix ellipsometry is often used to measure the Mueller matrix elements in scattered light [21, 27, 28]. In this case, multi-channel ellipsometers are utilized.

9.3.6 Imaging Ellipsometry

Many thin films encountered in practice exhibit an area non-uniformity in their thickness, optical constants and other quantities (e.g. parameters of the boundary roughness). If this non-uniformity is sufficiently large, then the results of optical characterization obtained by the standard ellipsometric methods described above are misrepresented (see Chap. 10). For the optical characterization of these non-uniform

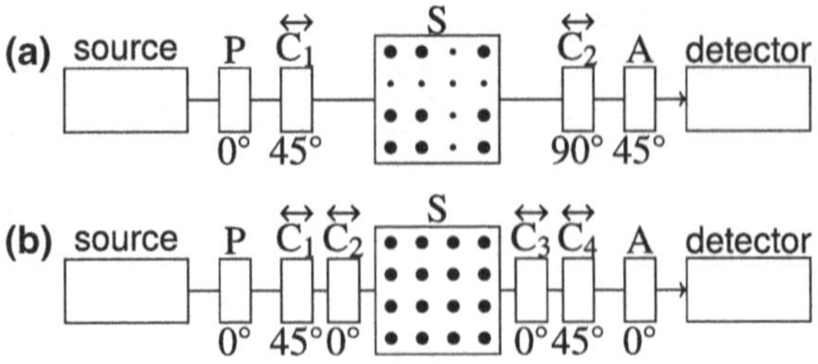

Fig. 9.6 Dual-phase-modulation Mueller-matrix ellipsometers: P, C, S and A denote the polarizers, compensators, samples and analyzers, respectively. The arrows indicate the modulated compensators and the angles specify the fixed positions of elements owing to the plane of incidence

thin films, imaging spectroscopic ellipsometry (ISE) is suitable. In this ellipsometry, the CCD cameras are utilized as detectors. Small areas on the samples correspond to individual pixels of the CCD camera used. This means that local ellipsometric quantities can be measured for these small areas. If it is possible to assume that the small areas on the samples are uniform in optical quantities, one can apply the formulae valid for uniform thin films at processing the local experimental data measured by the individual pixels. After processing these local experimental data one obtains distributions (maps) of the quantities characterizing the non-uniformity of the thin film samples. The ISE can be applied by means of all the ellipsometric techniques presented above. Of course it can also be achieved by using the other ellipsometric techniques, such as the technique based on rotating compensator.

The ISE has been applied in practice to the optical characterization of many layered systems. For example, the ISE based on the PCSA configuration with a rotating compensator was utilized for the optical characterization of a SiO_2/Si nanofilm stepped layer pattern in the visible range [29]. In [30], it was shown that the ISE is a versatile tool for investigating the uniformity and spectral dispersion functions with extreme lateral resolution for ITO-films. Monochromatic imaging ellipsometry was even used within in-situ mode in [31].

Within the imaging ellipsometry, there is a potentially promising method based on simultaneous processing of all the experimental data corresponding to the individual pixels. This method can be called multipixel ellipsometry. It is a modification of the multisample methods of spectroscopic ellipsometry and spectrophotometry utilized for a simultaneous processing of experimental data measured on different samples differing in values of some parameters (e.g. in thicknesses of the films, see e.g. [32–34]). The multisample methods improve the results of optical characterization because the correlations among the parameters sought are reduced or removed. The same effect can be achieved by multipixel ellipsometry since the local data measured by the individual pixels correspond to different values of some parameters sought (e.g. different local thicknesses). However, within the multisample modifications, several samples of the same layered system must be used while the multipixel ellipsometry can be applied to a single non-uniform sample because the local experimental data for all the pixels are obtained simultaneously. This represents a considerable advantage of this method. Note that the multipixel method was successfully applied to the optical characterization of non-uniform ZnSe thin films by means of imaging spectroscopic photometry in papers [35, 36] (see also Chap. 5).

9.4 Applications

In this section, examples of applying the matrix formalisms in optics of thin films are presented.

9.4.1 *Approximation of Reflection Coefficients of Inhomogeneous Layers*

One of the applications of the matrix formalism for isotropic layered systems is a derivation of the approximate formula for the reflection coefficients of isotropic inhomogeneous layers. It will be assumed that the inhomogeneous layer has a profile of the refractive index represented by a continuous function $\hat{n}(z)$ of the coordinate z corresponding to the axis perpendicular to the parallel boundaries. The inhomogeneous layer is surrounded by the ambient and the substrate. An approximation of this inhomogeneous layer is represented by a layered system consisting of a sufficiently large number of thin homogeneous layers [37]. The system transfer matrix is expressed by (9.36).

Due to the large number of thin films in the layered system, the refractive indices of the media adjacent to the inner boundaries are mutually very close. Then the matrices $\hat{\mathbf{W}}_j$ in (9.32) can be expressed as follows:

$$\hat{\mathbf{W}}_j = \mathbf{I} + \hat{\boldsymbol{\delta}}_j, \qquad \text{where} \qquad \hat{\boldsymbol{\delta}}_j = \begin{pmatrix} 0 & -\frac{\Delta \hat{Y}_j}{2\hat{Y}_j} \\ -\frac{\Delta \hat{Y}_j}{2\hat{Y}_j} & 0 \end{pmatrix}. \tag{9.85}$$

The symbol $\mathbf{I}$ denotes the unit matrix and $\Delta \hat{Y}_j = \hat{Y}_j - \hat{Y}_{j-1}$. The system transfer matrix can then be written as

$$\hat{\mathbf{P}} = \hat{\mathbf{W}}_1 \hat{\mathbf{Z}} \hat{\mathbf{W}}_{L+1}, \qquad \text{where} \qquad \hat{\mathbf{Z}} = \hat{\mathbf{U}}_1(\hat{\mathbf{I}} + \hat{\boldsymbol{\delta}}_2)\hat{\mathbf{U}}_2 \cdots (\hat{\mathbf{I}} + \hat{\boldsymbol{\delta}}_L)\hat{\mathbf{U}}_L. \tag{9.86}$$

The Fresnel reflection coefficients are calculated from the elements of the matrix $\hat{\mathbf{Z}}$ using the formula (9.37) as

$$\hat{r} = \frac{\hat{r}_1 \hat{Z}_{11} + \hat{r}_1 \hat{r}_{L+1} \hat{Z}_{12} + \hat{Z}_{21} + \hat{r}_{L+1} \hat{Z}_{22}}{\hat{Z}_{11} + \hat{r}_{L+1} \hat{Z}_{12} + \hat{r}_1 \hat{Z}_{21} + \hat{r}_1 \hat{r}_{L+1} \hat{Z}_{22}}, \tag{9.87}$$

where $\hat{Z}_{ij}$ are the elements of the matrix $\hat{\mathbf{Z}}$. The symbols $\hat{r}_1$ and $\hat{r}_{L+1}$ denote the Fresnel reflection coefficients for boundaries that separate the inhomogeneous layer from the ambient and substrate, respectively. If the elements of the matrices $\hat{\boldsymbol{\delta}}_j$ are sufficiently small it is reasonable to organize the terms in (9.86) into a perturbation series as:

$$\hat{\mathbf{Z}} = \prod_{a=1}^{L} \hat{\mathbf{U}}_a + \sum_{j=2}^{L} \left(\prod_{a=1}^{j-1} \hat{\mathbf{U}}_a \right) \hat{\boldsymbol{\delta}}_j \left(\prod_{b=j}^{L} \hat{\mathbf{U}}_b \right) + \sum_{k=3}^{L} \sum_{j=2}^{k-1} \left(\prod_{a=1}^{j-1} \hat{\mathbf{U}}_a \right) \hat{\boldsymbol{\delta}}_j \left(\prod_{b=j}^{k-1} \hat{\mathbf{U}}_b \right) \hat{\boldsymbol{\delta}}_k \left(\prod_{c=k}^{L} \hat{\mathbf{U}}_c \right) + \cdots, \tag{9.88}$$

where the number of $\hat{\boldsymbol{\delta}}_j$ matrices in each term determines the perturbation order. The dots at the end represent the terms consisting of matrix products with more than two $\hat{\boldsymbol{\delta}}_j$ matrices. It is convenient to denote the products of $\hat{\mathbf{U}}_a$ matrices (9.35) by the symbol $\hat{\mathbf{U}}_{j,k}$ defined as:

$$\hat{\mathbf{U}}_{j,k} = \prod_{a=j}^{k-1} \hat{\mathbf{U}}_a = \begin{pmatrix} \mathrm{e}^{-\mathrm{i}(\hat{X}_k - \hat{X}_j)} & 0 \\ 0 & \mathrm{e}^{\mathrm{i}(\hat{X}_k - \hat{X}_j)} \end{pmatrix}, \quad \text{where} \quad \hat{X}_j = \sum_{a=1}^{j-1} \hat{\phi}_a. \tag{9.89}$$

Using this notation, the perturbation series (9.88) can be written as:

$$\hat{\mathbf{Z}} = \hat{\mathbf{U}}_{1,L+1} + \sum_{j=2}^{L} \hat{\mathbf{U}}_{1,j} \hat{\boldsymbol{\delta}}_j \hat{\mathbf{U}}_{j,L+1} + \sum_{k=3}^{L} \sum_{j=2}^{k-1} \hat{\mathbf{U}}_{1,j} \hat{\boldsymbol{\delta}}_j \hat{\mathbf{U}}_{j,k} \hat{\boldsymbol{\delta}}_k \hat{\mathbf{U}}_{k,L+1} + \cdots. \tag{9.90}$$

In the limit of an infinite number of layers $L \to \infty$, the discrete indices in (9.90) and (9.89) become continuous variables and the sums are replaced by integrals. The result of this limit is given by:

$$\hat{\mathbf{Z}} = \hat{\mathbf{U}}(0, h) + \int_0^h \hat{\mathbf{U}}(0, z) \hat{\boldsymbol{\delta}}(z) \hat{\mathbf{U}}(z, h) \, \mathrm{d}z + \int_0^h \int_0^{z_2} \hat{\mathbf{U}}(0, z_1) \hat{\boldsymbol{\delta}}(z_1) \hat{\mathbf{U}}(z_1, z_2) \hat{\boldsymbol{\delta}}(z_2) \hat{\mathbf{U}}(z_2, h) \, \mathrm{d}z_1 \mathrm{d}z_2 + \cdots \tag{9.91}$$

with h being the thickness of the inhomogeneous layer. The matrix functions $\hat{\boldsymbol{\delta}}(z)$ and $\hat{\mathbf{U}}(z_1, z_2)$ are given as:

$$\hat{\mathbf{U}}(z_1, z_2) = \begin{pmatrix} \mathrm{e}^{-\mathrm{i}(\hat{X}(z_2) - \hat{X}(z_1))} & 0 \\ 0 & \mathrm{e}^{\mathrm{i}(\hat{X}(z_2) - \hat{X}(z_1))} \end{pmatrix}, \quad \hat{\boldsymbol{\delta}}(z) = \begin{pmatrix} 0 & -\frac{\hat{Y}'(z)}{2\hat{Y}(z)} \\ -\frac{\hat{Y}'(z)}{2\hat{Y}(z)} & 0 \end{pmatrix}, \tag{9.92}$$

where the symbol $\hat{Y}'(z)$ denotes the derivative of the admittance $\hat{Y}(z)$ with respect to the coordinate z and the function $\hat{X}(z)$ is defined as:

$$\hat{X}(z) = \frac{2\pi}{\lambda}\int_0^z \hat{n}(z')\cos\hat{\psi}(z')\,\mathrm{d}z' = \frac{2\pi}{\lambda}\int_0^z \sqrt{\hat{n}^2(z') - n_0^2\sin^2\varphi}\,\mathrm{d}z', \tag{9.93}$$

where $\hat{\psi}(z)$ denotes the refraction angle in the inhomogeneous layer. The expression for the admittance $\hat{Y}(z)$ is different for the p- and s-polarization:

$$\hat{Y}(z) = \frac{\hat{n}(z)}{\cos\hat{\psi}(z)} = \frac{\hat{n}^2(z)}{\sqrt{\hat{n}^2(z) - n_0^2\sin^2\varphi}} \quad \text{for p-polarization,} \tag{9.94}$$

$$\hat{Y}(z) = \hat{n}(z)\cos\hat{\psi}(z) = \sqrt{\hat{n}^2(z) - n_0^2\sin^2\varphi} \quad \text{for s-polarization.} \tag{9.95}$$

Performing the matrix products in (9.91), the result up to the second order is given as follows:

$$\hat{\mathbf{Z}} = \begin{pmatrix} \mathrm{e}^{-\mathrm{i}\hat{X}(h)}(1+\hat{D}_1) & -\mathrm{e}^{\mathrm{i}\hat{X}(h)}\hat{I}_2 \\ -\mathrm{e}^{-\mathrm{i}\hat{X}(h)}\hat{I}_1 & \mathrm{e}^{\mathrm{i}\hat{X}(h)}(1+\hat{D}_2) \end{pmatrix} \tag{9.96}$$

where the first order corrections are given by the integrals

$$\hat{I}_1 = \int_0^h \frac{\hat{Y}'(z)}{2\hat{Y}(z)}\mathrm{e}^{2\mathrm{i}\hat{X}(z)}\,\mathrm{d}z, \qquad \hat{I}_2 = \int_0^h \frac{\hat{Y}'(z)}{2\hat{Y}(z)}\mathrm{e}^{-2\mathrm{i}\hat{X}(z)}\,\mathrm{d}z, \tag{9.97}$$

and the second order corrections are given by the integrals

$$\begin{aligned} \hat{D}_1 &= \int_0^h\int_0^{z_2} \frac{\hat{Y}'(z_1)}{2\hat{Y}(z_1)}\frac{\hat{Y}'(z_2)}{2\hat{Y}(z_2)}\mathrm{e}^{-2\mathrm{i}\hat{X}(z_1)+2\mathrm{i}\hat{X}(z_2)}\,\mathrm{d}z_1\mathrm{d}z_2, \\ \hat{D}_2 &= \int_0^h\int_0^{z_2} \frac{\hat{Y}'(z_1)}{2\hat{Y}(z_1)}\frac{\hat{Y}'(z_2)}{2\hat{Y}(z_2)}\mathrm{e}^{2\mathrm{i}\hat{X}(z_1)-2\mathrm{i}\hat{X}(z_2)}\,\mathrm{d}z_1\mathrm{d}z_2. \end{aligned} \tag{9.98}$$

The Fresnel reflection coefficients are calculated from the elements of the matrix (9.96) by means of the formula (9.87).

The foregoing formulae for the reflection coefficients of the inhomogeneous layers take into account the approximation up to the second order, i.e. the integrals of the first and second orders occur in these formulae. The similar formulae for higher order approximations, i.e. for the higher order integrals, can be derived in the same way. Note that the derivation above was based on the calculations presented in [38].

9.4.2 Uniaxial Anisotropic Layer

In this section we will use the Yeh matrix formalism introduced in Sect. 9.2.4 to calculate the Fresnel reflection coefficients for the system consisting of an anisotropic layer surrounded by isotropic ambient and isotropic substrate. Only the simplest case of uniaxial anisotropy with the optical axis perpendicular to the boundaries will be considered in the medium forming the layer.

The coordinate system will be chosen such that the x–z plane will be the plane of incidence and the boundaries will be parallel to the x–y plane. The dielectric tensor describing the anisotropic film is given as:

$$\hat{\boldsymbol{\varepsilon}} = \begin{pmatrix} \hat{\varepsilon}_{\mathrm{o}} & 0 & 0 \\ 0 & \hat{\varepsilon}_{\mathrm{o}} & 0 \\ 0 & 0 & \hat{\varepsilon}_{\mathrm{e}} \end{pmatrix}. \tag{9.99}$$

In the anisotropic media, the wave equation is solved using (9.44)–(9.46). The equation (9.46) simplifies, in the case of uniaxial anisotropy, to

$$\begin{pmatrix} k_0^2\hat{\varepsilon}_{\mathrm{o}} - \hat{k}_z^2 & 0 & \hat{k}_x\hat{k}_z \\ 0 & k_0^2\hat{\varepsilon}_{\mathrm{o}} - \hat{k}_x^2 - \hat{k}_z^2 & 0 \\ \hat{k}_z\hat{k}_x & 0 & k_0^2\hat{\varepsilon}_{\mathrm{e}} - \hat{k}_x^2 \end{pmatrix} \begin{pmatrix} \hat{E}_x \\ \hat{E}_y \\ \hat{E}_z \end{pmatrix} = 0, \tag{9.100}$$

where $\hat{k}_x = k_0 \sin\varphi$. The condition of the vanishing determinant of the matrix in (9.100), which ensures that this equation has nontrivial solution, leads to the following quartic equation for $\hat{k}_z$:

$$k_0^2(k_0^2\hat{\varepsilon}_{\mathrm{o}} - \hat{k}_x^2 - \hat{k}_z^2)(k_0^2\hat{\varepsilon}_{\mathrm{o}}\hat{\varepsilon}_{\mathrm{e}} - \hat{\varepsilon}_{\mathrm{o}}\hat{k}_x^2 - \hat{\varepsilon}_{\mathrm{e}}\hat{k}_z^2) = 0. \tag{9.101}$$

The solution of (9.100) and (9.101) gives the following wavevectors and polarization vectors of the electric and magnetic fields

$$\begin{aligned}
\hat{\boldsymbol{k}}_1 &= (\hat{k}_x, 0, \hat{k}_{\mathrm{e},z}), & \hat{\boldsymbol{p}}_1 &= \hat{\alpha}^{-1}(\hat{\varepsilon}_{\mathrm{e}}\hat{k}_{\mathrm{e},z}, 0, -\hat{\varepsilon}_{\mathrm{o}}\hat{k}_x), & \hat{\boldsymbol{q}}_1 &= \hat{\alpha}^{-1}(0, k_0^2\hat{\varepsilon}_{\mathrm{o}}\hat{\varepsilon}_{\mathrm{e}}, 0), \\
\hat{\boldsymbol{k}}_2 &= (\hat{k}_x, 0, -\hat{k}_{\mathrm{e},z}), & \hat{\boldsymbol{p}}_2 &= \hat{\alpha}^{-1}(\hat{\varepsilon}_{\mathrm{e}}\hat{k}_{\mathrm{e},z}, 0, \hat{\varepsilon}_{\mathrm{o}}\hat{k}_x), & \hat{\boldsymbol{q}}_2 &= \hat{\alpha}^{-1}(0, -k_0^2\hat{\varepsilon}_{\mathrm{o}}\hat{\varepsilon}_{\mathrm{e}}, 0), \\
\hat{\boldsymbol{k}}_3 &= (\hat{k}_x, 0, \hat{k}_{\mathrm{o},z}), & \hat{\boldsymbol{p}}_3 &= (0, 1, 0), & \hat{\boldsymbol{q}}_3 &= (-\hat{k}_{\mathrm{o},z}, 0, \hat{k}_x), \\
\hat{\boldsymbol{k}}_4 &= (\hat{k}_x, 0, -\hat{k}_{\mathrm{o},z}), & \hat{\boldsymbol{p}}_4 &= (0, 1, 0), & \hat{\boldsymbol{q}}_4 &= (\hat{k}_{\mathrm{o},z}, 0, \hat{k}_x),
\end{aligned} \tag{9.102}$$

where

$$\hat{k}_{\mathrm{e},z} = \sqrt{k_0^2\hat{\varepsilon}_{\mathrm{o}} - \hat{\varepsilon}_{\mathrm{o}}/\hat{\varepsilon}_{\mathrm{e}}\hat{k}_x^2}, \qquad \hat{k}_{\mathrm{o},z} = \sqrt{k_0^2\hat{\varepsilon}_{\mathrm{o}} - \hat{k}_x^2},$$
$$\hat{\alpha} = \sqrt{k_0^2\hat{\varepsilon}_{\mathrm{o}}\hat{\varepsilon}_{\mathrm{e}}^2 + (\hat{\varepsilon}_{\mathrm{o}}^2 - \hat{\varepsilon}_{\mathrm{o}}\hat{\varepsilon}_{\mathrm{e}})\hat{k}_x^2}. \tag{9.103}$$

In (9.102) the indices 1 and 2 correspond to the right- and left-going extraordinary waves and the indices 3 and 4 correspond to the right- and left-going ordinary waves. In the case of uniaxial anisotropy considered here, the extraordinary and ordinary waves correspond to the p- and s-polarized waves, respectively, i.e. with the electric field parallel or perpendicular to the plane of incidence. Note that because the quartic equation (9.101) has four different roots, the polarization vectors are determined uniquely up to a sign. If some of the roots were degenerated (double roots), then there would be infinitely many choices of polarization vectors corresponding to these degenerate roots. The double roots are encountered, for example, in isotropic media or for waves propagating along the optical axes in anisotropic media. In isotropic media there are always two double roots and it is customary to choose the polarization vectors such that they correspond to the p- and s-polarized waves.

The matrices $\hat{\mathbf{D}}$ (9.58) and $\hat{\mathbf{T}}$ (9.54) are given as:

$$\hat{\mathbf{D}} = \begin{pmatrix} \hat{\varepsilon}_e\hat{k}_{e,z}/\hat{\alpha} & \hat{\varepsilon}_e\hat{k}_{e,z}/\hat{\alpha} & 0 & 0 \\ k_0^2\hat{\varepsilon}_o\hat{\varepsilon}_e/\hat{\alpha} & -k_0^2\hat{\varepsilon}_o\hat{\varepsilon}_e/\hat{\alpha} & 0 & 0 \\ 0 & 0 & 1 & 1 \\ 0 & 0 & -\hat{k}_{o,z} & \hat{k}_{o,z} \end{pmatrix}, \quad \hat{\mathbf{T}} = \begin{pmatrix} e^{-i\hat{\phi}_e} & 0 & 0 & 0 \\ 0 & e^{i\hat{\phi}_e} & 0 & 0 \\ 0 & 0 & e^{-i\hat{\phi}_o} & 0 \\ 0 & 0 & 0 & e^{i\hat{\phi}_o} \end{pmatrix}, \tag{9.104}$$

where $\hat{\phi}_e = h\hat{k}_{e,z}$, $\hat{\phi}_o = h\hat{k}_{o,z}$ and the symbol h denotes the thickness of the film. The Yeh matrix of the whole system (9.60) is equal to

$$\hat{\mathbf{X}} = \hat{\mathbf{D}}_0^{-1}\hat{\mathbf{D}}\hat{\mathbf{T}}\hat{\mathbf{D}}^{-1}\hat{\mathbf{D}}_S, \tag{9.105}$$

where the matrices $\hat{\mathbf{D}}_0$ and $\hat{\mathbf{D}}_S$ corresponding to the isotropic ambient and substrate are given by (9.67). It is convenient to calculate the product of matrices $\hat{\mathbf{D}}\hat{\mathbf{T}}\hat{\mathbf{D}}^{-1}$ in the middle of (9.105) separately:

$$\hat{\mathbf{D}}\hat{\mathbf{T}}\hat{\mathbf{D}}^{-1} = \begin{pmatrix} \cos\hat{\phi}_e & -i(k_0\hat{Y}_e)^{-1}\sin\hat{\phi}_e & 0 & 0 \\ -ik_0\hat{Y}_e\sin\hat{\phi}_e & \cos\hat{\phi}_e & 0 & 0 \\ 0 & 0 & \cos\hat{\phi}_o & i(k_0\hat{Y}_o)^{-1}\sin\hat{\phi}_o \\ 0 & 0 & ik_0\hat{Y}_o\sin\hat{\phi}_o & \cos\hat{\phi}_o \end{pmatrix}, \tag{9.106}$$

where $\hat{Y}_e$ and $\hat{Y}_o$ are the admittances corresponding to the extraordinary and ordinary waves:

$$\hat{Y}_e = \frac{k_0\hat{\varepsilon}_o}{\hat{k}_{e,z}}, \qquad \hat{Y}_o = \frac{\hat{k}_{o,z}}{k_0}. \tag{9.107}$$

Note that the 2×2 blocks on the diagonal are very similar to interference matrices introduced in (9.39). All the matrices (9.106), (9.67) in the product (9.105) have a block diagonal structure with nonzero elements only in the the 2×2 blocks on the diagonal. Therefore, the Yeh matrix of the whole system must have the same block diagonal structure and the nonzero elements of the Yeh matrix (9.105) are:

$$\hat{X}_{11} = \frac{1}{2}\frac{\hat{k}_0}{\hat{k}_{0,z}}\frac{\hat{k}_{S,z}}{\hat{k}_S}\left[\cos\hat{\phi}_e\left(1+\frac{\hat{Y}_{S,p}}{\hat{Y}_{0,p}}\right)-\mathrm{i}\sin\hat{\phi}_e\left(\frac{\hat{Y}_e}{\hat{Y}_{0,p}}+\frac{\hat{Y}_{S,p}}{\hat{Y}_e}\right)\right],$$
$$\hat{X}_{12} = \frac{1}{2}\frac{\hat{k}_0}{\hat{k}_{0,z}}\frac{\hat{k}_{S,z}}{\hat{k}_S}\left[\cos\hat{\phi}_e\left(1-\frac{\hat{Y}_{S,p}}{\hat{Y}_{0,p}}\right)-\mathrm{i}\sin\hat{\phi}_e\left(\frac{\hat{Y}_e}{\hat{Y}_{0,p}}-\frac{\hat{Y}_{S,p}}{\hat{Y}_e}\right)\right],$$
$$\hat{X}_{21} = \frac{1}{2}\frac{\hat{k}_0}{\hat{k}_{0,z}}\frac{\hat{k}_{S,z}}{\hat{k}_S}\left[\cos\hat{\phi}_e\left(1-\frac{\hat{Y}_{S,p}}{\hat{Y}_{0,p}}\right)+\mathrm{i}\sin\hat{\phi}_e\left(\frac{\hat{Y}_e}{\hat{Y}_{0,p}}-\frac{\hat{Y}_{S,p}}{\hat{Y}_e}\right)\right],$$
$$\hat{X}_{22} = \frac{1}{2}\frac{\hat{k}_0}{\hat{k}_{0,z}}\frac{\hat{k}_{S,z}}{\hat{k}_S}\left[\cos\hat{\phi}_e\left(1+\frac{\hat{Y}_{S,p}}{\hat{Y}_{0,p}}\right)+\mathrm{i}\sin\hat{\phi}_e\left(\frac{\hat{Y}_e}{\hat{Y}_{0,p}}+\frac{\hat{Y}_{S,p}}{\hat{Y}_e}\right)\right],$$
$$\hat{X}_{33} = \frac{1}{2}\left[\cos\hat{\phi}_o\left(1+\frac{\hat{Y}_{S,s}}{\hat{Y}_{0,s}}\right)-\mathrm{i}\sin\hat{\phi}_o\left(\frac{\hat{Y}_o}{\hat{Y}_{0,s}}+\frac{\hat{Y}_{S,s}}{\hat{Y}_o}\right)\right],$$
$$\hat{X}_{34} = \frac{1}{2}\left[\cos\hat{\phi}_o\left(1-\frac{\hat{Y}_{S,s}}{\hat{Y}_{0,s}}\right)-\mathrm{i}\sin\hat{\phi}_o\left(\frac{\hat{Y}_o}{\hat{Y}_{0,s}}-\frac{\hat{Y}_{S,s}}{\hat{Y}_o}\right)\right],$$
$$\hat{X}_{43} = \frac{1}{2}\left[\cos\hat{\phi}_o\left(1-\frac{\hat{Y}_{S,s}}{\hat{Y}_{0,s}}\right)+\mathrm{i}\sin\hat{\phi}_o\left(\frac{\hat{Y}_o}{\hat{Y}_{0,s}}-\frac{\hat{Y}_{S,s}}{\hat{Y}_o}\right)\right],$$
$$\hat{X}_{44} = \frac{1}{2}\left[\cos\hat{\phi}_o\left(1+\frac{\hat{Y}_{S,s}}{\hat{Y}_{0,s}}\right)+\mathrm{i}\sin\hat{\phi}_o\left(\frac{\hat{Y}_o}{\hat{Y}_{0,s}}+\frac{\hat{Y}_{S,s}}{\hat{Y}_o}\right)\right], \qquad (9.108)$$

where the admittances of the ambient $\hat{Y}_{0,s}$, $\hat{Y}_{0,p}$ and the substrate $\hat{Y}_{S,s}$, $\hat{Y}_{S,p}$ are defined by (9.28). The meaning of the symbols $\hat{k}_0$, $\hat{k}_{0,z}$, $\hat{k}_S$ and $\hat{k}_{S,z}$ is the same as in (9.67) with the subscripts 0 and S corresponding to the ambient and substrate, respectively. The Fresnel reflection coefficients are calculated as (9.64):

$$\hat{r}_{pp} = \frac{\hat{r}_{1,p}+\hat{r}_{2,p}e^{2\mathrm{i}\hat{\phi}_e}}{1+\hat{r}_{1,p}\hat{r}_{2,p}e^{2\mathrm{i}\hat{\phi}_e}}, \qquad \hat{r}_{ss} = \frac{\hat{r}_{1,s}+\hat{r}_{2,s}e^{2\mathrm{i}\hat{\phi}_o}}{1+\hat{r}_{1,s}\hat{r}_{2,s}e^{2\mathrm{i}\hat{\phi}_o}}, \qquad \hat{r}_{ps} = \hat{r}_{sp} = 0, \qquad (9.109)$$

where the Fresnel reflection coefficients $\hat{r}_{1,p}$, $\hat{r}_{1,s}$ for the boundary between the ambient and anisotropic film and the Fresnel reflection coefficients $\hat{r}_{2,p}$, $\hat{r}_{2,s}$ for the boundary between the anisotropic film and substrate are defined as:

$$\hat{r}_{1,p} = \frac{\hat{Y}_{0,p}-\hat{Y}_e}{\hat{Y}_{0,p}+\hat{Y}_e}, \quad \hat{r}_{2,p} = \frac{\hat{Y}_e-\hat{Y}_{S,p}}{\hat{Y}_e+\hat{Y}_{S,p}}, \quad \hat{r}_{1,s} = \frac{\hat{Y}_{0,s}-\hat{Y}_o}{\hat{Y}_{0,s}+\hat{Y}_o}, \quad \hat{r}_{2,s} = \frac{\hat{Y}_o-\hat{Y}_{S,s}}{\hat{Y}_o+\hat{Y}_{S,s}}. \qquad (9.110)$$

The foregoing formulae for the Fresnel reflection coefficients are in coincidence with those presented in the monograph [1].

9.4.3 Reflection and Transmission of Light by Transparent Slabs Covered with Layered Systems

Often, one needs to carry out optical characterization of the layered systems placed on substrates formed by transparent slabs with parallel boundaries. Generally, it is possible to assume that the layered systems are present on both the lower and upper boundaries of the slabs. The presence of the transparent slabs makes the optical characterization of the layered systems more complicated because of the reflections inside the slabs (see Fig. 9.7). If the slabs are sufficiently thick (compared to coherence length of light), the inner reflections in the slab must be described as incoherent superposition of light intensities. This can be achieved using the Stokes–Mueller matrix formalism. In this formalism, the reflection and transmission of light by the slab covered with layered systems are described by the Mueller matrices **R** and **T**. These matrices contain the information about the reflectance and transmittance as well as the information concerning the ellipsometric quantities. The Mueller matrices **R** and **T** can be expressed as a sum of Mueller matrices corresponding to the paths of the individual beams in the following way:

$$\mathbf{R} = \mathbf{R}_1 + \mathbf{T}_1'\mathbf{U}'\mathbf{R}_2\mathbf{U}\left[\sum_{n=0}^{\infty}(\mathbf{R}_1'\mathbf{U}'\mathbf{R}_2\mathbf{U})^n\right]\mathbf{T}_1\,, \tag{9.111}$$

$$\mathbf{T} = \mathbf{T}_2\mathbf{U}\left[\sum_{n=0}^{\infty}(\mathbf{R}_1'\mathbf{U}'\mathbf{R}_2\mathbf{U})^n\right]\mathbf{T}_1\,, \tag{9.112}$$

where $\mathbf{R}_j$, $\mathbf{R}_j'$ and $\mathbf{T}_j$, $\mathbf{T}_j'$ are the Mueller matrices describing the reflection and transmission of light by the layered systems placed on the upper ($j = 1$) and lower ($j = 2$) boundaries of the slab, respectively. The quantities without the primes correspond to the light wave incident from the top while the primed quantities correspond to the light wave incident from the bottom (see Fig. 9.7). The matrices **U** and $\mathbf{U}'$ describe the influence of the slab if the light wave travels from top to bottom and in the opposite direction, respectively. The infinite sums of matrices in (9.111) and (9.112) can be calculated as follows:

$$\sum_{n=0}^{\infty}\mathbf{S}^n = \mathbf{I} + \mathbf{S} + \mathbf{SS} + \cdots = (\mathbf{I} - \mathbf{S})^{-1}\,, \tag{9.113}$$

where $\mathbf{S} = \mathbf{R}_1'\mathbf{U}'\mathbf{R}_2\mathbf{U}$.

In the simplest case, the slab is formed by an isotropic medium. The matrices **U** and $\mathbf{U}'$ are then identical and proportional to the unit matrix, i.e.

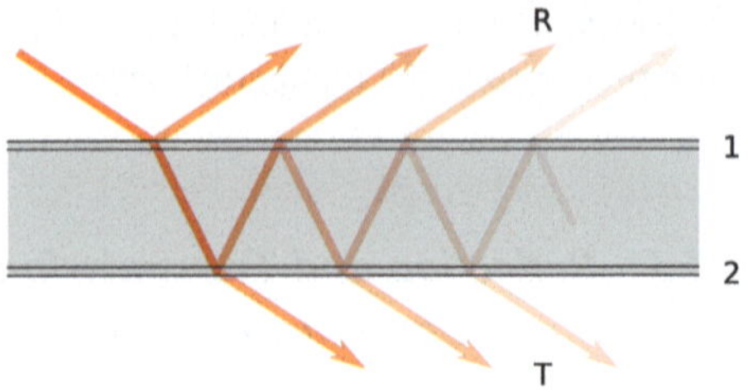

Fig. 9.7 Schematic diagram of multiple reflections inside the slab

$$\mathbf{U} = \mathbf{U}' = U\mathbf{I}, \qquad U = \exp\left[-2k_0 \mathrm{Im}\left(\hat{n}_S \cos\hat{\psi}_S\right) h_S\right], \tag{9.114}$$

where $\hat{n}_S$ is the complex refractive index of the medium forming the slab, h_S is the thickness of the slab and $\hat{\psi}_S$ is the complex refraction angle in the slab. The scalar quantity U describes the attenuation of the waves passing through the slab. If the slab is anisotropic then the expressions for the matrices $\mathbf{U}$ and $\mathbf{U}'$ are more complicated. In general, they are mutually different.

In practice, the stress caused by the layered systems placed onto the boundaries of isotropic slabs often induces weak artificial anisotropy in these slabs. It is reasonable to assume that this weak anisotropy has a negligible influence onto the values of the Fresnel reflection and transmission coefficients of the boundaries of the slab. However, this anisotropy has a non-negligible influence on the phases of waves propagating inside slabs. In a special case of anisotropy with each principal axis lying either in the plane of incidence or perpendicular to the plane of incidence, this results in a phase retardation between the p- and s-polarized waves. This phase retardation can be described by means of the Mueller matrix $\mathbf{C}(\delta)$ corresponding to the special Mueller matrix (9.23) for transmitted light with $\hat{t}_p = 1$ and $\hat{t}_s = e^{i\delta}$, where δ is a small retardation angle. The matrices $\mathbf{U}$ and $\mathbf{U}'$ are then given as:

$$\mathbf{U} = \mathbf{U}' = U\mathbf{C}(\delta), \qquad \text{where} \qquad \mathbf{C}(\delta) = \begin{pmatrix} 1 & 0 & 0 & 0 \\ 0 & 1 & 0 & 0 \\ 0 & 0 & \cos(\delta) & -\sin(\delta) \\ 0 & 0 & \sin(\delta) & \cos(\delta) \end{pmatrix}. \tag{9.115}$$

Of course, the retardation angle δ depends on the elements of the dielectric tensor describing the anisotropy, wavelength of light, incidence angle and the thickness of the slab.

The formulae (9.111) and (9.112) describe only the depolarization caused by multiple reflections inside the slab. Often it is also necessary to take into account the depolarization caused by other effects discussed in Sect. 9.2.2. In this case the sums in (9.111) and (9.112) must be performed first and the formula (9.21) is then applied to the resultant matrices $\mathbf{R}$ and $\mathbf{T}$. The situation is relatively simple if the special form of the Mueller matrix described by (9.23) and (9.24) can be employed (e.g. for optically isotropic systems).

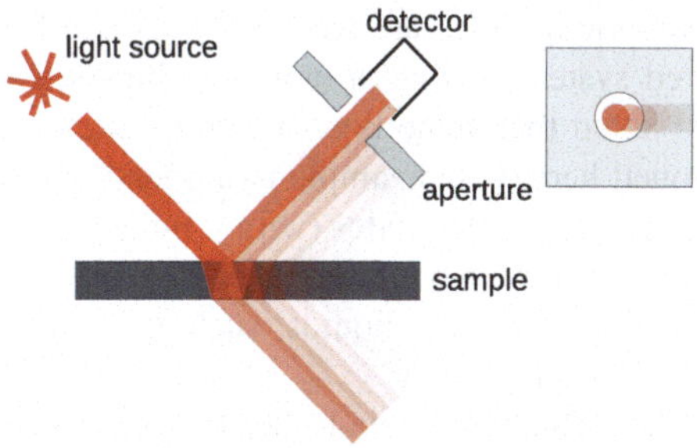

Fig. 9.8 Schematic diagram of experimental setup in which only part of reflected light beams is registered by the detector

It is evident from Figs. 9.7 and 9.8 that the outgoing light beams corresponding to different paths of light inside the slab are mutually shifted. If the slab is sufficiently thick, these shifts can not be neglected and the finite size of the detector aperture must be taken into account. Since each term in the sums in (9.111) and (9.112) corresponds to a particular path of light in the sample, it is possible to incorporate this phenomenon by multiplying each of these terms by a factor that expresses the fraction of the light intensity registered by the detector for a given outgoing beam.

It should be noted that an influence of depolarization, including an induced anisotropy and thickness non-uniformity in the transparent slab, must often be taken into account when performing the optical characterization of the layered systems covering these slabs. Of course, the influence of the finite size of the detector apertures of the ellipsometers discussed above must also be considered.

9.5 Conclusion

This chapter briefly describes principles of ellipsometry needed for the optical characterization of the layered systems.

In the first part of this chapter, the matrix formalisms usable in ellipsometry of layered systems are presented. These matrix formalisms allow us to formulate the theoretical results in an efficient and concise form. The Jones and Stokes–Mueller matrix formalisms are used to describe the theoretical principles of ellipsometric measurements. Two other matrix formalisms applicable to calculation of optical quantities of layered systems are also presented. One of them is applicable to the systems consisting of isotropic layers and the other formalism, called the Yeh matrix formalism, is applicable also to systems containing anisotropic layers.

In the second part of this chapter, the basic types of ellipsometry, i.e. conventional ellipsometry, generalized ellipsometry and Mueller-matrix ellipsometry, together with three frequently utilized ellipsometric techniques, i.e. null ellipsometry, rotating analyzer (polarizer) ellipsometry and phase-modulated ellipsometry, are introduced. The other ellipsometric techniques such as the rotating compensator ellipsometry, oscillating-analyzer ellipsometry, return-path ellipsometry, etc., were not described, even though they are frequently applied for the optical characterization of the layered systems. Modern imaging ellipsometry was also briefly discussed. It should be noted that some types of ellipsometry that are used in practice were not mentioned here. One example is immersion ellipsometry employed in monochromatic or spectroscopic modes (within immersion ellipsometry the values of the refractive index of the ambient are varied by means of immersion of the samples into various non-absorbing liquids, such as acetone, toluene, chloronaphthalene, etc.), see e.g. [39–41].

The chapter concludes by the presentation of three selected applications of the matrix formalisms introduced in the first part, i.e. by deriving the approximate expressions for the reflection coefficients of the isotropic inhomogeneous layer, the formulae for the reflection coefficients of the uniaxial anisotropic layer placed onto the isotropic substrate and the formulae for the reflection and transmission of light by transparent slabs covered with layered systems. These examples illustrate the practical use of the corresponding matrix formalisms.

Acknowledgements The work on this chapter was supported by the project LO1411 (NPU I) funded by Ministry of Education Youth and Sports of the Czech Republic.

References

1. R.M.A. Azzam, N.M. Bashara, *Ellipsometry and Polarized Light* (North-Holland, Amsterdam, 1977)
2. I. Ohlídal, D. Franta, in *Progress in Optics*, vol. 41, ed. by E. Wolf (Elsevier, Amsterdam, 2000), pp. 181–282
3. H. Fujiwara, *Spectroscopic Ellipsometry: Principles and Applications* (Wiley, Chichester, 2007)
4. D. Goldstein, *Polarized Light, Revised and Expanded* (Dekker, New York, 2003)
5. M. Born, E. Wolf, *Principles of Optics*, 3rd edn. (Pergamon Press, Oxford, 1965)
6. E. Garcia-Caurel, R. Ossikovski, M. Foldyna, A. Pierangelo, B. Drévillon, A. De Martino, in *Ellipsometry at the Nanoscale*, ed. by M. Losurdo, K. Hingerl (Springer, Berlin, 2013), pp. 31–143. Chap. 2
7. S. Huard, *Polarization of Light* (Wiley, New York, 1997)
8. R.A. Chipman, in *Handbook of Optics*, vol. 2, ed. by M. Bass (McGraw Hill, New York, 1995), pp. 22.1–22.37. Chap. 22
9. G.E. Jellison, in *Handbook of Ellipsometry*, ed. by H.G. Tompkins, E.A. Irene (William Andrew, New York, 2005), pp. 237–296. Chap. 3
10. A. Aiello, J.P. Woerdman, Linear algebra for Mueller calculus (2004), arXiv:math-ph/0412061
11. Z. Knittl, *Optics of Thin Films* (Wiley, London, 1976)
12. P. Yeh, Surf. Sci. **96**, 41 (1980)
13. D.W. Berreman, J. Opt. Soc. Am. **62**, 502 (1972)
14. M. Schubert, Phys. Rev. B **53**, 4265 (1996)
15. M. Schubert, B. Rheinländer, J.A. Woollam, B. Johs, C.M. Herzinger, J. Opt. Soc. Am. A **13**, 875 (1996)
16. M. Schubert, in *Handbook of Ellipsometry*, ed. by H.G. Tompkins, E.A. Irene (William Andrew, New York, 2005), pp. 637–717. Chap. 9
17. D.E. Aspnes, Opt. Commun. **8**, 222 (1973)

18. J.M.M. de Nijs, A.H.M. Holtslag, A. Hoeksta, A. van Silfhout, J. Opt. Soc. Am. A **5**, 1466 (1988)
19. O. Acher, E. Bigan, B. Drévillon, Rev. Sci. Instrum. **60**, 65 (1989)
20. L. Halagačka, K. Postava, M. Foldyna, J. Pištora, Phys. Status Solidi A **205**, 752 (2008)
21. R.M.A. Azzam, in *Polarization: Measurement, Analysis, and Remote Sensing*. Proceedings of SPIE, vol. 3121 (1997), pp. 396–405
22. P.S. Hauge, J. Opt. Soc. Am. **68**, 1519 (1978)
23. D.H. Goldstein, R.A. Chipman, J. Opt. Soc. Am. A **7**, 693 (1990)
24. P.S. Hauge, Surf. Sci. **96**, 108 (1980)
25. R.C. Thompson, J.R. Bottiger, E.S. Fry, Appl. Opt. **19**, 1323 (1980)
26. A. Laskarakis, S. Logothetidis, E. Pavlopoulou, M. Gioti, Thin Solid Films **455–456**, 43 (2004)
27. R.M.A. Azzam, Opt. Lett. **11**, 270 (1986)
28. R. Azzam, Thin Solid Films **234**, 371 (1993)
29. Y.H. Meng, Y.Y. Chen, C. Qi, L. Liu, G. Jin, Phys. Status Solidi C **5**, 1050 (2008)
30. M. Vaupel, M. Vinnichenko, Phys. Status Solidi C **5**, 1137 (2008)
31. L. Jin, Y. Wakako, K. Takizawa, E. Kondoh, Thin Solid Films **571**, 532 (2014)
32. D. Franta, I. Ohlídal, Acta Phys. Slov. **50**, 411 (2000)
33. D. Franta, L. Zajíčková, I. Ohlídal, J. Janča, K. Veltruská, Diam. Relat. Mater. **11**, 105 (2002)
34. I. Ohlídal, D. Franta, E. Pinčík, M. Ohlídal, Surf. Interface Anal. **28**, 240 (1999)
35. D. Nečas, I. Ohlídal, D. Franta, M. Ohlídal, V. Čudek, J. Vodák, Appl. Opt. **53**, 5606 (2014)
36. D. Nečas, J. Vodák, I. Ohlídal, M. Ohlídal, A. Majumdar, L. Zajíčková, Appl. Surf. Sci. **350**, 149 (2015)
37. R. Jacobsson, in *Progress in Optics*, vol. 5, ed. by E. Wolf (Elsevier, Amsterdam, 1966), pp. 247–286
38. M. Kildemo, O. Hunderi, B. Drévillon, J. Opt. Soc. Am. A **14**, 931 (1997)
39. K. Vedam, R. Rai, F. Lukeš, R. Srinivasan, J. Opt. Soc. Am. **58**, 526 (1968)
40. I. Ohlídal, M. Líbezný, Surf. Interface Anal. **16**, 46 (1990)
41. C. Zhao, P. Lefebvre, E. Irene, Thin Solid Films **313–314**, 286 (1998)

Part III
Characterization of Defective and Corrugated Coatings

Chapter 10
Optical Characterization of Thin Films Exhibiting Defects

Ivan Ohlídal, Martin Čermák and Jiří Vohánka

Abstract In this chapter the influence of the main defects on the optical characterization of thin films is described. These defects are random roughness of boundaries, thickness non-uniformity, optical inhomogeneity corresponding to refractive index profiles, overlayers and transition layers. The theoretical approaches and the formulae for the corresponding optical quantities of the thin films exhibiting these defects are presented. The attention is concentrated on the ellipsometric parameters and reflectance of these thin films belonging to the specular reflection. The selected numerical examples illustrating the influence of the defects are introduced. Several experimental examples of the optical characterization of the thin films with the defects are also shown. The discussion of both the numerical and experimental results is carried out too.

10.1 Introduction

Thin film systems occurring in practice exhibit various defects. These defects often influence the optical properties and optical characterization of these systems in a non-negligible way. Results of their characterization can be misrepresented if the defects are neglected. Therefore, it is necessary to employ methods allowing us to include these defects into optical characterization of thin films. Including the defects requires their incorporation into structural models of the corresponding thin film systems and using a suitable theoretical approach mathematically describing the influence of these defects. The creation of correct structural and dispersion models of the layered systems is the basic condition for their reliable and precise optical characterization. The five most frequent defects occurring in practice are: random surface (boundary) roughness, area non-uniformity, optical inhomogeneity represented by profiles of the complex refractive index across the films, transition layers and overlayers.

I. Ohlídal (✉) · M. Čermák · J. Vohánka
Department of Physical Electronics, Faculty of Science, Masaryk University,
Kotlářská 2, 611 37 Brno, Czech Republic
e-mail: ohlidal@physics.muni.cz

O. Stenzel and M. Ohlídal (eds.), *Optical Characterization of Thin Solid Films*, Springer Series in Surface Sciences 64,
https://doi.org/10.1007/978-3-319-75325-6_10

In this chapter, both the theoretical and experimental features of the methods for the optical characterization of the thin films with defects mentioned above will be presented. Several typical examples of the thin films exhibiting the defects or their combination will be utilized for experimental illustration.

10.2 Quantities for the Optical Characterization

For the optical characterization of thin films, spectroscopic ellipsometry and spectroscopic photometry are used most frequently. Therefore, the spectral dependencies of ellipsometric parameters and spectral dependencies of reflectance or transmittance of these films are employed for this purpose. In the specular direction the reflectances are defined as follows:

$$R_q = \hat{r}_q \hat{r}_q^*, \quad q = p, s, \quad R = \hat{r}\hat{r}^*, \tag{10.1}$$

where $\hat{r}_\mathrm{p}$ and $\hat{r}_\mathrm{s}$ are the complex Fresnel reflection coefficients for p- and s-polarizations at the oblique incidence, respectively. The symbol R denotes the reflectance at normal incidence ($\hat{r} = \hat{r}_\mathrm{p} = \hat{r}_\mathrm{s}$). In specular reflection the ellipsometric parameters are defined by means of the ellipsometric ratio $\hat{\varrho} = \hat{r}_\mathrm{p}/\hat{r}_\mathrm{s}$ in the following way [1, 2]:

$$\hat{\varrho} = \tan \psi \mathrm{e}^{\mathrm{i}\Delta}, \tag{10.2}$$

where ψ and Δ are the azimuth and the phase change, respectively. Ellipsometric parameters ψ and Δ are measured within null ellipsometry while $\tan \psi$ and $\cos \Delta$ are measured within rotating-analyzer ellipsometry and rotating-polarizer ellipsometry. Within phase-modulated ellipsometry, the associated ellipsometric parameters I_s, I_c and I_n are measured. They are defined as [1, 3]

$$I_\mathrm{s} = -\mathrm{i}\frac{\hat{r}_\mathrm{p}\hat{r}_\mathrm{s}^* - \hat{r}_\mathrm{p}^*\hat{r}_\mathrm{s}}{|\hat{r}_\mathrm{s}|^2 + |\hat{r}_\mathrm{p}|^2}, \quad I_\mathrm{c} = \frac{\hat{r}_\mathrm{p}\hat{r}_\mathrm{s}^* + \hat{r}_\mathrm{p}^*\hat{r}_\mathrm{s}}{|\hat{r}_\mathrm{s}|^2 + |\hat{r}_\mathrm{p}|^2}, \quad I_\mathrm{n} = \frac{|\hat{r}_\mathrm{s}|^2 - |\hat{r}_\mathrm{p}|^2}{|\hat{r}_\mathrm{s}|^2 + |\hat{r}_\mathrm{p}|^2}. \tag{10.3}$$

Note that the other ellipsometric quantities can also be utilized for the optical characterization of thin films. In transmitted light, the ellipsometric parameters and transmittances are defined by means of the Fresnel transmission coefficients of the corresponding thin film systems (see e.g. [4, 5]).

It should also be noted that in this chapter, we will deal with the systems consisting of optically isotropic thin films and substrates. It is also necessary to emphasize that the foregoing equations are usable for the optical characterization of thin film systems only if the optical response of the system, i.e. the relation between the incident and outgoing waves, can be described with the help of the Fresnel reflection and transmission coefficients in an unambiguous way. In some cases, the relation between the incident and outgoing waves cannot be completely described by the

Fresnel reflection and transmission coefficients. For example, if the sample exhibits area non-uniformity, which will be discussed in Sect. 10.4 of this chapter, or in the case of samples which depolarize light (see Chap. 9).

10.3 Random Roughness of Thin Film Boundaries

Random roughness of boundaries is a defect which occurs in most thin films encountered within fundamental research, applied research and industrial applications. The choice of the theoretical approaches applicable for incorporation of this roughness into the formulae expressing the measured optical quantities of these films depends on the relations between the linear dimensions of roughness and wavelengths of incident light.

The utilized theoretical approaches enabling us to include random roughness into the optical characterization of thin films will be presented in this section. These approaches cover the whole domain of random roughness having a practical meaning in the thin film characterization. Note that homogeneous and isotropic random roughness will be assumed in this chapter.

10.3.1 Effective Medium Approximation

If the inequalities $\sigma \ll \lambda$ and $T \ll \lambda$ are fulfilled then the effective medium approximation (EMA) can be used to include this fine random roughness into the formulae of the optical quantities (λ, σ and T are the wavelength, the RMS value of the heights of irregularities and autocorrelation length of roughness, respectively). This fine roughness is often called random microroughness.

Within the EMA, random microroughness of surfaces or boundaries is replaced by a homogeneous fictitious layer (HFL) with a certain effective thickness and effective dielectric function (or effective optical constants). The effective dielectric function can be calculated by the Lorentz-Lorenz (LL), Maxwell Garnet (MG) and Bruggeman (BR) formulae [6]. These formulae have the same generic formula [6]:

$$\frac{\hat{\bar{\varepsilon}} - \hat{\varepsilon}_{\mathrm{h}}}{\hat{\bar{\varepsilon}} + 2\hat{\varepsilon}_{\mathrm{h}}} = \sum_{j=1}^{N} p_j \frac{\hat{\varepsilon}_j - \hat{\varepsilon}_{\mathrm{h}}}{\hat{\varepsilon}_j + 2\hat{\varepsilon}_{\mathrm{h}}}, \tag{10.4}$$

where $\hat{\bar{\varepsilon}}, \hat{\varepsilon}_{\mathrm{h}}, \hat{\varepsilon}_j$ $(j = 1, 2, \ldots, N)$ are the dielectric functions of the effective medium, host medium, and inclusions of types j in the host medium, respectively, and p_j represent volume fractions of materials of inclusions of types j in the total volume (the volume fraction is the relative part of the corresponding inclusion volume in the host medium volume related to the total volume). The symbol N denotes the total number of these inclusions. The dielectric function of the effective medium will be referred to as the effective dielectric function below.

The generic formula (10.4) was derived for mixtures of bulk solids under the assumptions of spherical inclusion geometry and dipole interaction only [6–8]. Due to the spherical inclusion geometry, the depolarization factor was taken in the value of $1/3$ at deriving the generic formula. Neither of these two assumptions is rigorously satisfied by random roughness. This fact is a weak point of the EMA at its application to randomly microrough surfaces and thin film boundaries. In spite of this fact, the formulae implied by this generic formula (10.4) are mostly utilized for expressing the optical response from randomly microrough surfaces and boundaries (see e.g. [6–8]). A change of the depolarization factor from value of $1/3$, i.e. a change from the spherical inclusion symmetry, is not generally significant [6]. In the MG approximation one medium is selected as a host medium in (10.4) and other media are considered to be inclusions into this host medium. Therefore, the MG approximation exhibits the ambiguity in selecting the host medium for the mixtures containing two bulk media since the roles of the host medium and inclusion can be interchanged. In bulk applications, this ambiguity was removed by Bruggeman [9] who suggested replacing $\hat{\varepsilon}_{\mathrm{h}}$ with $\hat{\bar{\varepsilon}}$, i.e. letting the effective medium itself act as host medium. Removing of the ambiguity is probably the main reason why the Bruggeman formula derived by the above mentioned replacement in (10.4) is mostly used to express optical quantities of the bulk mixtures and randomly microrough surfaces and boundaries. Note that the LL approximation assuming vacuum as a host medium is used to describe microrough surfaces only exceptionally. Ellipsometric measurements of smooth and rough surfaces are mostly performed in air. The majority of these surfaces are then covered with various overlayers. These overlayers are usually very thin, i.e. they exhibit thickness in nanometric scale (see e.g. [1, 10–24]). Nevertheless, they must be taken into account in ellipsometric studies because ellipsometry is extremely sensitive to them (overlayers need not only be considered under special conditions, e.g. at ellipsometric measurements in ultrahigh vacuum). If the overlayers are very thin, it is reasonable to model them by the identical thin films (ITF). The ITF exhibits the upper and lower randomly rough boundaries that are identical from both the geometrical and statistical points of view (see [1, 12, 13, 25–27]). Microroughness of both the boundaries is the same as that of the surface on which the overlayer is placed. Moreover, it is assumed that the overlayers are continuous, i.e. overlayers with island character and volume defects are not taken into account. It is also assumed that the overlayers are optically homogeneous together with their substrates. When the Bruggeman approximation is employed, the effective dielectric function is given as follows (see (10.4)):

$$p_0 \frac{\varepsilon_0 - \hat{\bar{\varepsilon}}}{\varepsilon_0 + 2\hat{\bar{\varepsilon}}} + p_{\mathrm{O}} \frac{\hat{\varepsilon}_{\mathrm{O}} - \hat{\bar{\varepsilon}}}{\hat{\varepsilon}_{\mathrm{O}} + 2\hat{\bar{\varepsilon}}} + p_{\mathrm{S}} \frac{\hat{\varepsilon}_{\mathrm{S}} - \hat{\bar{\varepsilon}}}{\hat{\varepsilon}_{\mathrm{S}} + 2\hat{\bar{\varepsilon}}} = 0, \qquad (10.5)$$

where p_0, p_O and p_S, are the volume fractions of ambient, overlayer and substrate, respectively. The symbols ε_0, $\hat{\varepsilon}_O$ and $\hat{\varepsilon}_S$ denote the dielectric functions of ambient, overlayer and substrate, respectively.

By means of the following equations, it is possible to define fictitious thicknesses corresponding to the volume fractions of the ambient, overlayer and substrate, i.e.

$$p_0 = h_0^H / h_F^H, \qquad p_O = h_O^H / h_F^H, \qquad p_S = h_S^H / h_F^H, \tag{10.6}$$

where $h_F^H = h_0^H + h_O^H + h_S^H$ is the thickness of the HFL. If the model of the ITF is used then the volume fractions of the ambient and substrate are usually chosen to be the same, i.e. $h_0^H = h_S^H$.

Equation (10.5) leads to a cubic equation with regard to the unknown effective dielectric function $\hat{\bar{\varepsilon}}$. The roots $\hat{\bar{\varepsilon}}_k$, $(k = 0, 1, 2)$ of this cubic equation can be found using the known Cardano's method. This method allows to find three roots of the cubic equation from which only one has a physical meaning. The finding of the solution of the cubic equation together with choice of the physically-meaningful root is a non-trivial task. It seems that to avoid this complication, some researchers used the model of the system of randomly microrough surface covered with very thin overlayer consisting of two layers. The lower layer is the HFL describing microroughness with the effective dielectric function corresponding to a mixture of vacuum voids and surface material. The upper layer then consists of a continuous and homogeneous overlayer. For example, this special model was utilized for a slightly rough silicon surface covered with very thin SiO_2 overlayer in [28]. However, such a model is not completely correct since the material of the overlayer must also be incorporated into the fictitious layer describing random microroughness.

There is a further problematic point with the EMA based on the HFL. Specifically, it is evident that a distribution of volume occupied by the material forming the randomly rough surface changes along the normal direction to the mean plane of this surface. Results of several researchers indicate this fact. For example, Aspnes et al. [6] showed that the fictitious layer representing random microroughness of the upper boundary of an amorphous silicon film did not correspond to a homogeneous layer, but it exhibited a clear density gradient. Therefore, microroughness of this silicon sample was approximated by two HFL mutually differing in thickness and an effective dielectric function. Aspnes et al. fitted also the experimental data of this rough silicon sample by the inhomogeneous fictitious layer (IFL) with the dielectric function profiles calculated by means of the volume fraction profiles for periodically rough surfaces corresponding to triangular ridges, hemicylindrical ridges, pyramids, hemicylindrical pyramids and hemispherical geometry. They employed the theoretical results derived for the volume fraction profiles of the periodic models of roughness discussed above in [29]. After processing the experimental ellipsometric data of the amorphous silicon sample, they found that the best fit was obtained for the hemispherical geometry.

If the statistical approach is utilized for replacing random microroughness by the IFL, the profiles of the volume fractions of the individual materials are determined by a one-dimensional distribution of probability density of the heights of roughness irregularities $w_1(z)$ of the overlayer boundaries. It holds that [30]:

$$p_S(z) = \int_z^{\infty} w_1(z')\mathrm{d}z'. \tag{10.7}$$

For the Gaussian distribution $w_1(z)$ is given by

$$w_1(z) = \frac{1}{\sqrt{2\pi}\sigma^{\mathrm{I}}} \exp\left(-\frac{z^2}{2(\sigma^{\mathrm{I}})^2}\right). \tag{10.8}$$

The symbols σ^{I} and $p_S(z)$ denote the RMS value of the heights of the roughness irregularities and function describing the volume fraction profile of the substrate (surface) material, respectively. The symbol $w_1(z)$ denotes the Gaussian one-dimensional probability density of heights of roughness irregularities corresponding to the lower boundary of the overlayer. Function $p_O(z)$ describing the volume fraction profile of the overlayer material is given as

$$p_O(z) = \int_z^{\infty} w_1(z' - h_O^{\mathrm{I}})\mathrm{d}z' - \int_z^{\infty} w_1(z')\mathrm{d}z' \tag{10.9}$$

where $w_1(z' - h_O^{\mathrm{I}})$ is the Gaussian one-dimensional distribution of probability density corresponding to the upper boundary and h_O^{I} is the mean thickness of the overlayer. The function $p_0(z)$ describing the volume fraction profile of the ambient is given as follows:

$$p_0(z) = \int_{-\infty}^{z} w_1(z' - h_O^{\mathrm{I}})\mathrm{d}z' = 1 - p_S(z) - p_O(z). \tag{10.10}$$

After inserting functions $p_S(z)$, $p_O(z)$ and $p_0(z)$ into the Bruggeman formula, one obtains the corresponding profile of the effective dielectric function of the IFL.

The Fresnel reflection coefficients for the HFL are calculated using the well-known formula for the system of a smooth substrate covered with a single layer (see e.g. [2–5]). In the case of the IFL, several procedures can be used to calculate its reflection coefficients (see Sect. 10.5).

After assessing numerical and experimental results available in literature, the advantages of the EMA approaches can be summarized in the following way:

- The mathematical formalism concerning the EMA approaches is very simple and its utilization is quite easy in the processing of the optical data. This is why the EMA approaches are employed very frequently in practice.

- In spite of the fact that the replacement of the random microroughness by the fictitious layer is unphysical the EMA approaches allow to partially exclude the influence of this random microroughness.
- The overlayer thickness is determined correctly and precisely if the RMS value of the heights or RMS value of the slopes of random microroughness are very small. One can even determine the spectral dependence of the refractive index of the overlayer if a simple dispersion model is used.

The limitation of the EMA approaches are as follows:

- The EMA approaches can not describe roughness correctly. In these approaches, random microroughness is described by the effective parameters such as the effective RMS values of the heights and effective thicknesses that do not describe the real properties of this microroughness.
- Within the EMA approaches one can not take into account the scattering of light caused by random microroughness. Consequently, the EMA approaches are applicable rather for ellipsometry since the ellipsometric quantities are less sensitive to the influence of scattering losses in the specular direction than the intensity quantities such as reflectance.
- The expressions for the effective dielectric function used in the EMA approaches are based on the formulae derived on the basis of the assumptions that are not completely fulfilled for random microroughness.
- It is very difficult to estimate the conditions under which the EMA approaches give reasonable results.

10.3.2 Rayleigh–Rice Theory

The Rayleigh–Rice theory (RRT) is one of perturbation theories applicable for describing interaction of light with randomly rough surfaces and thin films with randomly rough boundaries [31–33]. It is a second-order perturbation theory that can be used to derive formulae for quantities describing specularly reflected, transmitted and scatered light. To our experience, the RRT represents a very efficient theoretical approach for optical characterization of the randomly rough surfaces and thin films if the following relations are fulfilled: $\sigma \ll \lambda$ and $\tan \beta_0 \lesssim 0.1$, The roughness corresponding to the foregoing inequalities is referred to as slight roughness. Within this perturbation theory, the Fresnel reflection coefficients $\hat{r}_q$ ($q = \mathrm{p, s}$) of slightly randomly rough surfaces are given as follows [33, 34]:

$$\hat{r}_q = \hat{r}_q^{(0)} + \Delta \hat{r}_q, \tag{10.11}$$

where $\hat{r}_q^{(0)}$ are the reflection coefficients of the corresponding smooth surfaces and $\Delta \hat{r}_q$ represent the corrections in specularly reflected light calculated as

$$\Delta \hat{r}_q = \int_{-\infty}^{\infty} \int_{-\infty}^{\infty} \hat{f}_q(K_x, K_y) W(K_x', K_y) \mathrm{d}K_x \mathrm{d}K_y. \tag{10.12}$$

The symbols K_x, K_y are the spatial frequencies of the roughness.

$$K_x' = K_x - (2\pi/\lambda) n_0 \sin\varphi,$$

where φ is the incidence angle on the mean plane of the surface and $n_0 = \sqrt{\varepsilon_0}$ is the refractive index of the ambient. The function $\hat{f}_q(K_x, K_y)$ is a complicated function of K_x, K_y, λ, φ and optical constants of media forming the system. The power spectral density function (PSDF) $W(K_x', K_y)$ can be expressed by the Gaussian function

$$W(K_x', K_y) = \frac{\sigma^2 T^2}{4\pi} \mathrm{e}^{-(K_x'^2 + K_y^2) T^2/4}. \tag{10.13}$$

Note that the PSDF can be expressed by other functions of K_x and K_y. The corrections to the Fresnel reflection coefficients of the multilayer system with slightly randomly rough boundaries are given as follows [33]:

$$\Delta \hat{r}_q = \sum_{i=1}^{L+1} \sum_{j=1}^{L+1} \int_{-\infty}^{\infty} \int_{-\infty}^{\infty} \hat{f}_{ij,q}(K_x, K_y) W_{ij}(K_x', K_y) \mathrm{d}K_x \mathrm{d}K_y, \tag{10.14}$$

where $L+1$ is number of boundaries. The symbols $W_{ij}(K_x', K_y)$ and $\hat{f}_{ij,q}$ denote the mutual PSDF of the ith an jth boundaries and functions dependent on K_x, K_y, λ, φ and optical constants of media forming the system, respectively.

The system consisting of a substrate covered by a single layer with rough boundaries is the most frequently occurring in practice. The expressions for $\Delta \hat{r}_q$ of this system are presented in detail in [33, 34].

Similar formulae can be derived for light transmitted by the single surfaces and systems with rough boundaries. The RRT can also be utilized for expressing formulae for intensity of scattered light (see e.g. [33]).

In [35, 36], the formulae for the RRT presented in [33] are modified for small autocorrelation lengths of slight random roughness. Within this modification, it is assumed that the autocorrelation length is much smaller than the wavelength of incident light. For randomly microrough surfaces, this small autocorrelation length modification is utilized for a comparison to the EMA. However, this modification must be used with caution since the utilization of the RRT is based on the assumption that the roughness slopes are also relatively small. Therefore, the relation $T \ll \lambda$ restricts the use of this RRT modification for very small values of σ ensuring the validity of relation $\tan\beta_0 \lesssim 0.1$. In other words, the validity of relation $T \ll \lambda$ can imply that the microrough surfaces fulfilling this relation are practically smooth and so the use of this modification of the RRT is irrelevant for such surfaces. This statement is evident, for example, for the Gaussian roughness because $\tan\beta_0 = \sqrt{2}\sigma/T$ [1, 12, 13, 37].

The RRT can also be used for other surfaces such as self-affine surfaces and rough surfaces exhibiting fractal character [35, 36, 38].

The approaches based on physical principles, such as the RRT, provide a more suitable alternative to the EMA for the optical characterization of randomly microrough and slightly rough surfaces covered with very thin overlayers. This statement is also implied by results presented in literature (see e.g. [30]). Of course, the RRT is also more suitable than the EMA for the layered systems with microrough and slightly rough boundaries.

In [36], it is stated that the model of the randomly rough surfaces characterized by a Gaussian PSDF is unphysical. This statement is generally not correct. In practice, it is often possible to encounter real randomly rough surfaces or boundaries exhibiting this Gaussian PSDF [39–43].

There are several other perturbation theories applicable for describing the interaction of light with slightly randomly rough surfaces and boundaries (e.g. Green's function perturbation theory [44, 45]). A survey of the perturbation theories concerning the randomly rough surfaces and multilayer systems is presented in Ogilvy's monograph [45] and paper [25].

10.3.3 Scalar Diffraction Theory

If the relations $\sigma \lesssim \lambda$ and $T \gg \lambda$ are fulfilled, the scalar diffraction theory (SDT) can be used to derive formulae for calculating the values of the optical quantities of the corresponding randomly rough surfaces and thin films. Although the theoretical consideration require the inequality $T \gg \lambda$, the real applicability of the SDT extends even to the situations, in which a weaker condition $T \gtrsim \lambda$ is fulfilled. The randomly rough surfaces and boundaries fulfilling inequality $T \gg \lambda$ are locally smooth surfaces and boundaries because at all their points they can be approximated by the corresponding tangent planes (see e.g. [12, 22]). The roughness corresponding to the foregoing inequalities is often referred to as the moderate roughness. For the rough surface the starting point of this theory is given by the Helmholtz-Kirchhoff integral (HK) [46, 47], i.e.

$$\hat{E}_q(\boldsymbol{r}_Q) = \frac{1}{4\pi}\iint_S \left[\hat{E}_q(\boldsymbol{r}_B)\frac{\partial \hat{G}(\boldsymbol{r}_Q - \boldsymbol{r}_B)}{\partial n} - \hat{G}(\boldsymbol{r}_Q - \boldsymbol{r}_B)\frac{\partial \hat{E}_q(\boldsymbol{r}_B)}{\partial n}\right] \mathrm{d}S, \tag{10.15}$$

where $\hat{E}_q(\boldsymbol{r}_B)$ and/or $\hat{E}_q(\boldsymbol{r}_Q)$ is the local electric field at a point B on the rough surface and/or the electric field at point Q in the far zone. The symbol $\partial/\partial n$ represents a directional derivative with respect to the normal of the rough surface (the illuminated part of this surface is denoted as S). The function $\hat{G}$ is expressed as

$$\hat{G}(\boldsymbol{r}_Q - \boldsymbol{r}_B) = \frac{1}{|\boldsymbol{r}_Q - \boldsymbol{r}_B|} e^{-\mathrm{i}k_0|\boldsymbol{r}_Q - \boldsymbol{r}_B|}, \tag{10.16}$$

where k_0 is the absolute value of the wave vector of the light wave propagating in the ambient and $|\boldsymbol{r}_Q - \boldsymbol{r}_B|$ denotes the distance of point Q from an arbitrary point B on the rough surface ($\boldsymbol{r}_Q$ and $\boldsymbol{r}_B$ denote the radius vectors of these points). The Kirchhoff approximation is employed for expressing the local electric field [47] i.e.

$$\hat{E}_q(\boldsymbol{r}_B) = (1 + \hat{r}_q^{(l)})\hat{E}_{0q}(\boldsymbol{r}_B), \tag{10.17}$$

where $\hat{r}_q^{(l)}$ are the local Fresnel reflection coefficients on the rough surface and $\hat{E}_{0q}(\boldsymbol{r}_B)$ is given as

$$\hat{E}_{0q}(\boldsymbol{r}_B) = \hat{A}_{0q} e^{-\mathrm{i}\boldsymbol{k}_0 \boldsymbol{r}_B}, \tag{10.18}$$

where $\hat{A}_{0q}$ and $\boldsymbol{k}_0$ are the amplitude and wave vector of the incident monochromatic plane wave corresponding to p- and s-polarizations. After several mathematical operations one obtains the following equation for the Fresnel reflection coefficients of the randomly rough surface $\hat{r}_q$, which corresponds to the specular direction (see e.g. [12, 22]):

$$\hat{r}_q = \int_{-\infty}^{\infty}\int_{-\cot\varphi}^{\cot\varphi}\int_{-\infty}^{\infty} \hat{r}_q^{(l)}(z, z_x, z_y) e^{\mathrm{i}uz} w(z, z_x, z_y) \mathrm{d}z \mathrm{d}z_x \mathrm{d}z_y, \tag{10.19}$$

where $u = -(4\pi/\lambda)n_0 \cos\varphi$, z denotes the values of the random function $\eta(x, y)$ describing the rough surface, z_x and z_y represent the values of the derivatives of the random function $\eta(x, y)$ with respect to x and y coordinates within the mean plane, respectively. The symbol $w(z, z_x, z_y)$ denotes the three-dimensional distribution of the random functions $\eta(x, y)$, $\eta_x(x, y)$ and $\eta_y(x, y)$. Note that (10.19) takes into account the presence of shadowing among roughness irregularities (see [22, 48]). In the case of a surface, the local reflection coefficient is independent on z, and therefore, only the dependence $\hat{r}_q(z_x, z_y)$ must be included in (10.19). If the random function $\eta(x, y)$ is statistically independent on its derivatives, i.e. the heights of the roughness are statistically independent on the slopes of the roughness, then the formula (10.19) can be written as:

$$\hat{r}_q = \hat{A}\int_{-\infty}^{\infty}\int_{-\cot\varphi}^{\cot\varphi} \hat{r}_q^{(l)}(z_x, z_y) w_2(z_x, z_y) \mathrm{d}z_x \mathrm{d}z_y, \tag{10.20}$$

where

$$\hat{A} = \int_{-\infty}^{\infty} \mathrm{e}^{\mathrm{i}uz} w_1(z) \mathrm{d}z, \tag{10.21}$$

where the symbol $w_2(z_x, z_y)$ denotes the two-dimensional probability density of the values of the derivatives of $\eta(x, y)$.

The same formula is true for the single ITF. Of course, the expression for $\hat{r}_q^{(l)}(z_x, z_y)$ is different from that for the rough surface (see [12, 22]). Formula (10.20) with the corresponding expression for $\hat{r}_q^{(l)}(z_x, z_y)$ is evidently true for a multilayer system formed by the homogeneous ITFs. If the multilayer systems consists of homogeneous non-identical films, such as films with correlated boundaries or statistically independent boundaries, the formula (10.20) is not valid. For such complicated rough multilayer system, only formulae corresponding to negligible slopes at the boundaries have been formulated so far. In multilayer systems with correlated boundaries and negligible boundary slopes, the reflection coefficients can be written as:

$$\hat{r}_q = \int_{-\infty}^{\infty} \cdots \int_{-\infty}^{\infty} \hat{r}_q^{(l)}(z_1, \ldots, z_{L+1}) \mathrm{e}^{\mathrm{i}uz_1} w(z_1, \ldots, z_{L+1}) \mathrm{d}z_1 \ldots \mathrm{d}z_{L+1}, \quad (10.22)$$

where L is the number of films in the system, z_j denotes the values of the random function describing the jth boundary and the symbol $w(z_1, \ldots, z_{L+1})$ denotes the $L+1$ dimensional distribution function of random functions $\eta_1(x, y), \ldots, \eta_{L+1}(x, y)$. The approximate recursive formulae for reflection and transmission coefficients based on (10.22) were presented in [1, 49–51]. The exact formulae for the reflection coefficients of the system with the randomly rough boundaries exhibiting negligible slopes were derived in [52]. If the distribution function $w(z_1, \ldots, z_{L+1})$ is given by the $(L+1)$-dimensional Gaussian distribution then the following formula is valid for normal incidence [52]:

$$\hat{r}_q = \hat{r}_1 \exp\left(-\frac{u^2\sigma_1^2}{2}\right) + \sum_{p=1}^{\infty} \sum_{b=1}^{\min(L,p)} \sum_{\mathbf{m}} \exp\left(\mathrm{i}\sum_{j=1}^{b} m_j \hat{X}_j\right) \hat{Q}_b(\mathbf{m}) \hat{H}_b(\mathbf{m}), \quad (10.23)$$

where $\hat{X}_j = 4\pi \hat{n}_j \overline{h}_j/\lambda$, and the symbol $\overline{h}_j$ denotes the mean thickness of the jth film. The remaining quantities are expressed in this way:

$$\hat{Q}_b(\mathbf{m}) = \hat{t}_1 \hat{t}_1' \hat{r}_1'^{\,m_1-1} \hat{r}_{b+1}^{m_b} \prod_{j=2}^{b} \sum_{o_j=1}^{\min(m_j, m_{j-1})} \binom{m_{j-1}}{o_j} \binom{m_j - 1}{o_j - 1} \hat{r}_j^{m_{j-1}-o_j} (\hat{t}_j \hat{t}_j')^{o_j} \hat{r}_j'^{\,m_j - o_j},$$

$$\hat{H}_b(\mathbf{m}) = \exp\left(-\frac{1}{2} \sum_{i=1}^{b+1} \sum_{j=1}^{b+1} \hat{D}_i \hat{D}_j S_{i,j}\right),$$

$$\hat{D}_j = \frac{4\pi}{\lambda} \left(m_j \hat{n}_j - m_{j-1} \hat{n}_{j-1}\right),$$

$$S_{i,j} = \langle \eta_i \eta_j \rangle = \sigma_i \sigma_j C_{i,j},$$

where σ_j is the RMS value of the heights of irregularities of the jth boundary and $C_{i,j}$ is the cross-correlation coefficient between the ith and jth boundaries with $C_{i,i} = 1$. The elements m_j of the vector $\mathbf{m} = (m_1, m_2, \ldots, m_L)$ express how many times the light passes through the layer j downwards, which is the same number it passes upwards. The summation over $\mathbf{m}$ is carried out for all sets of

$$m_j \geq 1 \text{ for } 1 \leq j \leq b, \quad m_j = 0 \text{ for } b < j, \quad \text{and} \quad m_1 + m_2 + \cdots + m_L = p.$$

The symbol p represents the total 'length' of the path of light through the multilayer system. The number $b \leq L$ denotes the path depth which is equal to the highest index of the media, through which the path passes. The summation over $\mathbf{m}$ is expressed as

$$\sum_{\mathbf{m}} = \sum_{m_1=1}^{p-b+1-M_0} \sum_{m_2=1}^{p-b+2-M_1} \cdots \sum_{m_{b-1}=1}^{p-1-M_{b-2}},$$

where $M_j = \sum_{l=1}^{j} m_l$ and $M_0 = 0$. In addition we define $m_0 = 1$, $m_b = p - M_{b-1}$ and $m_j = 0$ for $j > b$. The reflection and transmission coefficients of jth boundary are denoted as $\hat{r}_j, \hat{r}'_j, \hat{t}_j, \hat{t}'_j$ respectively. The coefficients without primes correspond to light incident from the top while the primed coefficients correspond to light incident from the bottom ($\hat{r}'_j = -\hat{r}_j$, $\hat{t}_j\hat{t}'_j = 1 - \hat{r}_j^2$). The symbol o_j expresses how many times the light:

- visits the subsystem from medium j downward,
- passes through the $(j-1)/j$ boundary from the side of medium $j-1$,
- passes through the $(j-1)/j$ boundary from the side of medium j.

The first point should be understood as follows: the entire path can be divided into segments contained within media from 0 to $j-1$ and segments contained within media from j downward. The symbol o_j is the number of the latter segments. For all paths and all layer systems for $j > b$, it holds that $o_j = m_j = 0$. For details see [52].

Similar formulae for the Fresnel transmission coefficients of the randomly rough multilayer systems can also be derived if the boundary slopes are neglected.

Eastman [53] employed the SDT for deriving the formulae for reflection and transmission coefficients of the multilayer system with randomly rough boundaries exhibiting the negligible slopes as well. His approach is based on a matrix formalism in which the random functions describing boundary roughness were incorporated into phase matrices. The reflection and transmission coefficients, which are given by ratios of the corresponding system matrix elements, were then expanded into a second order Taylor series with respect to the random functions. The statistical mean values of the optical quantities were then calculated using this Taylor series. This means that the Eastman's approach is also an approximate approach. Similar matrix approach based on the SDT was utilized for calculating the changes in specular reflectance and transmittance of multilayer systems exhibiting randomly rough boundaries in paper of Carniglia [54]. The formulae describing the diffuse scattering near the specular beams are also presented in this paper. In [55] the SDT was employed for predicting the angular distribution of light scattered from multilayer system with randomly rough boundaries. In [54] and [55] it is shown that scattered light can be used to characterize the layered systems with rough boundaries. Formulae describing light scattering from randomly rough surfaces and thin film systems derived using the SDT are also presented in [25, 45, 47, 56].

In literature, the approaches based on the vector diffraction theory (VDT) are presented for the derivation of formulae expressing the optical quantities of randomly rough surfaces. The starting point of the VDT approaches is the Stratton-Chu-Silver integral [57]. The formulae for the reflection coefficients of randomly rough surfaces are presented in papers [1, 58] (in these papers randomly rough surfaces fulfilling relations among λ, σ and T presented in this section are studied). The interaction of light with considerably or very rough surfaces can be described within the VDT as well, for example, in [45, 56, 58–60].

10.4 Area Non-uniformity of Thin Films

In this chapter the emphasis will be put on the area non-uniformity of the thin films corresponding to the non-uniformity in thickness. It has to be pointed out that there is a principal difference between surface roughness and thickness non-uniformity in optics of thin films. This principal difference is implied by the characteristic lateral scales of both the defects. The characteristic lateral scale of the random roughness is from nanometers to tens of microns (such the surfaces correspond to the EMA, RRT, SDT and VDT approaches). Therefore, the coherent formalism must be employed to describe the influence of this roughness on the optical quantities. On the other hand, the characteristic lateral scale of thickness non-uniformity is several orders of magnitude longer i.e. from millimeters to tens of centimeters. This implies that the thickness non-uniformity must be described by the incoherent formalism.

The necessity of using the incoherent formalism can be proved by the following mathematical evidence performed for the normal reflectance of a single thin film with the thickness non-uniformity [61]. The light beam incident on the homogeneous thin film non-uniform in thickness has finite dimensions. Therefore, it is possible to use the SDT for expressing the intensity of reflected light from this film. Within the SDT, the electric field $\hat{E}$ of the reflected light at point (x_0, y_0) of the detector is given as follows [62]:

$$\hat{E}\left(x_0, y_0\right) = \mathrm{i}C \int\limits_S \exp\left[\mathrm{i}k_0\left(\overline{L}(x, y) - 2\xi(x, y)\right)\right]\hat{r}(x, y)\mathrm{d}x\mathrm{d}y, \qquad (10.24)$$

where $C = A_0/(\lambda L_0)$, A_0 is the amplitude of the incident wave, L_0 is the distance between the center of the illuminated spot on the film and detector, $k_0 = 2\pi/\lambda$ is the wave number, $\hat{r}(x, y)$ is the local Fresnel reflection coefficient of the non-uniform film on the upper boundary, $\overline{L}(x, y)$ is the distance of the point in the detector plane from the point in the illuminated light spot, $\xi(x, y)$ is the local thickness deviation from the mean thickness and S denotes the area of the light spot on the upper boundary of the film.

Hence, the intensity $I(x_0, y_0)$ at the point of the detector is given as

$$I(x_0, y_0) = C^2 \int\limits_{S}\int\limits_{S} \exp\left[ik_0\left(\overline{L} - \overline{L}' - 2\xi + 2\xi'\right)\right]\hat{r}\hat{r}'^{*}\mathrm{d}x'\mathrm{d}y'\mathrm{d}x\mathrm{d}y, \quad (10.25)$$

where quantities calculated at primed coordinates are marked with primes for brevity. The total detected intensity I is then given in the following way

$$I = \int\limits_{S_D} I(x_0, y_0)\mathrm{d}x_0\mathrm{d}y_0, \quad (10.26)$$

where S_D denotes the detector area. Owing to the spatial distribution of the intensity diffracted within the Fresnel diffraction [46] and the fact that the areas of the detectors of spectrophotometers and irradiated spots on the films are sufficiently large, it is possible to write instead

$$I = \int\limits_{-\infty}^{\infty}\int\limits_{-\infty}^{\infty} I(x_0, y_0)\mathrm{d}x_0\mathrm{d}y_0. \quad (10.27)$$

Substituting (10.25) into the preceding formula and reversing the order of integrations over thin film and detector planes, one obtains

$$I = C^2 \int\limits_{S}\int\limits_{S} \exp[-2ik_0(\xi - \xi')]\hat{r}\hat{r}'^{*}\left[\int\limits_{-\infty}^{\infty}\int\limits_{-\infty}^{\infty} \exp\left[ik_0\left(\overline{L} - \overline{L}'\right)\right]\mathrm{d}x_0\mathrm{d}y_0\right]\mathrm{d}x'\mathrm{d}y'\mathrm{d}x\mathrm{d}y. \quad (10.28)$$

Keeping only the terms corresponding to the Fresnel approximation in the expansion of $\overline{L} - \overline{L}'$ [46], the integral in square brackets can be written as follows:

$$\exp\left[ik_0\frac{x^2 - x'^2 + y^2 - y'^2}{2L_0}\right]\int\limits_{-\infty}^{\infty} \exp\left[\frac{ik_0}{L_0}\left(x' - x\right)x_0\right]\mathrm{d}x_0 \int\limits_{-\infty}^{\infty} \exp\left[\frac{ik_0}{L_0}\left(y' - y\right)y_0\right]\mathrm{d}y_0. \quad (10.29)$$

By taking into account the expression for the inverse Fourier transform of the Dirac's delta function and performing the resulting trivial integration. one obtains the following final formula for the normal reflectance R of the homogeneous film non-uniform in thickness:

$$R = \frac{I}{SA_0^2} = \frac{1}{S}\int\limits_{S} R(x, y)\mathrm{d}x\mathrm{d}y, \quad (10.30)$$

where $R(x, y) = \hat{r}(x, y)\hat{r}^{*}(x, y)$ is the local reflectance of this non-uniform thin film. Similar formulae are valid for oblique incidence of light and for other photometric quantities corresponding to thin films with thickness non-uniformity.

The area non-uniformity along the substrates of thin films is a defect present in many thin films prepared by various technologies. For example, different plasma chemical technologies produce thin films exhibiting this area non-uniformity. It is known that disregarding the area non-uniformity when processing experimental data can lead to misrepresented results due to the deformation of the reflectometric and ellipsometric spectra. The thickness non-uniformity is the most common type of area non-uniformity occurring in practice. However, only a few papers have been devoted to the optical characterization of thin films non-uniform in thickness using spectrophotometry [63–67] or spectroellipsometry [68–71]. Moreover, these papers dealt with studies of a special case of wedge-shaped thickness non-uniformity and a special position of the rectangular light spot owing to the thickness gradient. This special case corresponds to uniform distribution of thickness, i.e. a thickness distribution density is constant inside a certain interval and zero outside. In the case of a general thickness non-uniformity, the use of the special formulae corresponding to the above-mentioned special case is not justified. The possibility of using a general thickness distribution in ellipsometry is briefly mentioned in [72] but without any specific examples or applications. The formulae for the optical quantities of the thin films exhibiting general thickness non-uniformity will be presented below.

It will be assumed that the non-uniform thin films and substrates are optically homogeneous, the ambient is non-absorbing, the boundaries of the non-uniform films are smooth and the thickness of these films vary sufficiently gradually along the substrates, i.e. the films are locally uniform.

The local reflection coefficient $\hat{r}(x, y)$ at normal incidence is given as follows:

$$\hat{r}(x, y) = \frac{\hat{r}_1 + \hat{r}_2 \exp\left[\mathrm{i}\hat{X}(x, y)\right]}{1 + \hat{r}_1 \hat{r}_2 \exp\left[\mathrm{i}\hat{X}(x, y)\right]}, \tag{10.31}$$

where $\hat{r}_1$ and $\hat{r}_2$ are the Fresnel reflection coefficients on the upper and lower film boundary, respectively, and $\hat{X}(x, y)$ denotes the local phase-shift angle at point (x, y). It holds that

$$\hat{r}_1 = \frac{n_0 - \hat{n}_1}{n_0 + \hat{n}_1}, \quad \hat{r}_2 = \frac{\hat{n}_1 - \hat{n}_S}{\hat{n}_1 + \hat{n}_S}, \quad \hat{X}(x, y) = \frac{4\pi}{\lambda} \hat{n}_1 h(x, y), \tag{10.32}$$

where $\hat{n}_1$ and $\hat{n}_S$ are the complex refractive indices of the film and substrate, respectively, and $h(x, y)$ is the local film thickness.

Since the local reflectance $R(x, y)$ is a function of the local film thickness only, the formula (10.30) can be written in the form of an integral over this local thickness, i.e.

$$R = \int R(h)\varrho(h)\mathrm{d}h, \tag{10.33}$$

where $\varrho(h)$ is the distribution of local thicknesses. It can easily be shown that $\varrho(h)$ is given as follows [73]:

$$\varrho(h) = \frac{1}{S} \int_{C_h} \frac{\mathrm{d}l}{|\mathrm{grad}\, h|}, \tag{10.34}$$

where the integration is performed over curves C_h of constant thicknesses h (contour lines) and the symbols $\mathrm{d}l$ and $\mathrm{grad}\, h$ denote the length element and gradient of the function $h(x, y)$ at a given point of the curve, respectively.

For wedge-shaped thin film thickness and an elliptic illuminated spot the distribution density is [73]

$$\varrho\,(h) = \begin{cases} \dfrac{2}{\pi a^2}\left[a^2 - \left(h - \overline{h}\right)^2\right]^{1/2} & \text{for} |h - \overline{h}| \leq a, \\ 0 & \text{otherwise}, \end{cases} \tag{10.35}$$

where $\overline{h}$ is the mean thickness within the illuminated spot and $\overline{h} - a$ and $\overline{h} + a$ are the minimum and maximum thicknesses, respectively.

The expression for $\varrho(h)$ for some other simple geometrical forms of the upper boundaries of the thin films non-uniform in thickness are presented in [73]. If the thickness non-uniformity of the thin films is sufficiently small, one can perform the Taylor expansion of $R(h)$ around the mean thickness $\overline{h}$ and then consider only a few terms at the beginning of the series of this expansion. This Taylor expansion is as follows:

$$R(h) = \sum_{m=0}^{\infty} \frac{1}{m!} R^{(m)}(\overline{h})(h - \overline{h})^m, \tag{10.36}$$

where $R^{(m)}(\overline{h})$ is the mth derivative of $R(h)$ calculated in the mean thickness $\overline{h}$.

After substituting this expansion into (10.33) and exchanging the order of integration and summation, the following formula is obtained

$$R = R(\overline{h}) + \sum_{m=2}^{\infty} \frac{1}{m!} R^{(m)}(\overline{h})\mu_m, \tag{10.37}$$

where μ_m denotes the mth central moment of ϱ, i.e.

$$\mu_m = \int_{-\infty}^{\infty} \varrho(h)(h - \overline{h})^m \mathrm{d}h. \tag{10.38}$$

If only the first two terms of the Taylor expansion are considered, one can write

$$R = R(\overline{h}) + \frac{\sigma_{\mathrm{t}}^2}{2} R''(\overline{h}), \tag{10.39}$$

where $\sigma_t = \sqrt{\mu_2}$ is the RMS value of the thickness distribution and $R''(\overline{h})$ denotes the second derivative of $R(h)$ calculated in $h = \overline{h}$. Equation (10.39) shows that in this approximation, the normal reflectance is independent on the shape of the non-uniformity. The higher-orders terms in the expansion can also be taken into account. However, the non-uniformities occurring in practice exhibit the thickness distributions that are often symmetric or nearly symmetric. In this case the third order term is close to zero and the fourth order term must be considered. This implies that (10.39) often represents approximation accurate up to σ_t^3.

The polarization states of the light waves are completely described by their Stokes vectors whose components are given by the total intensity of light I_0 and $I_{\uparrow}$, $I_{\leftarrow}$, $I_{\nwarrow}$, $I_{\nearrow}$, $I_{\circlearrowright}$ and $I_{\circlearrowleft}$ representing intensities transmitted by ideal polarizers transmitting the linearly polarized light along the axis tilted with respect to the plane of incidence by 0, $\pi/2$, $-\pi/4$, $\pi/4$ and left and right circularly polarized light, respectively [2, 69] (see also Chap. 9). The influence of the sample on the polarization state of the incident wave is given by the unnormalized Mueller matrix. For a non-depolarizing sample this matrix can be expressed as follows [74, 75]:

$$\mathbf{M} = \begin{pmatrix} \left(|\hat{r}_p|^2 + |\hat{r}_s|^2\right)/2 & \left(|\hat{r}_p|^2 - |\hat{r}_s|^2\right)/2 & 0 & 0 \\ \left(|\hat{r}_p|^2 - |\hat{r}_s|^2\right)/2 & \left(|\hat{r}_p|^2 + |\hat{r}_s|^2\right)/2 & 0 & 0 \\ 0 & 0 & \mathrm{Re}\left(\hat{r}_p\hat{r}_s^*\right) & \mathrm{Im}\left(\hat{r}_p\hat{r}_s^*\right) \\ 0 & 0 & -\mathrm{Im}\left(\hat{r}_p\hat{r}_s^*\right) & \mathrm{Re}\left(\hat{r}_p\hat{r}_s^*\right) \end{pmatrix}, \tag{10.40}$$

where $\hat{r}_p$ and $\hat{r}_s$ are the complex Fresnel reflection coefficients of the sample for p and s polarizations respectively ($|\hat{r}_p|$ and $|\hat{r}_s|$ are the modules of these reflection coefficients).

If the thin film corresponding to the sample is non-uniform, the matrix $\mathbf{M}$ is dependent on the local position within this film, where x and y are the Cartesian coordinates in the mean plane of the upper boundary. In this case one must consider that the intensity detected by the detector corresponds to the value integrated over the detector area (see formula (10.30)). Then the elements of the unnormalized Mueller matrix $\overline{\mathbf{M}}$ of the non-uniform film are expressed as

$$\overline{\mathbf{M}} = \frac{1}{\mathcal{S}} \int_{\mathcal{S}} \mathbf{M}(x, y)\mathrm{d}x\mathrm{d}y. \tag{10.41}$$

If the thin film exhibits thickness non-uniformity one can write $\mathbf{M}(x, y) = \mathbf{M}(h(x, y))$. Therefore, (10.41) can be rewritten as follows:

$$\overline{\mathbf{M}} = \int \varrho(h)\mathbf{M}(h)\mathrm{d}h. \tag{10.42}$$

For the wedge-shaped non-uniformity and elliptic illuminated spot the density $\varrho(h)$ can be expressed as [76]:

$$\varrho(h) = \frac{1}{2\pi\sigma_{\mathrm{t}}^2(\varphi)}\left[4\sigma_{\mathrm{t}}^2(\varphi) - (h-\overline{h})^2\right]^{1/2}. \tag{10.43}$$

The RMS value of thickness differences within the illuminated spot σ_t depends on the angle of incidence φ as follows:

$$\sigma_{\mathrm{t}}^2(\varphi) = \sigma_0^2\left(\frac{\cos^2\alpha}{\cos^2\varphi} + \sin^2\alpha\right), \tag{10.44}$$

where σ_0 is the RMS value for the normal incidence and α is the angle between the thickness gradient direction and the plane of incidence. The form of density (10.43) enables efficient evaluation of the integrals in (10.42) using a Chebyshev–Gauss quadrature of the second kind (for details see [76]). If the non-uniformity shape deviates from the ideal wedge this integration method is not sufficient. In that case, it is possible to parametrize the thickness $h(x, y)$ assuming a more general form, e.g. a higher order polynomial, but this approach has two disadvantages. The first disadvantage is a large number of non-uniformity parameters that have to be introduced. The second disadvantage is that the possibility of straightforward use of a Gaussian quadrature is lost as the thickness density form becomes dependent on the non-uniformity parameters and also on the angle of incidence φ. In our paper [76] we therefore proposed a different approach. Since small changes of the form of $\varrho(h)$ have only a weak influence on the measured optical quantities (see [73, 76]), the formula (10.43) for the density is retained. Only the dependence $\sigma_t^2(\varphi)$ is expressed with the help of a general polynomial in $1/\cos^2(\varphi)$ and the same type of dependence on φ is assumed also for the mean thickness $\overline{h}(\varphi)$. The mathematical procedures enabling us to perform the foregoing steps are described in [76]. As the result of these procedures the two following equations are obtained:

$$\overline{h}(\varphi) = \overline{h}_{00} + \frac{\overline{h}_{10}}{\cos^2\varphi} + \frac{\overline{h}_{20}}{\cos^4\varphi} + \cdots \tag{10.45}$$

and

$$\sigma_{\mathrm{t}}^2(\varphi) = s_0 + \frac{s_1}{\cos^2\varphi} + \frac{s_2}{\cos^4\varphi} + \cdots, \tag{10.46}$$

where $\overline{h}_{00}, \overline{h}_{10}, \overline{h}_{20}, \ldots, s_0, s_1, s_2, \ldots$ are used as the parameters instead of the geometrical parameters of non-uniformity. The foregoing equations are then substituted into (10.43).

The associated ellipsometric parameters I_{s}, I_{c}, I_{n} measured within the phase modulated ellipsometry occur in the normalized Mueller matrix $\mathbf{M}_{\mathrm{n}}$ defined as [77]

$$\mathbf{M}_{\mathrm{n}} = \frac{1}{M_{00}}\mathbf{M} = \begin{pmatrix} 1 & -I_{\mathrm{n}} & 0 & 0 \\ -I_{\mathrm{n}} & 1 & 0 & 0 \\ 0 & 0 & I_{\mathrm{c}} & I_{\mathrm{s}} \\ 0 & 0 & -I_{\mathrm{s}} & I_{\mathrm{c}} \end{pmatrix} \tag{10.47}$$

where M_{00} is the reflectance of the sample for the given angle of incidence.

From the foregoing, it is clear that the associated ellipsometric parameters $\overline{I_s}$, $\overline{I_c}$ and $\overline{I_n}$ corresponding to the single thin film non-uniform in thickness are given as follows:

$$\overline{I_s} = -i\frac{\langle \hat{r}_p\hat{r}_s^* \rangle - \langle \hat{r}_p^*\hat{r}_s \rangle}{\langle |\hat{r}_s|^2 \rangle + \langle |\hat{r}_p|^2 \rangle}, \quad \overline{I_c} = \frac{\langle \hat{r}_p\hat{r}_s^* \rangle + \langle \hat{r}_p^*\hat{r}_s \rangle}{\langle |\hat{r}_s|^2 \rangle + \langle |\hat{r}_p|^2 \rangle}, \quad \overline{I_n} = \frac{\langle |\hat{r}_s|^2 \rangle - \langle |\hat{r}_p|^2 \rangle}{\langle |\hat{r}_s|^2 \rangle + \langle |\hat{r}_p|^2 \rangle}. \quad (10.48)$$

The angle brackets denote the mean values of the corresponding quantities calculated using the density distribution (see (10.42)). The values of the parameters $\overline{h}_{00}$, $\overline{h}_{10}$, $\overline{h}_{20}$, ..., s_0, s_1, s_2, ... are then sought at processing the experimental data within the optical characterization of the thin films non-uniform in thickness.

The influence of non-uniformity of thin films in the optical constants has not been studied so far. In principle the incorporation of this type of area non-uniformity can be performed in a similar way as for the films with thickness non-uniformity. As for investigating thin films exhibiting non-uniformity in the optical constants, only mapping spectroscopic ellipsometry with a light microspot was utilized for assessing whether the single films of $SiO_xC_yH_z$ were non-uniform in the optical constants (see [78]).

The influence of the area non-uniformity on the optical quantities of the multilayer systems has not been studied so far.

10.5 Inhomogeneity of Thin Films Represented by Refractive Index Profiles

Within the majority of thin film technologies, great emphasis is put on achieving homogeneous thin films. Nevertheless, specific applications in thin film optics require to create inhomogeneous thin films whose refractive indices vary continuously in a prescribed manner across these films along the axis perpendicular to the parallel boundaries [79]. In that case, the refractive indices of the films form profiles described by continuous functions of the coordinate along this axis, i.e. the coordinate denoted z. Such an inhomogeneity can not be considered to be a defect. However, if the aim is to achieve the homogeneous thin films, refractive index profiles can be seen as a defect of these films.

10.5.1 Exact Solutions

It will be assumed below that inhomogeneous thin films are uniform along the substrates and that the boundaries are smooth and flat without any transition layers (substrates are optically homogeneous). This means that the functions of z which describe the refractive index profiles are not dependent on the corresponding Cartesian coordinates x and y lying in the plane of the upper boundary. Theoretical approaches applicable for describing the interaction of light with such inhomogeneous thin films are presented in [4, 5, 80–83]. Unfortunately, exact solutions in analytical forms are known only for several profiles of the refractive index $n(z)$ of the inhomogeneous thin films. For normal incidence of light the profiles corresponding to these exact solutions are introduced in [79]. They are as follows:

$$n^2(z) = n_U^2 - \frac{z}{h}\left(n_U^2 - n_L^2\right), \tag{10.49}$$

$$n^2(z) = \left[\frac{1}{n_U^4} - \frac{z}{h}\left(\frac{1}{n_U^4} - \frac{1}{n_L^4}\right)\right]^{-1/2}, \tag{10.50}$$

$$n(z) = \frac{n_U n_L}{n_L - \frac{z}{h}\left(n_L - n_U\right)}. \tag{10.51}$$

In the equations above, the symbols n_U, n_L and h denote the refractive index at the upper boundary, refractive index at the lower boundary and thickness, respectively. The formulae for the reflection and transmission coefficients corresponding to these exact solutions are presented in [79]. For the oblique incidence there is an exact solution for the following exponential profile

$$n(z) = n_U \left(\frac{n_L}{n_U}\right)^{z/h}. \tag{10.52}$$

The refractive index profile expressed in (10.52) can also be written as

$$n(z) = n_U \exp\left(az\right), \qquad \text{where} \qquad a = \frac{1}{h}\ln\frac{n_L}{n_U}. \tag{10.53}$$

The formulae for the reflection coefficients corresponding to this exact solution are presented in [4, 79, 84, 85] and the formulae for the transmission coefficients are introduced in [4, 79]. For the oblique angle of incidence the exact solution is also known for the Rayleigh profile expressed as

$$\frac{1}{n(z)} = \frac{1}{2}\left(\frac{1}{n_U} + \frac{1}{n_L}\right) + \left(\frac{z}{h} - \frac{1}{2}\right)\left(\frac{1}{n_L} - \frac{1}{n_U}\right). \tag{10.54}$$

The formulae for the reflection coefficients corresponding to this exact solution are presented in [86].

The utilization of the exact solutions is rather limited in practice because most real inhomogeneous thin films exhibit refractive index profiles considerably differing from those corresponding to these exact solutions. Moreover, the refractive index profiles given by (10.49)–(10.51) do not have exact solutions at the oblique incidence. Therefore, approximate methods enabling us to derive approximate formulae for the optical quantities of the inhomogeneous thin films are needed. The majority of them are applicable for an arbitrary refractive index profile. The most important approximate methods will be presented bellow. Some of these approximate methods have been employed in [87–93].

10.5.2 WKBJ Approximation

If the gradient of the profile is very small, it is possible to apply the Wentzel–Kramers–Brillouin–Jeffreys (WKBJ) approximation (see e.g. [1, 4, 81, 94]). The reflection coefficients corresponding to the WKBJ approximation are given as

$$\hat{r}_q = \frac{\hat{r}_{1q} + \hat{r}_{2q}\exp\left[\mathrm{i}\hat{\overline{X}}(h)\right]}{1 + \hat{r}_{1q}\hat{r}_{2q}\exp\left[\mathrm{i}\hat{\overline{X}}(h)\right]}, \tag{10.55}$$

where

$$\hat{r}_{1\mathrm{s}} = \frac{n_0 \cos\varphi - \hat{n}_\mathrm{U}\cos\hat{\psi}_1}{n_0 \cos\varphi + \hat{n}_\mathrm{U}\cos\hat{\psi}_1}, \quad \hat{r}_{2\mathrm{s}} = \frac{\hat{n}_\mathrm{L}\cos\hat{\psi}_2 - \hat{n}_\mathrm{S}\cos\hat{\psi}_\mathrm{S}}{\hat{n}_\mathrm{L}\cos\hat{\psi}_2 + \hat{n}_\mathrm{S}\cos\hat{\psi}_\mathrm{S}},$$

$$\hat{r}_{1\mathrm{p}} = \frac{n_0 \cos\hat{\psi}_1 - \hat{n}_\mathrm{U}\cos\varphi}{n_0 \cos\hat{\psi}_1 + \hat{n}_\mathrm{U}\cos\varphi}, \quad \hat{r}_{2\mathrm{p}} = \frac{\hat{n}_\mathrm{L}\cos\hat{\psi}_\mathrm{S} - \hat{n}_\mathrm{S}\cos\hat{\psi}_2}{\hat{n}_\mathrm{L}\cos\hat{\psi}_\mathrm{S} + \hat{n}_\mathrm{S}\cos\hat{\psi}_2},$$

$$\hat{\overline{X}}(h) = \frac{4\pi}{\lambda}\int_0^h \sqrt{\hat{n}^2(z) - n_0^2 \sin^2\varphi}\,\mathrm{d}z.$$

The symbols $\hat{r}_{1q}$, $\hat{r}_{2q}$ ($q = \mathrm{s}, \mathrm{p}$) and $\hat{\overline{X}}(h)$ denote the Fresnel reflection coefficients of the upper boundary, lower boundary and phase-shift angle of the inhomogeneous thin film, respectively. Furthermore, from the Snell's law it follows that $n_0 \sin\varphi = \hat{n}_\mathrm{U}\sin\hat{\psi}_1 = \hat{n}(z)\sin\hat{\psi}(z) = \hat{n}_\mathrm{L}\sin\hat{\psi}_2 = \hat{n}_\mathrm{S}\sin\hat{\psi}_\mathrm{S}$, where the symbols $\hat{\psi}_1$, $\hat{\psi}_2$, $\hat{\psi}_\mathrm{S}$ and $\hat{\psi}(z)$ represent the refraction angle at the upper boundary, lower boundary, substrate and variable refraction angle inside the inhomogeneous film, respectively. Equation (10.55) can be derived easily if one realizes that it is possible to approximate the inhomogeneous film by a multilayer system containing a large number of sub-layers with the property that adjacent sub-layers exhibit very small differences in the refractive indices. The formula (10.55) corresponds to neglecting all internal reflections on the inner boundaries belonging to the sub-layers and the limit for the number of dividing sub-layers going to infinity. A similar formula can be derived for the transmission coefficients of this film.

10.5.3 Approximation by Multilayer Systems

If the gradient of the refractive index profile of the inhomogeneous thin film is too large to utilize the WKBJ approximation, it is possible to use an approximation based on replacing this film by the multilayer system containing a sufficient number of sub-layers, as mentioned above. Therefore, in this approximation, the reflection coefficients $\hat{r}_q$ of the inhomogeneous film are expressed by the formulae for a multilayer system. For this purpose, the matrix formalism or recursive formalism [4, 5] can be employed. By using a sufficient number of sub-layers in this system, it is possible to achieve practically any precision in the calculations of the reflection coefficients of any inhomogeneous thin film. Thus, by employing this approximation, it is possible to calculate the values of the reflection coefficients even for the inhomogeneous thin films exhibiting large gradients of the refractive index profiles with required precision. The same statements are true for the transmission coefficients of the inhomogeneous thin films.

10.5.4 Approximation Based on Recursive Formulae

This approximation was presented in paper of Kildemo et al. [95]. It is based on recursive formulae for the reflection coefficients of a multilayer system replacing the inhomogeneous thin film. By subdividing the inhomogeneous film into more and more sub-layers and applying a recursive process, one can see that the reflection coefficients can be expressed in terms of single, double, triple and higher-order sums. In the limit of an infinite number of sub-layers, the sums occurring in the recursive formulae are replaced by integrals. Then the following formulae are obtained for inhomogeneous thin films [95]:

$$\hat{r}_q = \frac{\hat{r}_{1q} + \hat{I}_{1q} + \hat{r}_{1q}\hat{I}_{2q}\hat{E}_q + \hat{r}_{1q}\hat{D}_{1q} + \hat{D}_{2q}\hat{E}_q + \hat{T}_{1q} + \hat{r}_{1q}\hat{T}_{2q}\hat{E}_q + \cdots + \hat{E}_q}{1 + \hat{r}_{1q}\hat{I}_{1q} + \hat{I}_{2q}\hat{E}_q + \hat{D}_{1q} + \hat{r}_{1q}\hat{D}_{2q}\hat{E}_q + \hat{r}_{1q}\hat{T}_{1q} + \hat{T}_{2q}\hat{E}_q + \cdots + \hat{r}_{1q}\hat{E}_q}, \tag{10.56}$$

where

$$\hat{I}_{1q} = \int_0^h \hat{f}_q(z)\exp\left[\mathrm{i}\hat{\overline{X}}(z)\right]\mathrm{d}z, \quad \hat{I}_{2q} = \int_0^h \hat{f}_q(z)\exp\left[-\mathrm{i}\hat{\overline{X}}(z)\right]\mathrm{d}z,$$

$$\hat{E}_q = \hat{r}_{2q}\exp(\mathrm{i}\hat{\overline{X}}(h)), \quad \hat{D}_{1q} = \int_0^h \int_0^y \hat{f}_q(z)\hat{f}_q(y)\exp\left[\mathrm{i}\hat{\overline{X}}(y) - \mathrm{i}\hat{\overline{X}}(z)\right]\mathrm{d}z\mathrm{d}y,$$

$$\hat{D}_{2q} = \int_0^h \int_0^y \hat{f}_q(z)\hat{f}_q(y)\exp\left[\mathrm{i}\hat{\bar{X}}(z) - \mathrm{i}\hat{\bar{X}}(y)\right] \mathrm{d}z\mathrm{d}y,$$

$$\hat{f}_q(z) = \frac{1}{2\hat{Y}_q(z)} \frac{\mathrm{d}\hat{Y}_q(z)}{\mathrm{d}z}, \quad \hat{\bar{X}}(z) = \frac{4\pi}{\lambda} \int_0^z \sqrt{\hat{n}^2(z') - n_0^2 \sin^2 \varphi}\, \mathrm{d}z',$$

and the optical admittances $\hat{Y}_q(z)$ are given as

$$\hat{Y}_q(z) = \begin{cases} \hat{n}(z) \cos \hat{\psi}(z), & \mathrm{s-polarization} \\ \hat{n}(z)/\cos \hat{\psi}(z), & \mathrm{p-polarization} \end{cases}.$$

The symbols $\hat{T}_{1q}$ and $\hat{T}_{2q}$ denote triple integrals that can be expressed by means of double integrals $\hat{D}_{1q}$ and $\hat{D}_{2q}$ using the recursive formulae presented in [95]. It is also stated, in the aforementioned paper, that it is sufficient for many refractive index profiles to include only the single integrals in (10.56), i.e.

$$\hat{r}_q = \frac{\hat{r}_{1q} + \hat{I}_{1q} + \hat{r}_{1q}\hat{I}_{2q}\hat{E}_q + \hat{E}_q}{1 + \hat{r}_{1q}\hat{I}_{1q} + \hat{I}_{2q}\hat{E}_q + \hat{r}_{1q}\hat{E}_q}. \tag{10.57}$$

In [95], it is moreover shown that the formula (10.56) containing single, double and triple integrals is sufficient for the inhomogeneous thin films containing complicated profiles with large gradients. Note that if the terms comprising integrals can be neglected the WKBJ approximation is obtained.

It should be pointed out that the approximate formulae for the reflection coefficients of inhomogeneous thin films can also be derived using the matrix formalism (for details see Chap. 9).

10.5.5 Runge–Kutta Methods

These numerical methods are suitable for solving generic ordinary differential equations [96]. Therefore, they can also be utilized for solving the system of ordinary differential equations describing the propagation of monochromatic plane waves in media inhomogeneous along one axis. These methods are often used as reference methods for the comparison with results achieved by approximate methods for optical quantities of inhomogeneous thin films (see e.g. [95]).

10.6 Overlayers and Transition Layers

Overlayers and transition layers are unwanted homogeneous or inhomogenous thin films occurring in various layered systems. Their thicknesses are mostly in nanometers. These layers can exhibit smooth or rough boundaries (overlayers with randomly microrough boundaries modeled by the ITF are taken into account in Sect. 10.3.1).

Overlayers and transition layers can be considered as defects of thin film systems. They are unwanted products of thin film technologies or other miscellaneous processes. Overlayers often originate on surfaces of solids or the uppermost boundaries of thin film systems as native oxide layers or adsorption layers because of an influence of environmental media on these objects (see Sect. 10.3.1). Transition layers often appear at boundaries between substrates and films or at boundaries between adjacent films in layered systems. An example of their origin is mutual diffusion of materials of thin films adjacent to the boundary. Because of relatively small thickness of overlayers and transition layers it is considerably difficult to perform their optical characterization. This statement is true, in particular, if the optical characterization of overlayers and transition layers must be carried out together with the characterization of the other films occurring in the multilayer system studied. If overlayers and transition layers are modeled by the homogeneous thin films then the well known matrix or recursive formalisms can be used (see e.g. [4, 5]). If these layers are modeled by the inhomogeneous thin films the formulae corresponding to the procedures presented in Sect. 10.5 have to be used except for the WKBJ approximation which can not usually be utilized (overlayers and transition layers usually exhibit large profile gradients). If the thicknesses of overlayers and transition layers are substantially smaller than the wavelength of incident light ($h \ll \lambda$), the Drude approximation can be employed to calculate the Fresnel reflection coefficients of these inhomogeneous layers. After applying the procedures presented in [5, 97] one obtains the following formulae for overlayers placed onto substrates:

$$\hat{r}_s = \frac{n_0 \cos\varphi - \hat{n}_S \cos\hat{\psi}_S + ik_0\left(hn_0\hat{n}_S \cos\varphi \cos\hat{\psi}_S - \int_0^h \hat{n}^2(z)\cos^2\hat{\psi}(z)\mathrm{d}z\right)}{n_0 \cos\varphi + \hat{n}_S \cos\hat{\psi}_S + ik_0\left(hn_0\hat{n}_S \cos\varphi \cos\hat{\psi}_S + \int_0^h \hat{n}^2(z)\cos^2\hat{\psi}(z)\mathrm{d}z\right)},$$

$$\hat{r}_p = \frac{n_0 \cos\hat{\psi}_S - \hat{n}_S \cos\varphi + ik_0\left(n_0\hat{n}_S \int_0^h \cos^2\hat{\psi}(z)\mathrm{d}z - \cos\varphi\cos\hat{\psi}_S \int_0^h \hat{n}^2(z)\mathrm{d}z\right)}{n_0 \cos\hat{\psi}_S + \hat{n}_S \cos\varphi + ik_0\left(n_0\hat{n}_S \int_0^h \cos^2\hat{\psi}(z)\mathrm{d}z + \cos\varphi\cos\hat{\psi}_S \int_0^h \hat{n}^2(z)\mathrm{d}z\right)}. \tag{10.58}$$

In the foregoing equations, the symbol $\hat{n}(z)$ denotes the variable refractive index inside the overlayer. The same equations are valid for the reflection coefficients of the overlayers appearing at the uppermost boundaries and for inner transition layers inside the thin film systems. However, the refractive indices and refraction angles in (10.58) must be changed accordingly. The Drude approximation was mainly utilized for the optical description of overlayers (see e.g. [5, 98]).

Within the Drude's approximation the similar equations can be derived for the Fresnel transmission coefficients of overlayers and transition layers.

If the overlayers and transition layers cover randomly rough surfaces or boundaries, the approximation corresponding to the IFL and HFL can be used (see Sect. 10.3.1).

In spite of difficulties in optical characterization of the overlayers and transition layers, many works devoted to this problem were published (see e.g. [1, 10–24, 41, 99–107]).

10.7 Numerical Examples

In Fig. 10.1 the spectral dependencies of the normal reflectance calculated by means of the SDT (see Sect. 10.3.3) for a three-layer system with rough boundaries are introduced for illustration (see (10.23) for $L = 3$). This three-layer system is formed by the triple layer $Si_3N_4/SiO_2/Si_3N_4$ on a silicon single crystal substrate. The thicknesses of these layers were 150 nm (top Si_3N_4 layer), 130 nm (middle SiO_2 layer) and 100 nm (bottom Si_3N_4 layer). Optical constants of all materials were taken from standard tables [108–110]. The boundary roughness of this three-layer system is described by ten parameters, i.e. by ten independent elements of matrix **S** or, equivalently, by four σ_j and six C_{jk}. It was assumed that the kth layer adds an independent roughness contribution $\gamma_k(x, y)$ of the same RMS value σ on top of the $(k + 1)$th layer. The heights of irregularities were, therefore, related:

$$\eta_k(x, y) = \eta_{k+1}(x, y) + \gamma_k(x, y) \quad \text{for} \quad 1 \leq k \leq L, \tag{10.59}$$

where

$$\langle \eta_{L+1}(x, y)\eta_{L+1}(x, y)\rangle = \sigma_{L+1}^2, \tag{10.60}$$

$$\langle \gamma_k(x, y)\gamma_j(x, y)\rangle = \delta_{jk}\sigma^2, \tag{10.61}$$

$$\langle \gamma_k(x, y)\eta_{L+1}(x, y)\rangle = 0. \tag{10.62}$$

The symbol σ_{L+1} denotes the RMS of heights of the lowermost boundary. The symbol δ_{jk} is the Kronecker delta. This results in the following values of elements of the matrix **S**:

$$S_{j,k} = \sigma_{L+1}^2 + (L + 1 - \max(j, k))\sigma^2. \tag{10.63}$$

Values σ_{L+1} =10 nm and σ =5 nm were used in the calculation.

For comparison, numerical quadrature and ray tracing methods were also utilized for calculating the spectral reflectance of the system under study. From Fig. 10.1, one can see that the calculated spectral dependencies of reflectance are practically identical for all the three approaches employed, i.e. for the approach based on the series formula (10.23), numerical quadrature and ray tracing methods. Moreover, the

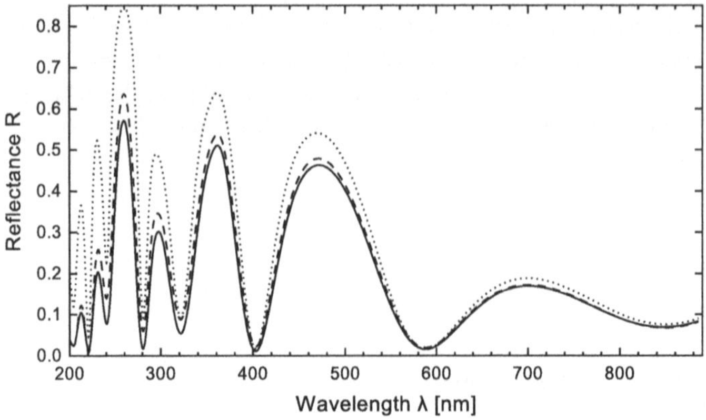

Fig. 10.1 The solid line corresponds to result obtained by series, integration or ray tracing when calculated to a sufficient precision. The dashed line represents the approximate method [1, 49–51]. The dotted line corresponds to reflectance calculated for smooth boundaries

approximate formulae presented in paper [1, 49–51] and formulae for the smooth three layer system were used to calculate the spectral dependencies of the reflectance. From Fig. 10.1, it is further seen that the approximate formulae provide the reflectance values that are relatively close to the values obtained by the series formula, numeric quadrature and ray tracing method. Nevertheless, deviations up to several percent are noticeable in the short-wavelength region. It is also evident that the formulae for the corresponding system with smooth boundaries provide inadequate results over the entire spectral range of interest (see Fig. 10.1).

In Fig. 10.2, the calculated spectral reflectance of SiO_2 thin films with a wedge-shaped non-uniformity for several selected values of σ_t are introduced for illustration. From this figure it is seen that with increasing the value of σ_t the contrast of the reflectance extrema, i.e. maxima and minima, is decreasing.

In Fig. 10.3, the comparison of the spectral dependencies of the normal reflectance for the non-uniform thin film and thin film with randomly rough boundaries is performed. It is apparent that the influence of the thickness non-uniformity on the one hand and the influence of the boundary roughness on the other hand are strongly different.

For illustration, the spectral dependencies of the reflectance for normal incidence are also introduced for two inhomogeneous thin films with a linear profile of the dielectric function in Fig. 10.4. It is observed that, for the inhomogeneous thin films, the positions of the maxima and minima changed compared to the homogeneous film as a consequence of the change in their optical thickness. The values of the reflectance at maxima and minima of both the inhomogeneous films changed relatively to the homogeneous film. The maxima of the inhomogeneous thin film with higher refractive index at the upper boundary are higher than the maxima for the homogeneous film. If the refractive index of the inhomogeneous thin film is higher at the lower boundary then the maxima are lower than those for the homogeneous

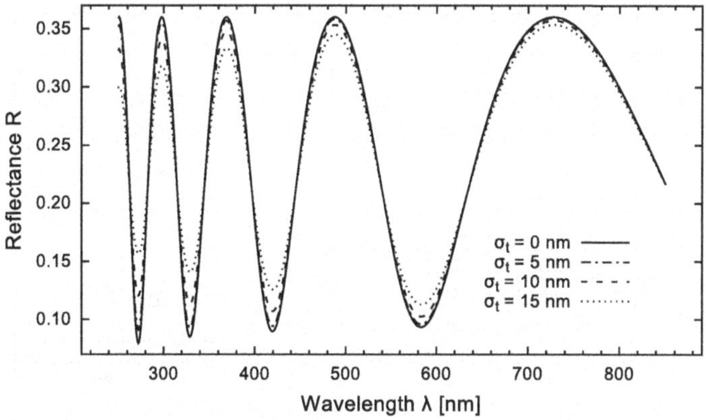

Fig. 10.2 Calculated spectral reflectances R of SiO_2 films with a wedge-shaped non-uniformity for several selected values of σ_t: the spectral dependence of the refractive index of the SiO_2 film is of $n_{SiO_2} = A_1 + A_2/\lambda^2$, where $A_1 = 1.4478$ and $A_2 = 3621\ nm^2$, the substrate refractive index is chosen for simplicity with the value of 4 within the entire spectral range and the mean thickness of the film is assumed to be 500 nm for all the films

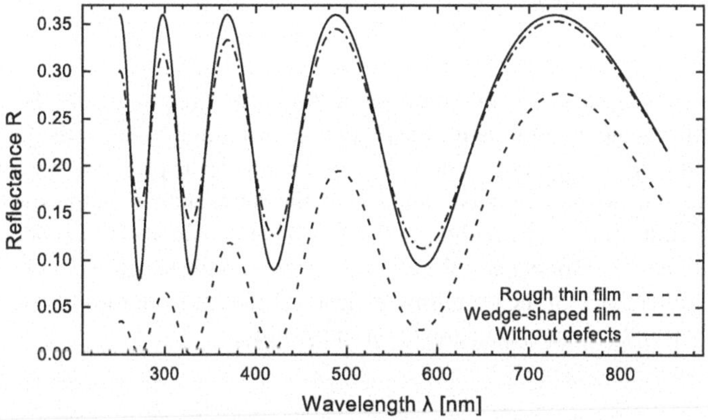

Fig. 10.3 Calculated spectral dependencies of the normal reflectance R of the rough thin film and wedge-shaped non-uniform film. The spectral reflectance of the rough thin film is calculated using the formula (10.23) for $L = 1$ under assumption that $C_{12} = 0$, $\sigma_1 = 20$ nm, $\sigma_2 = 25$ nm, the mean thickness $\overline{h}_1 = 500$ nm. The spectral reflectance of the non-uniform thin film was calculated for $\sigma_t = 15$ nm. The curve corresponding to the thin film without defects is plotted for comparison. The same optical constants as in Fig. 10.2 were used

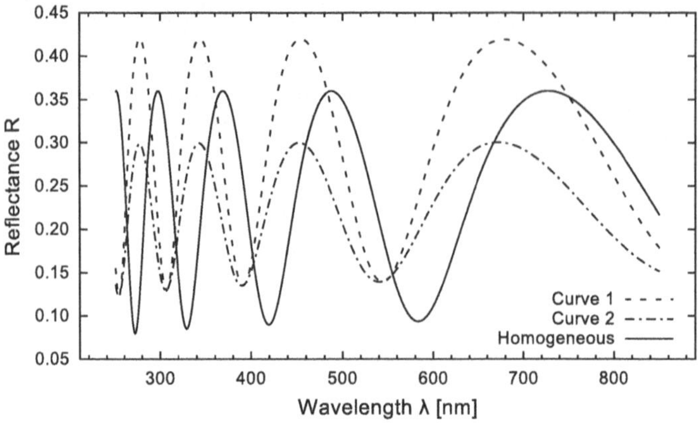

Fig. 10.4 Calculated spectral dependencies of the normal reflectance R of two inhomogeneous thin films with linear profile of the dielectric function $n(z)^2 = n_{\mathrm{U}}^2 - (z/d)(n_{\mathrm{U}}^2 - n_{\mathrm{L}}^2)$. The curve 1 corresponds to thin film with $n_{\mathrm{U}} = 1.4478 + 3621\,\mathrm{nm}^2/\lambda^2$, $n_{\mathrm{L}} = 1.24 + 2600\,\mathrm{nm}^2/\lambda^2$ and the curve 2 corresponds to $n_{\mathrm{U}} = 1.24 + 2600\,\mathrm{nm}^2/\lambda^2$, $n_{\mathrm{L}} = 1.4478 + 3621\,\mathrm{nm}^2/\lambda^2$. The refractive index of the substrate and thickness of the film are $n = 4$ and $h = 500\,\mathrm{nm}$, respectively. The homogeneous thin film is for $n_{\mathrm{U}} = n_{\mathrm{L}} = 1.4478 + 3621\,\mathrm{nm}^2/\lambda^2$

film. The minima of both the inhomogeneous thin films are practically the same and they are higher than the minima for the homogeneous film. Owing to the fact that the gradient of the inhomogeneity of both the films is small, the previous statements can also be obtained by mathematical analysis of the reflectance formula for the inhomogeneous thin film corresponding to the WKBJ approximation (see Sect. 10.5.2). One can see that the influence of the inhomogeneity on the spectral dependencies of the normal reflectance of the thin films is clearly different from those corresponding to boundary roughness and thickness non-uniformity.

10.8 Experimental Examples

This section presents four examples of the optical characterization of the thin film systems exhibiting defects discussed above.

10.8.1 Slightly Randomly Rough Surface Covered with Very Thin Overlayer

First, we show an example of the optical characterization of a sample with slight random roughness of silicon single crystal substrate covered with native oxide layer (NOL) using the EMA and RRT approaches (the NOL represents the homogeneous

Table 10.1 The values of the parameters determined using the RRT and EMA. The roughness parameter values determined by AFM are shown for comparison

	RRT	AFM		EMA IFL
sample with smooth boundaries				
$h_{\rm O}$ [nm]	3.31 ± 0.05		$h_{\rm O}$ [nm]	3.5 ± 0.1
$\chi_{\rm ell}$	5.04		$\chi_{\rm ell}$ [nm]	4.96
rough sample				
$h_{\rm O}$ [nm]	3.68 ± 0.05		$h_{\rm O}^{\rm I}$ [nm]	3.50 ± 0.08
σ [nm]	5.48 ± 0.01	4.83 ± 0.02	$\sigma^{\rm I}$ [nm]	0.85 ± 0.03
T [nm]	52.06 ± 0.09	47.8 ± 0.5		
$\chi_{\rm ell}$	3.93		$\chi_{\rm ell}$ [nm]	12.88
dispersion parameters				
B_1	1.41 ± 0.01		B_1	1.37 ± 0.02
B_2 [nm^2]	3032 ± 160		B_2 [nm^2]	2199 ± 212

overlayer). The rough silicon surfaces were prepared by anodic oxidation under constant voltage followed by dissolution of the grown oxide layers. The NOLs were created by the interaction of air and the rough surfaces. The experimental data were composed from the data corresponding to variable angle spectroscopic ellipsometry and spectroscopic reflectometry applied at near-normal incidence. The sample of smooth silicon single crystal surface covered with the NOL was also measured and the experimental data for both the samples were processed simultaneously, i.e. the multi-sample method was utilized [111–114]. If the EMA was utilized for the optical characterization only the ellipsometric data were used while both the ellipsometric and reflectometric data were employed within the RRT. The results of the optical characterization by the RRT and EMA approaches are presented in Table 10.1. In the table the symbols σ and T denote the RMS value of the heights and autocorrelation length found by the RRT (see Sect. 10.3.2). The symbol $h_{\rm O}$ denotes the NOL thickness. The symbols $\sigma^{\rm I}$ and $h_{\rm O}^{\rm I}$ denote the effective RMS value of the heights and effective NOL thickness corresponding to the EMA approach based on the IFL if the Gaussian distribution and Bruggeman formula are used (see Sect. 10.3.1). The refractive index of the NOL was modeled using the Cauchy formula $n_{\rm NOL} = B_1 + B_2/\lambda^2$ with parameters B_1 and B_2 common for both the samples. The optical constants of the silicon single crystal were taken from literature [115]. The quantity $\chi_{\rm ell}$ expressing the quality of the fit of the ellipsometric data is also presented. It is seen that the RRT achieves much better fit of the ellipsometric data than the EMA approach for the rough sample. It should be noted that the RMS value of the heights determined using the EMA approach represents clearly an effective quantity. The results in Table 10.1 support the theoretical statements introduced in Sect. 10.3.2.

In Fig. 10.5 the spectral dependence of the measured associated ellipsometric parameter $I_{\rm n}$ together with the differences of this parameter from the theoretical values calculated using the RRT and EMA approaches is shown. It is evident that these differences are smaller for the RRT than for the EMA approach. In Fig. 10.6 the measured spectral dependence of the relative reflectance of the slightly rough

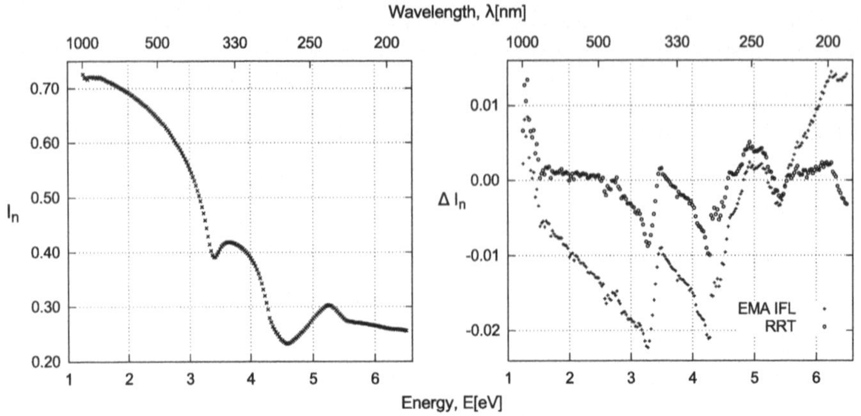

Fig. 10.5 Spectral dependence of the associated ellipsometric parameter I_n measured for the incidence angle of 60° (left) and the corresponding differences $\Delta I_n = I_n^{theoretical} - I_n^{experimental}$ between the experimental and theoretical values corresponding to the RRT and EMA based on the IFL (right)

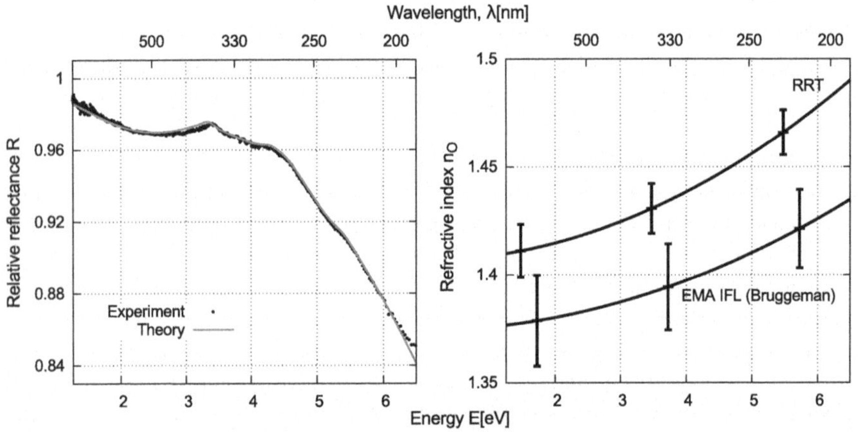

Fig. 10.6 Spectral dependence of the measured near-normal relative reflectance R of the rough sample and corresponding theoretical curve calculated by the RRT (left). Spectral dependencies of the refraction index n_O of the NOL determined by the RRT and EMA based on the IFL (right)

sample together with the fit performed by the RRT is plotted. The relative reflectance was measured with the smooth sample used as a reference sample. Furthermore, the spectral dependencies of the refractive index of the NOL determined by the RRT and EMA approaches are introduced in this figure. A relatively good agreement between both the dependencies is observed. From the previous, it is possible to deduce that the RRT is much better in describing slightly randomly rough surfaces covered with the NOLs than the EMA approach.

10.8.2 Thin Film with Thickness Non-uniformity, Boundary Roughness and Overlayer

The method of optical characterization of a selected zinc selenide (ZnSe) thin film will be presented. It is a thin film deposited by the molecular beam epitaxy on gallium arsenide (GaAs) single crystal substrate exhibiting three defects, specifically the thickness non-uniformity, roughness of the upper boundary and homogeneous overlayer. This method is based on processing of single- and variable-angle spectroscopic ellipsometry[1] and near-normal spectroscopic reflectometry. For describing the optical properties of the ZnSe film and overlayer, the universal dispersion model is employed [116] (see also Chap. 3). The optical constants of the GaAs substrate were also determined using the universal dispersion model applied to experimental data obtained for wafers of GaAs covered with native overlayers.

As for the structural model, it is assumed that the thickness non-uniformity of the ZnSe film is wedge-shaped which was indicated visually. Based on atomic force microscopy (AFM) studies, it was revealed that the random roughness of the upper boundary exhibits a wide interval of spatial frequencies. Therefore, this roughness is modeled by the combination of the RRT and SDT. In this combination, the SDT describes the roughness components corresponding to locally smooth roughness (low spatial frequencies) and the RRT describes the roughness components exhibiting high and moderate spatial frequencies. The overlayer is represented by an ITF. Then the reflection coefficients $\hat{r}_q$ of the ZnSe film are expressed by the formula (10.19) in which the distribution function $w(z, z_x, z_y)$ corresponds to the random function $\eta(x, y)$ describing the low spatial frequencies of the roughness [39]. The local reflection coefficients $\hat{r}_q^{(l)}(z, z_x, z_y)$ correspond to a wedge-shaped film, locally representing the ZnSe film covered with a thin overlayer. The slopes of this local wedge-shaped films are given by the derivatives z_x and z_y and the thickness of the film is $h_f + z$, where h_f is the mean thickness of the ZnSe film [39]. It is assumed, based on the AFM results, that these slopes are very small, i.e. $\tan \beta_0 \lesssim 0.01$. The reflection and transmission coefficients of this overlayer are calculated using the RRT with the PSDF corresponding to components of the roughness in the high and moderate range of spatial frequencies. Within the SDT, the Gaussian distribution for the heights and slopes of the upper boundary was utilized. The PSDF given by the Gaussian function (10.13) was used in the RRT. In this approximation, the integration over z in (10.19) can be performed independently on z_x, z_y with the result expressed as a finite series. The resulting expression for $\hat{r}_q$ contains a quadruple integral (double integral over z_x, z_y and double integral from the RRT). The Gauss quadrature was found to be efficient for calculating the double integral over z_x, z_y. The double integral corresponding to the RRT could also be performed quite efficiently by a specially designed numerical integration method.

Subsequently, the measured optical quantities, i.e. the reflectance R and the associated ellipsometric parameters $\overline{I}_s$, $\overline{I}_c$, $\overline{I}_n$, were calculated by means of (10.33) and

[1]The VUV ellipsometric data were measured for single incidence angle (70°) and in the IR, visible and UV range the ellipsometric data were measured in the variable-angle mode (55–75°).

(10.48), respectively. Thus, these optical quantities were calculated by averaging over the local thickness distribution corresponding to the wedge-shaped non-uniformity within the irradiated spot on the sample (for details see [39]).

Within the processing of the experimental data, it was found that the value of the autocorrelation length T_L corresponding to low spatial frequencies could not be determined with a sufficient accuracy. This is a consequence of a low sensitivity of the ellipsometric quantities to this autocorrelation length (reflectance at normal incidence is not sensitive to this quantity at all). This is why the value of T_L was fixed at the value determined by AFM . Owing to the convolution between the roughness and the tip of the AFM apparatus, it is possible to expect that the values of the autocorrelation length determined by AFM rather correspond to T_L than to the autocorrelation length T_H characterizing high and moderate spatial frequencies [117].

In Table 10.2, the values of the structural parameters related to the thickness non-uniformity are introduced. These values had to be determined for the individual spectral ranges and instruments separately since the irradiated spots were different for each instrument. Moreover, it was not possible to ensure the same position of the spots on the samples in each instruments. The values of the structural parameters of the roughness and overlayer are also introduced in this table. These structural parameters are common for all instruments and spectral ranges. The symbols σ_L and σ_H denote the RMS values of the heights corresponding to low spatial frequencies and to high and moderate spatial frequencies, respectively. The values of the roughness parameters determined by AFM are presented for comparison. The overlayer thickness is evidently larger than the expected value and the value determined in [40]. The typical thickness of the overlayers is roughly equal to 4 nm for the ZnSe films. Therefore, one can expect that the remaining roughly 6 nm in the thickness of the overlayer corresponds to microroughness manifested into the increase of this overlayer thickness. This microroughness could be incorporated into the structural model of the ZnSe film by means of the EMA approaches. The fact that this microroughness is not included in the total RMS value $\sigma_T = (\sigma_H^2 + \sigma_L^2)^{1/2}$ also explains why the

Table 10.2 The values of the structural parameters related to the thickness non-uniformity (left) and structural parameters of the roughness and overlayer (right). The spectral ranges are denoted as MIR (medium infrared), NIR (near infrared), VIS (visible), UV (ultraviolet), VUV (vacuum ultraviolet). The symbol h_O denotes the overlayer thickness and symbol σ_T denotes the total RMS value of the heights calculated as $\sigma_T = (\sigma_H^2 + \sigma_L^2)^{1/2}$. The values are rounded such that the errors are in the last digit

	h_f [nm]	σ_0 [nm]
ell,NIR	1199.5	32.8
ell,NIR-UV	1204.7	6.8
ell,VIS-VUV	1211.7	16.8
R,MIR	1196.6	32.9
R,NIR	1196.8	16.9
R,VIS-UV	1200.9	14.8

		AFM
h_O [nm]	10.59	
σ_T [nm]	7.29	7.45
σ_H [nm]	4.79	
T_H [nm]	56.77	
σ_L [nm]	5.5	7.45
T_L [nm]	608(fixed)	608
$\tan\beta_{0,L}$	0.013	0.017

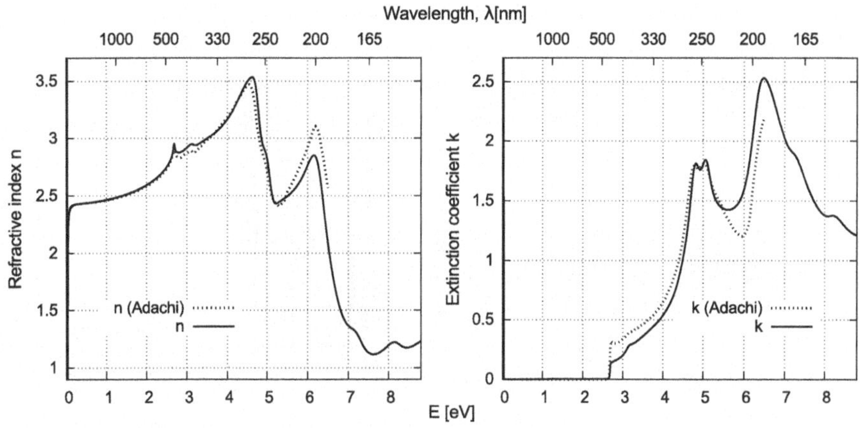

Fig. 10.7 Spectral dependencies of the refractive index *n* and extinction coefficient *k* of the selected ZnSe film

value determined by AFM is larger than the value σ_T determined from the optical characterization (because of the convolution mentioned above the RMS values of the heights determined by AFM are usually smaller than those determined optically, see e.g. [41, 117]).

In Fig. 10.7 the determined spectral dependencies of the optical constants of the selected ZnSe thin film are depicted. The optical constants of the ZnSe single crystal taken from Adachi [118] are introduced for comparison. An excellent agreement between the selected experimental data and their fits is seen from Fig. 10.8. This example of optical characterization shows that it is possible to perform the successful optical characterization of thin films exhibiting several defects if the combination of several theoretical approaches and several types of the experimental data measured in a wide spectral range are employed.

10.8.3 Inhomogeneous Thin Film

The optical characterization of the inhomogeneous thin film of non-stoichiometric silicon nitride prepared by plasma enhanced chemical vapor deposition onto silicon single crystal wafer is selected for illustrating the optical characterization of the thin films exhibiting the profiles of the refractive index. The optical characterization of this silicon nitride film was performed using phase-modulated variable angle spectroscopic ellipsometry. The profile of the complex refractive index was modeled by the linear dependence of the dielectric function, i.e. by the formula (10.49). The dispersion model of the silicon nitride film with the imaginary part of the dielectric function ε_i was used as follows [89]:

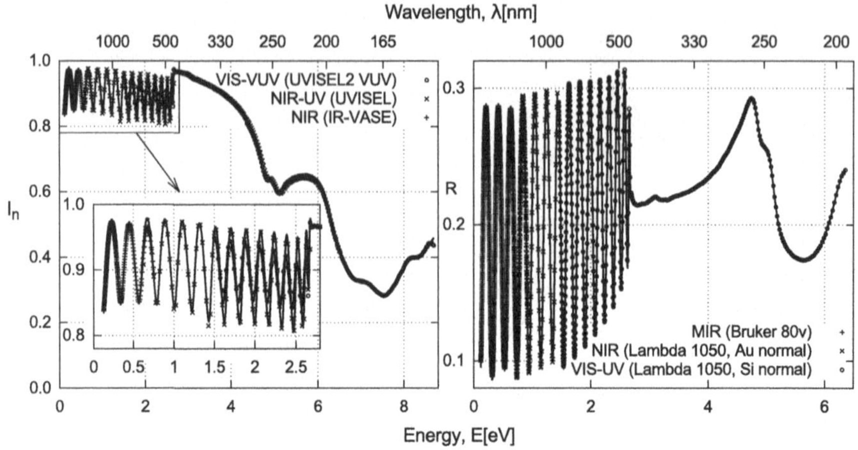

Fig. 10.8 Agreement between the experimental data and their fits for the ellipsometric parameter I_n measured at the incidence angle 70° (left) and the near-normal reflectance R (right)

$$\varepsilon_i(E) = \frac{32Q^2(E - E_g)^2(E_h - E)^2}{(E_h - E_g)^5 E^2} \Pi(E_g, E_h; E), \tag{10.64}$$

where E_g, E_h and Q are the band gap energy, maximum energy of transitions and parameter proportional to the density of electrons. The function Π is defined as:

$$\Pi(a, b; x) = \begin{cases} 1 & a \leq x \leq b, \\ 0 & \text{otherwise.} \end{cases} \tag{10.65}$$

The real part of the dielectric function was determined using the Krammers–Kronig transform [119]. It was necessary to search six dispersion parameters describing the spectral dependencies of the dielectric functions at upper and lower boundaries. The thickness value had to be searched together with these dispersion parameters.

Two structural models of the silicon nitride film were employed. The first model assumed a refractive index profile and fine random roughness on the upper boundary of this film (RPF model). The latter model assumed a refractive index profile and the overlayer on the upper boundary of the film (OPF model). The roughness of the upper boundary was included into the formulae for the associated ellipsometric parameters by means of the RRT. The incorporation of the overlayer represented by a homogeneous thin film was performed using the matrix algorithm.

The best fits of the experimental data were achieved for both the structural models. One could not distinguish between these models because their influence on the experimental data was practically identical. It was also impossible to distinguish between them using AFM . The RMS value of the heights of the roughness was 1.6 nm and the autocorrelation length was 6.0 nm. The thickness value of the overlayer was determined as 1.9 nm. The thickness of the silicon nitride film was found in the value of

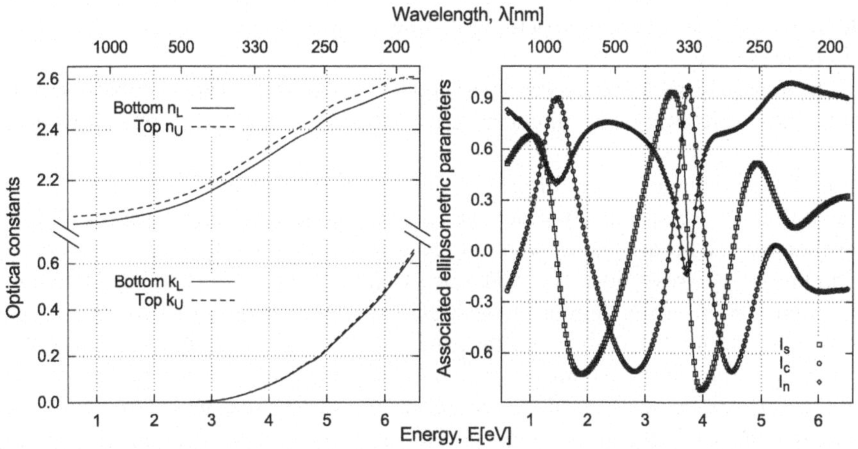

Fig. 10.9 Spectral dependencies of the optical constants for the upper n_U, k_U and lower n_L, k_L boundaries of the silicon nitride film (left). Spectral dependencies of the measured associated ellipsometric parameters I_s, I_c, I_n at angle of incidence of 65° and their fits (right)

114.8 nm for the RPF and 113.9 nm for the OPF. It is probable that both the defects, i.e. fine roughness and overlayer, are partially present onto the upper boundary of the silicon nitride film under study.

The spectral dependencies of the optical constants of the investigated silicon nitride film are plotted for the upper and lower boundary in Fig. 10.9. These spectral dependencies are indistinguishable for the RPF and OPF models. From this figure, it is evident that the differences between the upper and lower boundaries are larger for the refractive indices n_U and n_L than for the extinction coefficients k_U and k_L. In Fig. 10.9 the experimental values of the associated ellipsometric parameters are introduced together with their fits. One can see an excellent agreement between the experimental data and their fits. This implies that the structural and dispersion models of this silicon nitride thin film characterized were used in a correct way.

10.8.4 Transition Layers

The two examples concerning the optical characterization of the layered systems covered with overlayers are introduced in Sects. 10.8.1 and 10.8.2. Therefore, we will focus on the examples of optical characterization of the layered systems with transition layers here.

Transition layers between the silicon single crystal surfaces and their NOLs or thermally grown oxide films are studied in literature most frequently. This is due to their technological importance and also the fact that this is the simplest semiconductor-oxide layered system. This is why the results concerning two examples of optical characterization of the transition layers of the system mentioned above will be presented.

The transition layers between silicon single crystal substrates and thermally grown SiO_2 films were studied by monochromatic ellipsometry in [120]. First, the ideal model of the system Si–SiO_2 was assumed, i.e. assuming no defects. The dependence of the refractive index values of the SiO_2 films was measured as a function of their thicknesses within a relatively wide interval. Some anomalies were observed for this dependence. These anomalies were removed if the ideal model was improved by including weak stress-induced double refraction in the SiO_2 films and by including very thin non-absorbing homogeneous transition layers between silicon substrates and the SiO_2 films. After processing the experimental data by the improved model, the following values of the parameters characterizing the transition layers were found for the selected sample: the thickness of the transition layer was 0.6 nm and the refractive index was 2.8 at $\lambda = 546.1$ nm. It was also found that the transition layers were thicker for SiO_2 films grown at lower temperatures than for SiO_2 films grown at higher temperatures. The refractive index values of the transition layers were independent on temperature of the growth of the SiO_2 films (for details see [120]).

In [105, 106] in-situ studies of the transition layers between silicon single crystal surfaces and their thermal oxide films were performed using spectroscopic ellipsometry. The transition layers were again modeled by the homogeneous thin films. The ellipsometric data were processed by three structural models of the system. The best results were obtained for the model consisting of four media: silicon substrate, transition layer, SiO_2 thin film and ambient. It was assumed that the SiO_2 films contain voids with certain volume fractions. The mixing of the voids and SiO_2 was described by the Bruggeman formula. The weak stress-induced birefringence of SiO_2 films was also incorporated as in [120]. The dielectric functions of the transition layers were calculated for a physical and chemical mixtures of Si and SiO_2. The physical mixtures corresponded to optically identifiable separate regions of amorphous Si and SiO_2, i.e. microroughness. This mixing was described by the Bruggeman formula. The chemical mixtures corresponded to combining silicon and oxygen on atomic scale. The complex dielectric function corresponding to chemical mixing was calculated by means of the model developed in [121]. Within the four-media model, the chemical mixtures of the transition layers gave better fits of the ellipsometric data than the physical mixtures The best fit of the experimental data was achieved for the transition layer thickness of 0.7 ± 0.2 nm and average stoichiometry of $Si_{0.8\pm0.1}(SiO_2)_{0.2\mp0.1}$. The transition layer thickness value determined in this paper is in agreement with that found in [120]. However, it is necessary to note that in several papers the transition layers at the boundaries between silicon single crystal substrates and thermal SiO_2 thin films were not observed (see e.g. [122]).

The transition layers modeled by the inhomogeneous thin films were found at optical characterization of thin films of zirconia (ZrO_2) prepared by vacuum evaporation onto silicon single crystal substrates in [88]. In this paper it was observed that these ZrO_2 thin films exhibit certain refractive index profiles. In their optical characterization this profile was modeled by the following function:

$$n(z,\lambda) = n_{\rm L}(\lambda)p(z) + n_{\rm U}(\lambda)[1 - p(z)], \tag{10.66}$$

where the function $p(z)$ is expressed as

$$p(z) = -\frac{c + (1-c)\mathrm{e}^{-b}}{1-\mathrm{e}^{-a-b}}\mathrm{e}^{-az/h} + \frac{1-c+c\mathrm{e}^{-a}}{1-\mathrm{e}^{-a-b}}\mathrm{e}^{-b(1-z/h)} + c, \tag{10.67}$$

where a, b, c are parameters of the profile and h is the thickness of the inhomogeneous ZrO_2 film. The function $p(z)$ varies continuously from the value $p(z) = 0$ for $z = 0$ to the value $p(z) = 1$ for $z = h$. The spectral dependencies of the refractive indices $n_{\rm U}(\lambda)$ and $n_{\rm L}(\lambda)$ were given by the Cauchy formula as follows:

$$n_{\rm U}(\lambda) = A_{\rm U} + \frac{B_{\rm U}}{\lambda^2}, \qquad n_{\rm L}(\lambda) = A_{\rm L} + \frac{B_{\rm L}}{\lambda^2}. \tag{10.68}$$

All the parameters of the model, i.e. a, b, c, h, $A_{\rm U}$, $B_{\rm U}$, $A_{\rm L}$ and $B_{\rm L}$, were sought within the optical characterization. For determining these parameters the combined method of variable angle spectroscopic ellipsometry and near-normal spectroscopic reflectometry was utilized. The matrix formalism was used to calculate the spectral dependencies of ellipsometric parameters and reflectance. The following values of parameters were determined for the selected sample: $a = 60 \pm 14$, $b = 5.82 \pm 0.30$, $c = 0.560 \pm 0.066$, $h = 321.96 \pm 0.26\,\mathrm{nm}$, $A_{\rm U} = 1.610 \pm 0.078$, $B_{\rm U} = (2.91 \pm 0.28) \times 10^4\,\mathrm{nm}^2$, $A_{\rm L} = 2.21225 \pm 0.0061$, $B_{\rm L} = (6.16 \pm 0.98) \times 10^3\,\mathrm{nm}^2$. The refractive index profile of the selected ZrO_2 film and the spectral dependencies of the boundary refractive indices are plotted in Fig. 10.10. The refractive index n_{150} corresponding to depth of 150 nm is introduced as well. The spectral dependence of the refractive index of the ZrO_2 thin film determined by Chindaudom and Vedam [123] is depicted for comparison. From Fig. 10.10 one can see that the substantial part of the ZrO_2 film exhibits almost constant refractive index which corresponds to homogeneous thin film. The region adjacent to the boundary between the silicon substrate and ZrO_2 film can be considered as a transition layer with the refractive index profile (the depth of this region is about 70 nm). The region of the ZrO_2 film adjacent to the ambient corresponds to the overlayer also exhibiting refractive index profile (the depth of this region is about 10 nm). It is probable that the columnar structure of the ZrO_2 thin films is responsible for the existence of both the overlayers and the transition layers (as for columnar structure of thin films see e.g. [124–127]). The larger packing density of columns close to the substrate corresponds to the transition layer while the the overlayer mainly corresponds to microroughness formed by tops of columns (packing density of the columnar thin film is defined as a ratio of the volume occupied by the columns to the total volume of this film).

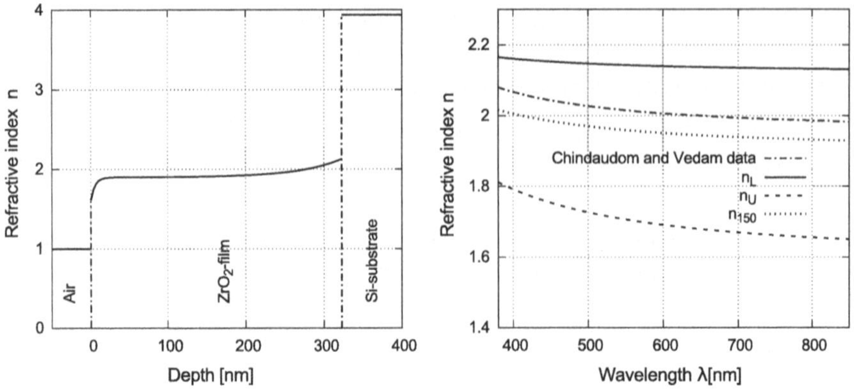

Fig. 10.10 The results concerning the selected ZrO_2 thin film: the refractive index profile for $\lambda = 600\,\text{nm}$ (left), the spectral dependencies of the refractive indices (right)

10.9 Closing Remarks

In practice it is possible to encounter other defects of thin films than those presented here. These defects can also influence the results of the optical characterization. One of these defects is a columnar structure of thin films usually accompanied with their porosity. This defect causes the dependence of the optical properties of such the films on temperature and environmental medium because of the capillary condensation in the pores. This can substantially change the properties of multilayer systems fabricated in optics industry that are utilized in many scientific and commercial instruments. Therefore, many studies of an influence of this defect on the optical properties of the columnar thin films have been published so far. Results concerning the optical characterization of columnar thin films are presented, for example, in [124–126, 128–131].

Mechanical stress inside films can induce a non-negligible artificial optical anisotropy. This artificial anisotropy influences the polarization states of light waves reflected or transmitted by the thin films characterized and, therefore, this defect affects the ellipsometric measurements of such films. The photometric quantities such as reflectance and transmittance of these films are also influenced by this defect. For the successful optical characterization of thin films with artificial anisotropy, the choice of the model of this anisotropy is important (see e.g. [132]). The influence of the induced artificial anisotropy of transparent slabs caused by the inner stress must often be taken into account at optical characterization of thin film systems covering these slabs.

An important defect is caused by the existence of local volume inhomogeneities such as voids and inclusions of materials differing from the host material of the films studied. If the linear dimensions of these volume inhomogeneities are substantially smaller than the wavelength of light the EMA formulae can be used to describe their influence on the optical properties and optical characterization of these films (see e.g. [105, 106, 133]).

The defects such as boundary roughness and volume inhomogeneities can cause scattering of light. The dependencies of the flux of scattered light on various parameters (e.g. angles describing the direction of scattered light) can also be utilized for optical characterization of thin films with these defects (see e.g. [134–136]).

10.10 Conclusion

This chapter describes the influence of the most important thin film defects, i.e. random boundary roughness, thickness non-uniformity, optical inhomogeneity in the form of refractive index profile, overlayers and transition layers, on the optical characterization of the films. The theoretical approaches enabling us to derive the formulae for the optical quantities of thin film systems exhibiting the defects mentioned above are presented. We focused on the quantities corresponding to specularly reflected light from these systems. Similar formulae and discussion can be presented for the optical quantities corresponding to transmitted light under assumption that the substrates are transparent. In the theoretical part of this chapter the influence of the defects on spectral reflectance of the selected thin film systems is illustrated by means of several numerical examples. These numerical examples make it easier to understand the influence of the defects from the practical point of view in photometry.

In the latter part of this chapter, the selected experimental examples of the optical characterization of the thin films are shown. The examples are chosen so that they include all the defects discussed in the theoretical part. The examples indicate that even the thin films exhibiting the combined defects can be successfully characterized.

From both the parts of this chapter, it can be implied that the neglecting of the defects mentioned above can cause the substantial misrepresentation of the results of the optical characterization of the thin film systems exhibiting these defects. Therefore, it is necessary to incorporate them into the structural models of such thin film systems in practice, in spite of the complications arising from this step. In this chapter it is also indicated that many sophisticated methods are available for the optical characterization of thin films with defects.

Acknowledgements The work on this chapter was supported by the project LO1411 (NPU I) funded by Ministry of Education Youth and Sports of the Czech Republic.

References

1. I. Ohlídal, D. Franta, in *Progress in Optics*, vol. 41, ed. by E. Wolf (Elsevier, Amsterdam, 2000), pp. 181–282
2. R.M.A. Azzam, N.M. Bashara, *Ellipsometry and Polarized Light* (North-Holland, Amsterdam, 1977)

3. H. Fujiwara, *Spectroscopic Ellipsometry: Principles and Applications* (Wiley, England, 2007)
4. Z. Knittl, *Optics of Thin Films* (Wiley, London, 1976)
5. A. Vašíček, *Optics of Thin Films* (North-Holland, Amsterdam, 1960)
6. D.E. Aspnes, J.B. Theeten, F. Hottier, Phys. Rev. B **20**, 3292 (1979)
7. D. Stroud, Phys. Rev. B **12**, 3368 (1975)
8. R. Landauer, AIP Conf. Proc. **40**, 2 (1978)
9. D.A.G. Bruggeman, Ann. Phys. **416**(636) (1935)
10. G.H. Bu-Abbud, D.L. Mathine, P. Snyder, J.A. Woollam, D. Poker, J. Bennett, D. Ingram, P.P. Pronko, J. Appl. Phys. **59**, 257 (1986)
11. G. Yu, T. Soga, J. Watanabe, T. Jimbo, M. Umeno, Jpn. J. Appl. Phys. **36**, 2829 (1997)
12. I. Ohlídal, F. Lukeš, Opt. Acta **19**, 817 (1972)
13. I. Ohlídal, F. Lukeš, K. Navrátil, Surf. Sci. **45**, 91 (1974)
14. S. Logothetidis, J. Appl. Phys. **65**, 2416 (1989)
15. F. Lukeš, Optik **31**, 83 (1970)
16. Y. Gaillyová, Thin Solid Films **155**, 217 (1987)
17. J.R. Blanco, P.J. McMarr, Appl. Opt. **30**, 3210 (1991)
18. D. Franta, I. Ohlídal, P. Klapetek, A. Montaigne Ramil, A. Bonanni, D. Stifter, H. Sitter, Thin Solid Films **468**, 193 (2004)
19. D. Franta, I. Ohlídal, P. Klapetek, A. Montaigne Ramil, A. Bonanni, D. Stifter, H. Sitter, J. Appl. Phys. **92**, 1873 (2002)
20. D. Franta, I. Ohlídal, P. Klapetek, A. Montaigne-Ramil, A. Bonanni, D. Stifter, H. Sitter, Acta Phys. Slov. **53**, 95 (2003)
21. R. Dahmani, L. Salamanca-Riba, N.V. Nguyen, D. Chandler-Horowitz, B.T. Jonker, J. Appl. Phys. **76**, 514 (1994)
22. I. Ohlídal, D. Nečas, J. Mod. Opt. **55**, 1077 (2008)
23. M. Šiler, I. Ohlídal, D. Franta, A.M. Ramil, A. Bonanni, D. Stifter, H. Sitter, J. Mod. Opt. **52**, 583 (2005)
24. K. Navrátil, I. Ohlídal, F. Lukeš, Thin Solid Films **56**, 163 (1979)
25. I. Ohlídal, K. Navrátil, M. Ohlídal, in *Progress in Optics*, vol. 34, ed. by E. Wolf (Elsevier, Amsterdam, 1995), pp. 249–331
26. I. Ohlídal, K. Navrátil, F. Lukeš, J. Opt. Soc. Am. **61**, 1630 (1971)
27. J. Bauer, L. Biste, D. Bolze, Phys. Status Solidi A **39**, 173 (1977)
28. L. Spanos, Q. Liu, E.A. Irene, T. Zettler, B. Hornung, J.J. Wortman, J. Vac. Sci. Technol. A **12**, 2653 (1994)
29. C.A. Fenstermaker, F.L. McCrackin, Surf. Sci. **16**, 85 (1969)
30. I. Ohlídal, J. Vohánka, M. Čermák, D. Franta, Appl. Surf. Sci. **419**, 942 (2017)
31. J.W.S.B. Rayleigh, *The Theory of Sound* (Macmillan and Company, New York, 1878)
32. S.O. Rice, Commun. Pure Appl. Math. **4**, 351 (1951)
33. D. Franta, I. Ohlídal, J. Mod. Opt. **45**, 903 (1998)
34. D. Franta, I. Ohlídal, Opt. Commun. **248**, 459 (2005)
35. A. Yanguas-Gil, H. Wormeester, Relationship Between Surface Morphology and Effective Medium Roughness (Springer, Berlin, 2013), pp. 179–202
36. A. Yanguas-Gil, B.A. Sperling, J.R. Abelson, Phys. Rev. B **84**, 085402 (2011)
37. B. Levin, *Theoretical Principles of Statistical Radio Engineering* (Sovyetskoe Radio, Moscow, 1974). In Russian
38. A.L. Barabási, H.E. Stanley, *Fractal Concepts in Surface Growth* (Cambridge University Press, Cambridge, 1995)
39. I. Ohlídal, D. Franta, D. Nečas, Appl. Surf. Sci. **360**, 28 (2016)
40. D. Nečas, I. Ohlídal, D. Franta, M. Ohlídal, V. Čudek, J. Vodák, Appl. Opt. **53**, 5606 (2014)
41. D. Franta, I. Ohlídal, P. Klapetek, Mikrochim. Acta **132**, 443 (2000)
42. P.J. Chandley, Opt. Quantum Electron. **8**, 329 (1976)
43. I. Ohlídal, K. Navrátil, Appl. Opt. **24**, 2690 (1985)
44. A.A. Maradudin, D.L. Mills, Phys. Rev. B **11**, 1392 (1975)

45. J.A. Ogilvy, *Theory Of Wave Scattering From Random Rough Surfaces* (A. Hilger, Philadelphia, 1991)
46. M. Born, E. Wolf, *Principles of Optics*, 3rd edn. (Pergamon Press, Oxford, 1965)
47. P. Beckmann, A. Spizzichino, *The Scattering of Electromagnetic Waves from Rough Surfaces* (Pergamon, London, 1963)
48. I. Ohlídal, D. Franta, D. Nečas, Thin Solid Films **571**, 695 (2014)
49. I. Ohlídal, J. Opt. Soc. Am. A **10**, 158 (1993)
50. I. Ohlídal, F. Vižd'a, M. Ohlídal, Opt. Eng. **34**, 1761 (1995)
51. I. Ohlídal, F. Vižd'a, J. Mod. Opt. **46**, 2043 (1999)
52. D. Nečas, I. Ohlídal, Opt. Express **22**, 4499 (2014)
53. J.M. Eastman, in *Physics of Thin Films*, ed. by G. Hass, M.H. Francombe (Academic Press, New York, 1978), pp. 167–226
54. C.K. Carniglia, Opt. Eng. **18**, 104 (1979)
55. J.M. Zavislan, Appl. Opt. **30**, 2224 (1991)
56. A.G. Voronovich, *Wave Scattering from Rough Surfaces* (Springer, Berlin, 1994)
57. S. Silver, *Microwave Antenna Theory and Design* (McGraw-Hill, New York, 1949)
58. I. Ohlídal, F. Lukeš, Opt. Commun. **7**, 76 (1973)
59. M. Sancer, I.E.E.E. Trans, Antennas Propag. **17**, 577 (1969)
60. Y.Q. Jin, J. Appl. Phys. **63**, 1286 (1988)
61. D. Nečas, Optical characterisation of non-uniform thin films. Ph.D. thesis, Masaryk University, Brno (2013)
62. I. Ohlídal, Appl. Opt. **19**, 1804 (1980)
63. T. Pisarkiewicz, T. Stapinski, H. Czternastek, P. Rava, J. Non-Cryst, Solids **137**, 619 (1991)
64. T. Pisarkiewicz, J. Phys. D Appl. Phys. **27**, 690 (1994)
65. E. Márquez, J.M. González-Leal, R. Jiménez-Garay, S.R. Lukic, D.M. Petrovic, J. Phys. D Appl. Phys. **30**, 690 (1997)
66. E. Márquez, P. Nagels, J. González-Leal, A. Bernal-Oliva, E. Sleeckx, R. Callaerts, Vacuum **52**, 55 (1999)
67. M. Török, Opt. Acta **32**, 479 (1985)
68. U. Richter, Thin Solid Films **313–314**, 102 (1998)
69. G.E.J. Jr., J.W. McCamy, Appl. Phys. Lett. **61**, 512 (1992)
70. S. Pittal, P.G. Snyder, N. Ianno, Thin Solid Films **233**, 286 (1993)
71. K. Roodenko, M. Gensch, H. Heise, U. Schade, N. Esser, K. Hinrichs, Infrared Phys. Technol. **49**, 39 (2006)
72. U. Rossow, Thin Solid Films **313–314**, 97 (1998)
73. D. Nečas, I. Ohlídal, D. Franta, J. Opt. A: Pure Appl. Opt. **11**, 045202 (2009)
74. K. Forcht, A. Gombert, R. Joerger, M. Köhl, Thin Solid Films **302**, 43 (1997)
75. R. Ossikovski, M. Kildemo, M. Stchakovsky, M. Mooney, Appl. Opt. **39**, 2071 (2000)
76. D. Nečas, I. Ohlídal, D. Franta, J. Opt. **13**, 085705 (2011)
77. G.E. Jellison, in *Handbook of Ellipsometry*, ed. by H.G. Tompkins, E.A. Irene (William Andrew, New York, 2005), pp. 237–296. chap. 3
78. D. Nečas, I. Ohlídal, D. Franta, V. Čudek, M. Ohlídal, J. Vodák, L. Sládková, L. Zajíčková, M. Eliáš, F. Vižd'a, Thin Solid Films **571**, 573 (2014)
79. R. Jacobsson, in *Physics of Thin Films*, vol. 8, ed. by G. Hass (Academic Press, New York, 1975), pp. 51–98
80. F. Abelès, in *Progress in Optics*, ed. by E. Wolf (Elsevier, Amsterdam, 1963), pp. 249–288
81. R. Jacobsson, in *Progress in Optics*, ed. by E. Wolf (Elsevier, Amsterdam, 1966), pp. 247–286
82. L. Brekhovskikh, *Waves in Layered Media*, Applied mathematics and mechanics (Academic Press, New York, 1980)
83. J. Wait, *Electromagnetic Waves in Stratified Media*, International series of monographs on electromagnetic waves (Pergamon Press, Oxford, 1962)
84. S.F. Monaco, J. Opt. Soc. Am. **51**, 280 (1961)
85. M.J. Minot, J. Opt. Soc. Am. **66**, 515 (1976)
86. J. Lekner, Physica A **116**, 235 (1982)

87. B. Sheldon, J.S. Haggerty, A.G. Emslie, J. Opt. Soc. Am. **72**, 1049 (1982)
88. D. Franta, I. Ohlídal, P. Klapetek, P. Pokorný, M. Ohlídal, Surf. Interface Anal. **32**, 91 (2001)
89. D. Nečas, D. Franta, I. Ohlídal, A. Poruba, P. Wostrý, Surf. Interface Anal. **45**, 1188 (2013)
90. T. Lohner, K.J. Kumar, P. Petrik, A. Subrahmanyam, I. Bársony, J. Mater. Res. **29**, 1528 (2014)
91. D. Li, A. Goullet, M. Carette, A. Granier, J. Landesman, Mater. Chem. Phys. **182**, 409 (2016)
92. M.J. Minot, J. Opt. Soc. Am. **67**, 1046 (1977)
93. D. Franta, I. Ohlídal, Surf. Interface Anal. **30**, 574 (2000)
94. C.K. Carniglia, J. Opt. Soc. Am. A **7**, 848 (1990)
95. M. Kildemo, O. Hunderi, B. Drévillon, J. Opt. Soc. Am. A **14**, 931 (1997)
96. W. Press, S. Teukolsky, W. Vetterling, B. Flannery, *Numerical Recipes in C: The Art of Scientific Computing* (Cambridge University Press, Cambridge, 1992)
97. P. Drude, Wied. Ann. **43**, 126 (1891)
98. L. Tronstad, Trans. Faraday Soc. **31**, 1151 (1935)
99. F. Lukeš, Surf. Sci. **30**, 91 (1972)
100. T. Saitoh, K. Mori, in *Proceedings of 1994 IEEE 1st World Conference on Photovoltaic Energy Conversion - WCPEC (A Joint Conference of PVSC, PVSEC and PSEC)*, vol. 2 (1994), pp. 1648–1651
101. H. Ono, T. Ikarashi, K. Ando, T. Kitano, J. Appl. Phys. **84**, 6064 (1998)
102. Y.J. Cho, Y.W. Lee, H.M. Cho, I.W. Lee, S.Y. Kim, J. Appl. Phys. **85**, 1114 (1999)
103. T. Easwarakhanthan, P. Alnot, Surf. Interface Anal. **26**, 1008 (1998)
104. G. Jungk, C. Jungk, T. Grabolla, Phys. Status Solidi B **215**, 731 (1999)
105. D.E. Aspnes, J.B. Theeten, J. Electrochem. Soc. **127**, 1359 (1980)
106. D.E. Aspnes, J.B. Theeten, Phys. Rev. Lett. **43**, 1046 (1979)
107. X. Chen, J.M. Gibson, J. Vac. Sci. Technol. A **17**, 1269 (1999)
108. H.R. Philipp, in *Handbook of Optical Constants of Solids*, ed. by E.D. Palik, vol. 1 (Academic Press, New York, 1985), pp. 749–763
109. H. Philipp, in *Handbook of Optical Constants of Solids*, ed. by E.D. Palik (Academic Press, Burlington, 1997), pp. 771–774
110. C.M. Herzinger, B.D. Johs, W.A. McGahan, J.A. Woollam, W. Paulson, J. Appl. Phys. **83**, 3323 (1998)
111. D. Franta, I. Ohlídal, Acta Phys. Slov. **50**, 411 (2000)
112. D. Franta, L. Zajíčková, I. Ohlídal, J. Janča, K. Veltruská, Diam. Relat. Mater. **11**, 105 (2002)
113. I. Ohlídal, M. Líbezný, Surf. Interface Anal. **16**, 46 (1990)
114. I. Ohlídal, M. Líbezný, Surf. Interface Anal. **17**, 171 (1991)
115. D. Franta, A. Dubroka, C. Wang, A. Giglia, J. Vohánka, P. Franta, I. Ohlídal, Appl. Surf. Sci. **421**, 405 (2017)
116. D. Franta, D. Nečas, I. Ohlídal, Appl. Opt. **54**, 9108 (2015)
117. P. Klapetek, I. Ohlídal, Ultramicroscopy **94**, 19 (2003)
118. S. Adachi, *Optical Properties of Crystalline and Amorphous Semiconductors: Materials and Fundamental Principles* (Kluwer, Boston, 1999)
119. D. Franta, D. Nečas, L. Zajíčková, Opt. Express **15**, 16230 (2007)
120. E. Taft, L. Cordes, J. Electrochem. Soc. **126**, 131 (1979)
121. D.E. Aspnes, J.B. Theeten, J. Appl. Phys. **50**, 4928 (1979)
122. I. Ohlídal, D. Franta, E. Pinčík, M. Ohlídal, Surf. Interface Anal. **28**, 240 (1999)
123. P. Chindaudom, K. Vedam, in *Physics of Thin Films*, ed. by K. Vedam, vol. 19 (Elsevier, Amsterdam, 1994), pp. 191–247
124. E. Ritter, Appl. Opt. **15**, 2318 (1976)
125. H. Pulker, E. Jung, Thin Solid Films **9**, 57 (1972)
126. M. Harris, H. Macleod, S. Ogura, E. Pelletier, B. Vidal, Thin Solid Films **57**, 173 (1979)
127. P.J. Martin, H.A. Macleod, R.P. Netterfield, C.G. Pacey, W.G. Sainty, Appl. Opt. **22**, 178 (1983)
128. I. Ohlídal, K. Navrátil, Thin Solid Films **74**, 51 (1980)
129. K. Kinosita, M. Nishibori, J. Vac. Sci. Technol. **6**, 730 (1969)

130. O. Stenzel, S. Wilbrandt, N. Kaiser, M. Vinnichenko, F. Munnik, A. Kolitsch, A. Chuvilin, U. Kaiser, J. Ebert, S. Jakobs, A. Kaless, S. Wüthrich, O. Treichel, B. Wunderlich, M. Bitzer, M. Grössl, Thin Solid Films **517**, 6058 (2009)
131. O. Stenzel, *The Physics of Thin Film Optical Spectra*, Surface sciences (Springer, Berlin, 2005)
132. D. Franta, D. Nečas, I. Ohlídal, Thin Solid Films **519**, 2637 (2011)
133. I. Ohlídal, D. Franta, Acta Phys. Slov. **50**, 489 (2000)
134. C. Amra, J.H. Apfel, E. Pelletier, Appl. Opt. **31**, 3134 (1992)
135. C. Amra, C. Grèzes-Besset, L. Bruel, Appl. Opt. **32**, 5492 (1993)
136. A. Duparré, J. Ferre-Borrull, S. Gliech, G. Notni, J. Steinert, J.M. Bennett, Appl. Opt. **41**, 154 (2002)

Chapter 11
Scanning Probe Microscopy Characterization of Optical Thin Films

Petr Klapetek

Abstract Scanning Probe Microscopy is a technique very frequently used for thin film surfaces measurements at different stages of their preparation and/or characterization. It provides information about the surface morphology and can also be used to measure locally various physical quantities. In this chapter we discuss typical quantities measured and evaluated in the field of thin films, which basically consists of roughness characterization and film thickness determination. The basics of instrumentation, related metrology and techniques to perform measurements on large area to obtain enough statistical information about the sample properties are discussed as well.

11.1 Introduction

Using a term Scanning Probe Microscopy (SPM) measurement techniques for solid surfaces characterization that have a common basis - use of a very sharp probe that is scanned in a close proximity to the studied sample. Various interactions can be used for preserving the small gap between probe and sample, e.g. keeping the interatomic forces or tunneling current constant. In top of the local probe position various other channels of information can be stored. A typical product of any SPM measurement is therefore a map of surface topography, eventually coupled to a map of some other physical quantity. The spatial resolution can go the sub-atomic range and in special cases (especially in UHV applications) atomic lattice can be imaged. After more than twenty years of development, Scanning Probe Microscopy had evolved into a widely used tool for surface characterization that is used in many fields of science and technology.

Optical thin films are only one of many classes of samples that are being routinely measured using SPM techniques. There are numerous resources about the SPM techniques itself, so in this chapter we want to focus on practical aspects of measurement

P. Klapetek (✉)
Czech Metrology Institute, Okružní 31, 63800 Brno, Czech Republic
e-mail: pklapetek@cmi.cz

O. Stenzel and M. Ohlídal (eds.), *Optical Characterization of Thin Solid Films*, Springer Series in Surface Sciences 64,
https://doi.org/10.1007/978-3-319-75325-6_11

of thin films, based on many years of experiences in SPM measurements for various groups working on the field of optics [1–3]. As for any type of samples, also here are some typical tasks, user wishes and instrument limitations and these will be discussed here. These are almost entirely related to dimensional measurements. Note that this selection represents simple and typical tasks in thin films analysis, not the full spectrum of possibilities that SPM can offer - using various advanced techniques we could e.g. measure film thermal, electrical, mechanical or even optical properties. However these advanced measurements are still far from anything like "routine analysis" and we refer to specific literature about SPM for more details [4, 5].

The most frequent routine measurement in the field of thin films is related to surface roughness. Roughness originates from nearly any technological operation while preparing thin films and its knowledge is important to separate its influence on optical properties during optical characterization (where surface and interface roughness is one of the important uncertainty sources). As all the SPM techniques provide surface topography data (see the example of a rough surface on Fig. 11.1a) it is relatively easy to evaluate the roughness statistical parameters. The only limitation is in scanned area and density of measured points, which limits the range of spatial frequencies accessible by SPM measurements; these problems can be partly overcame by large area measurements and/or advanced sampling techniques as demonstrated.

Apart of statistical properties, there are also dimensional quantities that can be evaluated directly from the data, like thickness of the film where its edge is exposed (e.g. due to some masking procedure or due to the film delamination). This is also frequently requested measurand as it can give an information about the sample thickness that is independent on any optical phenomena, and therefore not affected e.g. by the refractive index profile across the film. An example of data related to such SPM task is shown in Fig. 11.1b.

To present the basic aspects of the above mentioned surface measurements using SPM techniques, we first summarize the technological principles of SPM, focusing

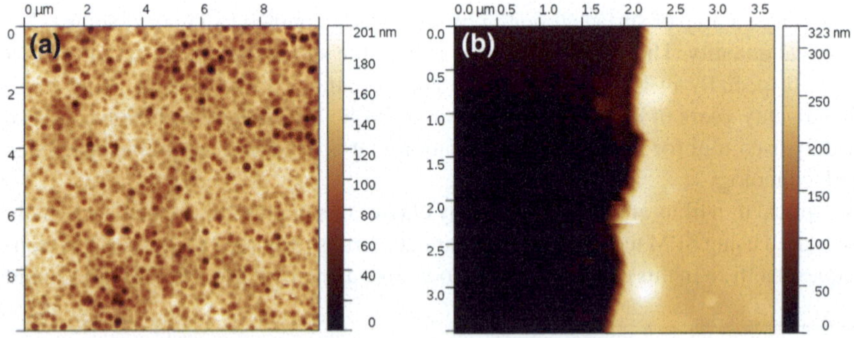

Fig. 11.1 Examples of typical SPM data measured on thin films: **a** rough surface where roughness parameters are evaluated, **b** film edge where film thickness is evaluated mostly

on the metrology of SPM devices and its capabilities for dimensional measurements as the absolute majority of the requests in the field of optical thin films is to give some quantitative numbers about dimensional properties of the sample. Then we review typical data processing related to extraction of the relevant information from the measured data.

11.2 Instrumentation

Scanning probe microscope operation is based on use of a very sharp probe (often called "tip") that is scanned across the sample surface, using some probe-sample interaction to keep the probe-sample distance constant via a feedback loop. The most often used interaction is force, monitored using a optical lever techniques, like that shown in a sketch of a possible SPM design in Fig. 11.2. Based on the interactions that are observed, probe selection and scanning regimes we can distinguish some basic types of SPM measurement regimes:

- Atomic Force Microscopy (AFM) - based on attractive or repulsive forces the topography is measured, often coupled with analysis of mechanical properties if samples are soft enough.
- Scanning Tunneling Microscopy and Spectroscopy (STM, STS) - based on tunneling current the sample topography and electronic properties are measured.

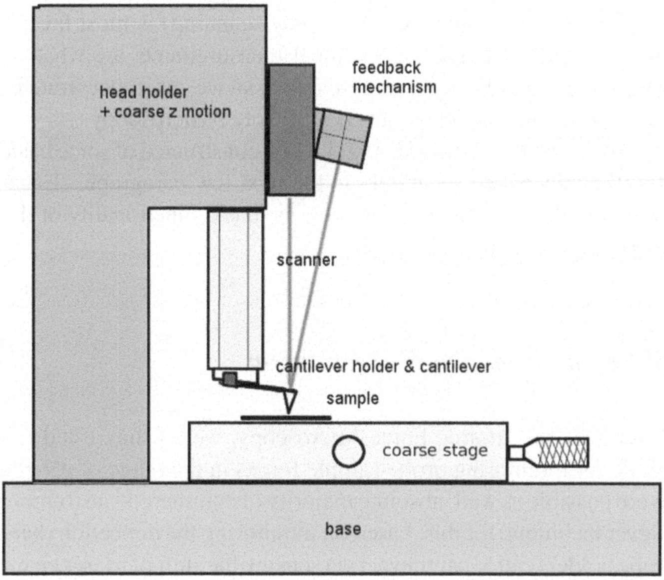

Fig. 11.2 Schema of a typical scanning probe microscope, using laser deflection for the feedback

- Conductive Atomic Force Microscopy (cAFM) - in addition to AFM the probe-sample current is measured, using conductive probe and applied voltage.
- Magnetic Force Microscopy (MFM) - above the topography scan another layer(s) of information about magnetic stray field are determined using a magnetic probe.
- Scanning Thermal Microscopy (SThM) - using a local heater and temperature senor thermal properties or temperature of the sample are measured.
- Scanning Near Field Optical Microscopy (SNOM) - using nanoscale light source, either based on aperture or on a scattering metallic probe the sample is illuminated to overcome the diffraction limit in classical optical microscopy.
- Tip Enhanced Raman Spectroscopy (TERS) - using plasmon resonance and related field amplification at sharp metallic tip apex the local Raman spectra are recorded with nanoscale resolution.

As mentioned in the Introduction, even if the advanced techniques had already matured to some level of reliability, obtaining quantitative results of local electrical, thermal, magnetic or optical properties is still not trivial task and cannot be considered as a standard measurement applicable on all the possible samples, without special attention to different caveats and without deeper understanding of the individual probe-sample interactions and their sensing mechanisms. This is caused partly by complexity of all the interactions observed in SPM techniques, partly by lack of good reference samples and intensive research is done in many academic labs and at instrument manufacturers to improve this state. However, so far, for measurements with metrological traceability (result can be related to a reference through a documented unbroken chain of calibrations), absolute majority of measurements is related to dimensions of some objects on the sample surface. This is also where the microscope principle (acquisition of data by scanning) is most favorable to get the traceability. Luckily enough, dimensional measurements are what is typically requested by people working with thin film optics so we can concentrate here on the very basic technique only, which is the Atomic Force Microscopy.

The AFM, already demonstrated in Fig. 11.2, is constructed of some basic building blocks that will be discussed separately in the next few paragraphs. Even if details can vary from manufacturer to manufacturer, the basic functionality of the building blocks is nearly the same in all the cases.

11.2.1 Probe and Feedback Mechanism

As we discuss here the Atomic Force Microscopy, we are interested in detection system suitable for monitoring probe-sample forces in the range of pN-μN. Even if alternatives are possible as well, absolute majority of commercial instruments is using the optical lever technique for this, based on monitoring the deflection (bending) of a very soft probe holder (called cantilever) via sensing the shift of a laser beam reflected from this holder. The spring constant of the cantilever is in range of 0.01–100 N/m (suited for different scanning regimes) and the cantilever dimensions are in range

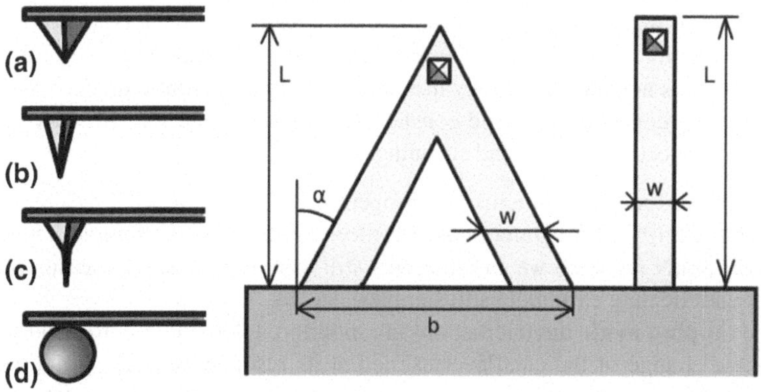

Fig. 11.3 Left: SPM probe geometries: **a** contact mode silicon nitride probe, **b** tapping mode silicon probe, **c** Focused Ion Beam cut supersharp probe, **d** colloidal probe for accurate force measurements. Right: SPM cantilever V shape and I shape geometry

of few hundreds of micrometers. Two typical cantilever geometries are shown in Fig. 11.3. Probe is integrated into the cantilever (being manufactured together with it), forming a pyramid or cone protruding from the cantilever, with length of few micrometers and apex radius of around 10 nm.

A low power laser diode is used to form the beam for cantilever deflection monitoring; the beam is reflected from area near to the cantilever apex and hits the position sensitive detector located relatively far from the cantilever, which assures the deflection signal magnification. The position sensitive detector can be formed e.g. by a quadrant photodiode, which provides four signals from four individual sensors. Then the signal used for feedback loop is calculated via hardware or software tools as follows:

$$I_{TB} = \frac{(I_{TL} + I_{TR} - I_{BL} - I_{BR})}{(I_{TL} + I_{TR} + I_{BL} + I_{BR})} \tag{11.1}$$

where I_{TL}, I_{TR}, I_{BL}, I_{BR}, are top-left, top-right, bottom-left and bottom-right photodiode currents.

If this signal is kept constant during the scan, using a proportional-integral-derivative (PID) feedback loop, we obtain a constant force image, so the position data from the scanning system can be directly treated as the measured sample topography. If the signal is only monitored and sample is not moving in the z direction (the "height" direction), we get constant height image which can be converted into topography using the cantilever deflection sensitivity information, however this second approach is not very frequently used in commercial instruments.

Even if we concentrate on the constant force mode only, we still have several possible variants of detection and scanning regimes:

- In the **contact mode** the microscope is operating in the repulsive forces regime, so in fact it is acting like a gramophone needle, sliding across the sample. The forces are still in nN range so we can treat this still as non-destructive measurement if we are not dealing extremely soft samples.
- In the **tapping mode** the interactions are monitored dynamically from amplitude or phase change of the cantilever excited at its resonant frequency, microscope operates at the border of repulsive forces and attractive forces (e.g. van der Waals) and measurement is really non-contact unless a feedback loop fault happens.
- Some of the **advanced regimes** use combination of both - e.g. individual measurements and evaluation of force-distance curves for both keeping the force constant, but also detecting the sample mechanical properties and optimizing the contact force and feedback loop parameters. This is also direction which most of the manufacturers follow at present in their novel instruments.

11.2.2 Scanners

To realize the probe-sample motion, some kind of positioning mechanism has to be used. This is called "scanner" and can be used to move either the probe or the sample, depending on the microscope construction. The absolute majority of instruments is based on use of piezoelectric elements as scanners, using the very small displacements that can be generated when a voltage is applied on some material like lead zirconate titanate (PZT). Scanners are manufactured as stacked elements (forming a rod), moving in one direction only or as tubes, bending in all the directions. If only voltage applied on the scanner would be used as a position information we would be facing many principal problems of the piezoceramic scanners (creep, hysteresis, aging, etc.), so nowadays nearly all the instruments are equipped by a feedback loop which connects the scanner voltage to some independently measured displacement signal. From metrology point of view, this would be ideally an interferometer, however most frequently a strain gauge is used which, together with driving circuit, is a simple strain-voltage transducer. Any sensor is used and scanner is "closed loop", it leads to dramatic increase of accuracy over the older "open loop" systems with no position feedback. A schematic drawing of different stages with piezoceramic actuators is shown in Fig. 11.4a.

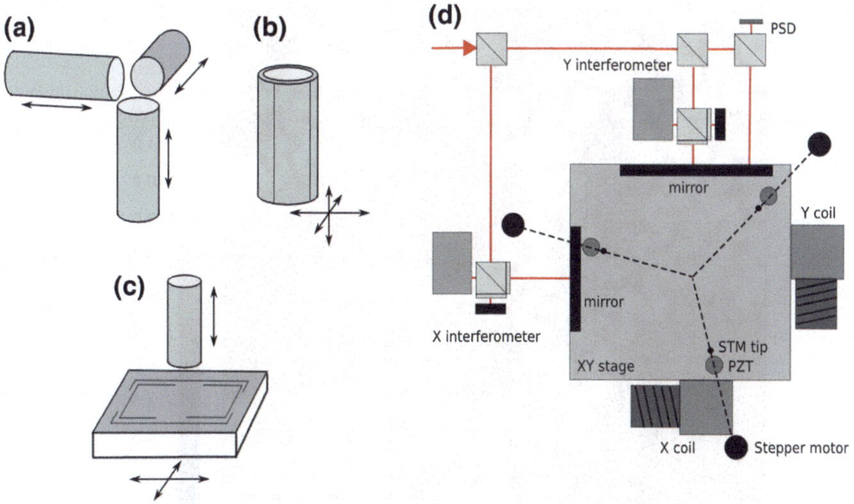

Fig. 11.4 Scanner schematics: **a–c** piezoceramic actuators in different configurations for XYZ motion and **d** a voice coils based XY motion system

As we are dealing optical thin films, we are often facing a need of performing measurements on a larger area, that would be more comparable to areas covered by optical instruments. Large area Scanning Probe Microscopy is relatively small field of research, but has very broad potential in metrology of thin films and optical surfaces. An increase of the range of SPMs above about $100 \times 100\,\mu\text{m}^2$ is still complicated and very rare as we need to solve various limitations of the SPM architecture. Piezoceramic scanners have only limited displacement per unit length and cannot be scaled infinitely without introducing significant systematic errors. Even if we use some lever technique all the problems are not solved. To reach a spatial resolution in order of nanometers for scans in a centimeter range using a voltage–displacement transducer need a 24 bit digital to analog converter and appropriate signal to noise ratio in the other parts of the electronics leading from the digital part to the transducer (scanner) e.g. in the high voltage amplifiers for piezoelectric actuators. Not only scanners, but also sensors are problematic in larger scanning ranges as the most frequent sensors - strain gauges - are limited to smaller displacements. A possible solution is using a combination of voice coils (as actuators) with interferometers (as sensors). Such system, schematically shown in Fig. 11.4d, is capable of performing measurements on areas up to square centimeters. This actuation principle is used only in very special instruments so far, however from the point of thin films metrology it offers many novel possibilities. As an example, in Fig. 11.5 we show a measurement on a delaminated thin films sample performed using a large area SPM [6].

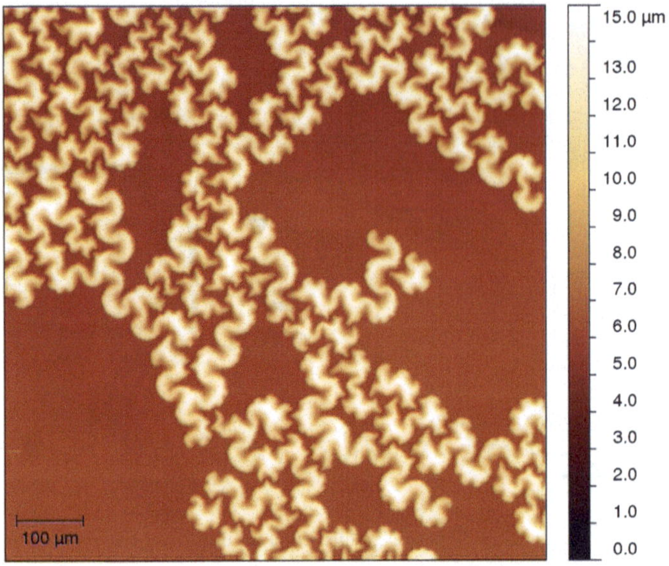

Fig. 11.5 Delaminated sample large area measurements performed using voice-coil based scanner

11.3 Metrological Traceability

Many of the SPM measurements in thin film analysis are done for reference purposes - to determine the film thickness, thickness variations or surface roughness and having the system properly calibrated is therefore important. From metrology point of view this means to have the system traceable - all the components characterized by some other etalons that form a chain of measurements up to primary definitions of the respective physical units. The uncertainties of the calibration procedures, together with all the other uncertainties related to the particular measurement can be then used to establish an uncertainty budget for each data provided by the instrument (a nice example of this procedure can be seen in [7]). This is however very rare; most of the instruments used in academic institutions are not properly calibrated and no special effort is given to uncertainty analysis. As a result, the data obtained from SPM measurements are by different users either taken as infinitely correct or as absolute unreliable, which both is clearly wrong. It is therefore important to follow at least basic steps of various SPM parts calibration and to understand the basic uncertainty sources and basic systematic errors.

11.3.1 Scanning System Calibration

The key step for bringing some metrology to the SPM measurements is to calibrate the scanning system and to do it periodically. Regarding traceability, there are two classes of instruments:

- Metrology SPMs are built typically by metrology institutes and are serving as an etalon for providing traceability for other microscopes. They are equipped by stabilized lasers that provide the traceability to length units and they are designed to operate with maximum possible accuracy, often at costs of low speed or limited range of possible measurement regimes. Traceability of these instruments is based on calibration of the laser and estimation of all the uncertainty sources.
- Commercial SPMs are designed and built to provide the best possible user experience, including large number of scanning regimes, speed, automated cantilever adjustment, etc., however they are not equipped by sensors that could be independently calibrated. Traceability needs to be provided via calibration samples, that user typically needs to purchase separately, namely if they should be really traceable. As calibration samples, gratings are used for lateral distances, for the z axis, step height standards that can be evaluated according to ISO 5436-1 are available.

For getting the right set of calibration samples or for getting the samples itself traceable, SPM probes manufacturers offer various solutions, or you can ask your National Metrology Institute.

11.3.2 Cantilever Stiffness Calibration

If we measure dimensional quantities, at first sight we don't need to have our cantilever stiffness calibrated, at least if we assume that we are measuring in the constant force regime and cantilever serves only as a null sensor. Cantilever calibration is however necessary when we want to know at which force the measurement is performed or we want to set this force to some value, or when we are performing measurements using some advanced automated measurement regime (e.g. ScanAsyst by Bruker), or when we want to estimate some other properties in parallel to dimensional measurements.

First of all, one can not believe the stiffness values provided by SPM tips manufacturer, as these values are only very coarse and evaluated for whole batches of cantilevers at best; more usually these are only some desired values that can differ from the real stiffness by tens of percents or even order of magnitude. There are several ways how to perform the cantilever stiffness calibration:

- Based on dimensional measurements of the cantilever shape we can get only a coarse estimate — probably similar to what the manufacturer provided as a nominal value. Assuming that the thickness of cantilever is t_c and the other geometric constants are defined in Fig. 11.3, we can then use the following equations [8], to determine the cantilever stiffness for a cantilever of 'I' shape

$$k_c = \frac{E w_c t_c^3}{4 L_c^3} \tag{11.2}$$

and similarly

$$k_c = \frac{E w_c t_c^3}{2 L_c^3} \tag{11.3}$$

for cantilever of a 'V' shape (under assumption that it is treated as two parallel beams as a simplest approximation). As the bulk mechanical properties that we would use for evaluation of the stiffness from cantilever dimensions differ from the real values of thin films forming the cantilever, we can't expect higher accuracy than tens of percents [9, 10].

- If we already have a reference cantilever with known stiffness, we can use it as a tool to characterize an unknown cantilever. If we are able to mount the reference cantilever in a way that we can press the unknown cantilever towards reference cantilever (or vice versa), we can estimate the stiffness of our cantilever from deflection of the reference one and its stiffness; this can lead to a uncertainty between 10 and 30 % [9, 11].
- We can measure the free resonant frequency of the cantilever, and its dimensions. This method was reported to give results with an accuracy below 7% when the dimensions are measured in SEM, which was tested by comparison of method performance on multiple different instruments [12]. Under standard conditions it was assumed that the typical uncertainty of this method is in order of 15–20% [9, 10].
- In many of the software packages the power spectrum method is used, based on measurement of the thermal fluctuations of the cantilever, treated as a harmonic oscillator. For this method a 5–25% uncertainty was reported in the literature [10, 13], depending on cantilever type and measurement conditions.
- Using special device (e.g. a nanoindenter) or a special samples constructed for these purposes, like MEMS we can obtain the best possible results like 1 % [14–16], however this is quite complex and costly approach.

Some National Metrology Institutes provide cantilever stiffness calibration as a service as well, mostly via the special devices route, however it is not so frequent comparing to calibration of the dimensional reference standards.

11.3.3 Apex Radius Calibration

The most common problem related to the SPM probe and related to all the SPM techniques, is an artifact known as tip-sample convolution. This effect is typically

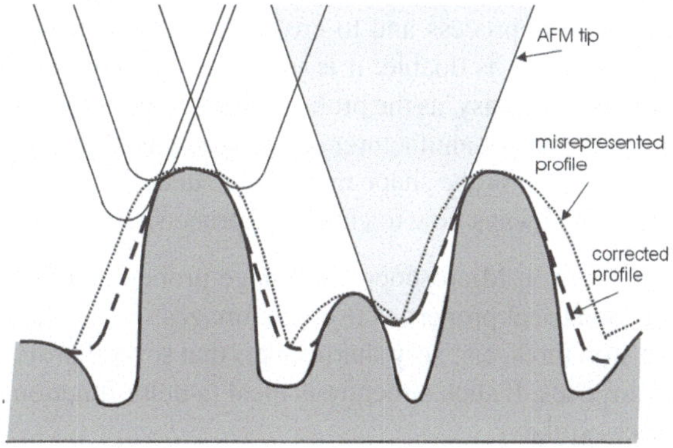

Fig. 11.6 Tip convolution and surface reconstruction schematics: misrepresented profile corresponds to the dilation morphological operation and corrected profile to the surface reconstruction algorithm

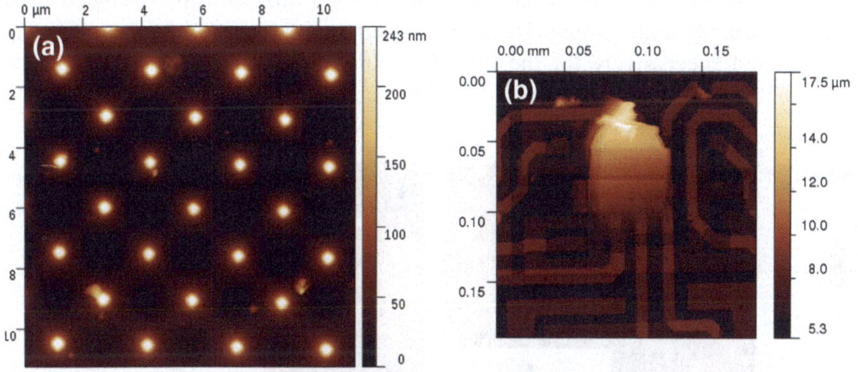

Fig. 11.7 Probe imaged on a microchip surface: **a** intentionally, multiple tips imaged, **b** unintentionally, showing even the cantilever end

seen in data when we use a bad (blunt) probe and its characteristic sign are repeated patterns seen on the image. It can be also observed when we measure a very sharp spike which in fact images the tip, as shown in Fig. 11.7, either obtained intentionally on sample designed for probe imaging or unintentionally on a complicated sample. An illustration of the process how the tip-sample convolution affects the SPM results is shown in Fig. 11.6. We can see that at some points the surface structures are only distorted while at some other points the structures are completely hidden from the tip. There are morphological operations that can describe these processes mathematically [17] — dilation, erosion and surface reconstruction, and under different names these are available in many software packages for SPM data processing.

To completely recover the true surface in the presence of tip-sample convolution artifacts is hard as part of the information might be missing. This is namely true for measurements on samples with steep slopes or pores [18]. As there are many algorithms to simulate the process and to create surface reconstruction at least in parts of the sample where it is doable, it is important to know the probe shape as good as possible. This is not easy, as the probe shape is in principle unknown already when it is taken from the box (manufacturer only claims what is the maximum radius) and during the scanning the probe shape may evolve undefinitely.

There are few possible ways how to check the probe shape:

- Using Scanning Electron Microscope and image probe from different sides and evaluate main geometrical properties from the images.
- Imaging known structures, e.g. very sharp spikes that several probe manufacturers offer for these purposes. If such structure is ideal (a delta-function), the tip would be imaged at every spike.
- Using Blind tip estimation algorithm which searches for local slopes in data measured in a rough sample, assuming that statistically there is enough of sloppy parts on the sample.

An example of using the Blind tip estimation on different tips is displayed in Fig. 11.8, where small gallium spots on the surface served as nice objects for tip imaging and successively for the tip estimation algorithm. Presence of repeated

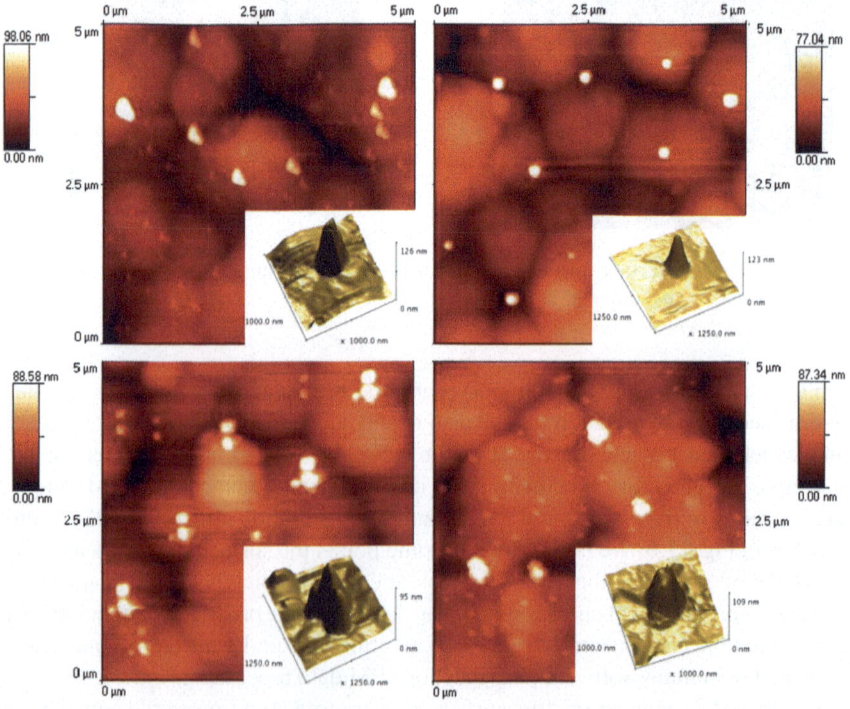

Fig. 11.8 Results of blind tip estimation routines together with the source data

patterns on the surface (as seen in most of the surfaces in Fig. 11.8 is usually a sign of tip convolution artifacts.

When we know the tip shape we can also try to recover parts of the sample affected by the tip-sample convolution. An algorithm called "surface reconstruction" can be used for these purposes. This can be performed only at positions where information was not lost, e.g. on positions where any part of the tip touched the surface and there were no multiple touches. Finally, a "certainty map" algorithm was designed to detect which parts can be reconstructed - where the tip touched the sample in a single point and where multiple touches happened and information is therefore lost.

11.4 Data Processing

11.4.1 Basic Tasks

The very basic operation that user needs to perform on the data after measurement is the mean plane or some polynomial background subtraction. This is often followed by some other data manipulation steps, like removal of defects in measurements (spikes, feedback loop faults, noisy lines). Except the mean plane subtraction case this is already a heavy data manipulation and if we want to determine our results with a known uncertainty, we should include some idea on how the data manipulation might affect the uncertainty, which can depend on how the user is experienced and similar factors that can be hardly estimated [19].

Direct dimensional quantities (width, height, depth, area, etc.) are typically evaluated from profiles that are extracted from the height field in some data processing software, like Gwyddion [20]. This includes typically measurements of height and lateral dimensions of various structures on a flat substrate, like semiconductor masks. It is important, whenever possible, to extract profiles that are then used for evaluation in the fast scanning direction as these are the least affected by drifts and similar error sources. Rotation of the sample to reach this state is always a benefit. When extracting profiles and evaluating results from the we need to have in mind the real pixel size (ratio between physical size and number of pixels) as obtained during the measurement. The pixel size affects the final uncertainty of the results and no interpolation can substitute the information that would be missing if the data would be measured too coarse. If a drift is not a big issue, having as large number of pixels as possible is always good if we do measurements for metrology purposes.

Besides direct measurements of some height or lateral values from profiles we can use also some other tools for dimensional measurements in an SPM data processing software. This includes e.g. a measurements of continuous areas, grains or particles, sometimes followed by a detection of the requested features and some statistical analysis. We refer to the user manual of the reader's favorite SPM data processing software, as the implementation of these tools can be different and there are no "standard" approaches that we could list here.

11.4.2 Roughness Analysis

Roughness is probably the most frequent measurand in thin films analysis via SPM techniques, as this is directly related to manufacturing processes performance and as it directly affects the optical quality of the thin film.

Atomic force microscopy data are usually represented as a two-dimensional array of datapoints of size $N \times M$, where N and M represent the number of rows and columns within the data field. A special case of data that are not-equidistant is discussed later in this chapter, but it is still very rare, so typically the data are equally spaced. The real size of this array will be here denoted as $L_x \times L_y$ where L_x and L_y are the sizes of axis x and y. The sampling interval (distance between two adjacent points within the scan) is denoted Δ. As mentioned above, in this section we will assume that the sampling interval is the same in both the x and y direction. As we discuss random roughness here, we assume that the surface height in a given point x,y can be described by a random function $\zeta(x, y)$ that has given statistical properties.

11.4.2.1 Probability Distribution of Heights and Angles

The most frequently used parameter for the description of statistical properties is probably a "roughness". This typically the means root-mean-square (RMS) value of height of surface irregularities, and is denoted as R_q or σ_{rms} (in novel standardization documents it is called S_q to distinguish it from single profile based roughness value).

If we start from the surface model given above, the roughness can be defined as:

$$\sigma_{rms}^2 = \int_{-\infty}^{\infty} z^2 w(z) \mathrm{d}z = \lim_{S\to\infty} \frac{1}{S} \int\int_S \zeta^2(x, y) \mathrm{d}x\mathrm{d}y, \tag{11.4}$$

where $w(z)$ is the one-dimensional distribution of the probability density of the random function $\zeta(x, y)$, z denotes the values of the heights of the irregularities of the AFM data and $S = L_x \times L_y$ represents the surface area of the AFM scan in the (x, y)-plane.

Similarly we can define the root-mean-square value of the slopes of the irregularities $\tan^2 \alpha_0$ can be written:

$$\tan^2 \alpha_0 = \int_{-\infty}^{\infty} z'^2 w(z') \mathrm{d}z' = \lim_{S\to\infty} \frac{1}{S} \int\int_S \zeta'^2(x, y) \mathrm{d}x\mathrm{d}y, \tag{11.5}$$

where

$$\zeta'(x, y) = \frac{\partial\zeta(x, y)}{\partial x} \text{ or } \frac{\partial\zeta(x, y)}{\partial y} \tag{11.6}$$

$w(z')$ represents the one-dimensional distribution of the probability density of the random function $\zeta'(x, y)$ and z' represents the values of function $\zeta'(x, y)$.

The above mentioned functions $w(z)$ and $w(z')$, are in practice calculated using the following formulas:

$$w(z) = \frac{\mathcal{N}(z, \delta z)}{NM\delta z} \tag{11.7}$$

and

$$w(z') = \frac{\mathcal{N}(z', \delta z')}{(N-1)M\delta z'}, \tag{11.8}$$

where the function $\mathcal{N}(z, \delta z)$ resp. $\mathcal{N}'(z, \delta z)$ gives the number of the values of z_{ij} resp. z'_{ij} that are within the interval $\langle z - \delta z/2, z + \delta z/2\rangle$ resp. $\langle z' - \delta z'/2, z' + \delta z'/2\rangle$.

The probability density of heights $w(z)$ defined here is also a very frequently used statistical result, used either directly (having the same shape as the "histogram"), sometimes it is also used in its cumulative form.

11.4.2.2 Autocorrelation Function

The height and slope quantities discussed above belong to the first-order statistical quantities, describing only the statistical properties of the individual points, and representing only statistical nature of the heights as they would be not connected somehow within the surface. However, for a complete description of the surface statistical properties it is necessary to use higher order functions that would include some information about lateral roughness properties. In practice this is namely the autocorrelation function (ACF), the power spectral density function (PSDF) and the height–height correlation function (HHCF).

The autocorrelation function is given by

$$G(\tau_x, \tau_y) = \int\int_{-\infty}^{\infty} z_1\, z_2\, w(z_1, z_2, \tau_x, \tau_y) \mathrm{d}z_1 \mathrm{d}z_2 =$$

$$\lim_{S\to\infty} \frac{1}{S} \int\int_S \xi(x_1, y_1)\xi(x_1 + \tau_x, y_1 + \tau_y)\mathrm{d}x_1\mathrm{d}y_1, \tag{11.9}$$

where z_1 and z_2 are the height values at points $[x_1, y_1]$, $[x_2, y_2]$, $\tau_x = x_1 - x_2$ and $\tau_y = y_1 - y_2$. The function $w(z_1, z_2, \tau_x, \tau_y)$ denotes the two-dimensional probability density of the random function $\xi(x, y)$ corresponding to points $[x_1, y_1]$, $[x_2, y_2]$ and the distance between these points $\tau = \sqrt{(\tau_x^2 + \tau_y^2)}$.

Within AFM measurements we usually evaluate the one-dimensional autocorrelation function determined only from profiles in the fast scanning axis (which is less affected by thermal and mechanical drifts) which can be evaluated from the discrete AFM data values as

$$G_x(\tau_x) = \frac{1}{N(M-m)} \sum_{l=1}^{N} \sum_{n=1}^{M-m} z_{n+m,l} z_{n,l} \tag{11.10}$$

where $m = \tau_x/\Delta$. Function is usually evaluated in a discrete set of values of τ separated by the sampling interval Δ, based on the nature of the measured data.

In order to get only some parameters describing roughness, not whole function we need to parametrize the functions somehow so we can fit them with some parametric model. For example, the one–dimensional autocorrelation function is often assumed to be Gaussian,

$$G_x(\tau_x) = \sigma_{rms}^2 \exp(-\tau_x^2/T^2); \tag{11.11}$$

where σ_{rms} and T are the root mean square deviation of the heights and the autocorrelation length, respectively.

Another frequently used model is the exponential form of autocorrelation function

$$G_x(\tau_x) = \sigma_{rms}^2 \exp(-\tau_x/T); \tag{11.12}$$

Note that in optical measurements (e.g. spectroscopic reflectometry, ellipsometry) the Gaussian autocorrelation function is usually expected to be in good agreement with the surface properties. However, some articles related to surface growth [21] and oxidation [22] usually assume that the exponential form is closer to the reality.

For illustration, in Fig. 11.9, the autocorrelation function computed for a simulated Gaussian model sample is plotted. The dataset used to evaluate it was generated using Gwyddion software Spectral synthesis module with the parameters $\sigma_{rms} = 20$ nm, $T = 100$ nm. The function is fitted by the Gaussian function given by (11.11). The results are $\sigma_{rms} = (20.17 \pm 0.08)$ nm, T = (102 ± 1) nm.

Note that the uncertainty of the function values is not the same for all the parameters τ_x. For larger values of τ_x the number of averaged components in (11.10) reduces dramatically. Thus we have to limit the function computation only to small values of τ_x.

In practice, the autocorrelation function is not very widely used (for its use see e.g. [23]); instead of it the power-spectral density function is mostly evaluated from the AFM data.

11.4.2.3 Height–Height Correlation Function

The two dimensional height–height correlation function can be written as

$$\begin{aligned} H(\tau_x, \tau_y) &= \int\int_{-\infty}^{\infty} (z_1 - z_2)^2 w(z_1, z_2, \tau_x, \tau_y) \mathrm{d}z_1 \mathrm{d}z_2 = \\ &= \lim_{S\to\infty} \frac{1}{S} \int\int_S (\xi(x_1, y_1) - \xi(x_1 + \tau_x, y_1 + \tau_y))^2 \mathrm{d}x_1 \mathrm{d}y_1, \end{aligned} \tag{11.13}$$

where z_1 and z_2 are height values at points $[x_1, y_1]$, $[x_2, y_2]$, $\tau_x = x_1 - x_2$ and $\tau_y = y_1 - y_2$. Function $w(z_1, z_2, \tau_x, \tau_y)$ denotes the two-dimensional probability density of random function $\xi(x, y)$ corresponding to points $[x_1, y_1]$, $[x_2, y_2]$ and the distance between these points $\tau = \sqrt{(\tau_x^2 + \tau_y^2)}$.

Similarly to the ACF, in practice the one-dimensional variant is used that can be evaluated from the discrete AFM data as

$$H_x(\tau_x) = \frac{1}{N(M-m)} \sum_{l=1}^{N} \sum_{n=1}^{M-m} (z_{n+m,l} - z_{n,l})^2 \tag{11.14}$$

where $m = \tau_x/\Delta$.

The height–height correlation function corresponding to a randomly rough surface with a Gaussian autocorrelation function is given by

$$H(\tau_x) = 2\sigma_{rms}^2 \left[1 - \exp\left(-\frac{\tau_x^2}{T^2}\right)\right], \tag{11.15}$$

where σ_{rms} and T denote the root mean square deviation of the heights and the autocorrelation length respectively. Similarly, height–height correlation function corresponding to a surface with an exponential autocorrelation function is

$$H(\tau_x) = 2\sigma_{rms}^2 \left[1 - \exp\left(-\frac{\tau_x}{T}\right)\right], \tag{11.16}$$

For the same sample as discussed in the previous paragraph, the corresponding T and σ_{rms} evaluated according to (11.15) are $\sigma_{rms} = (20.28 \pm 0.03)$ nm, T = (103.3 ± 0.82) nm, so we can see that this is an alternative for using autocorrelation function.

The height–height correlation function is used in practice for the correlation length evaluation and fractal dimension estimation, similarly as the power spectral density function. For details, see e.g. [23, 24].

11.4.2.4 Power Spectral Density Function

The two-dimensional power spectral density function can be written in terms of the Fourier transform of the autocorrelation function

$$W_j(K_x, K_y) = \frac{1}{4\pi} \int_{-\infty}^{\infty} G(\tau_x, \tau_y) \mathrm{e}^{-\mathrm{i}(K_x\tau_x + K_y\tau_y)} \mathrm{d}\tau_x \mathrm{d}\tau_y. \tag{11.17}$$

Similarly to the autocorrelation function, also here we usually evaluate the one-dimensional power spectral density function which is given by equation

$$W_1(K_x) = \int_{-\infty}^{\infty} W(K_x, K_y)\mathrm{d}K_y. \tag{11.18}$$

This function can be evaluated using the Fast Fourier transform as follows:

$$W_1(K_x) = \frac{2\pi}{NMh} \sum_{j=0}^{M-1} |\hat{P}_j(K_x)|^2, \tag{11.19}$$

where $\hat{P}_j(K_x)$ is the Fourier coefficient of the jth row, i.e.

$$\hat{P}_j(K_x) = \frac{h}{2\pi} \sum_{k=0}^{N-1} z_{kj} \exp(-\mathrm{i}K_x kh). \tag{11.20}$$

If we have chosen that a Gaussian autocorrelation function model is the right representation of our sample statistical properties, the corresponding PSDF is

$$W_1(K_x) = \frac{\sigma_{rms}^2 T}{2\sqrt{\pi}} \exp(-K_x^2 T^2/4) \tag{11.21}$$

For a surface with an exponential ACF we have

$$W_1(K_x) = \frac{\sigma_{rms}^2 T}{\pi} \frac{1}{1 + K_x{}^2 T^2}. \tag{11.22}$$

These functions can be therefore used for fitting the experimental data. To illustrate this, in Fig. 11.9, the power spectral density function for a Gaussian model sample (the same as in the previous section) is plotted. The function is fitted by the Gaussian PSDF given by (11.21). The results are $\sigma_{rms} = (20.1 \pm 0.2)$ nm, T $= (100 \pm 2)$ nm.

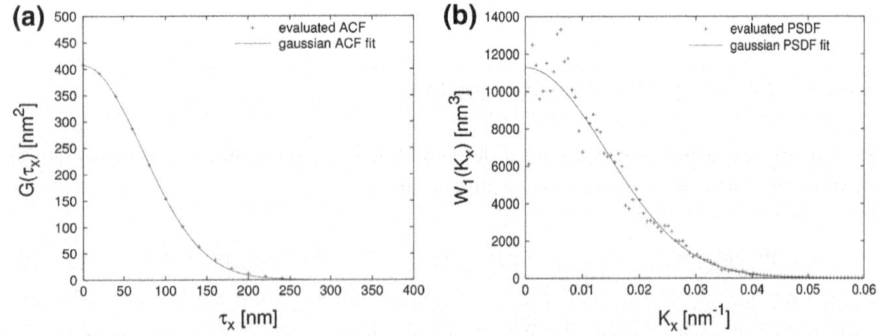

Fig. 11.9 Autocorrelation function (**a**) and Power Spectral Density Function (**b**) obtained on a rough sample and fitted by the analytical dependency based on assumption of Gaussian ACF of the roughness

This is a third way how to obtain the same statistical results. In practice, the PSDF is widely used for the correlation length determination and fractal analysis, see e.g. [25]. As seen from these three sections, any of the surface statistical properties functions is used we still need to fit it via some analytical model if we want to obtain some simple parameters that could be fed to some optical model. In thin film optics the Gaussian model is most frequently used, even if sometimes the thin film growth theories might favor other models. A choice of the fitting function should be considered as one of the uncertainty sources and a least some guess of its magnitude could be obtained by fitting the experimental data by few different models.

11.4.3 Step Height Analysis

Sometimes also step heights are evaluated from data measured on thin films. This is namely to estimate the film thickness. Even if very powerful optical methods exist for this, most of them are measuring optical thickness, which might be different from the "real" thickness as it includes the refractive index as well, which might be not correctly known. To prepare sample for thickness measurements via AFM is not easy as we need to have a steep edge at the surface. This can be obtained by masking some part of the sample during the deposition, however the deposition conditions can still prevent formation of a sharp edge.

Ideally, one would need to have two steps, forming a so called "step height sample" which can be evaluated using normative documents (ISO 5436-1) as shown in Fig. 11.10. On some samples this can be done by scratching the film, assuming that the substrate is hard enough to resist and that the film can be partly debonded easily. More often we are able to have only one step, so a similar approach (but not covered by any standard) can be used to fit only a single step, again skipping the part of the profile close to the step itself.

If the data are properly leveled we can also use histogram or similar height statistics to evaluate the step height from all the data, not only from profiles.

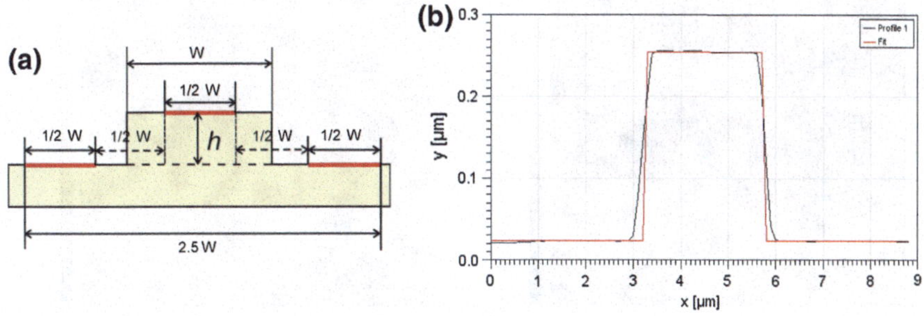

Fig. 11.10 Step height analysis schematics (**a**), and real data example (**b**)

11.5 Tip Sample Convolution Effect on Statistical Properties

As discussed earlier, tip-sample convolution is one of the biggest error sources in SPM measurement. If we characterize optical thin films, we very often want to see very small roughness as the films are of a very high quality, designed to scatter the light as low as possible. The issues of trying to measure features that are smaller than the probe radius are therefore a daily bread in optical thin films characterization. As an example, we show a measurement of the same area on a rough silicon sample (prepared by anodic oxidation and oxide dissolution [26]). We have used a fresh new probe designed for AFM measurements on sensitive samples (ScanasystAir by Bruker) and a relatively large probe that is used for Scanning Thermal Microscopy measurements (VITA-DM-GLA by Bruker). In Fig. 11.11 a comparison of the same areas is shown.

It can be expected that the tip convolution effects will not influence only dimensional measurements, but also the results of the statistical data treatment, values like roughness or autocorrelation length. While for direct measurements the effect of tip convolution on the data can be intuitively estimated (e.g. we see multiple tips imaged on the surface), for statistical algorithms we have no simple rule that could help us to see if the data were affected or not. The best solution is to simulate the effects for each concrete surface type that we measure. This work was performed earlier for columnar thin films [18] and for fractal-like rough surfaces [27] and it was found that the tip convolution leads to large suppression of higher spatial frequencies, which leads to decrease of the roughness and increase of the autocorrelation length. As an example, from the data presented in Fig. 11.11 we can evaluate the roughness value $\sigma_{rms} = 6.96$ nm for the sharp AFM probe and 5.68 nm for the bigger SThM probe. The autocorrelation length obtained under assumption of Gaussian ACF equals to 67 nm for the sharp AFM probe and 65 nm for the bigger SThM probe.

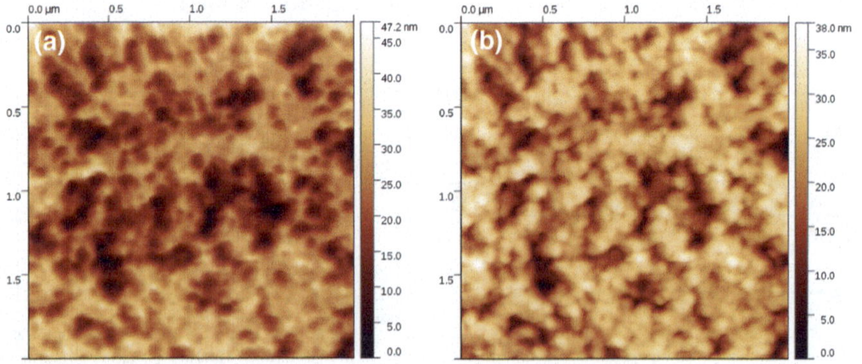

Fig. 11.11 Measurements of the same surface area by different probes: **a** sharp AFM probe, **b** larger radius thermal microscopy probe

11.6 Perspectives

Scanning probe microscopes are already quite well adapted for measurements of optical thin films and apart of lower noise, faster scanning and improved "easy-to-use" regimes there were not many substantial improvements in commercial devices in recent years if we discuss the dimensional measurements. The basic limit for standard SPMs is still the same — measurement time and measurement area. When roughness is analyzed we have always troubles with both, as we need to cover wide area of spatial frequencies and often we are not sure if increasing or decreasing the scanning area would not be a good step towards better result. Here we want to discuss some alternative measurement strategies, that are not yet commercially available, but could solve some of the problems in thin film roughness and defects analysis.

During scanning the data are usually collected in form of a regularly sampled matrix, which is a very simple approach from the point of their visualization and further data processing. If we are performing measurements on large areas, performing regular scans is not always an ideal approach. For example, if we want to determine the power spectral density function (see the Sect. 11.4 in wide range of spatial frequencies we need both dense sampling (to cover the high frequencies) and large area (to cover the low frequencies). These requirements leads to very long measurement times and significantly increased risk of loss of the probe. The problem of a suitable scanning algorithm appears also in the large area measurements data storage. If we need that our system moves over a millimeter or centimeter range it must be much faster than the present systems (typically with maximum speeds in hundreds of μm/s. Moreover, measuring over a millimeter or centimeter range with nanometer resolution would give us 10^{14} data points, which is not feasible both from the point of data acquisition and data processing. A possible solution is to use some non-equidistant (adaptive) sampling technique, measuring only the data points that are really needed in the data analysis phase. If we leave the concept of equally spaced points, we have plenty of possibilities how to arrange the scan path. As an example, we can perform measurement optimized on obtaining the maximum statistical information for spatial frequencies analysis in minimum time, or to reduce some typical errors in SPM measurements, like the anisotropy of the data caused by having fast and slow scan axis.

In our recent work [28] we have created a library for handling non-equidistant scan paths and here we present four different scan paths that are related to statistical analysis of roughness and therefore have high potential in optical thin films analysis. In Fig. 11.12 four different paths are shown:

- The space filling scan path, which is based on a finite-order approximation of the Hilbert space-filling curve [29–31] (Fig. 11.12a) which is another approach how to perform isotropic scanning with respect to the two Cartesian axes, suitable e.g. for roughness analysis as described above. It should be emphasized that in contrast to classical raster scan there is not preferred axis (fast scan axis). The path consists of short elements in x and y direction, changing the direction very frequently and going through the sample in a fractal-like pattern.

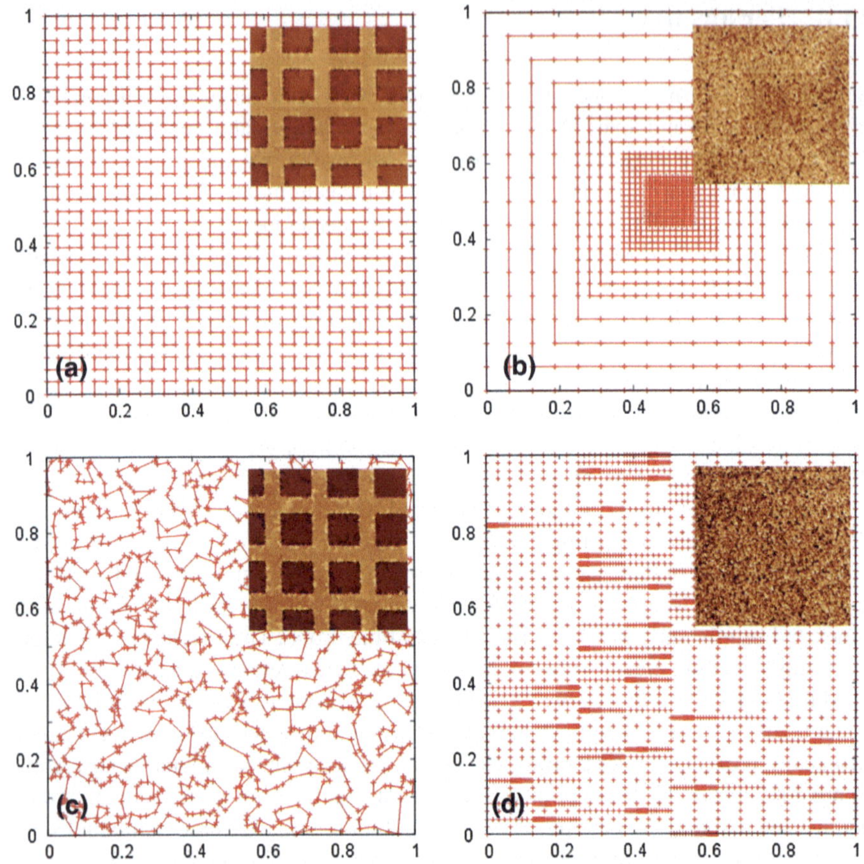

Fig. 11.12 Scan paths for rough samples measurements: **a** space filling scan path, **b** random scan path, **c** 2D octave scan path, **d** 1D octave scan path. Adapted from [28]

- The random scan path (Fig. 11.12b) is even more isotropic because the scanning directions are completely arbitrary, so even x and y direction is not preferred. This path exhibits no correlation between position and scanning direction and, but also uniformly covers the area. The random positions needed to generate this path are generated with uniform distribution in the requested region and then the traveling salesman problem [32] is approximately (partly) solved to construct the scan path that is effective for the tip motion.
- The two-dimensional octave scan path (Fig. 11.12c) is a variant of a scan path specially developed for the roughness measurements purposes, spanning the maximum range of spatial frequencies. It consists of a series of nested regular grids, each twice coarser than the smaller one inside, scanned in a spiral fashion. Therefore it is again isotropic, albeit with correlation between position and scanning direction. The motivation to create such path was to obtain as much of the roughness

statistical information (both height and lateral) as possible with a given number of measured points. This allows to have both dense information for high spatial frequencies and large area for low spatial frequencies in the resulting power spectral density function.

- One-dimensional octave scan path (Fig. 11.12d) is a simpler variant of the 2D octave scan path presented above, focusing only on the analysis of profiles along the fast scanning axis, from the point of typical microscope operation it is therefore more conventional way of measurement. This is very often performed in the field of roughness analysis, especially if the texture can be assumed isotropic. The path consists of straight lines along the fast axis, each formed by nested regular one-dimensional grids. Again, this scan path allows covering a wide range of spatial frequencies in a single measurement. However data are measured along one direction and we need to assume that the roughness is isotropic.

To show the potential improvements while using one of the scan paths, in Fig. 11.13 an example of one-dimensional octave scan path use for rough sample measurement is shown. A rough surface manufactured by Simetrics company was used for this analysis.

After the measurement based on the 1D octave scan path, the data were split to individual line profiles with different point spacing and a power spectral density function (PSDF) was calculated. We can see that using this approach we can span a large scale of spatial frequencies. The main benefit of the approach compared to e.g. individual scans with different resolutions, one after the other, is that all the data are collected at once, all the data post-processing, like drift correction or leveling is done at once (so there is no artificial tilt added for the high resolution and small range images). This approach also provides better statistical coverage of the surface as the detailed profiles are taken on different places of the scan area. It can be seen that as the information about high spatial frequencies comes from segments with smaller sampling steps in the octave scan, it not only covers a wider range of frequencies

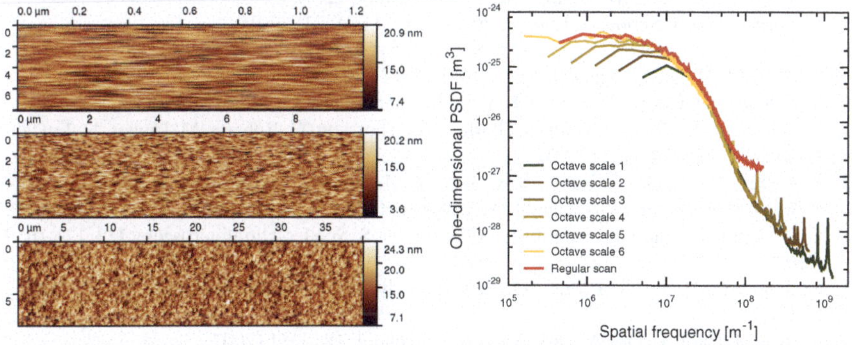

Fig. 11.13 Rough sample measurement using 1D octave scan: (left) extracted individual profiles at three selected scale levels, (right) PSDFs obtained from different levels compared with PSDF calculated from a regular scan with the same number of points. Reprinted from [28]

but is also considerably less affected by noise than the regular scan. To get the same spatial frequency coverage, the regular scan line would need 16 times more data points. More about the adaptive measurements and also the drift treatment in such data can be found in [28].

11.7 Conclusion

Scanning Probe Microscopy can be used as a complementary technique to the optical characterization of thin films. Majority of the measurements on this field is related to the surface roughness as this is a parameter that appears in many optical models. If we want to get the sample parameters correctly we need to spend some effort on making the instrument traceable, i.e. calibrated and with known uncertainty budget. Tasks that are typically solved using SPM in this scientific field might be considered as very basic, comparing to wide range of possibilities that SPM techniques offer, however even here are still open questions and novel research directions, like large area measurements or measurements covering better the surface statistical properties that were briefly discussed in this chapter.

References

1. D. Franta, I. Ohlídal, P. Klapetek, P. Pokorný, Surf. Interface Anal. **34**, 759 (2002)
2. P. Klapetek, I. Ohlídal, D. Franta, P. Pokorný, Surf. Interface Anal. **34**, 559 (2002)
3. P. Klapetek, I. Ohlídal, D. Franta, A. Montaigne-Ramil, A. Bonanni, D. Stifter, H. Sitter, Acta Phys. Slovaca **22**, 223 (2003)
4. P. Eaton, P. West, *Atomic Force Microscopy* (Oxford University Press, Oxford, 2010)
5. P. Klapetek, *Quantitative Data Processing in Scanning Probe Microscopy* (Elsevier, Amsterdam, 2012)
6. P. Klapetek, M. Valtr, M. Matula, Meas. Sci. Technol. **22**(9), 094011 (2011)
7. V. Korpelainen, J. Seppä, A. Lassila, Precis. Eng. **34**(4), 735 (2010)
8. H.J. Butt, B. Capella, M. Kappl, Surf. Sci. Rep. **59**, 1 (2005)
9. C. Clifford, M. Seah, Nanotechnology **16**(9), 1666 (2005)
10. M.S. Kim, J.H. Choi, J.H. Kim, Y.K. Park, Measurement **43**, 520 (2010)
11. N. Burnham, X. Chen, C. Hodges, G. Matei, E. Thoreson, C. Roberts, M. Davies, S. Tendler, Nanotechnology **14**(1), 1 (2003)
12. J. te Riet, A.J. Katan, C. Rankl, S.W. Stahl, A.M. van Buul, I.Y. Phang, A. Gomez-Casado, P. Schön, J.W. Gerritsen, A. Cambi, A.E. Rowan, G.J. Vancso, P. Jonkheijm, J. Huskens, T.H. Oosterkamp, H. Gaub, P. Hinterdorfer, C.G. Figdor, S. Speller, Ultramicroscopy **111**(12), 1659 (2011)
13. G. Matei, E. Thoreson, J. Pratt, D. Newell, N. Burnham, Rev. Sci. Instrum. **77**(8), 083703 (2006)
14. M. Kim, J. Choi, J. Kim, Y. Park, Meas. Sci. Technol. **18**, 3351 (2007)
15. P.J. Cumpson, C.A. Clifford, J. Hedley, Meas. Sci. Technol. **15**, 1337 (2004)
16. A. Campbellová, M. Valtr, J. Zůda, P. Klapetek, Meas. Sci. Technol. **22**, 094007 (2011)
17. J.S. Villarrubia, J. Res. Natl. Inst. Stand. Technol. **102**, 425 (1997)
18. P. Klapetek, I. Ohlídal, Ultramicroscopy **94**, 19 (2003)

19. D. Nečas, P. Klapetek, Meas. Sci. Technol. **28**(3), 034014
20. D. Nečas, P. Klapetek, Cent. Eur. J. Phys. **10**(1), 181 (2012)
21. S.M. Jordan, R. Schad, J.F. Lawler, D.J.L. Hermann, H. van Kempen, J. Phys, Condens. Matter **10**, L355 (1998)
22. T. Yoshinobu, A. Iwamoto, K. Sudoh, H. Iwasaki, J. Vac. Sci. Technol. B **13**(4), 1630 (1995)
23. H.N. Yang, Y.P. Zhao, A. Chan, T.M. Lu, G.C. Wang, Phys. Rev. B **56**, 4224 (1997)
24. R. Buzio, C. Boragno, F. Biscarini, F.B. de Mongeot, U. Valbusa, Nat. Mater. **2**, 233 (2003)
25. A. Mannelquist, N. Almqvist, S. Fredriksson, Appl. Phys. A **66**, 891 (1998)
26. J. Zemek, K. Olejnik, P. Klapetek, Surf. Sci. **602**(7), 1440 (2008)
27. P. Klapetek, I. Ohlídal, J. Bílek, Ultramicroscopy **102**, 5159 (2004)
28. P. Klapetek, A. Yacoot, P. Grolich, M. Valtr, D. Nečas, Meas. Sci. Technol. **28**(3), 034015 (2017)
29. D. Hilbert, Math. Ann. **38**, 459 (1891)
30. D.J. Abel, D.M. Mark, Int. J. Geogr. Inf. Syst. **4**(1), 21 (1990)
31. M. Bader, *Space-Filling Curves* (Springer, Berlin, 2012)
32. M.M. Flood, Op. Res. **4**(1), 61 (1956)

Chapter 12
Resonant Waveguide Grating Structures

Stefanie Kroker and Thomas Siefke

Abstract Resonant waveguide gratings are subwavelength structures that possess the ability to selectively reflect or transmit light in terms of wavelength, incidence angle and polarization state. They are of interest in a large variety of optical sensors and optoelectronic devices. Resonant waveguide gratings have also emerged as low-noise optical components in high-precision metrology, for example frequency stabilized laser systems for the realization of optical clocks or gravitational wave detectors. In these applications, Brownian thermal noise of optical coatings, sets a severe limitation to the feasible sensitivity. In this chapter, we will discuss the relevance of the mechanical loss of optical thin film coatings for Brownian thermal noise. We will present monolithic resonant waveguide gratings to circumvent the use of amorphous coatings to reduce thermal noise. First, we will introduce a method to characterize the mechanical loss of optical coatings and discuss its implications for high-precision metrology. Afterwards we will explain the working principle of resonant waveguide gratings. Then, several characterization techniques for the dimensional and optical characterization will be discussed and experimental results for monolithic waveguide gratings with one-dimensional and two-dimensional periodicity will be presented.

S. Kroker (✉)
LENA Laboratory for Emerging Nanometrology, Technische Universität Braunschweig, Pockelsstraße 14, 38106 Braunschweig, Germany
e-mail: s.kroker@tu-braunschweig.de

S. Kroker · T. Siefke
Physikalisch-Technische Bundesanstalt, Bundesallee 100, 38116 Braunschweig, Germany
e-mail: thomas.siefke@uni-jena.de

T. Siefke
Friedrich-Schiller-University Jena, Institute of Applied Physics, Albert-Einstein-Straße 15, 07745 Jena, Germany

O. Stenzel and M. Ohlídal (eds.), *Optical Characterization of Thin Solid Films*, Springer Series in Surface Sciences 64,
https://doi.org/10.1007/978-3-319-75325-6_12

12.1 Introduction

In the year 1985 Mashev and Popov observed for the first time anomalies in the zeroth order diffraction efficiency of dielectric coated gratings [1]. These sharp peaks in the reflectance spectra were identified as resonant excitations of guided waves. Since that time, such resonant waveguide gratings have attracted interest in numerous optical applications. They serve as bandstop [2–5] and bandpass filters [6, 7], compact mirrors in lasers [8], as elements for field enhancement in biosensors [9, 10] and light sources [11] as well as in nonlinear frequency converters [12, 13] or particle traps [14]. By using high-index materials, grating structures with two-dimensional periodicity and complex grating geometry, the spectral and angular properties of resonant waveguide gratings can be shaped in a large range [15–22].

During the past decade, resonant waveguide gratings have also emerged as low-noise optical components in high-precision metrology, for example frequency stabilized laser systems for the realization of optical clocks or gravitational wave (GW) detectors [23–25]. In these applications, Brownian thermal noise of optical components, sets a severe limitation to the sensitivity [26–30]. Gravitational wave detectors, for instance, need to sense relative length changes in the order of 10^{-21} and even below [31]. Therefore, highly sensitive laser interferometers are utilized [32, 33]. Figure 12.1 shows the schematic setup of such a Michelson interferometer as employed in the Advanced LIGO detectors (LIGO = Laser Interferometer Gravitational Wave Observatory). The first detection of gravitational waves in September 2015 [31] is not just another proof of Einstein's theory of general relativity [34] but also the beginning of a novel kind of astronomy, i.e. gravitational wave astronomy.

For GW astronomy, the sensitivity of current, second generation, GW detectors is planned to be enhanced by another order of magnitude in the next generation of detectors. A possible design of such a GW observatory, called Einstein Telescope [35, 36], has been developed in a European collaboration [37]. The Einstein Telescope will open a new window to the universe exploring new physical properties of astronomical objects and the universe itself. Due to their thermal noise, optical coatings are the critical part to reach the required sensitivity enhancement [38]. In the Advanced LIGO detectors the optical coatings components have been realized with multilayer stacks of silicon dioxide and tantalum pentoxide layers [39]. In future GW detectors operating at low temperatures, the amorphous coatings, of the end mirrors (end test masses) and cavity couplers (input test masses) stacks will be a limiting factor for the sensitivity.

In the following chapter, we will discuss the relevance of the mechanical loss, a measure for the internal friction, of optical thin film coatings for Brownian thermal noise. We will present monolithic resonant waveguide gratings to circumvent the use of amorphous coatings to reduce thermal noise. First, we will introduce a method to characterize the mechanical loss of optical coatings and discuss its implications for high-precision metrology. Afterwards we will explain the working principle of res-

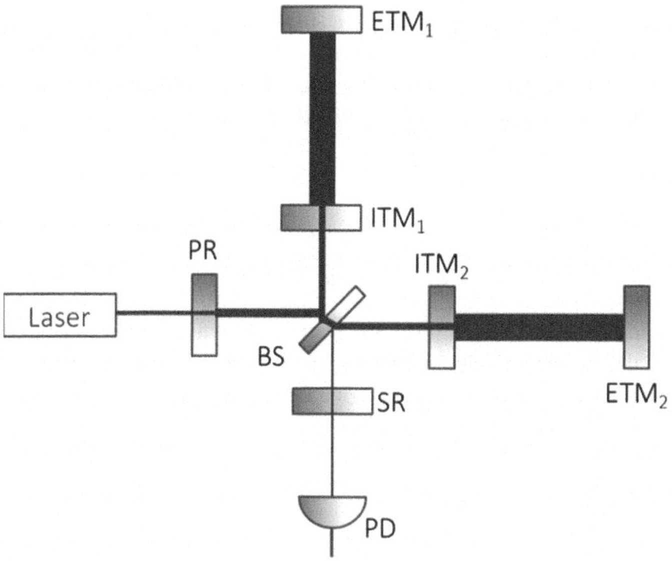

Fig. 12.1 Sketch of a typical Michelson-type gravitational wave interferometer with additional arm cavities, power recycling and signal recycling. The arm lengths are typically in the range of several kilometers. BS—beam splitter, ETM—end test mass, ITM—input test mass, PR—power recycling mirror, SR—Signal recycling mirror. The thickness of the laser beam indicates the beam power

onant waveguide gratings. Then, several characterization techniques for the dimensional and optical characterization will be discussed and experimental results for monolithic waveguide gratings with one-dimensional and two-dimensional periodicity will be presented.

12.2 Characterization of Mechanical Loss in Optical Coating Materials and Implications for Waveguide Gratings in Precision Metrology

Reflective elements in high-precision optical experiments commonly base on multiple beam interference realized by coating stacks of alternating high-index and low-index layers on optical substrates. As mentioned above, for gravitational wave detectors, the current coating materials of choice are silica (SiO_2) and tantala (Ta_2O_5) which are deposited on silica substrates [39]. These materials provide excellent optical properties, i.e. low absorption and scattering losses.

Brownian thermal noise in optical components originates from fluctuations of the sample surface [40]. The equipartition theorem states that every particle of finite temperature T exhibits a thermal energy of $k_B T/2$ per degree of freedom, where k_B is

Boltzmann's constant. Thus, the atoms of a solid body, e.g. of an optical component, are permanently in thermal motion adding a random phase to the light interacting with it. As a direct consequence of the fluctuation dissipation theorem thermal noise is related to the mechanical loss of the involved materials [41].[1] For a mirror substrate with a highly reflective coating stack of total thickness h the Brownian noise power spectral density $S_z(f, T)$ is [30]:

$$S_z(f, T) = \frac{2k_B T}{\pi^{3/2} f} \frac{1}{Yw} \left[\Phi_s(T) + \frac{h}{\sqrt{\pi} R} \left(\frac{Y_s}{Y_c} + \frac{Y_c}{Y_s} \right) \Phi_c(T) \right]. \qquad (12.1)$$

Y and Y' represent the Young's modulus of the substrate and the averaged Young's modulus of the coating stack, respectively. Φ_s and Φ_c are the related (temperature dependent) mechanical losses. T is the temperature, f the frequency and R the radius of the laser beam. Equation (12.1) contains four aspects to reduce Brownian thermal noise:

- Low temperature,
- low-loss materials,
- low coating thickness,
- large beam radii (require large substrates).

Due to their large mechanical loss, the optical coatings dominate Brownian thermal noise in highly reflective optical components. Thus, it is important to investigate relevant coating materials in terms of their loss. The mechanical loss can be measured by using ring-down techniques. Therefore, a mechanical resonance of a coated substrate is excited and the ring down time (the 1/e-decay of the initial oscillation amplitude) is measured. The mechanical loss is then given by:

$$\Phi = \frac{1}{\pi f_0 \tau}. \qquad (12.2)$$

Here, f_0 is the eigenfrequency of the mechanical resonance. Generally, the measured mechanical loss contains all mechanical dissipation processes in the system such as thermoelastic damping [43] or defect induced damping [44]. To characterize the mechanical loss of optical coatings it is necessary to minimize the influence of the substrate material and of potential suspension losses. This is achieved by the deposition of the related coating material on thin flexures. Loss measurements of the sample before (Φ_s) and after coating deposition (Φ_{total}) yield the mechanical loss Φ_c of the thin film coating [45]:

$$\Phi_{total} = \frac{1}{D} \left[(D-1)\,\Phi_s + \Phi_C \right] \qquad (12.3)$$

D is the dilution factor:

---

[1]The reader may be familiar with this relationship from the Johnson-Nyquist noise of electric resistors. Here thermal noise is also proportional to the electric resistance, i.e. the dissipation [42, 43].

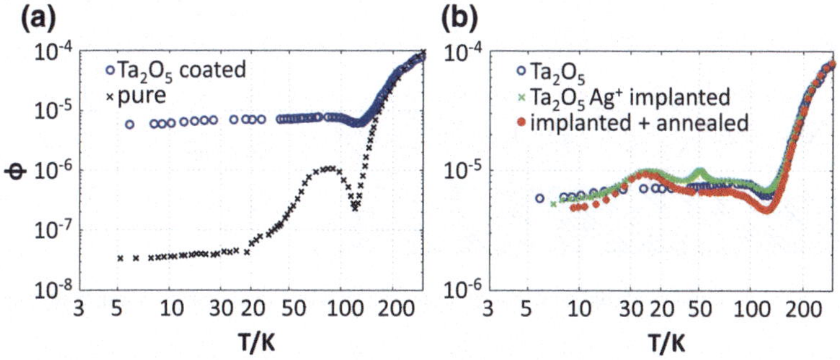

Fig. 12.2 **a** Temperature dependent measured mechanical loss of uncoated crystalline silicon flexure before and after coating with a tantala layer of 500 nm. **b** Mechanical loss of tantala coated silicon flexure after implantation and annealing [46]

$$D = \frac{E_C + E_S}{E_C}, \tag{12.4}$$

with the elastic energies E_C stored in the coating and E_S in the substrate, respectively. The dilution factor represents a volumetric weighting of the dissipation. It accounts for the fact, that only the fraction of elastic deformation energy which is stored in the coating (substrate) can be dissipated by loss mechanisms in the coating (substrate). The elastic energies and thus D depend on the shape of the mechanical mode which can be calculated by means of finite element tools like ANSYS or COMSOL [46, 47].

Figure 12.2a shows the temperature dependent mechanical loss for a pure silicon flexure and for the same flexure with a tantala coating. At low temperatures the amorphous coating increases the loss of the substrate by several orders of magnitude [43, 48, 49]. To reduce the mechanical loss of amorphous coating materials, techniques like ion doping with subsequent thermal treatment are investigated [45, 49, 50]. Figure 12.2b illustrates that these post-processing techniques allows for a loss reduction by about 20%. Beside an optimization of amorphous materials, crystalline materials such as silicon, sapphire or GaAs can outperform the mechanical loss of amorphous materials by several orders of magnitude [51]. This fact has raised interest in crystalline coating stacks based on Al_xGa_{1-x} As as an alternative to amorphous multilayer coatings. It has been demonstrated that these epitaxially grown stacks allow for a substantial reduction of thermal noise [52].

The third approach to minimize the detrimental effect of amorphous coatings, bases on the reduction of the coating thickness. In multilayer mirrors, a thickness reduction is unfavourable for a high reflectivity. Resonant waveguide gratings, instead, provide high reflectivity with a structured layer which is by a factor of about ten thinner than typical multilayer mirrors. As will be illustrated in the next section, resonant waveguide gratings can be implemented even completely monolithic without adding any lossy amorphous material. The monolithic nature enables an operation

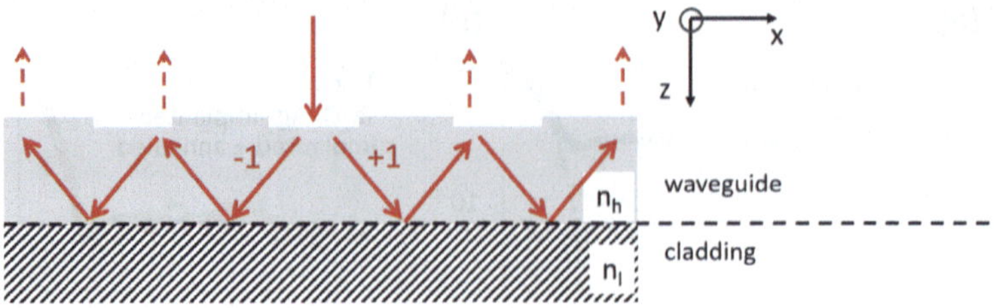

Fig. 12.3 Sketch of the working principle of a highly reflective resonant waveguide grating

at cryogenic temperatures without coating stress induce deformation. And, implemented in silicon, large sample sizes of up to 45 cm in diameter may be feasible by scaling up the highly developed silicon etching technology. Monolithic resonant waveguide gratings thus combine all necessary aspects for optical components with low Brownian thermal noise.

12.3 Functionality of Waveguide Gratings as High Reflectivity Mirrors

The effect of high reflectivity caused by resonant light coupling in dielectric structures was discovered and demonstrated for the first time in the 1980s by Mashev and Popov as well as by Golubenko [1, 2] with a configuration as shown in Fig. 12.3a. The grating structure leads to a perturbation of the waveguide. This perturbation allows an incident light wave propagating in z-direction to couple into the waveguide and to propagate along the horizontal (x-) direction. Usually the coupling of the incident wave is effected by the first diffraction orders [53]. To achieve this, for the grating period p the following condition needs to be fulfilled:

$$\frac{\lambda}{n_\mathrm{h} + \sin\varphi} < \mathrm{p} < \frac{\lambda}{n_\mathrm{l} + \sin\varphi}. \tag{12.5}$$

Here, φ is the incidence angle. The right part of the relation allows the structure to generate the first diffraction orders in the waveguide layer which are utilized to realize the coupling between the incoming wave and the waveguide modes. The left part ensures that the first diffraction orders cannot propagate in the cladding. With Rel. (12.5) it is also ensured that in free space only the propagation of the zeroth diffraction order is allowed. Thus, in an ideal structure no light would be lost in higher diffraction orders.

Due to the waveguide's perturbation the light is not perfectly guided but leaky. By adjusting the structural parameters (period, ridge width, thickness) it is possible to achieve constructive interference of the partial waves coupling out in upward direction. This effect of resonant light coupling is not limited to the configuration

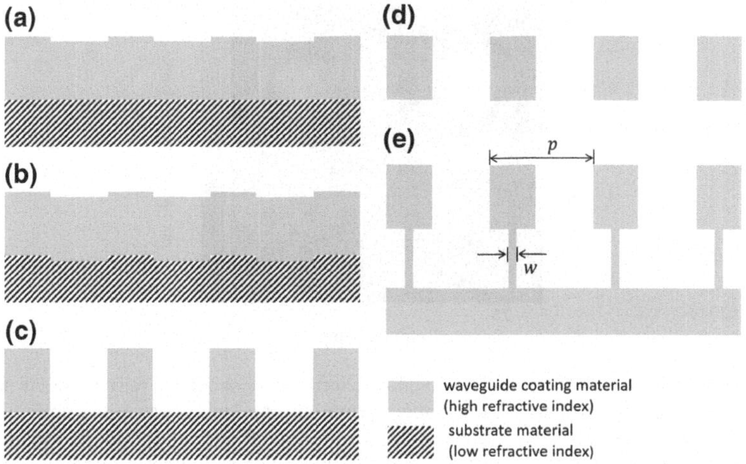

Fig. 12.4 Possible realizations of resonant waveguide gratings. **a** Configuration with structured waveguide layer. **b** Configuration with structured substrate. **c** Binary structure as waveguide grating. **d** Free standing waveguide grating structure. **e** Monolithic waveguide grating structure

shown in Fig. 12.3 but can be also extended to several other configurations, e.g. doubly corrugated waveguides [54] and even stand-alone or monolithic T-shape grating ridges [55]. An overview on typical configurations is given in Fig. 12.4. Although the structures in Fig. 12.4c–d do not exhibit complete waveguides, their high index grating layers are still able to confine light. A theoretical description of such strongly perturbed waveguide structures can be found in the works by Magnusson and coworkers [19–21]. An alternative to the model of perturbed waveguides and leaky modes was given by Lalanne and by Karagodsky et al. [56, 57]. They use Bloch modes to describe the behavior of strongly modulated waveguide gratings.

A benefit of the stand-alone and T-shape grating is their monolithic character. This avoids coating stress. In addition, it minimizes thermal noise of these elements which will be discussed in detail in the next section. In contrast to stand-alone structures T-shape gratings are not restricted to small sample sizes but scalable to virtually arbitrary areas. In these structures the low-index cladding layer is replaced by an effective low-index layer which prevents the light to couple from the waveguide layer to the high-index silicon substrate. The decoupling is the more the smaller the effective ridge width w of the supporting structure (see Fig. 12.4e). Figure 12.5 shows the rigorously calculated energy density distribution in a T-shaped monolithic mirror. The simulation was performed by means of RCWA (Rigorous Coupled Wave Analysis) [58]. The electromagnetic field is confined in the waveguide layer and merely penetrates into the support structure. However, in x-direction (in which the waveguide modes are meant to propagate, compare Fig. 12.3) the energy density is clearly confined in the material and not homogeneously distributed as one would expect from a waveguide mode. This may be understood by the following explanation: Due to the high index contrast between ridge (silicon, refractive index 3.48) and groove

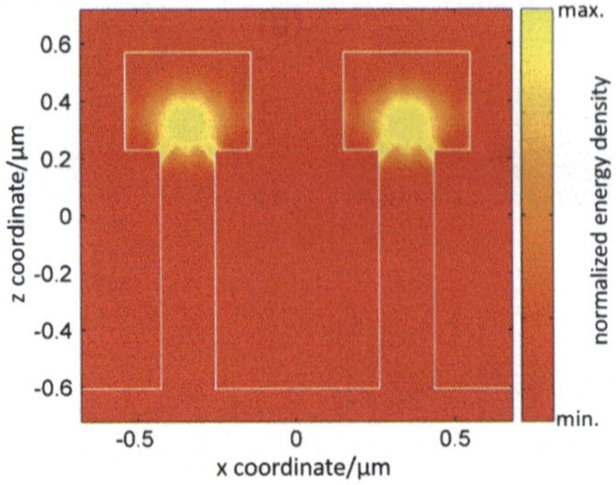

Fig. 12.5 Energy density distribution in a highly reflective nanostructured mirror. For the simulation the following parameters were used: grating period: 688 nm, depth of waveguide layer: 350 nm, duty cycle of waveguide layer: 0.56, depth of support structure: 800 nm, duty cycle of support structure: 0.25, wavelength: 1550 nm, incidence angle: 0°, polarization: transverse-magnetic

(free space, refractive index 1.0) the perturbation of the waveguide is strong. The strong perturbation leads to a very short propagation length of the waveguide modes of about just one period. The short propagation length manifests itself in the light confinement within the ridges of the upper part of the T-shape.

An optimum design of monolithic nanostructured mirrors necessitates high reflectivity with possible large fabrication tolerances as well as a good mechanical stability. A structure with a ridge width ratio of about 1:2 between the support structure and the upper layer is found to meet both requirements.

12.4 Fabrication of Monolithic Waveguide Gratings

The realization of nanostructured crystalline silicon mirrors bases on techniques which are well established in semiconductor industry. In a first step, the silicon substrate is coated with a chromium layer and a layer of electron beam sensitive resist (see Fig. 12.6a). For the former an ion beam deposition and for the latter a spin coating process is employed. Then, the subwavelength grating pattern is patterned in the resist by means of electron beam lithography. Within this process structures with periodicity in one dimension as well as in two dimensions can be realized. The development of the resist leads to the structured resist layer shown in Fig. 12.6b. After this step the structure is transferred into the chromium layer via an anisotropic (i.e. binary) etching process as illustrated in Fig. 12.6c. This layer then serves as a hard mask to finally structure the silicon substrate. Here, first an anisotropic process is applied to define the upper (binary) layer (Fig. 12.6d). The etching time is adjusted to match the groove depth of the design.

To finally achieve the T-shape structure, two strategies can be pursued: One possibility is, to cover the side walls with a thin chromium layer by coating the element

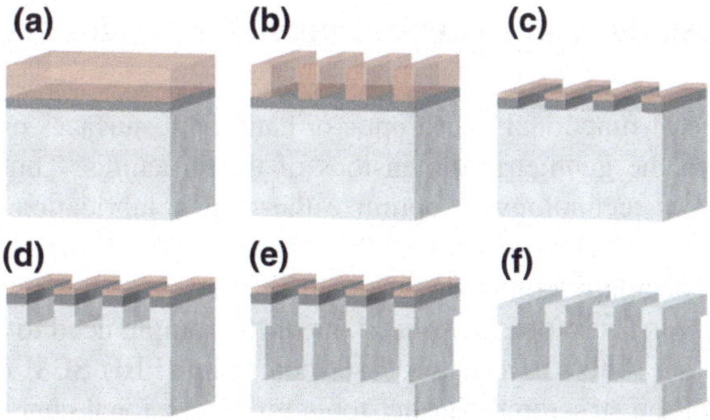

Fig. 12.6 Fabrication of nanostructured crystalline silicon waveguide gratings. **a** Substrate with chromium coating and resist layer on top. **b** Resist pattern after electron beam lithography and development. **c** Pattern transferred into the chromium layer by means of anisotropic dry etching. **d** Binary structure after anisotropic etching. **e** T-shape structure realized with an isotropic etching process and **f** Final mirror after removal of chromium and resist layer

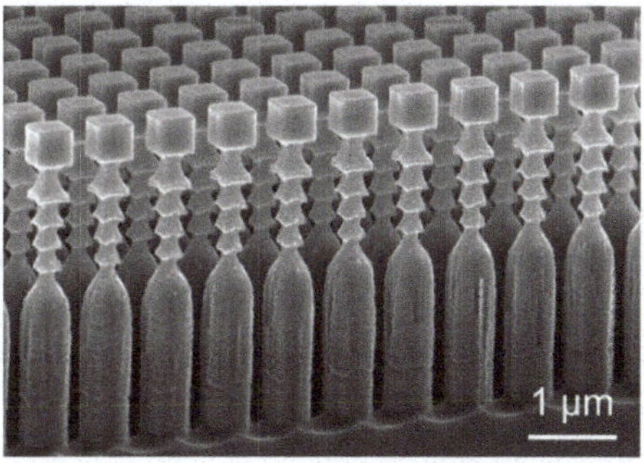

Fig. 12.7 Scanning electron image of a monolithic waveguide grating made of crystalline silicon with two-dimensional periodicity of a period of 720 nm

under an oblique angle [24]. This procedure protects the ridge side walls from further etching while the bottom of the groove is still accessible. Then, an isotropic (i.e. polydirectional) etching process enables the undercut of the upper structure to generate the thin supporting structure. This finally ends up in the structure illustrated in Fig. 12.6e, f, respectively. Another strategy to realize the undercut, is to apply a gas chopping process [19]. Therefore, the ratio of fluorine and carbon species in the reactive etching gas allows to alternatingly switch between (isotropic) etching and passivation of the ridge surface. Typical features of T-shape structures made by this process are corrugations of the supporting structure as shown in the scanning electron microscope image in Fig. 12.7. It should be mentioned that these features do not set an intrinsic limit to the feasible optical function if they are properly considered in the design of the structure.

12.5 Dimensional Characterization of Waveguide Gratings

The performance of functional micro-optic or nano-optic surfaces usually substantially depends on the geometric dimensions of the structures. Thus, dimensional metrology is a key technology to optimize the related fabrication processes and to understand the optical properties of the structures which may differ from the design values. A standard tool for dimensional characterization is scanning electron microscopy (SEM). Therefore, cross sections of the sample need to be prepared as shown in Fig. 12.7. Alternatively, a focused ion beam (FIB) SEM can be utilized to take images of the structures. Further tools for dimensional characterization are atomic force microscopy (AFM), helium ion microscopy (HIM) as well as optical techniques like scatterometry and ellipsometry [59–61]. Often a combination of several of these tools is necessary to retrieve reliable information about dimensional waveguide grating properties.

12.6 Optical Characterization of Waveguide Gratings

In the following sections, we will discuss techniques for the characterization of transmittance and reflectance in dependence of wavelength, incidence angle and polarization. It has already been mentioned that cryogenic temperatures will be beneficial for resonant waveguide gratings in high precision optical metrology. Therefore, the impact of temperature changes on the spectral properties of monolithic waveguide gratings is discussed as well.

12.6.1 Reflectance Measurements in a Cavity

The reflectivity of highly reflective waveguide gratings can be precisely measured by Cavity ring-down techniques which are described in Chap. 16. Alternatively, a resonator setup as illustrated in Fig. 12.8a can be utilized [24]. A conventional multilayer mirror serves as coupler with a transmittance of t_1^2. By measuring the cavity finesse F, the amplitude reflectivities $r_{12} = r_1 r_2$ of coupler end waveguide coating end mirror can be determined by:

$$r_{12} = r_1 r_2 = 2 - \cos\frac{\pi}{F} - \sqrt{\left(\cos\frac{\pi}{F} - 2\right)^2 - 1}. \tag{12.6}$$

The finesse F is the ratio of free spectral range (FSR) f_{FSR} and cavity linewidth Δf (full width at half maximum). The FSR results from the length of L the cavity by:

$$f_{FSR} = \frac{c}{2L}. \tag{12.7}$$

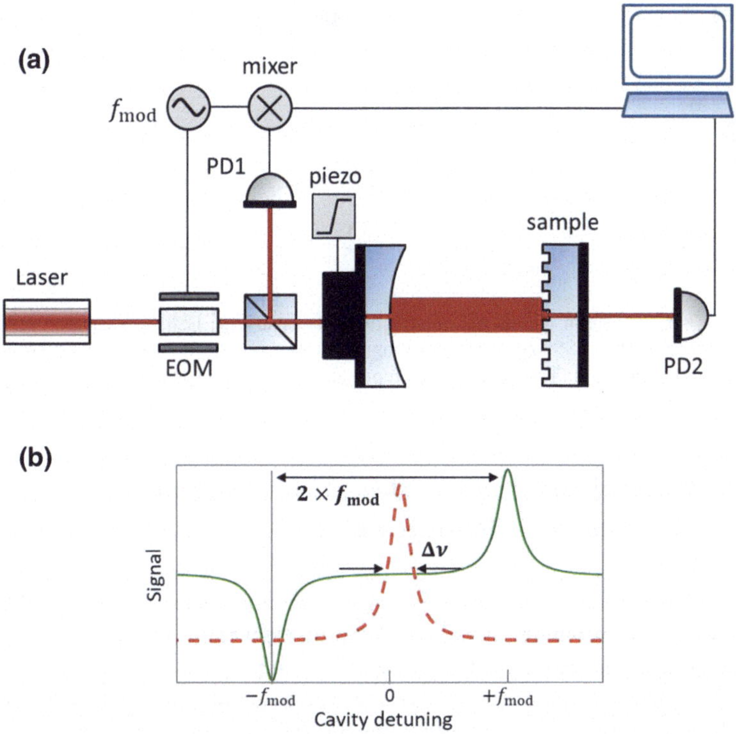

Fig. 12.8 **a** Schematic of reflectivity measurement setup using a cavity. **b** Signals in dependence of the cavity detuning. The frequency markers at $\pm f_{mod}$ serve calibrate the frequency range

The length L is changed by means of a piezo actuator attached to the input mirror. Detuning the cavity around the resonance leads to the typical Airy peak in the transmitted signal which is measured by photodiode 2 (PD2) (see Fig. 12.8b). To determine f, a frequency calibration is required. Therefore, the Pound-Drever-Hall (PDH) technique is used. By means of an electro-optical modulator (EOM) sidebands at $\pm f_{mod}$ around the cavity resonance frequency are generated. These sidebands serve as frequency markers. The cavity reflection which is detected in photodiode 1 (PD1) is then demodulated by a local oscillator. From the signal curves shown in Fig. 12.8b the reflectivity of the waveguide grating can be calculated. In their proof-of principle experiment Brückner et al. demonstrated a reflectivity of (99.79 ± 0.01) % at wavelength of 1550 nm [24].

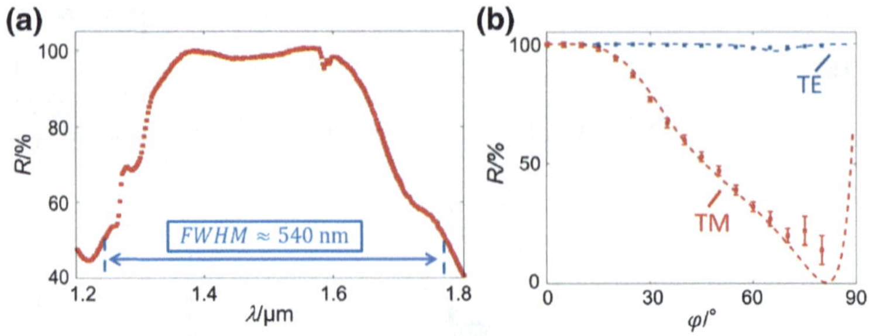

Fig. 12.9 Characterization results of the structure shown in Fig. 12.8. **a** Measured reflectance at an incidence angle of 8° and transverse magnetic (TM) polarization. **b** Angular spectrum at a wavelength of 1550 nm. The dashed lines represent the results of simulations performed with RCWA

12.6.2 Spectral and Angular Dependent Reflectance and Transmittance Measurements

The reflectivity of monolithic resonant waveguide gratings exhibits large tolerances with respect to wavelength and angle of the incident light. Figure 12.9 illustrates the characterization results of the 2D periodic structure shown in the SEM image of Fig. 12.7. The spectrum displayed in Fig. 12.9a was recorded with a commercial Perkin Elmer Lambda 950 spectrophotometer [62]. The reflectance reaches values close to unity in a broad wavelength range. This spectral broadband performance is caused by the large index contrast of the structure [15, 57]. Here, the full width at half maximum of more than 500 nm is readily competitive, if not superior, to the bandwidth of conventional multilayer based mirrors. Similar spectral results can be achieved with properly designed 1D periodic structure [24].

A feature that is unique for 2D structures and cannot be realized with their 1D counterparts is the extraordinary large angular tolerance [24]. Figure 12.9b compares the angular dependent reflectance for transverse magnetic (TM) and transverse electric (TE) polarization. The related measurement was performed using a fiber-coupled laser of 1550 nm wavelength and a rotation stage on which the sample was mounted. The experimental setup is shown in Fig. 12.10. Due to the structural symmetry, at normal incidence, i.e. 0°, the reflectances for both polarizations are identical within the limits of accuracy. With increasing incidence angle the TM reflectance decreases, whereas the TE reflectance remains close to unity ($\geq 98.2\%$) within the entire angular spectrum. In the simplified picture of a disturbed waveguide this behaviour can be understood the following way: In TE-polarization the only non-vanishing component of the electric field is perpendicular to the plane of incidence. Changing the angle of incidence within this plane does not substantially influence the coupling conditions of this field component to the waveguide. In the direction perpendicular to the plane of incidence the light still *sees* the same period as for normal incidence. For TM-polarization the field components in the plane of incidence have to be taken

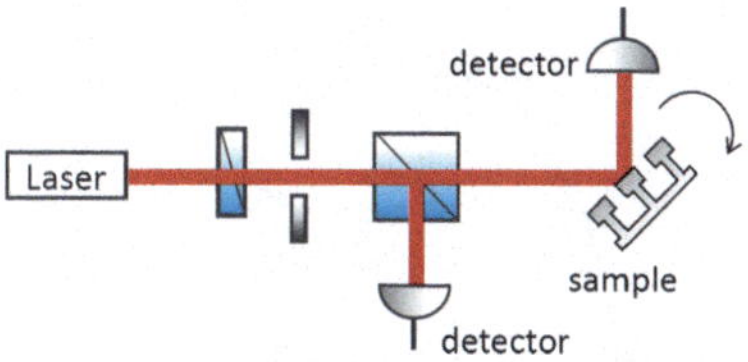

Fig. 12.10 Schematic set up for the angular dependent reflectivity characterization of waveguide grating. The beam splitter enables a relative measurement independent from power fluctuations of the laser light source

into account. In this case, a change of the angle leads to an increased effective grating period which detunes the structure from the high reflectivity. At incidence angles close to normal incidence Fresnel reflection dominates the reflection behaviour of the structure which again leads to an increase of the reflectance. In summary, with monolithic nanostructured mirrors large angular and spectral tolerances are accessible rendering the related elements robust towards adjustment errors.

12.6.3 Temperature Dependent Transmittance Measurements

The monolithic silicon waveguide coatings are intended for use at low temperatures. In contrast to conventional multilayer based solutions, their monolithic character enables cooling without additional stress induced by materials with different thermal expansion coefficients. Bringing monolithic silicon mirrors to cryogenic temperature leads to two effects: Firstly, thermal expansion changes the structural period, depth and ridge width. The thermal expansion coefficient of silicon at room temperature is 2.5×10^{-6}/K and reduces with decreasing temperature down to about 5×10^{-10}/K at a temperature of 10 K [63]. Thus, a relative change of the structural parameters is less than 2×10^{-4}/K and may be neglected. Secondly, the band gap of silicon changes from 1.12 eV at 300 K to 1.17 eV at 10 K [63, 64]. The temperature dependence of the band gap is responsible for a decrease of the refractive index at low temperatures. Hereby, the value of the index change is determined by the thermo-optic coefficient $\beta(T)$:

$$\Delta n(T_1) = \int_{T_0=300\,\mathrm{K}}^{T_1} \beta(T)\, dT \tag{12.8}$$

With the experimental data of $\beta(T)$ provided by Komma et al. [65] $\Delta n(T_1 = 10\,\mathrm{K})$ can be determined to be -0.03. This index reduction leads to a spectral blue shift of the reflectivity maximum. Since the material dispersion dn/dλ of crystalline silicon only accounts for 10^{-4} [66] its influence is much less than the temperature dependent refractive index change dn/dT. Figure 12.11a shows the transmittance spectra of the structure from Fig. 12.7 at room temperature and 10 K. A cryostat with optical

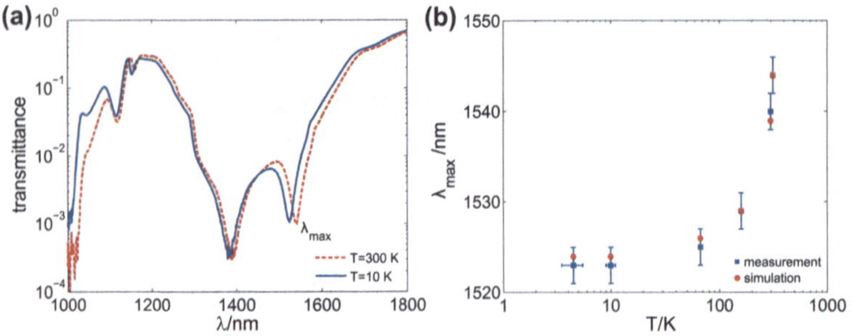

Fig. 12.11 **a** Measured transmittance spectra of the structure shown in Fig. 12.7. The blue shift of the room temperature spectrum is result of the refractive index reduction by 0.03. **b** Wavelength of maximum reflectance (minimum transmittance) of the right minimum in (**a**) in dependence of the temperature

windows was utilized for the characterization. By means of a halogen lamp and a monochromator setup the spectra were measured. Each of the spectra exhibits two transmittance minima. For the investigated structure, a wider supporting structure would reduce the transmittance between the two minima leading to broad wavelength range of low transmittance equivalent to a reflectance close to unity. In the spectra the transmittance minimum at a wavelength of about 1380 nm is more pronounced than the minimum at about 1540 nm. This fact may originate from a higher level of scattered light at shorter wavelengths as well as by a different interaction of the fields in the upper grating region with the supporting structure. The fact that the minima are located at a wavelength of less than 1550 nm can be attributed to a depth of the upper grating region which is less than the design value. Figure 12.11a implies that the shift of the short wavelength transmittance minimum (left transmittance minimum) is smaller than the shift of the minimum at larger wavelengths. However, the price for the larger temperature tolerance are stricter requirements for the accuracy of the upper grating layer thickness. Therefore, shifting this transmittance minimum to a wavelength of 1550 nm by adapting the structural parameters is not advisable.

From $\Delta n\,(T_1 = 10\,\text{K})$ the wavelength of minimum transmittance can be calculated by means of rigorous methods solving Maxwell's equations, for example RCWA [58]. Figure 12.11b shows the results of these simulations in comparison to the experimental data. The data illustrate that nanostructured crystalline mirrors needs be designed for a specific target temperature. Because of the decreasing $\beta\,(T)$ at low temperatures there is only a small spectral shift at temperatures below 100 K. The shift implies that the fabrication and pre-characterization of the mirrors at room temperature has to be performed at $\lambda_{max} + \Delta\lambda$ to account for spectral shifts at low temperatures. The value of $\Delta\lambda$ depends on the operation temperature as well as on the structural parameters. In applications close to room temperature the temperature dependent refractive index provides the possibility to precisely tune the wavelength of maximum reflectance (or minimum transmittance) to the desired wavelength.

12.7 Outlook

To overcome the limitation of Brownian thermal noise in future gravitational wave detectors the major challenge for all low-noise concepts will be the improvement of the mechanical, thermal and/or optical properties of the involved components. Ever improved properties require ever improved characterization methods. For waveguide gratings a major challenge will be to identify the origin of scattered light and to systematically mitigate it by optimized fabrication technology. For the characterization of scattered light, the techniques described in Chap. 14 can be applied. Due to the pronounced modulation of the electromagnetic field, in waveguide gratings defects and surface roughness are particularly critical in regions with high field intensities. The reduction of scattered light will be the key to enhance reflectivity and make waveguide gratings competitive to amorphous and crystalline multilayer based mirrors. Once this challenge is coped, a tenfold enhancement of sensitivity can be achieved in high-precision optical experiments [67].

Acknowledgements The sample preparation of the optical waveguide coatings for high-precision metrology was performed at the Institute of Applied Physics (Friedrich-Schiller University Jena). The support of T. Käsebier, M. Banasch, H. Schmidt, W. Gräf and W. Rockstroh in the sample preparation is acknowledged.

References

1. L. Mashev, E. Popov, Zero order anomaly of dielectric coated gratings. Opt. Commun. **55**, 377–380 (1985)
2. A. Sharon, D. Rosenblatt, A.A. Friesem, Narrow spectral bandwidths with grating waveguide structures. Appl. Phys. Lett. **69**, 4154 (1997)
3. D.L. Brundrett, E.N. Glytsis, T.K. Gaylord, Normal-incidence guided-mode resonant grating filters: design and experimental demonstration. Opt. Lett. **23**, 700–702 (1998)
4. Y. Wang, Y. Kanamori, J. Ye, H. Sameshima, K. Hane, Fabrication and characterization of nanoscale resonant gratings on thin silicon membrane. Opt. Express **17**, 4938–4943 (2009)
5. F. Brückner, S. Kroker, D. Friedrich, E.-B. Kley, A. Tünnermann, Widely tunable monolithic narrowband grating filter for near-infrared radiation. Opt. Lett. **36**, 436–438 (2011)
6. Y. Ding, R. Magnusson, Doubly resonant single-layer bandpass optical filters. Opt. Lett. **29** 1135–1137 (2004)
7. S. Steiner, S. Kroker, T. Käsebier, E.-B. Kley, A. Tünnermann, Angular bandpass filters based on dielectric resonant waveguide gratings. Opt. Express **20**, 22555–22562 (2012)
8. M.C.Y. Huang, Y. Zhou, J. Chang-Hasnain, A surface-emitting laser incorporating a high-index-contrast subwavelength grating. Nat. Photonics **1**, 119–122 (2007)
9. Y. Fang, A.M. Ferrie, N.H. Fontaine, J. Mauro, J. Balakrishnan, Resonant Waveguide Grating Biosensor for Living Cell Sensing. Biophys. J. **91**, 1925–1940 (2006)
10. H.N. Daghestani, B.W. Day, Theory and applications of surface plasmon resonance, resonant mirror, in *Resonant Waveguide Grating, and Dual Polarization Interferometry Biosensors*. Sensors, vol. 10, pp. 9630–9646 (2010)
11. I.-S. Chunga, Jesper Mørk, Silicon-photonics light source realized by III–V/Si-grating-mirror laser. Appl. Phys. Lett. **97**, 151113 (2010)
12. M. Siltanena, S. Leivo, P. Voima, M. Kauranen, Strong enhancement of second-harmonic generation in all-dielectric resonant waveguide grating. Appl. Phys. Lett. **91**, 111109 (2007)

13. A. Saari, G. Genty, M. Siltanen, P. Karvinen, P. Vahimaa, M. Kuittinen, M. Kauranen, Giant enhancement of second-harmonic generation in multiple diffraction orders from subwavelength resonant waveguide grating. Opt. Lett. **18**, 12298–12303 (2010)
14. R. Magnusson, Y. Ding, K. J. Lee, D. Shin, P.S. Priambodo, P.P. Young, A. Maldonado, Photonic devices enabled by waveguide-mode resonance effects in periodically modulated films, in Nano- and Micro-Optics for Information Systems. Proceedings of SPIE, vol. 5225, pp. 20–34 (2003)
15. C.F.R. Mateus, M.C.Y. Huang, Y. Deng, A.R. Neureuther, C.J. Chang-Hasnain, Ultrabroadband mirror using low-index cladded subwavelength grating. IEEE Photonics Technol. Lett. **16**, 518–520 (2004)
16. K.J. Lee, J. Curzan, M. Shokooh-Saremi, R. Magnusson, *Resonant wideband polarizer with single silicon layer*. Appl. Phys. Lett. **98**, 211112 (2011)
17. S. Kroker, T. Käsebier, E.-B. Kley, A. Tünnermann, Coupled grating reflectors with highly angular tolerant reflectance. Opt. Lett. **38**, 3336–3339 (2013)
18. S. Peng, G.M. Morris, Experimental demonstration of resonant anomalies in diffraction from twodimensional gratings. Opt. Lett. **21**, 549–551 (1996)
19. S. Kroker, T. Käsebier, S. Steiner, E.-B. Kley, A. Tünnermann, High efficiency two-dimensional grating reflectors with angularly tunable polarization efficiency. Appl. Phys. Lett. **102**, 161111 (2013)
20. Y. Ding, R. Magnusson, Resonant leaky-mode spectral-band engineering and device applications. Opt. Express **12**, 5661–5674 (2004)
21. R. Magnusson, M. Shokooh-Saremi, Physical basis for wideband resonant reflectors. Opt. Express **16**, 3456–3462 (2008)
22. Y. Ding, R. Magnusson, Use of nondegenerate resonant leaky modes to fashion diverse optical spectra. Opt. Express **12**, 1885–1891 (2004)
23. A. Bunkowski, O. Burmeister, D.Friedrich, K. Danzmann, R. Schnabel, High reflectivity grating waveguide coatings for 1064 nm. Class. Quant. Gravity **23**, 7279–7303 (2006)
24. F. Brückner, D. Friedrich, T. Clausnitzer, M. Britzger, O. Burmeister, K. Danzmann, E.-B. Kley, A. Tünnermann, R. Schnabel, Realization of a monolithic high-reflectivity cavity mirror from a single silicon crystal. Phys. Rev. Lett. **104**, 163903 (2010)
25. D. Friedrich, et al., Waveguide grating mirror in a fully suspended 10 meter Fabry-Perot cavity. Class. Quant. Gravity **19**, 14955–14963 (2011)
26. Y. Levin, Internal thermal noise in the LIGO test masses: a direct approach. Phys. Rev. D **57**, 659 (1998)
27. K. Numata, A. Kemery, J. Camp, Thermal-noise limit in the frequency stabilization of lasers with rigid cavities. Phys. Rev. Lett. **93**, 250602 (2004)
28. T. Kessler, C. Hagemann, C. Grebing, T. Legero, U. Sterr, F. Riehle, M.J. Martin, L. Che, J. Ye, A sub-40-mHz-linewidth laser based on a silicon single-crystal optical cavity. Nat. Photonics **6**, 687–692 (2012)
29. K. Jacobs, I. Tittonen, H.M. Wiseman, S. Schiller, Quantum noise in the position measurement of a cavity mirror undergoing Brownian motion. Phys. Rev. A **60**, 538 (1999)
30. G.M. Harry, et al., Thermal noise from optical coatings in gravitational wave detectors. Appl. Opt. **45**, 1569–1574 (2006)
31. B.P. Abbott et al. (LIGO Scientific Collaboration and Virgo Collaboration), Observation of gravitational waves from a binary black hole merger. Phys. Rev. Lett. **116**, 061102 (2016)
32. W.A. Edelstein, J. Hough, J.R. Pugh, W. Martin, Limits to the measurement of displacement in an interferometric gravitational radiation detector. J. Phys. E: Sci. Instrum. **11**, 710 (1978)
33. G.M. Harry. (for the LIGO Scientific Collaboration), Advanced LIGO: the next generation of gravitational wave detectors. Class. Quant Gravity **27**, 084006 (2010)
34. A. Einstein, Die Grundlage der allgemeinen Relativitätstheorie. Annalen der Physik. **7**, 769–822 (1916)
35. B. Sathyaprakash, et al., Scientific objectives of Einstein telescope. Class. Quant. Gravity **29**, 124013 (2012)

36. M. Punturo, et al., The Einstein Telescope: a third-generation gravitational wave observatory. Class. Quant. Gravity **27**, 194002 (2010)
37. The Einstein Telescope Science Team, Einstein gravitational wave Telescope conceptual design study, ET-0106C-10, http://www.et-gw.eu/etdsdocument, 2011
38. R. Nawrodt, S. Rowan, J. Hough, M. Punturo, F. Ricci, J.-Y. Vinet, Challenges in thermal noise for 3rd generation of gravitational wave detectors. Gen. Relat. Gravity **43**, 593–622 (2011)
39. L. Pinard, C. Michel, B. Sassolas, L. Balzarini, J. Degallaix, V. Dolique, R. Flaminio, D. Forest, M. Granata, B. Lagrange, N. Straniero, J. Teillon, G. Cagnoli, Mirrors used in the LIGO interferometers for first detection of gravitational waves. Appl. Opt. **56**, C11–C15 (2016)
40. G.E. Uhlenbeck, L.S. Ornstein, On the theory of the Brownian Motion. Phys. Rev. **36**, 823 (1930)
41. H.B. Callen, T.A. Welton, Irreversibility and generalized noise. Phys. Rev. **83**,34 (1951)
42. J.B. Johnson, Thermal agitation of electricity in conductors. Phys. Rev. **32**, 97 (1928)
43. H. Nyquist, Thermal agitation of electric charge in conductors. Phys. Rev. **32**, 100 (1928)
44. C. Zener, Internal friction in solids II. General theory of thermoelastic internal friction. Phys. Rev. **53**, 90 (1938)
45. D. Heinert, A. Grib, K. Haughian, J. Hough, S. Kroker, P. Murray, R. Nawrodt, S. Rowan, C. Schwarz, P. Seidel, A. Tünnermann, Potential mechanical loss mechanisms in bulk materials for future gravitational wave detectors. J. Phys: Conf. Ser. **228**, 012032 (2010)
46. C. Schwarz, *Private Communication*
47. M. Granata, E. Saracco, N. Morgado, A. Cajgfinger, G. Cagnoli, J. Degallaix, V. Dolique, D. Forest, J. Franc, C. Michel, L. Pinard, R. Flaminio, Mechanical loss in state-of-the-art amorphous optical coatings. Phys. Rev. D **93**, 012007 (2017)
48. M. Principe, I.M. Pinto, V. Pierro, R. DeSalvo, I. Taurasi, A.E. Villar, E.D. Black, K.G. Libbrecht, C. Michel, N. Morgado, L. Pinard, Material loss angles from direct measurements of broadband thermal noise. Phys. Rev. D **91**, 022005 (2015)
49. S. Gras, H. Yu, W. Yam, D. Martynov, M. Evans, Audio-band coating thermal noise measurement for advanced LIGO with a multi-mode optical resonator. Phys. Rev. D **95**, 022001 (2017)
50. I.W. Martin et al., Comparison of the temperature dependence of the mechanical dissipation in thin films of Ta2O5 and Ta2O5 doped with TiO2. Class. Quant. Gravity **26**, 155012 (2009)
51. R. Flaminio, J. Franc, C. Michel, N. Morgado, L. Pinard, B. Sassolas, A study of coating mechanical and optical losses in view of reducing mirror thermal noise in gravitational wave detectors. Class. Quant. Gravity **27**, 084030 (2010)
52. G.D. Cole, W. Zhang, M.J. Martin, J. Ye, M. Aspelmeyer, Tenfold reduction of Brownian noise in high-reflectivity optical coatings. Nat. Photonics **7**, 644–650 (2013)
53. A. Sharon, D. Rosenblatt, A.A. Friesem, Resonant grating-waveguide structures for visible and near-infrared radiation. J. Opt. Soc. Am. A 14, 2985–2993 (1997)
54. O. Stenzel, S. Wilbrandt, X. Chen, R. Schlegel, L. Coriand, A. Duparré, U. Zeitner, T. Benkenstein, C. Wächter, Observation of the waveguide resonance in a periodically patterned high refractive index broadband antireflection coating. Appl. Opt. **53**, 3147–3156 (2014)
55. F. Brückner, T. Clausnitzer, O. Burmeister, D.Friedrich, E.-B. Kley, K. Danzmann, A. Tünnermann, R. Schnabel, Monolithic dielectric surfaces as new low-loss light–matter interfaces. Opt. Lett. **33** 264–266 (2008)
56. P. Lalanne, J.P. Hugonin, P. Chavel, Optical properties of deep lamellar gratings: a Bloch mode insight. J. Lightwave Technol. **24**, 2442–2449 (2006)
57. V. Karagodsky, C.J. Chang-Hasnain, Physics of near-wavelength high contrast gratings. Opt. Express **20**, 10888–10895 (2012)
58. M.G. Moharam, T.K. Gaylord, Rigorous coupled-wave analysis of planar grating diffraction. J. Opt. Soc. Am. A **71**, 811–818 (1981)
59. S. Schröder, T. Herffurth, H. Blaschke, A. Duparré, Angle-resolved scattering: an effective method for characterizing thin-film coatings. Appl. Opt. **50**, C164–C171 (2011)
60. H. Gross, R. Model, M. Bär, M. Wurm, B. Bodermann, A. Rathsfeld, Mathematical modelling of indirect measurements in scatterometry. Measurement **39**, 782–794 (2006)

61. R.M.A. Azzam, N.M. Bashara, Generalized ellipsometry for surfaces with directional preference: application to diffraction gratings. J. Opt. Soc. Am. **46**, 1521–1523 (1972)
62. www.perkinelmer.com
63. http://www.ioffe.rssi.ru. Accessed 16 Jan 2017
64. J. Komma, C. Schwarz, G. Hofmann, D. Heinert, R. Nawrodt, Thermo-optic coefficient of silicon at 1550 nm and cryogenic temperatures. Appl. Phys. Lett. **101**, 041905 (2012)
65. http://refractiveindex.info/. Accessed 16 Jan 2017
66. H.Y. Fan, Temperature dependence of the energy gap in semiconductors. Phys. Rev. **82**, 900 (1951)
67. D. Heinert, S. Kroker, D. Friedrich, S. Hild, E.-B. Kley, S. Leavey, I. W. Martin, R. Nawrodt, A. Tünnermann, S. P Vyatchanin, K. Yamamoto, Calculation of thermal noise in grating reflectors, Phy. Rev. D **88**, 042001 (2013)

Chapter 13
Polarization Control by Deep Ultra Violet Wire Grid Polarizers

Thomas Siefke and Stefanie Kroker

Abstract Polarization is an inherent property of transversal electromagnetic light waves. Hence, the control of the polarization state is a fundamental requirement in many optical applications. Nowadays, optical measurement and fabrication technology strive to shorter wavelengths in the ultra violet to benefit from larger resolution and material specific electronic transitions used for analysis. Thanks to the progress of nano-technology it has become feasible to manufacture subwavelength devices such as wire grid polarizers for this wavelength regime. These elements offer a very large acceptance angle, large areas and can be integrated with other optical elements such as photo masks or image sensors. However, not only geometrical properties must be met, but also specific materials properties must be provided. In this chapter the principle concepts of polarizers basing on birefringence, reflection and dichroism are very briefly explained and their limitations are discussed. An overview of commercially available elements is given to set wire grid polarizer in a bigger picture and the characterization of polarizing elements is described. Further the working principle, structural and material requirements for wire grid polarizer are discussed in detail. The fabrication and design is presented. The transmittance spectra of fabricated elements exhibit resonances in the near ultra violet spectral region. It is discussed how these can be utilized to reconstruct the geometry and deduce the performance of the polarizers at much shorter, less accessible, wavelengths in the far ultra violet. Finally, a comparison of different materials for wire grid polarizers in the ultra violet wavelength range is presented.

T. Siefke (✉) · S. Kroker
Physikalisch-Technische Bundesanstalt, Bundesallee 100, 38116 Braunschweig, Germany
e-mail: thomas.siefke@uni-jena.de

S. Kroker
e-mail: s.kroker@tu-braunschweig.de

T. Siefke
Friedrich-Schiller-University Jena, Institute of Applied Physics, Albert-Einstein-Straße 15, 07745 Jena, Germany

S. Kroker
LENA Laboratory for Emerging Nanometrology, Technische Universität Braunschweig, Pockelsstraße 14, 38106 Braunschweig, Germany

O. Stenzel and M. Ohlídal (eds.), *Optical Characterization of Thin Solid Films*, Springer Series in Surface Sciences 64,
https://doi.org/10.1007/978-3-319-75325-6_13

13.1 Introduction

Light, a transversal electromagnetic wave, inherently possesses polarization properties—the mode and direction in which the electric field oscillates. Therefore, the generation and analysis of polarized light is a very important task in nowadays optical technology such as photo lithography [5, 6, 8], photo mask inspection [38, 39] or ellipsometry [30, 31]. These techniques benefit from increased resolution and material specific electronic transitions providing additional information. Thus, polarizers for shorter wavelengths are required. Compared to other possible solutions, wire grid polarizers offer large element sizes, very wide acceptance angles and can be easily integrated e.g. into the photo mask. Therefore, the further development of such elements will enable the future optical technology.

One of the first experimental investigations of the polarization properties of light were conducted by Etienne Louis Malus and published in the year 1808 [15]. Based on these experiments, the transversal character of light waves was further investigated by Augustin Fresnel and Thomas Young. Later, in the year 1861 James Clerk Maxwell published his famous set of equations, from which i.a. polarization properties can be concluded. This, in those days, controversial theory was experimentally supported by Heinrich Hertz in the year 1888 [13]. For his experiments with radio waves at a wavelength of 66 cm, Hertz developed the very first wire grid polarizer. This device consisted of a wooden frame where copper wires of 1 mm diameter were attached in a grid with a period of 3 cm.

During the following decades, this approach of a freestanding grid consisting of individual wires was extended to wavelengths in the far infrared. To circumvent the mechanical limitations regarding the grating period of this approach, George R. Bird and Maxfield Parrish Jr. presented the first wire grid polarizer fabricated by physical vapor deposition of gold or aluminum onto a polymer grating in year 1960 [3]. Fostered by improved nano-fabrication techniques the same approach of metallic gratings on a transparent substrate was further extended to the near ultra violet wavelength range [32–34]. Recently, aluminum wire grid polarizers with a period as small as 33 nm were fabricated by self-assembling diblock copolymers [25]. Unfortunately, however fabricating structures with very small periods is solely not sufficient to achieve a good optical performance. Efficient polarizers require materials with high absolute value of the relative permittivity and extinction coefficient. In metals this is provided by the strong interaction of light with free electrons. However, in the UV or at shorter wavelengths this interaction becomes weaker. Therefore, other materials such as wide bandgap semiconductors with direct band transitions are more suitable materials [27, 28].

In the following chapter, polarization properties in general are discussed, the most important commercially available polarizing elements are introduced and a characterization method is discussed. Then, the working principle of wire grid polarizer is explained and material requirements are deduced. Furthermore, the design of deep ultra violet wire grid polarizers is described and a fabrication route for such elements is discussed. Additionally, resonances occurring in the transmittance spectra are correlated to irregularities induced by the fabrication process deviations. Finally, the application ranges of wire grid polarizer consisting of different materials are compared according to measurements results from literature.

13.2 Polarization of Light

Light, as a transversal electromagnetic wave travelling along the z-axis, can be described by a superposition of two individual waves. According to a Cartesian decomposition the electric fields E_x and E_y can be defined as:

$$E_{\mathrm{x}}(z,t) = E_{0,\mathrm{x}} \cos(\omega t - kz), \tag{13.1}$$

$$E_{\mathrm{y}}(z,t) = E_{0,\mathrm{y}} \cos(\omega t - kz + \delta). \tag{13.2}$$

$E_{0,x}$, $E_{0,y}$ are the electric field amplitudes and δ is the phase between the two waves. Each polarization state can be described by the superposition of these two orthogonal waves. In Fig. 13.1a two orthogonal waves are oscillating in phase with the same amplitude. The superposition of these two waves leads to a linear polarized wave oscillation along the diagonal plane. By introducing a phase difference of $\pi/2$ between the two orthogonal waves so called circular polarized light can be achieved. The tip of the resulting electric field vector moves on a circle projected to the x-y-plane (see Fig. 13.1b)) [7, 12]. By the relation between the amplitudes and phase, arbitrary polarization states can be achieved. The denomination of the polarization states is summarized in Table 13.1.

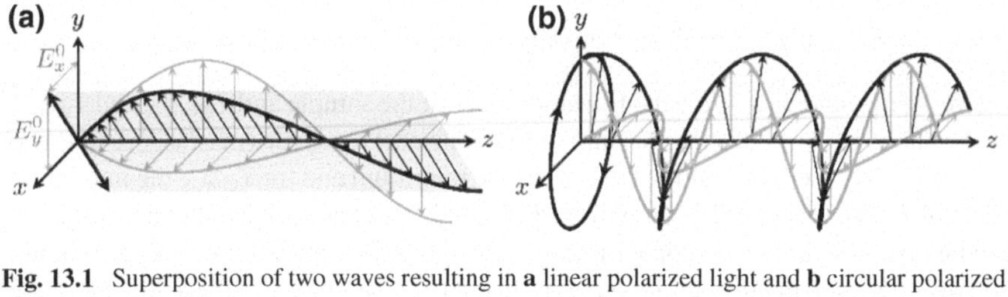

Fig. 13.1 Superposition of two waves resulting in **a** linear polarized light and **b** circular polarized light

Table 13.1 Polarization states

Polarization state	Condition
Linear	$E_{0,\mathrm{x}}$ and $E_{0,\mathrm{y}}$ arbitrary, $\delta = \pm 2\pi N$ for $N \in \mathbb{N}_0$, or equivalent $E_{0,\mathrm{x}} = 0$ or $E_{0,\mathrm{y}} = 0$
Circular	$E_{0,\mathrm{x}} = E_{0,\mathrm{y}}$, $\delta = \pi/2 \pm 2\pi N$ for $N \in \mathbb{N}_0$
Elliptical	$E_{0,\mathrm{x}}$, $E_{0,\mathrm{y}}$ and δ arbitrary

Many light sources emit light whose polarization state changes very fast and statistically. Such light is called natural light. The often used term "unpolarized" is somewhat misleading, since the light is actually polarized at a specific point in space and time. The duration in which the polarization state is not altered significantly is called polarization time [1], which may be much shorter than resolvable by a detector. Hence, the light appears to be not polarized. The here discussed wire grid polarizer are linear polarizers, i.e. the transmitted light is linear polarized.

13.3 Characterization of Polarizing Elements

Assuming an ideal linearly polarized beam passing through a ideal linear polarizer, it is possible to find an orientation where the maximum of the light is transmitted. Orthogonal to this orientation a minimum transmittance is found. The transmittances along the two orthogonal axes T_{trans} and T_{block} can be utilized for the characterization. The suppression of the transmittance along the undesired polarization direction can be expressed by the extinction ratio:

$$Er = \frac{T_{\text{trans}}}{T_{\text{block}}}. \tag{13.3}$$

Practically, the spectral characterization of a polarizer can be performed by using a spectrophotometer equipped with a reference polarizer, typically called analyzer (placed either in front or behind the sample). The measured transmittance T for different angles Θ between sample polarizer and analyzer is given by:

$$\begin{aligned} T =& 1/4(T_{\text{block,S}} + T_{\text{trans,S}})(\mathrm{T}_{\text{block,A}} + T_{\text{trans,A}}) \\ &+ 1/4 \cos(\Theta) \times (T_{\text{block,S}} - T_{\text{trans,S}})(T_{\text{block,A}} - T_{\text{trans,A}}), \end{aligned} \tag{13.4}$$

where $T_{\text{block,S}}$ and $T_{\text{trans,S}}$ are the transmittances of the sample and $T_{\text{block,A}}$ and $T_{\text{trans,A}}$ of the analyzer, respectively. Assuming that a perfect analyzer (i.e. $T_{\text{block,A}} = 0$ and $T_{\text{trans,A}} = 1$) is used, the value of $T_{\text{block,S}}$ is the minimum and for $T_{\text{trans,S}}$ the maximum value of $I(\Theta)$, respectively. However, if the analyzer is not perfect the measurement can be invalidated. For practical purposes the extinction ratio of the analyzer should be two orders of magnitude larger than that of the sample. Or alternatively, if the performance of the analyzer is known, (13.4) can be used to correct the results. Additionally, the polarization effects inherent to the device, e.g. due to reflection at internal mirrors must be considered.

13.4 Common Elements for Polarization Control

For polarization control, asymmetrical or anisotropic phenomena are required, in particular birefringence, dichroism and reflection are of technological importance.

Table 13.2 Comparison of several commercially available polarizers

	Wavelength range (nm)	Transmittance (%)	Extinction ratio	φ	Area	Cost
Birefringent prism Glan-Thompson	200–5000	>95	$> 10^6$	–	–	–
Rochon	130–7000	>95	$> 10^5$	–	–	–
Dichroic Glass	400–5000	25–40	10^4	o	+	+
Polymer	400–700	30–40	$> 10^3$	o	+	+
Wire grid	300–30000	50–80	$> 10^3$	+	o	o
Reflection	200–30000 (single wavelength)	>95	$> 10^3$	–	o	+

In the following section a brief overview of existing technological solutions and their respective capabilities is given and summarized in Table 13.2.

Birefringent materials have a refractive index which depends on the propagation direction and the polarization of the incident light. For the fabrication of polarizers uniaxial materials e.g. calcite, magnesium fluoride or α-BBO are typically utilized. Two prism of such a birefringent material are polished at suitable angles and attached to each other. At the interface, one of the orthogonal polarized components of the incident light experience either total internal reflection or is differently refracted than the other. Birefringent prism polarizers offer extremely high extinction ratios up to 10^6 and high transmittances. The application wavelength range is determined by the utilized material and can range from the far ultra violet to the far infrared. However, these elements are typically rather bulky and the angle of incidence is limited to a few 10°. Due to the limited available crystal size such components typically have clear apertures of a few 10 mm [7, 12].

Dichroic materials show an optical anisotropy of the extinction coefficient, comparable to birefringence. The absorption for polarized light oriented parallel to a certain axis is much larger than for orthogonal polarized light. Commercially, there are essentially two types of dichroic polarizer available, basing on polymer sheets and glass plates. They are fabricated by stretching either a thin foil of polymer [12] or metallic nanoparticle loaded glass sheets at elevated temperatures. Thereby an anisotropy is achieved. These can be very cheap elements with areas in the range of several square meters. However these components are temperatures sensitive (especially those basing on polymer), the extinction ratio and transmittance is limited. The application wavelength range is typically limited to the VIS to the mid-infrared spectral region.

Wire grid polarizers belong to this group of dichroic polarizers, their mode of operation will be subsequently discussed in detail. Commercially available wire grid polarizers offer application ranges from about 300 nm in the UV to 30 μm in the far infrared. The achievable extinction ratios are of about 10^4 at transmittances of

50–80%. A particular advantage of wire grid polarizers is their large acceptance angle covering almost the entire half space.

Reflection at an interface depends in general on the polarization of the incident light. At the Brewster's angle the reflection of p-polarized light drops to zero, hence the extinction ratio for the reflected light becomes, theoretically, infinity. However, this can only be achieved at a specific angle and wavelength. Hence these devices are very sensitive to misalignment or divergence of the incident beam. This can be partially circumvented by the utilization of multiple plates or by utilizing multiple thin films and allows for use in transmittance. Practically, this kind of polarizer overs extinction ratios of about 1000 and allows deviation of the incident angle of a few degrees [21].

13.5 Wire Grid Polarizers

Wire grid polarizers (WGPs) are nano-optical grating structures (see Fig. 13.2). The polarization direction of the incident light is denominated according to the orientation between the electrical field vector and the ridges. The parallel orientation is denominated as TE (transversal electric) and orthogonal as TM (transversal magnetic), respectively.

To avoid diffraction at the grating structure the period p has to obey the zero order grating condition (ZOG) (compare (12.5):

$$p < \frac{\lambda}{(n_{\text{sub}} + \sin\varphi)}, \tag{13.5}$$

where λ is the application wavelength, n_{sub} is the refractive index of the substrate and φ the angle of incidence (x-z plane, see Fig. 13.2). Technologically very small periods can be achieved [25], hence (13.5) can be fulfilled even for grazing incidence i.e. $\varphi \sim 90°$ in the visible and infrared spectral range. Therefore, elements with very large acceptance angles are feasible, rendering wire grid polarizer superior to other solutions.

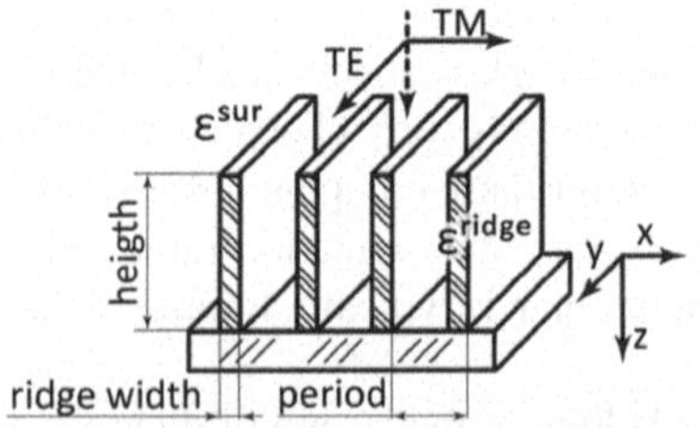

Fig. 13.2 Schematic wire grid polarizer. Incident light which is polarized parallel to the ridges is called transversal electric (TE), light which is orthogonal to the ridges is called transversal magnetic (TM)

To understand the working principle of WGPs the polarization dependent conditions for electric fields $\vec{E}$ [4] at the horizontal surfaces between the ridge material and the surrounding medium (here air or vacuum) can be considered:

$$\vec{E}_{\mathrm{TE}}^{\mathrm{ridge}} = \vec{E}_{\mathrm{TE}}^{\mathrm{vac}} \quad \text{for TE} \tag{13.6}$$

$$\varepsilon^{\mathrm{ridge}} \vec{E}_{\mathrm{TM}}^{\mathrm{ridge}} = \varepsilon^{\mathrm{vac}} \vec{E}_{\mathrm{TM}}^{\mathrm{sur}} \quad \text{for TM} \tag{13.7}$$

For TE-polarized light, the electric field is continuous at the interfaces (see (13.7). Therefore, the distortion of a plane wave incident on the WGP is small (see Fig. 13.3a). For a material with nonzero extinction coefficient the light is exponentially damped. This can be approximated by the Beer-Lambert law:

$$T_{\mathrm{TE}} = \mathrm{e}^{-\frac{4\pi k_{\mathrm{eff}} z}{\lambda}}. \tag{13.8}$$

For a wavelength λ the transmittance of TE polarized light becomes small if the ridge height z becomes large, or if the effective extinction coefficient k_{eff} of the grating becomes large. This means that for a wire grid polarizer a material with a large extinction coefficient must be utilized.

For TM polarized the dielectric displacement is continuous at the surfaces (see (13.7). Regarding the absolute value of the electric field 3 different cases can be distinguished.

(I) $\left|\boldsymbol{\varepsilon}^{\mathbf{ridge}}\right| \gg \left|\boldsymbol{\varepsilon}^{\mathbf{sur}}\right|$

If the absolute value of the permittivity of the ridges is much larger than that of the surrounding medium the electric field is mainly localized in the gap between the ridges. Hence, the transmittance of TM polarized light is very high (see Fig. 13.3b). This is the preferred operation mode for WGPs.

(II) $\left|\boldsymbol{\varepsilon}^{\mathbf{ridge}}\right| \gtrsim \left|\boldsymbol{\varepsilon}^{\mathbf{sur}}\right|$

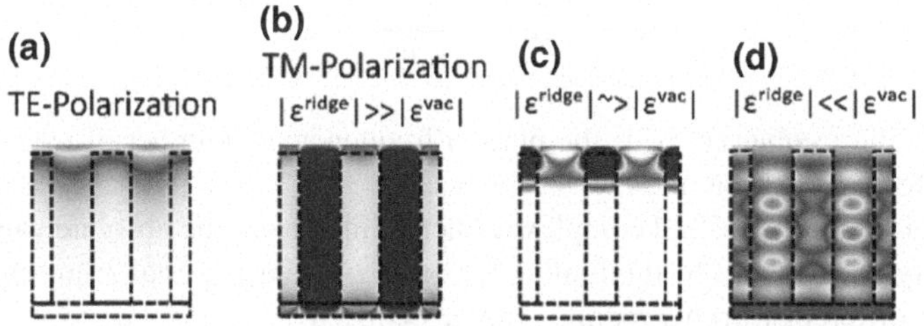

Fig. 13.3 Schematic electric field distribution in a wire grid polarizer. The dashed lines denote the position of the grating ridges and the substrate. **a** For TE polarized light and a grating material with nonzero extinction coefficient. **b** For a material whose absolute relative permittivity is much larger than that of the surrounding medium and **c** for a material whose absolute relative permittivity is much smaller than that of the surrounding medium

If the absolute value of the permittivity of the ridges approaches that of the surrounding medium the electric field penetrates into the ridges thus the transmittance of TM polarized light is considerably smaller. This operation mode is less expedient or may not be applicable at all.

(III) $\left|\varepsilon^{\text{ridge}}\right| \ll \left|\varepsilon^{\text{sur}}\right|$

If the absolute value of the permittivity of the ridges is much smaller than that of the surrounding medium the electric field is mainly localized in the ridges. Since the grating material must have a nonzero extinction coefficient the electric field is strongly absorbed (see Fig. 13.3c). The transmittance of TM polarized light is very low. Under special circumstances the transmittance of TM polarized light can become smaller than that of TE polarized light. This is than called inverse polarization effect. Typically this is only feasible in a very narrow spectral range of some ten nanometers [18].

The absolute value of the permittivity is coupled to the refractive index n and the extinction coefficient k:

$$|\varepsilon| = n^2 + k^2. \tag{13.9}$$

In summary, to achieve WGPs with a large extinction ratio and a large transmittance a material with a large extinction coefficient and a large absolute value of the permittivity is necessary.

In the visible and infrared wavelength range this can be easily achieved by using metals such as aluminum, gold or silver. In these wavelength regions, these materials can be appropriately described by the Drude model. Thereby the relative permittivity is expressed as [9]:

$$\varepsilon(\omega) = 1 - \frac{{\omega_{\mathrm{p}}}^2}{\left(\omega^2 + \mathrm{i}\gamma\omega\right)} \tag{13.10}$$

$$\omega_{\mathrm{p}} = \sqrt{\frac{n_{\mathrm{e}} e^2}{\varepsilon_0 m_{\mathrm{e}}}} \tag{13.11}$$

where ω is the frequenzy, ω_{p} is the plasma frequency, n_{e} number of free electrons and m_{e} effective electron mass.

Figure 13.4 the calculated complex refractive index and the absolute value of the permittivity are shown for aluminum. According to the material requirements for WGP four different working regimes can be deduced:

(A) For large wavelength in the visible and infrared wavelength regime both the extinction coefficient and the relative permittivity are large. Hence, this material is very well suited.
(B) Towards shorter wavelength in the near ultra violet, both the extinction coefficient and the relative permittivity decrease. Hence the optical performance of such WGPs decreases as well.

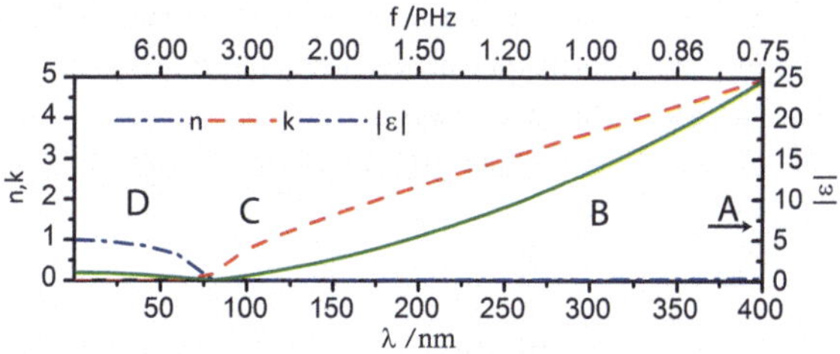

Fig. 13.4 Calculated refractive index and relative permittivity for aluminum according to the Drude model

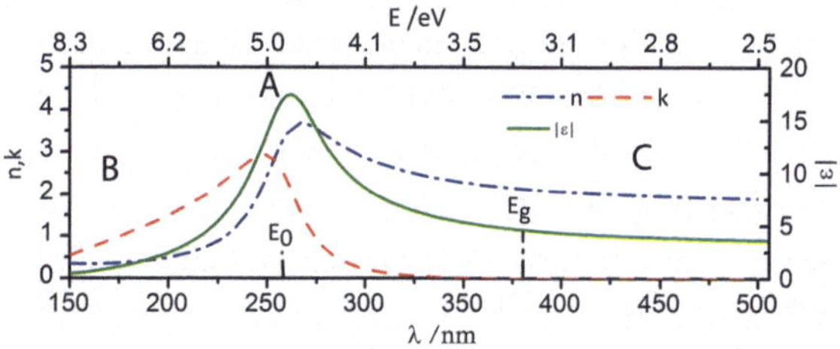

Fig. 13.5 Calculated complex refractive index according to the Tauc-Lorentz model (see 13.12)

(C) At wavelengths slightly larger than the plasma frequency the relative permittivity becomes smaller than that of the surrounding medium. Hence, an inverse WGP can possibly be achieved.

(D) For wavelength below the plasma frequency the extinction coefficient of the material becomes very small, hence such a material cannot be utilized for WGP.

From these considerations it can be concluded that the optical performance of WGPs consisting of Drude metals will diminish towards the ultra violet spectral range and will be not suitable below a certain wavelength. This behavior is experimentally observed for several materials such as iridium [34] or aluminum [23].

Especially, in the ultraviolet spectral range many materials show interband transitions. If the energy of an incident photon is larger than the bandgap energy of a material, electrons can be excited to a state in the conduction band. This absorption phenomenon becomes dominant for wide bandgap semiconductors such as titanium dioxide in the ultra violet. These interband absorption processes can be described by a Tauc-Lorentz model [14]

$$\mathrm{Im}\{\varepsilon\} = \frac{AE_0\Gamma\left(E - E_g\right)^2}{\left(E^2 - E_0^2\right)^2 + \Gamma^2 E^2}\frac{1}{E} \quad \text{for } E > E_g$$

$$\mathrm{Im}\{\varepsilon\} = 0 \quad \text{for } E \leq E_g \tag{13.12}$$

where E is the energy of the incident photon, E_g is the bandgap energy, E_0 the peak transition energy and A describes the strength of the oscillator. The real part of the permittivity can be obtained applying the Kramers–Kronig relation.

It is again possible to find several working regimes for WGPs as marked in Fig. 13.5.

(A) In the vicinity of the peak transition energy E_0 both the extinction coefficient and the relative permittivity are large. Therefore, the material is well suited for WGPs at this wavelength.
(B) At smaller wavelengths both the extinction coefficient and the relative permittivity are reduced. Thus the material is less suited at this wavelength, if applicable at all.
(C) For higher photon energies other transitions can usually be excited [24]. Hence, the relative permittivity does not drop below that of the surrounding vacuum and no inverse polarization effect can be observed.
(D) At smaller energies than the bandgap energy, interband transitions cannot be excited. The extinction coefficient is virtually zero. Hence this material cannot be applied as wire grid polarizer.

It can be concluded that wide bandgap semiconductors with a peak transition energy at the targeted application wavelength can be applied as material for WGPs. Noteworthy, the dominant loss mechanism for WGPs based on interband transition processes is absorption. Therefore incident TE polarized light is not reflected as for WGPs based on metals. This avoids the back reflection towards the incident beam which must otherwise be taken care of e.g. for laser applications.

13.6 Fabrication of Wire Grid Polarizers

For the fabrication WGPs, a variety of methods are available. Since small periods are necessary, self-aligned double pattering (SADP) processes are often applied [2, 29, 34] to relax the requirements on the lithography.

For the SADP process an initial layer stack is fabricated. First a phenolic resin is applied by spin coating and is then fully cured by subsequent thermal treatment. Afterwards a chromium layer is deposited by ion beam deposition. Finally, an electron beam lithography resist is applied by means of spin coating. This is then structured by character projection electron beam lithography (see Fig. 13.6a). This technique employs a patterned mask in the electron beam path e.g. with a grating structure, which is then printed at an area of several micrometer in one shot. This reduces the writing time by several orders of magnitude compared to serial electron beam lithography approaches [16]. This grating is then transferred into the chromium layer by ion beam etching. This patterned chromium layer serves as hard mask to enhance the selectivity for the subsequent reactive ion beam etching with oxygen for the phenolic resin layer (see Fig. 13.6b). This achieved polymer grating is then utilized as template onto which the target material for the WGP is deposited by means of atomic layer deposition [26, 32, 33] or ion beam deposition [29] (see Fig. 13.6c). Finally, the material on the horizontal surfaces is removed by ion beam etching, and the template grating is removed by reactive ion beam etching. Thereby the final WGP structure is achieved (see Fig. 13.6c) [27–29].

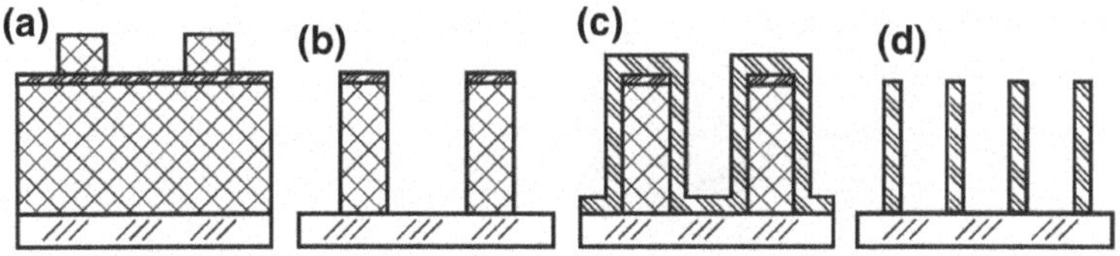

Fig. 13.6 Schematic fabrication process for a wire grid polarizer: **a** Fabrication of an initial layer stack and lithography **b** Patterning of the template grating **c** Deposition of the grating material and **d** Removal of the material on the horizontal surfaces as well as the template grating

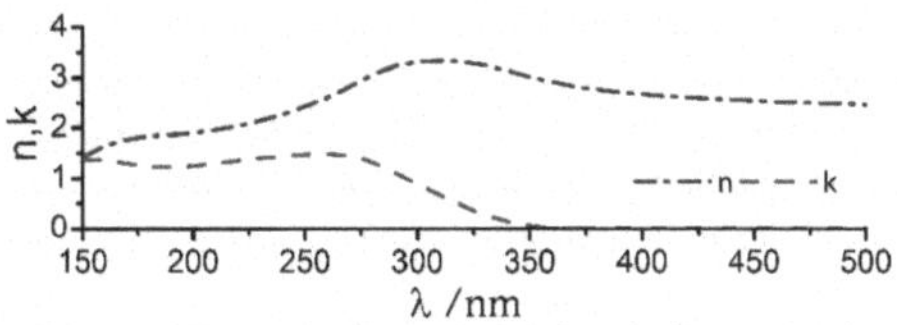

Fig. 13.7 Measured complex refractive index of titanium dioxide deposited by atomic layer deposition

13.7 Design and Characterization of Titania Wire Grid Polarizers

In the following section the design of WGPs is discussed exemplified at an element consisting of titanium dioxide. Usually the two optical target parameter extinction ratio and transmittance of TM polarized light must be met. Since, large transmittance and extinction ratio are contradictory, a tradeoff must be found. Therefore many commercially available wire grid polarizer are offered either as high transmittance or high contrast option.

It is known that the optical properties of the utilized materials have a great influence to the performance of the final element [29] and may strongly vary between different deposition techniques and tools [19]. Hence, the optical properties of the intended material deposited by the actual deposition process should be measured prior to the design. This can be accomplished by ellipsometry [10, 11] (see Chaps. 8 and 9). In Fig. 13.7 such measurement results for the complex refractive index for the here used titanium dioxide fabricated by atomic layer deposition [26] is shown.

To begin with, the period is determined. In a first place the zero order grating condition (see (13.5)) must by obeyed. Furthermore, fabrication technique and target period must be balanced. In this example a period of 104.5 nm was utilized.

Next, the ridge width and height are determined. As previously shown, a WGP can be understood as a form- birefringent thin film. The polarization dependent refractive indices can be approximated by the effective medium theory (EMT). Therefrom the transmittances can be calculated according to the Fresnel equation [20, 37]. While this approach is elegant and demonstrative it contains some simplification. For the effective medium approximation a period much smaller than that of the incident light

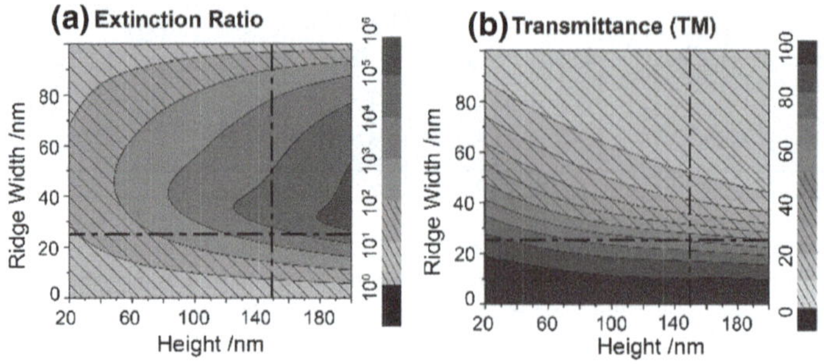

Fig. 13.8 Simulated dependency of the **a** extinction ratio and **b** transmittance of a titanium wire grid polarizer on the ridge width and height. The areas with an extinction ratio below 100 and a transmittance below 50% are hatched. The chosen target parameter ridge width of 26 nm and a height of 100 nm are marked. The simulation was performed at 248 nm wavelength and normal incidence

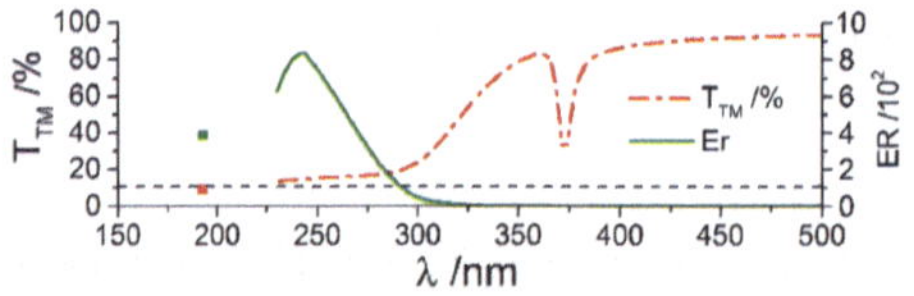

Fig. 13.9 Measured transmittance and extinction ratio of a titanium dioxide WGP. The dotted line marks an extinction ratio of 100. If the measured extinction ratio is above this level the wire grid polarizer is regarded as applicable, thereby the application ranges are deduced

is required [17]. Especially for DUV-WGPs this is hardly the case. Thus, numerical methods like rigorous coupled wave analysis (RCWA) are necessary [22]. By such a method the dependency of the extinction ratio and transmittance of TM polarized light on ridge width and height (see Fig. 13.2) is simulated (see Fig. 13.8). By comparison with the targeted parameters for the optical performance (here transmittance larger than 50% and extinction ratio lager 100) a set of geometrical parameters can be chosen. Here we chose a ridge width of 26 nm and a height of 150 nm.

According to this design titanium dioxide polarizer were fabricated. The measured optical performance is shown in Fig. 13.9. At 193 nm wavelength an extinction ratio of 385 was achieved. This is currently the largest reported value at this wavelength.

However, the measured values are below those expected from the design. The actual performance of a wire grid polarizer is strongly sensitive towards process induced geometrical deviations [29]. At the one hand this requires a close feedback between design and fabrication technique at the other hand necessitates a tight process control. This process control can be based on the evaluation of resonance features in the transmittance of TM polarized light, here arising at about 375 nm wavelength. The reasons for that are not equidistantly positioned and tilted ridges [27]. Figure 13.10a shows a STEM image of a fabricated titanium dioxide wire grid polarizer [28] where these features are visible.

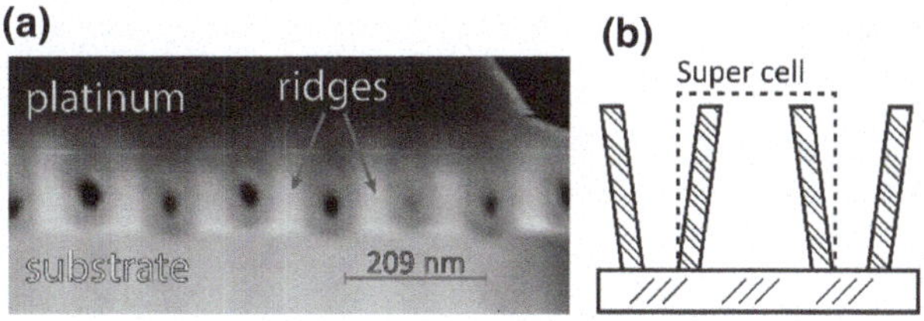

Fig. 13.10 **a** STEM image of a titanium dioxide wire grid polarizer. **b** Schematic of the distorted structure

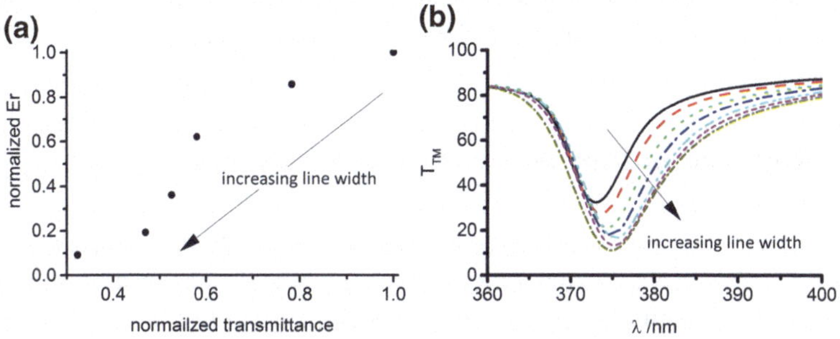

Fig. 13.11 **a** Correlation between depths of the dips in the transmittance of TM polarized light at about 375 nm and measured extinction ratio Er at 248 nm wavelength. **b** Measured transmittance for increasing linewidth from 76 nm to 95 nm

From these asymmetries results a supercell with the period of the initial template grating (see Fig. 13.10b). The period of this supercell of 209 nm obeys the ZOG (see (13.5), therefore the dip cannot be related to the occurrence of diffraction. But, since the condition:

$$\lambda \geq n_{\text{sub}}\Lambda \tag{13.13}$$

is fulfilled, guided mode resonance (for more information see Chap. 12) can occur for wavelength larger than 313 nm. According to the theory of perturbed waveguides the strength of the resonance, here the depth of the dip, is proportional to the perturbation i.e. the asymmetry. This asymmetry in turn leads to a reduction of the extinction ratio at shorter wavelengths [29]. Therefore, the depth of the dip can be correlated with the extinction ratio at the targeted wavelength. This was experimental verified by variation of the width of the initial template grating from 76 nm to 95 nm and evaluation of the extinction ratio at 248 nm and the minimum transmittance (see Fig. 13.11b) in the resonance (see Fig. 13.11a).

The evaluation of such resonances in the near ultra violet can be utilized to predict the performance of WGPs at much shorter wavelength. This is particular beneficial if the applications wavelength is in the far ultra violet where special devices and greater experimental effort is required. By evaluation of further properties of the

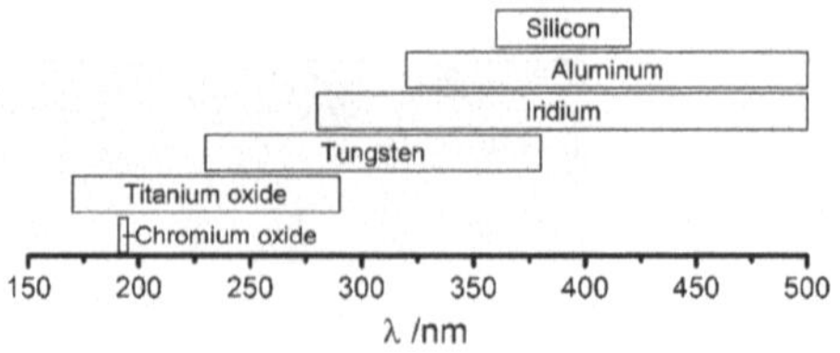

Fig. 13.12 Application range for wire grid polarizer consisting of different materials in the ultra violet spectral range [2, 27, 28, 32–35]

resonance such as depth, position, spectral width and angle dependence the original grating structure can be reconstructed in more detail [36]. This information can then be utilized to control and improve the fabrication technology.

13.8 Application Ranges for Different Materials

As discussed previously, the choice of the material for WGPs determines the optical performance and the wavelength range where a specific element can be utilized. To compare spectral application ranges of WGPs consisting of different materials we utilize an extinction ratio of 100 as criterion. If the extinction ratio is larger than this we regard the element as applicable. A suppression of the undesired polarization direction by two orders of magnitude seems to be appropriate for many purposes, although for a specific application other values may be more useful. However, a general conclusion can be drawn from this (Fig. 13.12).

13.9 Outlook

The application of wire grid polarizers has a long history in optics science. With the development of the modern fabrication and characterization technology it became feasible to fabricated elements for the infrared, visible and near ultraviolet spectral region. In the future, the accessible spectral range will be further shifted towards even shorter wavelength in the far ultraviolet spectral range. As demonstrated in this chapter, other materials have to be utilized than those which are currently used in commercially available elements.

Acknowledgements The authors would like to acknowledge the support by the German Science Foundation (project PolEx (KR4768/1-1) and NanoMet (GrK 1952/1)).

References

1. S. Andriy, R. Matthieu, T.F. Ari, T. Setälä, Polarization time of unpolarized light. Optica **4**(1), 64–70 (2017)
2. K. Asano, S. Yokoyama, A. Kemmochi, T. Yatagai, Fabrication and characterization of a deep ultraviolet wire grid polarizer with a chromium-oxide subwavelength grating. Appl. Opt. **53**, 2942–2948 (2014)
3. G.R. Bird, M. Parrish Jr, The wire grid as a near-infrared polarizer. J. Opt. Soc. Am. **9**(50), 886–891 (1960)
4. M. Born, E. Wolf, *Principles of Optics*, 7th(expanded) edn. (n.d.)
5. Y. Bourgin, T. Siefke, T. Käsebier, P. Genevée, A. Szeghalmi, E.-B. Kley, U.D. Zeitner, Double-sided structured mask for sub-micron resolution proximity i-line mask-aligner lithography. Opt. Express **23**(13), 16628–16637 (2015)
6. Y. Bourgin, T. Siefke, T. Käsebier, P. Genevée, A. Szeghalmi, E.-B. Kley, U.D. Zeitner, Double-sided diffractive photo-mask for sub-500 nm resolution proximity i-line mask-aligner lithography, in *SPIE Proceedings*, vol. 9426, 94260E (2015)
7. W. Demtröder, *Experimentalphysik 2*. (Springer Spektrum, Berlin, 2013)
8. D. Flagello, B. Geh, S. Hansen, M. Totzeck, Polarization effects associated with hyper-numerical-aperture (> 1) lithogrphy. J. Micro/Nanolith. MEMS MOEMS. **4**, 3 (2005)
9. M. Fox, *Optical Properties of Solid*, 2nd edn. (Oxford University Press, Oxford, 2010)
10. D. Franta, I. Ohlídal, Optical characterization of inhomogeneous thin films of ZrO2 by spectroscopic ellipsometry and spectroscopic reflectometry. Surf. Interface Anal. **30**, 574–579 (2000)
11. D. Franta, I. Ohlídal, P. Klapetek, P. Pokorný, Characterization of the boundaries of thin films of TiO2 by atomic force microscopy and optical methods. Surf. Interface Anal. **34**, 759–762 (2002)
12. E. Hecht, *Optik*. Oldenbourg: Oldenbourg Wissenschaftsverlag (2009)
13. H. Hertz, Ueber Strahlen elektrischer Kraft. *Koeniglich Preussische Akademie der Wissenschaften* (1888)
14. G.E. Jellison, F.A. Modine, Parameterization of the optical functions of amorphous materials in the interband. Appl. Phys. Lett. **69**, 371–373 (1996)
15. B. Kahr, K. Claborn, The lives of Malus and his bicentennial law. ChemPhysChem **9**, 43–58 (2008)
16. E.-B. Kley, H. Schmidt, U. Zeitner, M. Banasch, B. Schnabel, Enhanced E-beam pattern writing for nano-optics based on character projection, in *Proceedings of SPIE*, vol. 8352 (2012)
17. P. Lalanne, J.-P. Hugonin, High-order effective-medium theory of subwavelength gratings in classical mounting: application to volume holograms. JOSA A **15**, 1843 (1998)
18. A. Lehmuskero, B. Bai, P. Vahimaa, M. Kuittinen, Wire-grid polarizers in the volume plasmon region. Opt. Express **17**, 5481–5489 (2009)
19. A. Lehmuskero, M. Kuittinen, P. Vahimaa, Refractive index and extinction coefficient dependence of thin Al and Ir films on deposition technique and thickness. Opt. Express **15**(17), 10744 (2007)
20. Y.-L. Liao, Y. Zhao, Design of wire-grid polarizer with effective medium theory. Opt. Quant Electron. **46**, 641–647 (2014)
21. A. Macleod, Thin film polarizers and polarizing beam splitters. SVC Bulletin (Summer), 24–29 (2009)
22. M. Moharam, T. Gaylord, Rigorous coupled-wave analysis of metallic surface-relief gratings (1986)
23. MOXTEC. (n.d.). *ProFlux® UV Series Datasheet*
24. D.A. Papaconstantopoulos, *Handbook of the Band Structure of Elemental Solids*. Plenum Press, New York (1986)
25. V. Pelletier, K. Asakawa, M. Wu, D.H. Adamson, R.A. Register, P.M. Chaikin, Aluminum nanowire polarizing grids: Fabrication and analysis. Appl. Phys. Lett. **88**, 211114 (2006)
26. S. Ratzsch, E.-B. Kley, A. Tünnermann, A. Szeghalmi, Influence of the oxygen plasma parameters on the atomic layer deposition of titanium dioxide. Nanotechnology **26**, 024003 (2015)

27. T. Siefke, E.-B. Kley, A. Tünnermann, S. Kroker, Design and fabrication of titanium dioxide wire grid polarizer for the far ultraviolet spectral range, in *Proceedings of SPIE*, vol. 992706, 992706 (2016)
28. T. Siefke, S. Kroker, K. Pfeiffer, O. Puffky, K. Dietrich, D. Franta, T. Siefke, Materials pushing the application limits of wire grid polarizers further into the deep ultraviolet spectral range. Adv. Opt. Mater. (2016)
29. T. Siefke, D. Lehr, T. Weber, D. Voigt, E. Kley, A. Tünnermann, Fabrication influences on deep-ultraviolet tungsten wire grid polarizers manufactured by double patterning. Opt. Lett. **39**, 6434–6437 (2014)
30. H.G. Tompkins, E.A. Irene, *Handbook of Ellipsometry* (Springer, Norwich, NY, 2005)
31. H.G. Tompkins, T. Zhu, E. Chen, Determining thickness of thin metal films with spectroscopic ellipsometry for applications in magnetic random-access memory. JVSTA **16**, 1297–1302 (1998)
32. T. Weber, T. Kaesebier, E.-B. Kley, S. Babin, G. Glushenko, A. Szeghalmi, Application of double pattering technology to fabricate optical elements: Process simulation, fabrication, and measurement. J. Vac. Sci. Technol. B **30**(3), 2166 (2012)
33. T. Weber, T. Käsebier, M. Helgert, E.-B. Kley, A. Tünnermann, Tungsten wire grid polarizer for applications in the DUV spectral range. Appl. Opt. **51**, 3224–3227 (2012)
34. T. Weber, T. Käsebier, B. Kley, A. Tünnermann, Broadband iridium wire grid polarizer for UV applications. Opt. Lett. **36**, 445–447 (2011)
35. T. Weber, S. Kroker, T. Käsebier, B. Kley, A. Tünnermann, Silicon wire grid polarizer for ultraviolet applications. Appl. Opt. **53**, 8140–8144 (2014)
36. M. Wurm, F. Pilarski, B. Bodermann, A new flexible scatterometer for critical dimension metrology. Rev. Sci. Instrum. **81**, 023701 (2010)
37. P. Yeh, A new optical model for wire grid polarizer. Opt. Commun. **3**(26), 289–292 (1978)
38. A.M. Zibold, W. Harnisch, T. Scherübl, N. Rosenkranz, J. Greif, Using the aerial image measurement technique to speed up mask development for 193 nm immersion and polarization lithography, in *Proceedings of SPIE* (2004)
39. A. Zibold, U. Strössner, N. Rosenkranz, A. Ridley, R. Richter, W. Harnisch, A. Williams, First results for hyper NA scanner emulation from AIMS 45-193i, in *Proceedings of SPIE*, vol. 628312 (2006)

Part IV
Scatter and Absorption

Chapter 14
Roughness and Scatter in Optical Coatings

M. Trost and S. Schröder

Abstract The measurement of light scattering from optical components has received increased attention in the last decade. In addition to being a serious source of noise, light scattering can lead to a reduced optical throughput, degrade the imaging quality, or cause straylight in optical systems. On the other hand, the high sensitivity towards small imperfections and inhomogeneities makes light scattering measurements a powerful inspection tool. Measured light scattering is a good indicator of the surface quality and can be used to characterize the surface roughness or local defects. After an introduction of the main scattering quantities as well as their standardization, this chapter will focus on the instrumentation used to characterize light scattering and provide various application examples ranging from the light scattering based roughness characterization of supersmooth substrates to the roughness evolution of different multilayer coatings and its impact on the scattering properties.

Light scattering from optical components can critically affect and influence the performance of optical systems. An apt example for this is the diffusely reflected light from this white page or computer monitor [1, 2], which creates a homogenous, bright background, making it possible to easily read the black printed letters. Further examples are diffuse materials or rough surfaces which are used for a similar purpose in order to achieve a homogenous illumination in an optical system or to trap light inside a coating. This can for instance be used to increase the efficiency of thin film solar cells [3, 4].

But, even high quality optical components exhibit small imperfections and inhomogeneities that can lead to a reduced optical throughput, degrade the imaging quality, or cause straylight, which more often than not became an unexpected source of practical difficulties in optical systems [5].

M. Trost (✉) · S. Schröder
Fraunhofer Institute for Applied Optics and Precision Engineering IOF, Albert-Einstein-Str. 7, 07745 Jena, Germany
e-mail: Marcus.Trost@iof.fraunhofer.de

S. Schröder
e-mail: Sven.Schroeder@iof.fraunhofer.de

O. Stenzel and M. Ohlídal (eds.), *Optical Characterization of Thin Solid Films*, Springer Series in Surface Sciences 64,
https://doi.org/10.1007/978-3-319-75325-6_14

The origins for light scattering are manifold and range from residual surface roughness and contaminations [6–8] to local defects on the surface (scratches, pits, etc.) [9, 10], small cracks just below the surface (subsurface damage) [11, 12], or inclusions inside the bulk material [13, 14].

For thin film coatings, each interface inside the multilayer stack leads to light scattering [15–18]. Depending on the actual multilayer design and the roughness properties of all interfaces, these contributions can interfere destructively or constructively. Thus, the light scattering may be reduced or increased in certain scattering directions compared to a single highly reflective surface [19, 20]. But, also for coatings that exhibit no sharp interfaces between the layer materials, such as a Rugate coating, light scattering can occur from refractive index fluctuations [21].

The high sensitivity towards small imperfections on the other hand makes light scattering measurements a powerful inspection tool [6, 7, 22]. Because of the non-contact and robust data acquisition, the surface quality can often be advantageously characterized over large areas or even during the actual manufacturing process [23]. This allows controlling the ever increasing roughness and defect requirements of optical components.

This chapter is therefore intended to provide an insight into the scattering measurement principles for optical components and to give an overview about the scattering mechanism of substrates and multilayer coatings.

14.1 Definitions and Standards

Light scattering quantities for optical components can be divided into two major categories: Angle Resolved Scattering (ARS) and Total Scattering (TS). In the following, an overview about these two quantities as well as their standardization will be given.

The basic geometry and the nomenclature used for the definition of the specularly reflected and transmitted beams as well as light scattering are shown in Fig. 14.1.

The sample is illuminated at an angle of incidence φ_i and besides the reflected and transmitted beam at the angles φ_r and φ_t, part of the light is scattered into off-specular directions described by the azimuthal, ϕ_s, and polar, φ_s, scattering angles. The orientation of the sample with respect to the incoming beam is defined by the angle ϕ_i. All angles are measured with respect to the macroscopic sample normal.

14.1.1 Angle Resolved Scattering

The most important quantity in the discussion of scattering is the angle resolved scattering, ARS, which describes the relative amount of light scattered into a certain direction. This is achieved by normalizing the scattered light power ΔP_s into the solid angle $\Delta \Omega_\mathrm{s}$ to the incident power P_i and the solid angle [24]:

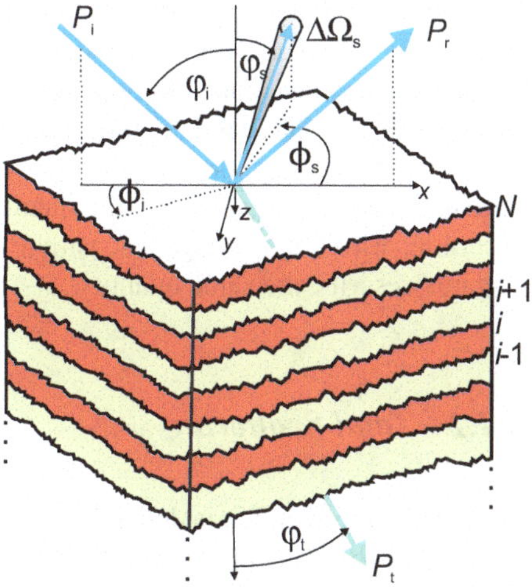

Fig. 14.1 Geometry for the definitions of specular quantities and scattering

$$\mathrm{ARS}\left(\varphi_{\mathrm{s}},\phi_{\mathrm{s}}\right)=\frac{\Delta P_{\mathrm{s}}\left(\varphi_{\mathrm{s}},\phi_{\mathrm{s}}\right)}{P_{\mathrm{i}}\Delta\Omega_{\mathrm{s}}}. \tag{14.1}$$

Besides the two scattering angles, the ARS also depends on the angle of incidence, the light wavelength, and the polarization of the incident and scattered light and is thus a multidimensional function.

Other common terms used to express angle resolved scattering are the Bidirectional Reflectance Distribution Function (BRDF) and the Bidirectional Transmittance Distribution Function (BTDF) referring to backscattering or forward scattering, respectively, or, more generally, the Bidirectional Scattering Distribution Function (BSDF). The main difference to the scattered intensity definition of the ARS is that these functions are radiometric quantities defined as the scattered radiance divided by the irradiance incident on a surface [7, 25]:

$$\mathrm{BSDF}\ (\varphi_{\mathrm{s}},\phi_{\mathrm{s}})=\frac{\mathrm{d}P_{\mathrm{s}}\,(\varphi_{\mathrm{s}},\phi_{\mathrm{s}})}{P_{\mathrm{i}}\mathrm{d}\Omega_{\mathrm{s}}\cos\varphi_{\mathrm{s}}}. \tag{14.2}$$

Because of the differential notation, the BSDF is only approximated when measurements are taken with a finite-diameter aperture. However, if the flux density is reasonably constant over the measuring aperture, actual measurements approximate the definition very good. Only for the combination of a large detector aperture and a focused specular or diffracted beam, the approximation becomes very poor.

Comparing (14.1) and (14.2), reveals that the BSDF and ARS are basically the same functions except for the cosine factor. The normalization to the incident light power and detector solid angle ensures that both quantities become independent of the actual measurement system.

Many fundamental aspects for measuring angle resolved scattering are described in the ASTM standard E 1392-90 [26]. The procedure was also verified in several round-robin experiments at different wavelengths [27, 28] but is restricted to opaque samples. Therefore, the ASTM standard ASTM E 2387-5 [29] has been established later that has a wider range of application.

Currently, an ISO standard procedure for angle resolved light scattering measurements is being developed by the international working group TC172/SC9/WG 6 of the international organization for standardization to meet the increased demands concerning wavelength ranges, sensitivity, flexibility, and practicability. All these standards describe the same quantity.

14.1.2 Total Scattering

The Total Scattering, TS, is defined as the light power scattered into the forward (f) or backward (b) hemisphere normalized to the incident light power [30]:

$$\mathrm{TS}_{\mathrm{b/f}} = \frac{P_{\mathrm{s,\,b/f}}}{P_{\mathrm{i}}}. \tag{14.3}$$

Practically, the TS can be measured with an Ulbricht/integrating sphere that simply collects the scattered light into the forward or backward hemisphere and uniformly spreads the light inside the sphere by using a diffusely reflecting coating. Part of the sphere can then be viewed with a recessed detector, which allows determining the total scattered light power $P_{\mathrm{s,b/f}}$. Alternatively, the TS can be determined with the help of a Coblentz sphere, which is basically a hemispherical mirror that collects the scattered light and images it onto the detector.

Both techniques are described in the international standard ISO13696 [30], which also suggests the angular acceptance range for the scattering angles ($\phi_{\mathrm{s}} = 0°\ldots 360°$, $\varphi_{\mathrm{s}} = \leq 2°\ldots \geq 85°$ [backward hemisphere] and $\varphi_{\mathrm{s}} = \leq 95°\ldots \geq 178°$ [forward hemisphere]). The specular beam is excluded in the calculation, which allows TS values to be treated as a loss factor similar to an absorption loss. The energy balance thus reads:

$$1 = R + T + A + \mathrm{TS}_{\mathrm{b}} + \mathrm{TS}_{\mathrm{f}}, \tag{14.4}$$

with the reflectance, R, and transmittance, T.

Another quantity used to describe the hemispherical scattering is the Total Integrated Scattering, TIS, which is defined as the ratio between the diffusely reflected light and the sum of the diffusely and specularly, P_{r}, reflected light [6, 7]:

$$\mathrm{TIS} = \frac{P_{\mathrm{s}}}{P_{\mathrm{s}} + P_{\mathrm{r}}} \tag{14.5}$$

It should be emphasized that TIS and TS are two different quantities although they are often mixed up with each other which can cause some confusion. For instance, several optical design software codes use the definition of TS but call it TIS.

The standard ASTM F 1048-87 [31] provides a procedure on how to measure TIS for opaque samples. One drawback is, however, that no angular ranges for the diffuse and specularly reflected light are given. It is thus difficult to distinguish at which angular position the specular beam transits into the diffuse scattering or how much diffuse scattering is actually included in the specular beam. These tricky questions are simply avoided in the ISO standard by defining specific angular acceptance ranges. This is essential when comparing light scattering measurements with reflectance and transmittance measurements.

If the diffuse scattering is small compared to the specularly reflected light, as it is usually fulfilled for high quality optics, TS_b and TIS can be converted into each other with the help of the sample's reflectance:

$$\mathrm{TIS} = \frac{P_\mathrm{s}}{P_\mathrm{s} + P_\mathrm{r}} \approx \frac{P_\mathrm{s}}{P_\mathrm{r}} = \frac{P_\mathrm{s}}{R P_i} = \frac{\mathrm{TS_b}}{R}. \tag{14.6}$$

Aside from measuring TS directly, the scattering loss can also be determined by integrating the ARS over the corresponding hemispheres:

$$\mathrm{TS}_{b/f} = \int_0^{2\pi} \int_{2^\circ/95^\circ}^{85^\circ/178^\circ} \mathrm{ARS}\,(\varphi_s, \phi_s) \sin\varphi_s \mathrm{d}\varphi_s \mathrm{d}\phi_s. \tag{14.7}$$

This allows the acceptance angles to be precisely controlled. It is also possible to determine the TS for arbitrary angles of incidence, which otherwise would require specially designed Ulbricht- or Coblentzspheres with different entrance and exit ports.

14.2 Theoretical Background

Scattering from imperfect surfaces has been intensively studied in the fields of acoustics, radiophysics [32], and optics [6, 7, 33]. In principle, Maxwell's equation can be used to connect the surface irregularities to the ARS. But, in most cases approximations are necessary in order to get practically useful results. The best known of these approximations are: (1) the diffraction based Kirchhoff [34] and (2) the perturbation based Rayleigh-Rice scattering theory [32, 35]. Despite their advanced age, these theories are still subject of a lively discussion [36, 37].

However, in the case of small deviations of the ideal surface profile and small angular scattering, both theories agree with each other as well as with experimental results very good [7, 38]. The main difference between both scattering theories is that the Kirchhoff approach can be used for rougher surfaces as well. But, it contains a paraxial (small angle) assumption that limits its ability to accurately calculate the scattering distribution at large scattering angles or large angles of incidence.

In contrast, the perturbation approach used in the Rayleigh-Rice scattering theory requires that the surface irregularities are small when compared to the light wavelength (rms-roughness, $\sigma \ll \lambda$). The main advantage, however, is that all calculations are vector based, which allows the scattering distribution to be determined at arbitrary angles of incidence and scattering. This high flexibility and the circumstance that most high quality optical components fulfill the small roughness requirement make the Rayleigh-Rice approach a powerful tool for predicting scattering from surface roughness and vice versa. The following paragraphs will thus focus on the Rayleigh-Rice scattering theory in more detail.

14.2.1 Light Scattering from a Single Rough Surface

The basic procedure for modeling light scattering from a rough surface is to solve Maxwell's equations for an ideally flat surface and replacing the interface roughness by plane surface current sheets that act as sources of the scattered light. If only first order terms are considered in the calculation, the following relationship between the ARS and the surface roughness is obtained [7, 33]:

$$\mathrm{ARS}\left(\varphi_{\mathrm{s}}, \phi_{\mathrm{s}}\right) = \frac{16\pi^2}{\lambda^4} \cos\varphi_{\mathrm{i}} \cos^2\varphi_{\mathrm{s}}\, Q \mathrm{PSD}\left(f_x, \mathrm{f}_y\right). \tag{14.8}$$

The ideal sample characteristics of the perfectly smooth surface, such as the refractive index of the ambient and substrate material, as well as the polarization of the incident and scattered light are described by the dimensionless factor Q, which can also be viewed as a generalized surface reflectance. For instance, for s-polarized incident light, Q can be approximated by the specular reflectances at the angles of incidence and scattering: $Q_{\mathrm{s-pol}} = \sqrt{R_{\mathrm{s-pol}}\left(\varphi_i\right) R_{\mathrm{s-pol}}\left(\varphi_s\right) \cos^2 \phi_{\mathrm{s}}}$.

The roughness properties are described by the power spectral density function, PSD, which is defined as the squared modulus of the Fourier transform of the surface topography, $z(x,y)$ [39, 40]:

$$\mathrm{PSD}\left(f_{\mathrm{x}}, f_{\mathrm{y}}\right) = \lim_{L_x, L_y \to \infty} \left| \int_{-\frac{L_y}{2}}^{\frac{L_y}{2}} \int_{-\frac{L_x}{2}}^{\frac{L_x}{2}} z(x, y)\, \mathrm{e}^{-2\pi \mathrm{i}\left(f_x x + f_y y\right)} \mathrm{d}x \mathrm{d}y \right|^2 \tag{14.9}$$

Another way to look at (14.8) and (14.9) is that the scattered intensity is essentially a Fourier spectrogram of the surface errors, which makes it convenient to discuss scattering in the frequency domain. The link between the scattering angles and the spatial frequencies, f_{x} and f_{y}, is given by:

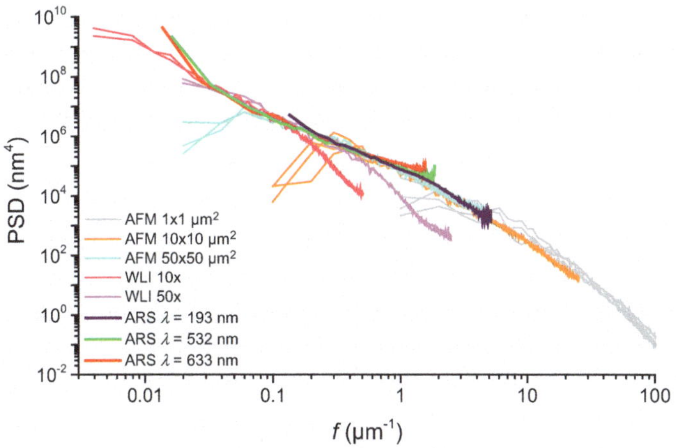

Fig. 14.2 Roughness spectra determined with different characterization techniques

$$f = \begin{pmatrix} f_x \\ f_y \end{pmatrix} = \begin{pmatrix} \frac{\sin\varphi_s \cos\phi_s - \sin\varphi_i}{\lambda} \\ \frac{\sin\varphi_s \sin\phi_s}{\lambda} \end{pmatrix}, \tag{14.10}$$

which can be viewed as a generalization of the conical grating equation for first-order diffraction for a grating with a period of $1/f_x$.

Stochastic processes such as polishing, etching, or thin film growth usually do not create a preferential direction and show an isotropic surface roughness [41, 42]. In this case, the PSD is nearly symmetric in $|f|$ and can be averaged over all azimuthal angles, enabling a compact visualization of the roughness spectrum. Another advantage of the PSD is that the roughness information from different characterization techniques, such as topography measurements performed with an atomic force microscope (AFM), white light interferometer (WLI), or profilometry, as well as light scattering measurements can be easily combined as shown in Fig. 14.2.

The actual accessible spatial frequency range of the different techniques is usually smaller than the theoretical limit set by the instrumental resolution and measurement area. For instance, at the lower end of the PSD determined by AFM or WLI, the curves tend to fall to smaller values because of a poor sampling. This can also occur at the upper end, because of a low pass behavior of the microscopy objective.

PSDs determined by light scattering measurements tend to overestimate the roughness spectra at their lower end because of the influence from the specular beam. Also at the upper end, the PSD usually increases in form of a sharp hook [36]. However, in the intermediate and thus largest part of the PSD, the curves from different characterization techniques overlap each other very good.

Changing the spatial frequency range covered by the light scattering measurements is possible by using different light wavelengths. This also illustrates that roughness components responsible for light scattering are not always the same for different light wavelengths. For instance, in the visible spectral range, only spatial frequencies up to $f \sim 2\ \mu\text{m}^{-1}$ contribute directly to the light scattering distribution. For shorter light wavelengths, the upper spatial frequency limit is higher.

Integrating the PSD allows determining the rms-roughness in the bandwidths relevant for the later application:

$$\sigma = \left[2\pi \int_{f_{min}}^{f_{max}} \text{PSD}\,(f)\, f \mathrm{d}f \right]^{\frac{1}{2}}. \tag{14.11}$$

By integrating (14.8) according to (14.7) a closed form solution for the TS can be retrieved. For normal incidence and under the assumption that the scattered light is concentrated in the vicinity of the specular reflex, this yields:

$$\text{TS}_b = R_0 \left(\frac{4\pi\sigma}{\lambda} \right)^2, \tag{14.12}$$

where the optical factor, Q, is approximated by the ideal Fresnel reflection of the surface, R_0.

Using scalar diffraction theory, the same result can be derived for smooth surfaces ($\sigma \ll \lambda$) as shown by H. Davies [43, 44]. In order to ease the mathematics, he used a Gaussian PSD function for the surface roughness. However, as (14.12) was obtained without any assumptions on the surface PSD, this is not necessarily required. It is interesting to note that the interim result,

$$\text{TS}_b = R_0 \left[1 - \mathrm{e}^{-\left(\frac{4\pi\sigma}{\lambda}\right)^2} \right] \tag{14.13}$$

in the paper of H. Davies is also valid for rough surfaces ($\sigma \gg \lambda$), as was later demonstrated by P. Beckmann and A. Spizzichino [34], assuming a Gaussian PSD function as well. However, in a recent publication [45], it was shown that for sinusoidal gratings, which obviously do not exhibit a Gaussian PSD, (14.13) provides a very good estimate even for large grating amplitudes by comparing the simulation results to rigorous calculations. Hence, for a large range of surface roughness values, (14.13) can be used to predict the scattering loss from single surfaces or highly reflective metal coatings.

14.2.2 Light Scattering from Thin Film Coatings

Going from one rough interface to multiple rough interfaces is now straight forward. For each interface, the scattered intensity can be calculated similarly to the single rough surface. The only difference is that the incident light has to propagate to the rough interface and the scattered light has to find its way back to the ambient media. This can however easily be achieved based on the ideal multilayer design, as the incident and scattered intensities are not significantly altered by the perturbation approach. The scattering distribution is then obtained by a superposition of the individual contribution from all interfaces and results in the following ARS for a thin film stack consisting of N layers [15–17]:

$$\mathrm{ARS}\left(\varphi_{\mathrm{s}}, \phi_{\mathrm{s}}\right) = \frac{1}{\lambda^4} \sum_{i=0}^{N} \sum_{j=0}^{N} F_i F_j^* \mathrm{PSD}_{i,j}\left(f_x, f_y\right). \tag{14.14}$$

Analogous to the single interface, all properties of the perfect multilayer and the conditions of illumination and observation (dielectric constants, multilayer design, and polarization) are described by the optical factors F_i. Interference between the scattered electric fields from the individual interfaces of the multilayer is considered by the cross-correlation PSDs ($i \neq j$), besides the individual interface PSDs ($i = j$).

14.2.3 Roughness Evolution of Multilayer Coatings

In total, $(N+1)^2$ PSDs are required to model the scattering from a multilayer. One way to determine them experimentally is by using cross-sectional transmission electron microscopy [20] or AFM after the deposition of different number of layers [46, 47]. This, however, usually results in the destruction of the sample or can be a tedious and time consuming task because of several deposition runs.

An alternative is given by modeling the thin film growth based on the substrate roughness and the thin film characteristics. Powerful approaches for this are: (1) the scaling of the surface roughness according to [48]:

$$\sigma_i = c i^{\beta} \tag{14.15}$$

and (2) the linear continuum model [49, 50].

Although very simple in its approach, a good agreement between experimental results and the first method can be achieved for columnar thin films [51, 52]. A nearly saturated roughness increase or a rapid roughening from one interface, i, to the next can be directly described with the help of the roughening exponent, β.

Dense polycrystalline or vitreous amorphous thin films, which can often be achieved by ion and plasma assisted deposition processes, such as ion beam sputtering, pulsed laser deposition, or magnetron sputtering [53], can be more advantageously described with the linear continuum model [21, 54–56]. This approach assumes that the interface roughness of the ith interface can be expressed by the sum of the intrinsic thin film roughness, $\mathrm{PSD}_{\mathrm{int},i}(f)$, and a replication of the underlying interface weighted by a local smoothing factor, $a_{\mathrm{rep},i}(f)$, in order to consider the mobility of the deposited particles:

$$\mathrm{PSD}_i\,(f) = \mathrm{PSD}_{\mathrm{int},i}\,(f) + a_{\mathrm{rep},i}\,(f)\,\mathrm{PSD}_{i-1}\,(f)\,. \tag{14.16}$$

The replication factor is essentially a low-pass filter that replicates the roughness components at low spatial frequencies while high spatial frequency components are decreased. This agrees with the physical picture that over a short distance the adatom mobility leads to a purely stochastic roughness, while large ranging roughness structures are replicated:

$$a_i\,(f) = \mathrm{e}^{-\sum_{\kappa_i} h_i \gamma_{\kappa_i} |2\pi f|^{\kappa_i}}\,. \tag{14.17}$$

The thickness of the ith layer is described by h_{i}, the relaxation rate is characterized by γ_{κ_i}, and the actual relaxation mechanism is denoted by κ_{i}. Different relaxation mechanisms such as viscos flow ($\kappa_i = 1$), evaporation-recondensation ($\kappa_i = 2$), bulk diffusion ($\kappa_i = 3$), and surface diffusion ($\kappa_i = 4$) can be directly associated with the different integer values of the relaxation parameter [48].

The intrinsic thin film roughness can be described by the same parameters with the addition of the minimal growth volume, Ω_i:

$$\mathrm{PSD}_{i,\mathrm{int}}\,(f) = \Omega_i \frac{1 - \mathrm{e}^{-2\sum_{\kappa_i} h_i \gamma_{\kappa_i} |2\pi f|^{\kappa_i}}}{2\sum_{\kappa_i} \gamma_{\kappa_i} |2\pi f|^{\kappa_i}}\,. \tag{14.18}$$

One way to determine the set of growth parameters for each layer material is to compare the experimental PSD of the multilayer top surface with the modeling results [54–56]. All interface PSDs are then automatically obtained by the recursion relation in (14.16).

In addition, also all cross-correlation PSDs can be modeled with this approach:

$$\mathop{\mathrm{PSD}}_{\substack{i,\,j\\ i<j}}\,(f) = a_j\,(f)\,a_{j-1}\,(f)\ldots \mathrm{a}_{i+1}\,(f)\,\mathrm{PSD}_i\,(f)\,. \tag{14.19}$$

Other common cross-correlation models that can be used in combination with the two presented roughness evolution approaches are:

- Partial correlation model: $\mathrm{PSD}_{i,j}\,(f) = \min\left[\mathrm{PSD}_i\,(f)\,, \mathrm{PSD}_j\,(f)\right]$,
- Uncorrelated model: $\mathrm{PSD}_{i,j}\,(f) = 0$ and the
- Fully correlated model: $\mathrm{PSD}_{i,j}\,(f) = \mathrm{PSD}_{\mathrm{top\ surface}}\,(f)$

Which one of these models is the most representative for a given multilayer coating depends to a large extend on the deposition parameters. For instance, high adatom mobilities or etching with a secondary ion source during the deposition process can destroy the correlation in favor of the uncorrelated model.

14.3 Instruments for Light Scattering Measurements

Measuring light scattering requires fairly sophisticated instruments as the scattered intensity is usually very small compared to the incident light power and can vary quickly by several orders of magnitude when approaching the specular beam directions. For a complete scattering analysis, also a large fraction of the scattering sphere has to be characterized. In the following, different setups for total scattering, angle resolved scattering, and compact scattering sensors are introduced.

14.3.1 Total Scattering Measurements

Instruments to measure TS (and TIS) have been developed in a number of laboratories [6, 40, 57–63] based on integrating/Ulbricht and Coblentz spheres. For the former, different coatings such as Spectralon (PTFE), barium sulfate, or rough gold are available, which allow performing these measurements in a wide spectral range from the ultraviolet to the infrared. Only below $\lambda = 200$ nm, no diffusely scattering coatings are available. Therefore, setups for shorter wavelengths require the use of a Coblentz sphere [64–66], which can be used at longer wavelengths as well.

For integrating spheres, the multiple scattering from the side walls and the small ratio between detector area and surface area of the sphere reduces the measurement signal. In addition, the large field of view of the detector is prone to Rayleigh scattering from air molecules, which limits the sensitivity of an Ulbricht sphere based setup.

The Coblentz sphere approach on the other hand provides a higher signal to the detector which usually offers a substantially higher sensitivity. However, some of this light is incident on the detector at very high angles, which tends to discriminate high angle scattering. For many samples this is not a problem or can be compensated with the help of a small integrating sphere in front of the detector.

According to the ISO standard 13696 [30], the Ulbricht and Coblentz sphere exhibit a small exit hole so that the radiation within 2° of the specular beam can leave the sphere without being detected as shown in Fig. 14.3.

14.3.2 Angle Resolved Scattering Measurements

For the measurement of the ARS or BSDF special photogoniometers, sometimes referred to as scatterometers, as well as fixed illumination and detection systems based on CCD or CMOS detectors have been developed in different laboratories [10, 19, 67–72]. The majority of the systems operate in the visible spectral range. But also specially designed instruments for the lithography wavelengths 13.5 nm [73, 74] and 193 nm [64, 66] as well as the infrared spectral range have been developed. With the

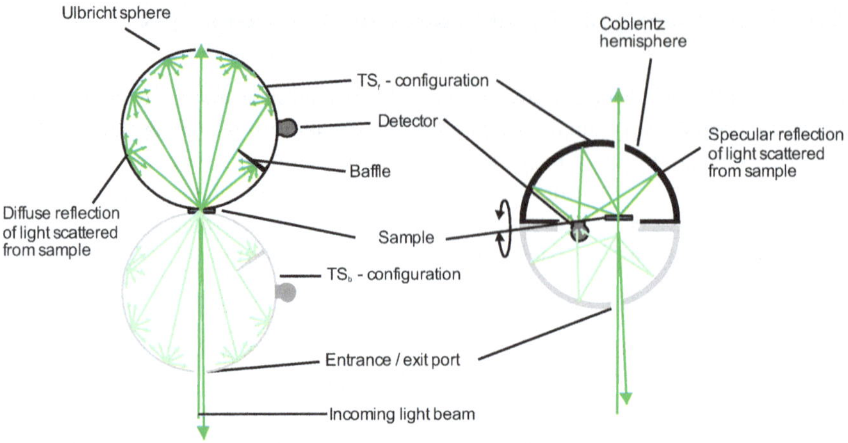

Fig. 14.3 Measurement principles for total scattering. Left: Ulbricht/integrating sphere based setup; right: Coblentz sphere based approach

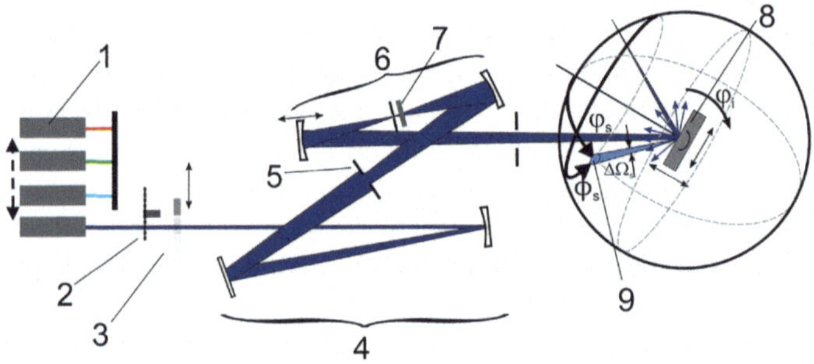

Fig. 14.4 Typical 3D scatterometer for angle resolved light scattering measurements

availability of tunable light sources with high light powers, also first instruments have been built that enable spectrally resolved light scattering measurements [19, 75].

In principle, all of these instruments can be divided in in-plane scatterometers that detect the scattered light in the plane of incidence ($\phi_s = 0°$) and 3D scatterometers which cover the entire scattering sphere. A schematic for most common instrument configuration in use is shown in Fig. 14.4.

Besides this classical setup, there exist also other measurement configurations. For instance, the source and the detector may be fixed and just the sample is rotated, which has the advantage of an easy mechanical design but complicates the analysis because the scattering and the incident angle are changed simultaneously [41].

Other approaches with a fixed detector include the variation of the illumination direction, which has the advantage that no long detector arms have to be used [75, 76]. In this case, the detector can also be rather bulky and incorporate additional optics such as a monochromator [77]. The sample is either fixed to the illumination direction or can be moved freely with respect to the incoming light beam. For the latter, the illumination spot size at the sample position changes for the different angles of incidence and thus leads to different measurement spots during a measurement. Therefore, these instruments usually incorporate a tunable slit in the illumination path to compensate this effect [75]. This can however introduce additional scattering and diffraction from the edges of the slit.

Following the beam path in the schematic sketch shown in Fig. 14.4, the light from the individual laser sources (1) passes a mechanical chopper (2) to enable lock-in amplification and noise suppression. Variable neutral density filters (3) are then used to adjust the incident light power so that the detector (9) operates in its linear response range, even for the large variations in signal powers between the incident light beam and low-level scattering.

In order to avoid detecting scattering from the instrument itself which is later reflected or transmitted by the sample, several beam preparation optics (4) are used: this includes an iris (5) to alter the beam size on the sample and a spatial filter (6), consisting of two focusing mirrors and a pinhole. If required, polarizers and wave plates (7) used to define the incident polarization are commonly placed just before the pinhole in order to reduce scattering from them as well.

The positions of one or both spatial filter mirrors can be changed in order to compensate for a possible curvature of the sample being tested (8), so that the pinhole is imaged over the sample onto the detector aperture. This enables light scattering measurements in close vicinity to the specular beam which is important for imaging optics. The typical spot size at the sample position is 1–5 mm.

Alternatively, the light can be focused onto the sample with a typical spot diameter of 100 μm in order to achieve a high lateral resolution on the sample but at the cost of a slightly reduced near angle measurement capability.

The irradiation position and the angle of incidence can be adjusted with the sample positioning system. The detector – usually a photomultiplier tube or a photodiode – can be scanned freely around the sample. The detector solid angle, Ω_s, is set by the distance between the detector and the sample as well as the diameter of the detector aperture. Aperture diameters between 0.1 and 5 mm are typically used, depending on the specific measurement task, sensitivity requirements, speckle reduction, and near angle limit.

By application of neutral density filters, more than 14 order of magnitude dynamic range can be achieved in the visible spectral range [69, 78]. This is sufficient to investigate samples ranging from superpolished substrates with rms roughness values below 0.1 nm to thin film coatings, optical materials, as well as nanostructured and machined surfaces. For high-end scatterometers, the lowest detectable signal is not necessarily set by electronic noise but by Rayleigh scattering from air molecules within the light beam that is partly viewed by the detector [79]. This is illustrated in Fig. 14.5 in more detail.

One way to minimize the unwanted influence of Rayleigh scattering from air molecules on the measurement signal is to reduce the detector field of view. However, it still has to be larger than the illumination spot at the sample position so that the entire scattered light from the sample is detected.

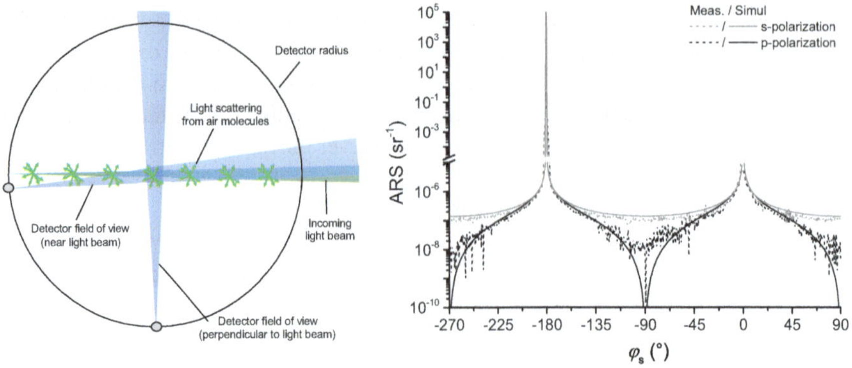

Fig. 14.5 Light scattering from air molecules. Left: Schematic illustrating the influence of the detector field of view on the measurement signal; right: exemplary instrument signature at λ = 405 nm in the plane of incidence

The characteristic shape of the instrument signature, which is an ARS measurement without a sample, results from the different length of the observed beam path and the incident polarization. For instance at $\varphi_s = -180°$, the detector looks along the entire beam path, while in the direction perpendicular to the incident beam, the observed beam length is confined to the width of detector field of view at the sample position. The air molecules behave as point polarizable dipoles because they are smaller than the light wavelengths. Therefore, a typical Hertzian radiation pattern can be observed for the two polarizations.

14.3.3 Compact Scattering Sensors

The large degrees of freedom for the sample positioning, illumination, and detection unit can quickly lead to more than 10 motorized axes for a laboratory system. A different approach is to keep all angles fixed which enables very compact scattering sensors, as shown in Fig. 14.6.

Similar to the laboratory based systems presented in the previous section, a spatial filter is used to minimize scattering from the sensor as much as possible. The scattered light from the sample is detected by a CMOS or CCD matrix array. Because of the limited size of these sensors, the angular detection range is restricted to a cone with an opening angle of a few degrees around the specular beam. Another drawback of this kind of detection is the higher noise compared to a lock-in based technique which limits the detectable ARS level to values $>10^{-4}$ sr^{-1} and thus above the Rayleigh scattering signal from air molecules. This is, however, often sufficient even for well-polished surfaces with rms roughness values down to 0.5 nm.

The matrix based detection allows for a quick data acquisition of less than 1 s per measurement position, which is a huge advantage compared to the point wise

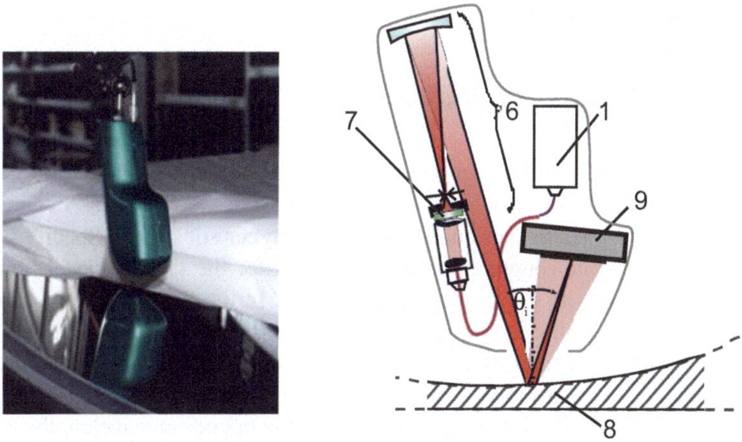

Fig. 14.6 Compact light scattering and roughness sensor *horos* [10]. Left: photograph of sensor during the characterization of 2 m telescope mirror, right: Schematic of sensor setup

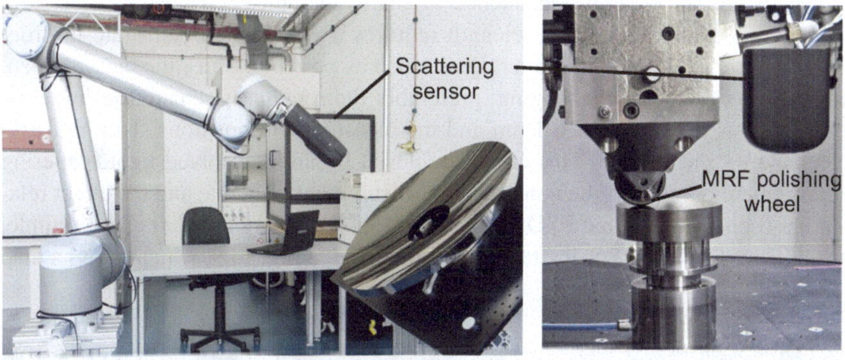

Fig. 14.7 Light scattering and roughness analysis with compact sensor. Left: sensor mounted to robotic arm during characterization of a primary mirror of a Ritchey-Chrétien-Cassegrain telescope. Right: implementation of sensor into magnetorheological finishing (MRF) machine

measurement approach of classical scatterometers. In combination with the compact size, this enables many different application scenarios. Two of them are shown in Fig. 14.7.

When light scattering is measured from a slightly vibrating sample, the scattering distribution is not significantly changed. Thus, the high sensitivity to detect even small surface irregularities is still maintained in harsh environments such as a polishing machine. This would be an unthinkable application scenario for conventional roughness metrology techniques. Furthermore, because of the large illumination spot of a few millimeters during the light scattering measurements, the roughness data obtained with the scattering sensor are very robust.

14.4 Application Examples

14.4.1 Light Scattering and Roughness of Substrates

The replication of the roughness properties of the substrate from one interface to the next in a multilayer stack is often responsible for most of the scattering. In particular the near angular scattering can quickly rise for rougher substrates. Specifying and analyzing the substrate roughness is thus essential for the fabrication of high-end optical components with good imaging properties. In the following paragraphs, it will be described how this can be achieved by light scattering measurements.

A first example is the collector mirror substrate shown in Fig. 14.8, which is used after the deposition of a molybdenum silicon multilayer as the first mirror in an extreme ultraviolet (EUV) lithography scanner at the operation wavelength 13.5 nm [80]. The substrate has a diameter of more than 660 mm and exhibits an elliptical shape. This allows the EUV light emitted from a laser produced plasma to be collected before it is redirected to further illumination and imaging optics [81, 82].

The strong wavelength dependence of scattered light of $1/\lambda^4$ (see (14.14)) in combination with the short wavelength requires a very smooth substrate in order to avoid optical losses from scattering as much as possible. Thus, the measurement requirements are quite challenging: preferably the roughness sensitivity should be below an rms roughness of 0.1 nm and the measurement positions shall be freely selectable on the surface. This can be achieved by angle resolved light scattering measurement because of the non-contact measurement principle and the direct relation between the ARS und PSD, as described in (14.8). By translating the sample, the entire surface can be characterized as illustrated in Fig. 14.9, which shows a map

Fig. 14.8 Roughness characterization of an EUV collector mirror substrate by light scattering measurements performed by the first author of this chapter

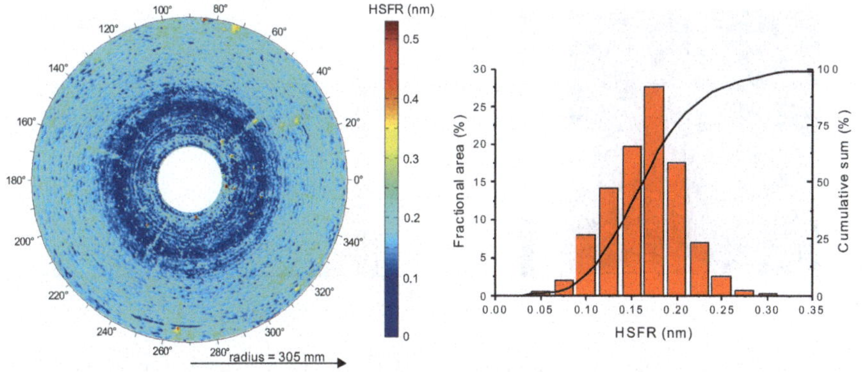

Fig. 14.9 Roughness characterization of an EUV collector mirror substrate. Left: Roughness map obtained from ARS measurements at 442 nm; right: corresponding histogram

of the rms-roughness between $f = 1\ \mu m^{-1}$ and $50\ \mu m^{-1}$, often referred to as high spatial frequency roughness (HSFR).

The roughness map consists of more than 34 000 individual roughness measurements and enables a detailed overview about the sample roughness and homogeneity; no severe defect areas and a homogenous surface finish that improves slightly towards the inner edge of the sample can be observed. Most of the substrate surface has a HSFR of better than 0.17 nm which is sufficient for the later application.

An alternative to the light scattering based roughness characterization are topography measurements performed with an AFM. However, the long measurement time for one AFM scan of 10–20 min quickly limits the number of measurements points. It would for instance require more than a year to generate a similar roughness map as shown in Fig. 14.9 with AFM in a non-stop operation mode. This shows the potential of the light scattering based roughness characterization in addition to its large flexibility to different sample geometries.

Another application example is given in Fig. 14.10, which shows the roughness map obtained from ARS measurements of an unstructured mask blank, which are known for their low surface roughness. In order to observe just the scattering from the front side, the sample was coated with a thin ruthenium layer.

The average HSFR is just 0.04 nm. Considering that the fundamental building blocks of the substrate – the silicon-oxygen tetrahedron – exhibits a distance of 0.16 nm and 0.26 nm between the individual ions, reveals that the topography of the mask blank was very close to an atomically flat surface.

Measuring such a remarkably low surface roughness with an AFM is also very challenging because the vertical instrument noise of an AFM is typically between 0.03 and 0.04 nm and thus just below the roughness of the sample. Hence, from a sensitivity point of view, also very smooth surfaces can be characterized by light scattering measurements. It should however be noted that for such low roughness values, Rayleigh scattering from air molecules has to be considered [83].

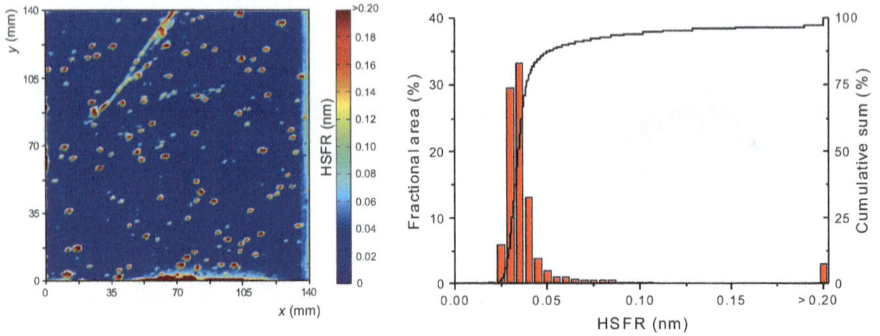

Fig. 14.10 Characterization of an unstructured mask blank. Left: roughness map obtained from ARS measurements at $\lambda = 405$ nm; right: corresponding histogram

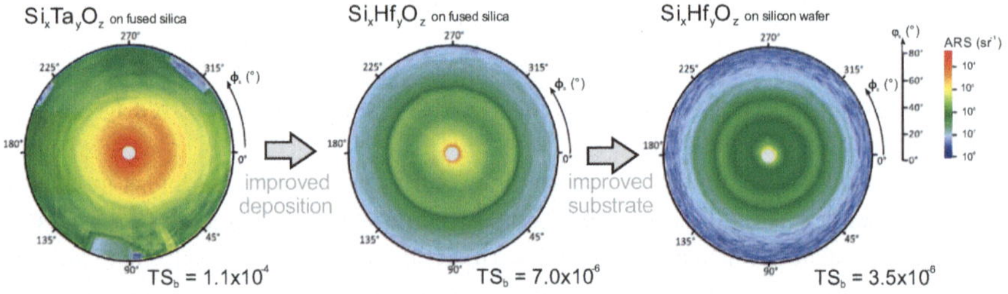

Fig. 14.11 Scattering characterization of rugate coating in an initial state and after the optimization of the multilayer deposition and substrate roughness

14.4.2 *Light Scattering from Multilayer Coatings*

Characterizing or modeling the light scattering of ready to use optical components, which is usually done after the coating deposition, is a prerequisite for simulating the stray light in entire optical systems. But, also the 3D-ARS and the TS data often provide a valuable feedback about the deposition process. This is exemplified in the following section in more detail by a variety of application examples.

14.4.2.1 Light Scattering from Rugate Coatings

In Fig. 14.11, exemplary 3D scattering plots are shown which illustrate the continuous optimization of a highly reflective rugate coating for $\lambda = 532$ nm towards low scattering.

The rugate coatings were deposited by reactive co-sputtering with a double ring magnetron with silicon as inner target and tantalum or hafnium as outer target [21]. The index profile during the deposition was achieved by varying the power ratio

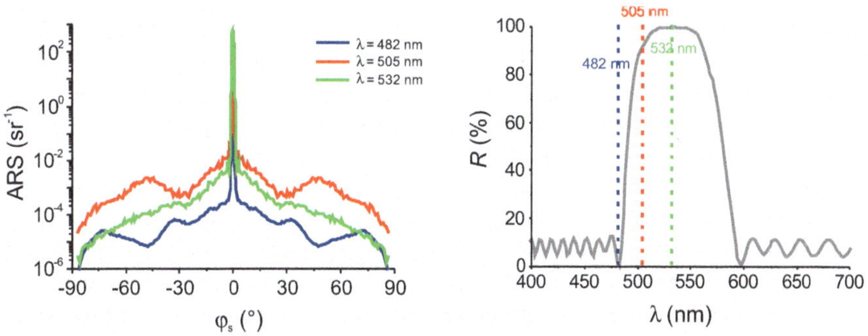

Fig. 14.12 Spectral scattering analysis of $Si_xTa_yO_z$ rugate coating. Left: angle resolved scattering measurements near the lower reflectance band edge. Right: Spectral reflectance plot

between the inner and outer discharge in a total of 30 cycles. The thickness of the $Si_xTa_yO_z$ and $Si_xHf_yO_z$ rugate coating is $h = 4.5$ μm.

By changing the outer target from tantalum to hafnium, the scattering loss is decreased by more than a factor of 15. This could be attributed to a reduced number of defects during the deposition process as a result of an optimized cleaning of the sputter source and the recipient as well as a more advantageous growth behavior of the thin film.

In a next step, the overall scattering could be reduced once more by a factor of two by improving the substrate roughness through an exchange of the substrate from fused silica ($\sigma = 0.14$ nm) to a silicon wafer ($\sigma = 0.10$ nm).

Another observation that can be made from these plots is that the 3D-ARS of the tantalum based rugate coating is slightly asymmetric although the angle of incidence for the scattering measurements is $\varphi_i = 0°$. This can be explained by an oblique thin film growth [20]. The 3D-ARS of the hafnium based rugate coating on the other hand exhibits a concentric ring with a high scattering intensity around $\varphi_s = 40°$. This is typical for resonant scattering within the thin film, often referred to as resonant scattering wings. Thus, even for coatings that do not exhibit classical interfaces, the replication of the substrate roughness through the thin film can influence the scattering characteristics.

The resonant scattering wings occur because of constructive interference of the scattering contributions from different depths of the coating. Thus, they depend on the coating design and illumination wavelength as illustrated in Fig. 14.12.

As discussed in Sect. 14.2.1, the scattering from a single interface directly depends on the sample reflectance at this wavelength. Therefore, it might be surprising that the wavelength with the highest scattering distribution does not correspond to the wavelength for which the rugate coating exhibits the highest reflectance. In fact, the most pronounced scattering can be observed at the lower edge of the reflectance band. The total scattering loss at $\lambda = 505$ nm is TS = 2.2% and thus a factor of 22 higher than at the central wavelength $\lambda = 532$ nm.

A simple explanation for this behavior can be found when considering that the incoming light has to (1) reach the irregularity or inhomogeneity in the film and (2) the scattered light has to propagate to the ambient media. Furthermore, if a reflective coating designed for a longer wavelength and normal incidence shall be used at shorter wavelengths, the mirror can be used under oblique incidence in order to maintain a high reflectance. From a Helmholtz reciprocity point of view, this directly corresponds to light scattering at large angles. Thus, light scattered within the coating can propagate through the thin film easily at large angles for light wavelengths below the design wavelength, which can increase the scattering for these wavelengths. This is a particular concern for filtering optics in combination with a broadband illumination.

14.4.2.2 Light Scattering from Highly Reflective Coatings

Resonant scattering is not just limited to Rugate coatings, but can also be observed for classical multilayer coatings, as shown in Fig. 14.13.

The coating is a quarter-wave multilayer consisting of aluminum fluoride and lanthanum fluoride with 20 periods. The design wavelength is $\lambda = 193$ nm. For a comparison, also a hypothetical, perfectly reflecting single surface with the same interface roughness is plotted. According to (14.8), this curve continuously increases according to $1/\lambda^4$ towards shorter wavelengths. This principle trend can also be observed for the multilayer stack with the addition of several resonant scattering peaks that are most pronounced around the band edges.

This drastic scattering enhancement is illustrated in Fig. 14.14 in more detail, which shows the spectral ARS and the electric field intensity, $|E|^2$, inside the multilayer stack on the same wavelength scale.

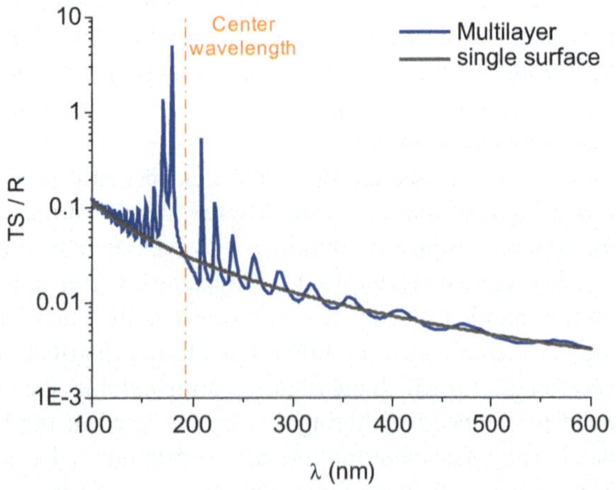

Fig. 14.13 Spectral scattering simulation of multilayer stack with fully correlated interfaces for $\lambda = 193$ nm

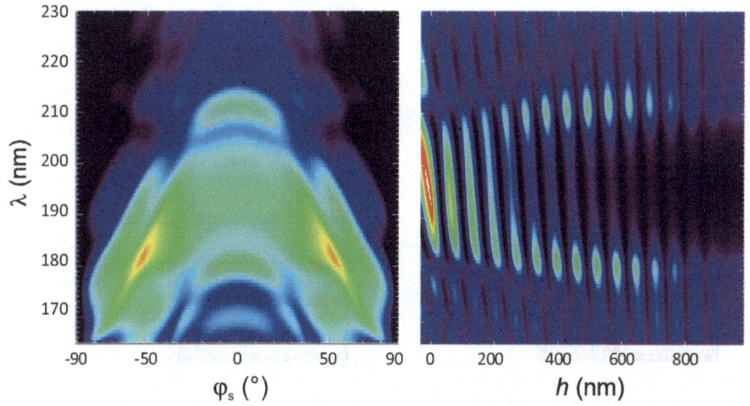

Fig. 14.14 Light scattering from quarter-stack for λ = 193 nm. Left: ARS as a function of the incident wavelength; right: field intensity inside the coating on the same wavelength scale

At λ = 193 nm, the field intensity is concentrated to the uppermost layers. In contrast, near the band edges of the coating at λ = 182 nm and 210 nm, the electric field propagates much deeper into the multilayer. Thus, even interfaces close to the substrate can cause significant scattering within the multilayer stack. At the upper band edge, this scattered light cannot propagate as easily outside the coating as for the lower band edge. Therefore, the spectral ARS diagram is narrower around λ = 210 nm compared to λ = 182 nm. For the latter wavelength also resonant scattering peaks around $\varphi_s = \pm$ 55 can be observed.

Considering that real coatings can exhibit small spectral shifts, caused by thickness variations during the deposition process or changes in the temperature or humidity, enhanced scattering can also occur for applications that use just a single fixed wavelength.

In addition to these scattering simulations with a fixed PSD for all interfaces, the scattering was also modeled based on topography measurements performed by atomic force microscopy of the substrate and multilayer top surface (see Fig. 14.15).

Starting with the substrate, the surface roughness increases quickly after 20 layer pairs. This can be modeled with the approach presented in (14.15). The ARS can then be calculated with the help of the partial correlation model. The results of this simulation are shown in Fig. 14.16.

The best fit to the experimental results can be achieved with $\beta = 1$, indicating a rapid roughening from one layer to the next. However, at large scattering angles there exists still some difference between the simulations and the experimental observations. This can be explained by small thickness variations within the multilayer stack, as illustrated in Fig. 14.16 right. The best agreement between the simulations and the measurements for the angular positions of the resonant scattering wings can be achieved for an average thickness variation of 3%.

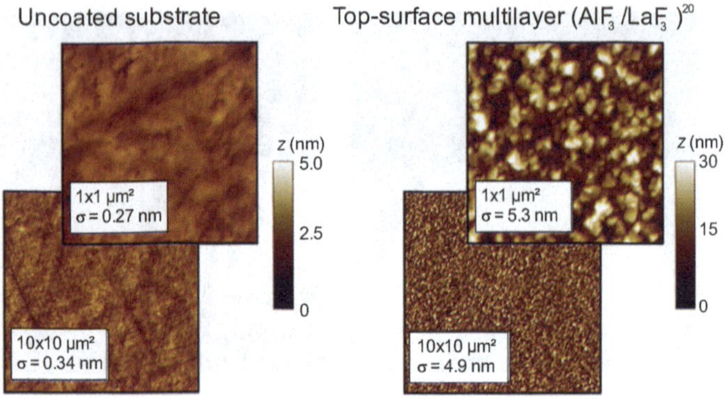

Fig. 14.15 AFM images of AlF_3/LaF_3 multilayer stack for $\lambda = 193$ nm. The roughness values in the insets are for the corresponding scan areas

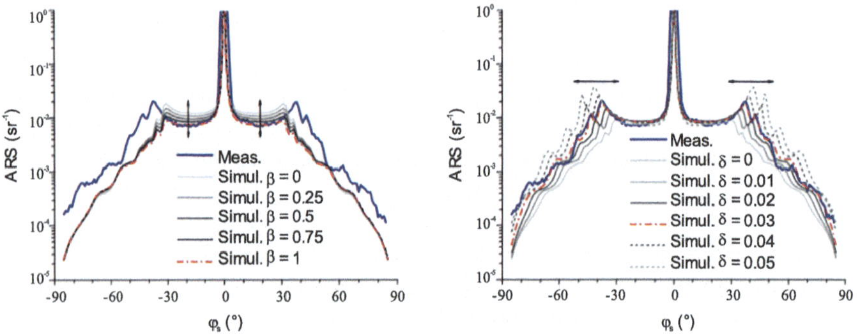

Fig. 14.16 ARS simulation of AlF_3/LaF_3 multilayer stack for $\lambda = 193$ nm. Left: Influence of the roughening exponent, β; right: effect of thin film thickness variations from the ideal multilayer design, δ

The roughness scaling approach has also been successfully used with other high index materials in combination with AlF_3 for the low index layer material inside a multilayer stack. Exemplary results obtained from AFM measurement after increasing number of layers are shown in Fig. 14.17.

For a slow or moderate roughness evolution inside the multilayer stack, an alternative to the scaling approach of the rms roughness is the linear continuum model. This is exemplified in Fig. 14.18 in more detail, which shows the interface PSDs after different number of layers of a highly reflective multilayer stack for 13.5 nm. The high index layer material is molybdenum and the low index material is silicon.

At low spatial frequencies, the substrate roughness is almost perfectly replicated by all layers. At high spatial frequencies, a continuous increase of the interface roughness from one layer to the next can be observed until the PSDs are on top of each other around $f = 100\ \mu m^{-1}$. This convergence can be explained by the

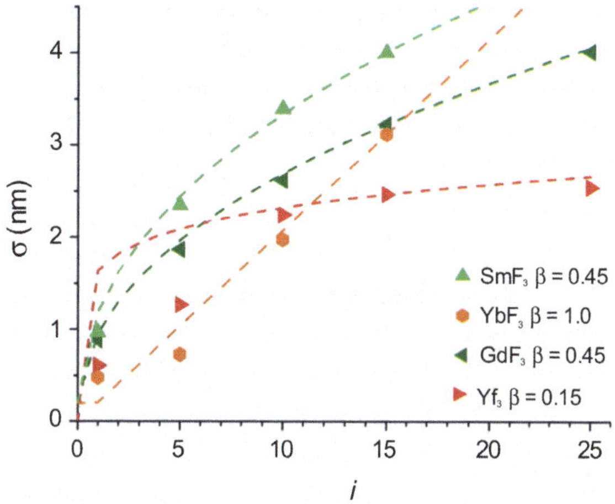

Fig. 14.17 Roughness evolution of quarter wave multilayer stacks with different high index materials in combination with AlF_3 as low index material. The rms-roughness values were obtained from AFM measurements in a scan area of $10 \times 10\ \mu m^2$

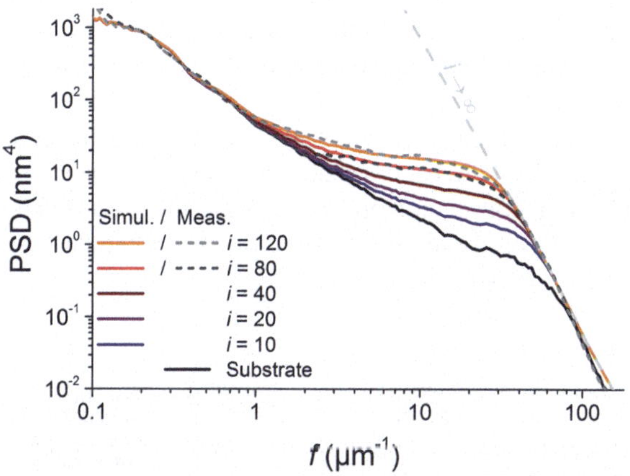

Fig. 14.18 Roughness evolution of Mo/Si multilayer. The gray dashed line indicates the equilibrium between intrinsic roughening and smoothing for an infinite number of layers

equilibrium between the roughness increase from intrinsic thin film roughness and the smoothing capabilities of the adatoms.

The roughness information from the linear continuum model can now be used to simulate the ARS as shown in Fig. 14.19.

The good agreement between measurement and simulations indicates that the combination of the linear continuum model with the Rayleigh-Rice scattering theory is a powerful approach when simulating scattering from multilayer coatings. The deviation close to the specular beam can be attributed to scattered light from the beam preparation optics of the measurement system.

A nice feature of the linear continuum model is that the different sources of roughness can be easily switched off in the simulation. Hence, a detailed analysis of the individual scattering contributions can be obtained, which could otherwise not be accessed by scattering measurements on real coatings. This shows that the dominant scattering source for the Mo/Si multilayer and the given substrate finish is intrinsic thin film roughness. Only close to the specularly reflected beam, the observed scattering can be almost entirely attributed to replicated substrate roughness. Hence, the imaging quality in an optical system critically depends on the substrate quality, while the overall scattering loss depends on the intrinsic thin films roughness and the used deposition parameters.

The detailed roughness evolution model can now also be used to predict the multilayer scattering for different substrate roughness values, as demonstrated in Fig. 14.20.

First, the substrate was characterized by angle resolved light scattering measurements as described in Sect. 14.4.1. The corresponding roughness map is shown in Fig. 14.20 left. Based on the local surface PSDs, the ARS could then be simulated. This also allows determining the impact of the scattering characteristics on the multilayer reflectance before the actual coating process. The results of this calculation are shown in Fig. 14.20 center. After the multilayer deposition, the final EUV reflectance was characterized at the application wavelength $\lambda = 13.5$ nm (see Fig. 14.20 right).

The predicted values and the actual reflectance measurements are in excellent agreement with each other. Both the average reflectance of 65% and the severe reflectance drop of more than 40% at the extended defect area are accurately predicted.

The simulation of the scattering and reflectance properties provides a valuable feedback and helps to ensure the final performance parameters already early during the entire fabrication process. For instance, at the extended defect area, the substrate could be easily re-polished before the coating process. This is not possible after the multilayer deposition.

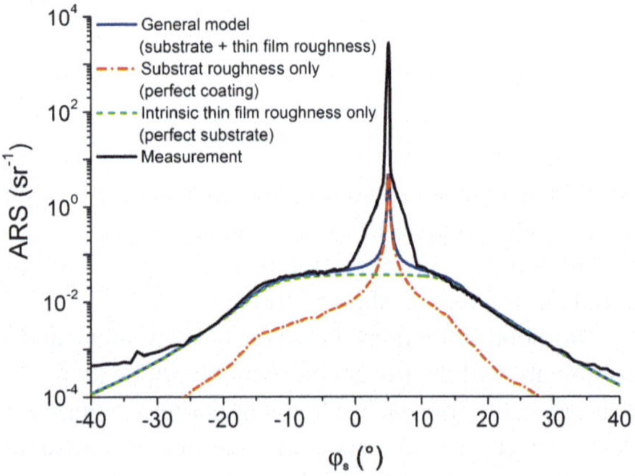

Fig. 14.19 Angle resolved light scattering of Mo/Si multilayer, measurements and modeling results

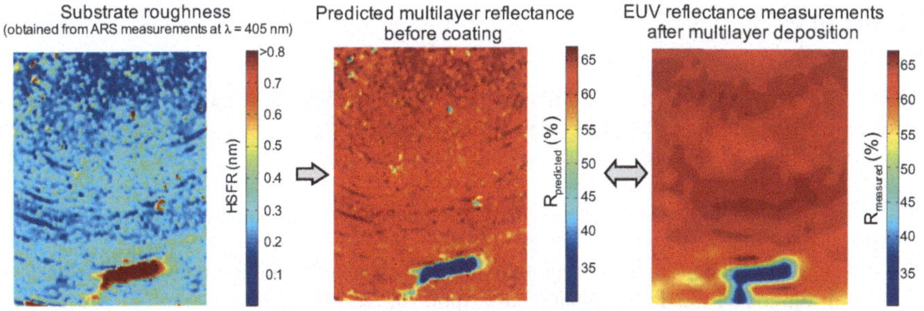

Fig. 14.20 Influence of substrate roughness on EUV reflectance of Mo/Si multilayer stack at $\lambda = 13.5$ nm. Left: roughness map obtained from ARS measurements at $\lambda = 405$ nm, center: prediction of multilayer reflectance prior to multilayer deposition; right: measured reflectance at application wavelength

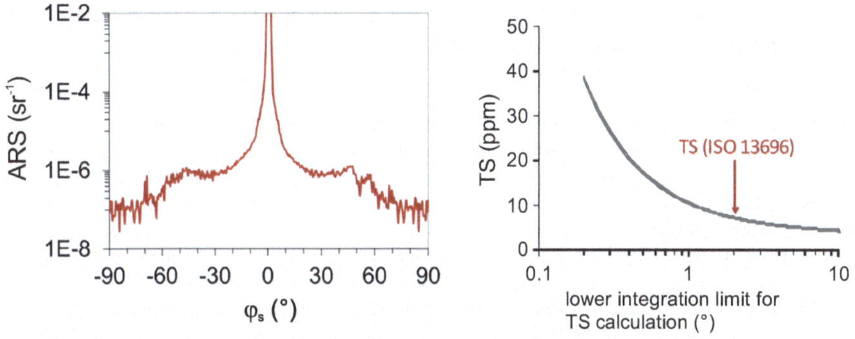

Fig. 14.21 Scattering analysis of low-loss mirror at $\lambda = 1064$ nm. Left: experimental ARS; right: TS when calculated from ARS for different lower integration limits

14.4.2.3 Energy Balance

Important for the analysis of the sample's reflectance or transmittance is that the angular acceptance range for the specular quantities are considered. For instance, in an international round robin experiment, dielectric mirrors made of silica and tantala with a reflectance of higher than 99.99% at $\lambda = 1064$ nm were characterized with respect to their losses and reflectance using spectrophotometry and laser ratiometry [84]. Substantial deviations between the reflectance values and thus the losses could be observed. This could be explained by the scattering characteristics as shown in Fig. 14.21.

Integrating the ARS according to the ISO 13696 between $\varphi_s = 2°$ and 85° leads to a scattering loss of TS = 7 ppm. However, substantially higher values up to 38 ppm can be obtained by varying the lower integration limit as illustrated in Fig. 14.21 right. This illustrates that considering the ranges of the acceptance angle is of crucial importance when measuring optical properties.

References

1. V. Džimbeg-Malcic, Ž. Barbaric-Mikocevic, K. Itric, Kubelka-Munk theory indescribing optical properties of paper (I). Tehn. vjesn. **18**, 117–124 (2011)
2. M.A. Hubbe, J.J. Pawlak, A.A. Koukoulas, Paper's appearance: a review. BioResour. **3**, 627–665 (2008)
3. E. Yablonovitch, G.D. Cody, Intensity enhancement in textured optical sheets for solar cells. IEEE Trans. Electron. Dev. **29**, 300–305 (1982)
4. M.A. Green, Lambertian light trapping in textured solar cells and light-emitting diodes: analytical solutions. Prog. Photovoltaics: Res. Appl. **10**, 235–241 (2002)
5. R.P. Breault, Control of stray light, in *Handbook of Optics Volume I Fundamentals, Techniques, and Design*, ed. by M. Bass (1995), pp. 38.1–38.35
6. J.M. Bennett, L. Mattsson, in *Introduction to Surface Roughness and Scattering* (Optical Society of America, Washington, 1989)
7. J.C. Stover, *Optical Scattering: Measurement and Analysis*, 3rd ed. (SPIE Press, Bellingham, 2012)
8. M.G. Dittman, Contamination scatter functions for stray-light analysis. Proc. SPIE **4774**, 99–110 (2002)
9. P.A. Bobbert, J. Vlieger, Light scattering by a sphere on a substrate. Physica **137A**, 209–242 (1986)
10. T. Herffurth, S. Schröder, M. Trost, A. Duparré, A. Tünnermann, Comprehensive nanostructure and defect analysis using a simple 3D light-scatter sensor. Appl. Opt. **52**, 3279–3287 (2013)
11. T.A. Germer, Angular dependence and polarization of out-of-plane optical scattering from particulate contamination, subsurface defects, and surface microroughness. Appl. Opt. **36**, 8798–8805 (1997)
12. M. Trost, T. Herffurth, D. Schmitz, S. Schröder, A. Duparré, A. Tünnermann, Evaluation of subsurface damage by light scattering techniques. Appl. Opt. **52**, 6579–6588 (2013)
13. E.L. Church, P.Z. Takacs, Surface scattering, in *Handbook of Optics: Fundamentals, Techniques, and Design*, ed. by M. Bass, 2nd ed. (McGraw-Hill, Inc., New York, 1995), pp. 7.1–7.14
14. A. Sentenac, H. Giovannini, M. Saillard, Scattering from rough inhomogeneous media: splitting of surface and volume scattering. J. Opt. Soc. Am. A **19**, 727–736 (2002)
15. P. Bousquet, F. Flory, P. Roche, Scattering from multilayer thin films: theory and experiment. J. Opt. Soc. Am. **71**, 1115–1123 (1981)
16. C. Amra, Light scattering from multilayer optics. I. Tools of investigation. J. Opt. Soc. Am. A **11**, 197–210 (1994)
17. J.M. Elson, J.P. Rahn, J.M. Bennett, Light scattering from multilayer optics: comparison of theory and experiment. Appl. Opt. **19**, 669–679 (1980)
18. P. Bussemer, K. Hehl, S. Kassam, Theory of light scattering from rough surfaces and interfaces and from volume inhomogeneities in an optical layer stack. Waves Random Media **1**, 207–221 (1991)
19. S. Schröder, D. Unglaub, M. Trost, X. Cheng, J. Zhang, A. Duparré, Spectral angle resolved scattering of thin film coatings. Appl. Opt. **53**, A35–A41 (2014)
20. M. Trost, T. Herffurth, S. Schröder, A. Duparré, A. Tünnermann, Scattering reduction through oblique multilayer deposition. Appl. Opt. **53**, A197–A204 (2014)
21. T. Herffurth, M. Trost, S. Schröder, K. Täschner, H. Bartzsch, P. Frach, A. Duparré, A. Tünnermann, Roughness and optical losses of rugate coatings. Appl. Opt. **53**, A351–A359 (2014)
22. S. Schröder, A. Duparré, Measurement of light scattering, transmittance, and reflectance, in *Laser-Induced Damage in Optical Materials*, ed. by D. Ristau. (CRC Press/Taylor & Francis Group, 2014), pp. 213–245
23. M. Trost, T. Herffurth, S. Schröder, A. Duparré, M. Beier, S. Risse, A. Tünnermann, N. Böwering, In situ and ex situ characterization of optical surfaces by light scattering techniques. Opt. Eng. **53**, 092013 (2014)

24. V. Rehn, V.O. Jones, J.M. Elson, J.M. Bennett, The role of surface topography in predicting scattering at grazing incidence from optical surfaces. Nucl. Instrum. Methods **172**, 307–314 (1980)
25. F.E. Nicodemus, J.C. Richmond, J.J. Hsia, I.W. Ginsberg, T. Limperis, Geometrical considerations and nomenclature for reflectance. Natl. Bur. Stand. Monogr. **160**, 1–52 (1977)
26. ASTM E 1392-90, *Standard Practice for Angle Resolved Optical Scatter Measurements on Specular or Diffuse Surfaces* (American Society for Testing and Materials, Philadelphia, 1990)
27. T.A. Leonard, M. Pantoliano, J. Reilly, Results of a CO2 BRDF round robin. Proc. SPIE **1165**, 444–449 (1989)
28. T.A. Leonard, P. Rudolph, BRDF round robin test of ASTM E1392. Proc. SPIE **1995**, 285–293 (1993)
29. ASTM E 2387-5, *Standard Practice for Goniometric Optical Scatter Measurements* (American Society for Testing and Materials, Philadelphia, 2011)
30. ISO 13696, *Optics and Photonics – Lasers and Laser Related Equipment – Test Methods for Radiation Scattered by Optical Components* (International Organization for Standardization, Geneva, Switzerland, 2002)
31. ASTM F 1048-87, *Standard Test Method for Measuring the Effective Surface Roughness of Optical Components by Total Integrated Scattering* (American Society for Testing and Materials, Philadelphia, 1999)
32. S.O. Rice, Reflection of electromagnetic waves from slightly rough surfaces. Commun. Pure Appl. Math. **4**, 351–378 (1951)
33. E.L. Church, H.A. Jenkinson, J.M. Zavada, Relationship between surface scattering and microtopographic features. Opt. Eng. **18**, 125–136 (1979)
34. P. Beckmann, A. Spizzichino, *The Scattering of Electromagnetic Waves from Rough Surfaces* (Pergamon Press, Oxford, 1963)
35. Lord Rayleigh, On the dynamical theory of gratings. Proc. R. Soc. Lond. Ser. A **79**, 399–416 (1907)
36. J.C. Stover, Experimental confirmation of the Rayleigh-Rice obliquity factor. Proc. SPIE **7792**, 77920J (2010)
37. N. Choi, J.E. Harvey, Numerical validation of the generalized Harvey-Shack surface scatter theory. Opt. Eng. **52**, 115103 (2013)
38. S. Schröder, A. Duparré, L. Coriand, A. Tünnermann, D.H. Penalver, J.E. Harvey, Modeling of light scattering in different regimes of surface roughness. Opt. Express **19**, 9820–9835 (2011)
39. J.M. Elson, J.M. Bennett, Calculation of the power spectral density from surface profile data. Appl. Opt. **34**, 201–208 (1995)
40. A. Duparré, J. Ferré-Borrull, S. Gliech, G. Notni, J. Steinert, J.M. Bennett, Surface characterization techniques for determining the root-mean-square roughness and power spectral densities of optical components. Appl. Opt. **41**, 154–171 (2002)
41. E.L. Church, H.A. Jenkinson, J.M. Zavada, Measurement of the finish of diamond-turned metal surfaces by differential light scattering. Opt. Eng. **16**, 360–374 (1977)
42. C. Amra, From light scattering to the microstructure of thin-film multilayers. Appl. Opt. **32**, 5481–5491 (1993)
43. H. Davies, The reflection of electromagnetic waves from a rough surface. Proc. IEEE **101**, 209–214 (1954)
44. H.E. Bennett, J.O. Porteus, Relation between surface roughness and specular reflectance at normal incidence. J. Opt. Soc. Am. **51**, 123–129 (1961)
45. J.C. Stover, S. Schröder, T.A. Germer, Upper roughness limitations on the TIS/RMS relationship. Proc. SPIE **8495**, 849503 (2012)
46. J.V. Grishchenko, M.L. Zanaveskin, Investigation into the correlation factor of substrate and multilayer film surfaces by atomic force microscopy. Crystallogr. Rep. **58**, 493–497 (2013)
47. S. Schröder, A. Duparré, A. Tünnermann, Roughness evolution and scatter losses of multilayers for 193 nm optics. Appl. Opt. **47**, C88–C97 (2008)
48. W.M. Tong, R.S. Williams, Kinetics of surface growth: phenomenology, scaling, and mechanisms of smoothing and roughening. Annu. Rev. Phys. Chem. **45**, 401–438 (1994)

49. M. Kardar, G. Parisi, Y.-C. Zhang, Dynamic scaling of growing interfaces. Phys. Rev. Lett. **56**, 889–892 (1986)
50. D.G. Stearns, Stochastic model for thin film growth and erosion. Appl. Phys. Lett. **62**, 1745–1747 (1993)
51. J. Ebert, H. Pannhorst, H. Küster, H. Welling, Scatter losses of broadband interference coatings. Appl. Opt. **18**, 818–822 (1979)
52. S. Schröder, H. Uhlig, A. Duparré, N. Kaiser, Nanostructure and optical properties of fluoride films for high-quality DUV/VUV optical components. Proc. SPIE **5963**, 59630R (2005)
53. K.H. Guenther, Revisiting structure zone models for thin film growth. Proc. SPIE **1324**, 2–12 (1990)
54. S. Schröder, T. Feigl, A. Duparré, A. Tünnermann, EUV reflectance and scattering of Mo/Si multilayers on differently polished substrates. Opt. Express **15**, 13997–14012 (2007)
55. R. Canestrari, D. Spiga, G. Pareschi, Analysis of microroughness evolution in X-ray astronomical multilayer mirrors by surface topography with the MPES program and by X-ray scattering. Proc. SPIE **6266**, 626613 (2006)
56. M. Trost, S. Schröder, T. Feigl, A. Duparré, A. Tünnermann, Influence of the substrate finish and thin film roughness on the optical performance of Mo/Si multilayers. Appl. Opt. **50**, C148–C153 (2011)
57. H.E. Bennett, Scattering characteristics of optical materials. Opt. Eng. **17**, 480–488 (1978)
58. K.H. Guenther, P.G. Wierer, J.M. Bennett, Surface roughness measurements of low-scatter mirrors and roughness standards. Appl. Opt. **23**, 3820–3836 (1984)
59. J.A. Detrio, S.M. Miner, Standardized total integrated scatter measurements of optical surfaces. Opt. Eng. **24**, 419–422 (1985)
60. D. Rönnow, E. Veszelei, Design review of an instrument for spectroscopic total integrated light scattering measurements in the visible wavelength region. Rev. Sci. Instrum. **65**, 327–334 (1994)
61. O. Kienzle, J. Staub, T. Tschudi, Description of an integrated scatter instrument for measuring scatter losses of "superpolished" optical surfaces. Meas. Sci. Technol. **5**, 747–752 (1994)
62. A. Duparré, S. Gliech, Non-contact testing of optical surfaces by multiple-wavelength light scattering measurement. Proc. SPIE **3110**, 566–573 (1997)
63. P. Kadkhoda, A. Mueller, D. Ristau, Total scatter losses of optical components in the DUV/VUV spectral range. Proc. SPIE **3902**, 118–127 (2000)
64. S. Gliech, J. Steinert, A. Duparré, Light-scattering measurements of optical thin-film components at 157 and 193 nm. Appl. Opt. **41**, 3224–3235 (2002)
65. M. Otani, R. Biro, C. Ouchi, Y. Suzuki, K. Sone, S. Niisaka, T. Saito, J. Saito, A. Tanaka, A. Matsumoto, Development of optical coatings for 157-nm lithography. II. Reflectance, absorption, and scatter measurement. Appl. Opt. **41**, 3248–3255 (2002)
66. S. Schröder, S. Gliech, A. Duparré, Measurement system to determine the total and angle-resolved light scattering of optical components in the deep-ultraviolet and vacuum-ultraviolet spectral regions. Appl. Opt. **44**, 6093–6107 (2005)
67. D.R. Cheever, F.M. Cady, K.A. Klicker, J.C. Stover, Design review of a unique complete angle scatter instrument (CASI). Proc. SPIE **818**, 13–20 (1987)
68. C.C. Asmail, C.L. Cromer, J.E. Proctor, J.J. Hsia, Instrumentation at the national institute of standards and technology for bidirectional reflectance distribution function (BRDF) measurements. Proc. SPIE **2260**, 52–61 (1994)
69. A. von Finck, T. Herffurth, S. Schröder, A. Duparré, S. Sinzinger, Characterization of optical coatings using a multisource table-top scatterometer. Appl. Opt. **53**, A259–A269 (2014)
70. A. von Finck, M. Trost, S. Schröder, A. Duparré, Parallelized multichannel BSDF measurements. Opt. Express **23**, 33493–33505 (2015)
71. C. Amra, D. Torricini, P. Roche, Multiwavelength (0.45–10.6 μm) angle-resolved scatterometer or how to extend the optical window. Appl. Opt. **32**, 5462–5474 (1993)
72. F.M. Cady, J.C. Stover, D.R. Bjork, M.L. Bernt, M.W. Knighton, D.J. Wilson, D.R. Cheever, A design review of a multiwavelength, three-dimensional scatterometer. Proc. SPIE **1331**, 201–208 (1990)

73. S. Schröder, T. Herffurth, M. Trost, A. Duparré, Angle-resolved scattering and reflectance of extreme-ultraviolet multilayer coatings: measurement and analysis. Appl. Opt. **49**, 1503–1512 (2010)
74. M.M. Barysheva, Y.A. Vainer, B.A. Gribkov, M.V. Zorina, A.E. Pestov, N.N. Salashchenko, N.I. Chkhalo, A.V. Shcherbakov, Investigation of supersmooth optical surfaces and multilayer elements using soft x-ray radiation. Tech. Phys. **58**, 1371–1379 (2013)
75. M. Zerrad, M. Lequime, C. Amra, Multimodal scattering facilities and modelization tools for a comprehensive investigation of optical coatings. Proc. SPIE **8169**, 81690K (2011)
76. M. Lequime, M. Zerrad, C. Deumié, C. Amra, A goniometric light scattering instrument with high-resolution imaging. Opt. Commun. **282**, 1265–1273 (2009)
77. C.C. Cooksey, M.E. Nadal, D.W. Allen, K.-O. Hauer, A. Höpe, Bidirectional reflectance scale comparison between NIST and PTB. Appl. Opt. **54**, 4006–4015 (2015)
78. S. Schröder, T. Herffurth, H. Blaschke, A. Duparré, Angle-resolved scattering: an effective method for characterizing thin-film coatings. Appl. Opt. **50**, C164–C171 (2011)
79. C.C. Asmail, J. Hsia, A. Parr, J. Hoeft, Rayleigh scattering limits for low-level bidirectional reflectance distribution function measurements. Appl. Opt. **33**, 6084–6091 (1994)
80. C. Wagner, N. Harned, EUV lithography: lithography gets extreme. Nat. Photonics **4**, 24–26 (2010)
81. N.R. Böwering et al., Performance results of laser-produced plasma test and prototype light sources for EUV lithography. J. Micro/Nanolithogr. MEMS MOEMS **8**, 041504 (2009)
82. M. Trost, S. Schröder, A. Duparré, S. Risse, T. Feigl, U.D. Zeitner, A. Tünnermann, Structured Mo/Si multilayers for IR-suppression in laser-produced EUV light sources. Opt. Express **21**, 27852–27864 (2013)
83. M. Trost, S. Schröder, C.C. Lin, A. Duparré, A. Tünnermann, Roughness characterization of EUV multilayer coatings and ultra-smooth surfaces by light scattering. Proc. SPIE **8501**, 85010F (2012)
84. A. Duparré, D. Ristau, Optical interference coatings 2010 measurement problem. Appl. Opt. **50**, C172–C177 (2011)

Chapter 15
Absorption and Fluorescence Measurements in Optical Coatings

Christian Mühlig

Abstract Absorption characterization of optical thin films and coatings has become one central task for the manufacturers e.g. to ensure stability in the production process, to verify functionalities and to understand possible performance changes and limitations during their use in e.g. high power laser applications. Set by this trend, numerous direct absorption measurement techniques have been developed in the last two decades which all—despite particular pro and cons—feature a high sensitivity. However, the different techniques possess remarkable differences regarding a universal and efficient absolute calibration procedure. After a survey of different measurement techniques with their calibration procedures, this chapter will focus on the laser induced deflection (LID) technique, its independent absolute calibration, particular measurement concepts and experimental results.

Increasing laser power for material processing, decreasing structures for semiconductor lithography—meeting the advanced requirements put high demands on state-of-the-art optical components. It turned out that amongst the key parameters in sophisticated laser applications, absorption is recently gaining more and more attention due to the undesired effects resulting from laser induced thermal lensing. Focus shifting, wave front deformation and depolarization are prominent critical issues which need to be taken into account in modern optic design to assure the target system performance.

When taking a look at optical coatings, measuring the spectral reflectance and transmittances is the most general way to investigate optical losses of thin films with a subsequent calculation of the extinction coefficient k. However, the thus calculated extinction coefficient may contain scatter contributions, because small absorption and scatter losses cannot be distinguished when only directed transmission and reflection spectra are measured (compare (2.4), Chap. 2). An increasing number of applications, however, demand for individual absorption and scattering data as a result of their dif-

C. Mühlig (✉)
Leibniz-Institute of Photonic Technology IPHT, Albert-Einstein-Straße 9, 07745 Jena, Germany
e-mail: christian.muehlig@leibniz-ipht.de

O. Stenzel and M. Ohlídal (eds.), *Optical Characterization of Thin Solid Films*, Springer Series in Surface Sciences 64,
https://doi.org/10.1007/978-3-319-75325-6_15

ferent possible interference on the optical system performance. In addition, separating the effects from bulk and coating/surface is essential to discriminate between different sources of absorption and scattering, respectively. Consequently, direct measuring of absorption properties in optical thin films and bulk materials turned out to be one challenge for optics manufacturers to secure or optimize stabile production processes, to establish particular optical functionalities and to understand potential performance limitations of the optics when used in highly sophisticated laser applications. Set by this trend, numerous direct absorption measurement techniques have been developed in the last two decades which all—despite particular pro and cons—feature a high sensitivity. Recently, the ability of absolute calibration has gained more and more attention in order to allow for absolute absorption data. Here, however, the different techniques possess remarkable differences regarding a universal and efficient procedure. After a survey of different measurement techniques with their calibration procedures, this chapter will focus on the laser induced deflection (LID) technique, its independent absolute calibration, particular measurement concepts and experimental results.

Along with direct absorption measurements, laser induced fluorescence (LIF) investigations can be a very useful tool to identify very sensitively defects and/or impurities responsible for the observed absorption properties. However, compared to investigations in bulk materials, there is rather a lack of data with respect to LIF measurements in optical coatings. This might be attributed to the fact that the excitation volume of thin films decreases by about 5 orders of magnitude in comparison to bulk samples, not accounting for the peculiarities of e.g. DUV measurement techniques. But, bearing in mind a presumably higher defect density in coatings and the high photon flux of e.g. excimer lasers, on the other hand, even the small excitation volume of coating samples can be considered to be an accessible range for experimental investigation.

15.1 Overview of Absorption Measurement Techniques and Absolute Calibration

When looking for potential ways to directly access the absorption of light inside bulk materials or optical coatings, it is worth considering the effect(s) of absorbed light. Figure 15.1 gives a scheme of possible processes. In the end it turned out that the heat generation and related effects (photo-thermal and photo-acoustic effects) have shown the largest potential for direct absorption measurement techniques. However, as seen from Fig. 15.1, it needs to be mentioned that for all those techniques only that part of the absorbed light is detected which is actually transferred into heat. In particular, this has to be taken into account when considering materials showing fluorescence with high fluorescence quantum efficiency.

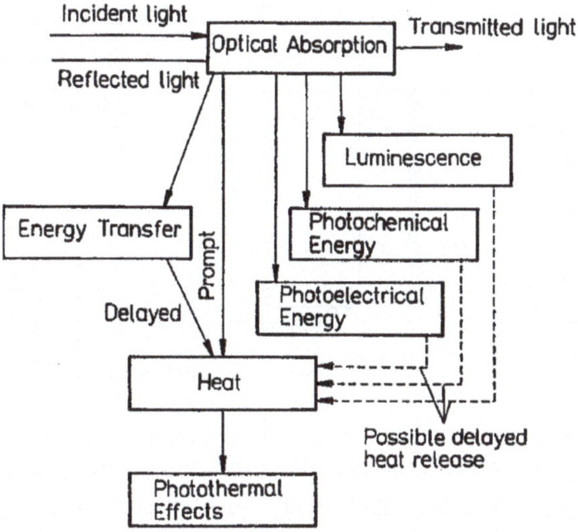

Fig. 15.1 Block diagram that indicates the possible consequences of optical absorption that leads to prompt or delayed heat production compared with other deexcitation channels (taken from [1] with kind permission)

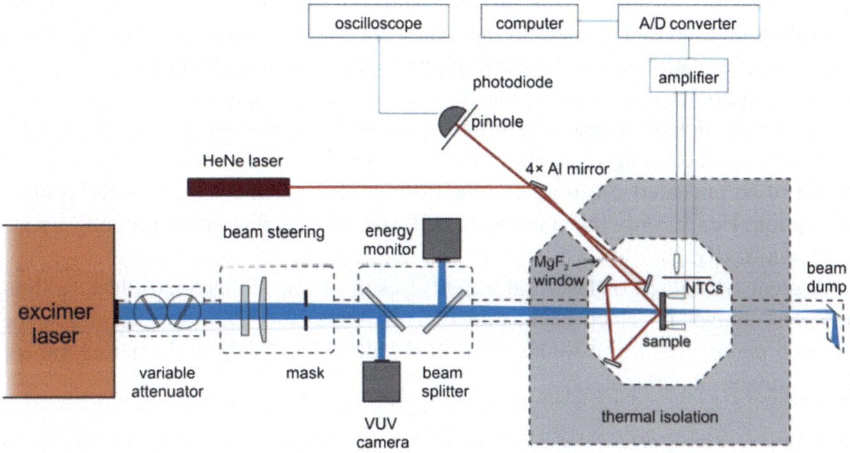

Fig. 15.2 Schematic diagram of the optical arrangement for the combined laser calorimetry (LC) and surface thermal lens (STL) technique (taken from [2] with kind permission)

15.1.1 Calorimetry

Longer in use than any other technique is the laser calorimetry. Here, the sample of investigation is irradiated by a small laser spot and the increasing temperature is measured directly at defined position on the sample surface (Fig. 15.2).

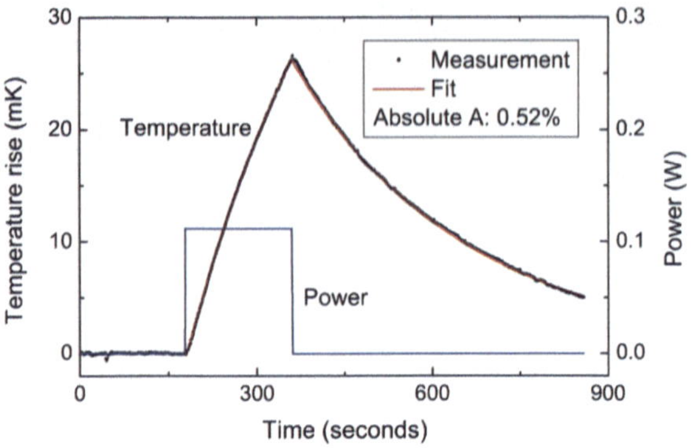

Fig. 15.3 Calorimetric signal obtained with a highly reflective mirror irradiated at 193 nm. The irradiation fluence is 0.52 J cm^2. The repetition rate of the laser is 57 Hz (taken from [2] with kind permission)

Figure 15.3 shows the typical result of a laser calorimetric measurement. When applying the laser power to the sample of investigation, its temperature rises until the irradiation stops. Then the sample cools down again. In general, both the temperature rising and falling can be fitted by corresponding models and used for data analysis according to the calorimetry ISO standard description (11551 [3]).

In order to distinguish between coating and substrate absorption, there are two approaches. First, samples with different substrate thickness but identical coating have to be measured. The measurement data are plotted against substrate thickness and the extrapolation to zero thickness serves for the coating/interface absorption. Since this approach is quite costly and the accuracy of this approach is strongly related to identical sample properties (substrate absorption and coating absorption), mostly a second approach is applied. Here, the result for the coated substrate is compared to that of an uncoated substrate and the difference is taken to calculate the coating absorption. Hence, only the substrate's bulk and surface absorption properties need to be identically.

A great advantage of laser calorimetry results from the easy absolute calibration. An "equivalent" electrical heating of a reference sample allows determining the required parameters for absolute absorption calculation, without the actual knowledge of material parameters.

15.1.2 *Photo-Thermal Techniques*

A large variety of absorption measurement techniques belong to photo-thermal techniques [4]. The majority of those techniques are using the so-called collinear pump-probe-configuration, i.e. pump and probe beams cross each other under a very small angle. In contrast to the laser calorimetry, the sample heating by pump beam absorption is detected indirectly by the induced changes in the probe beam characteristics.

In general, the common challenge of all photo-thermal techniques is the absolute calibration. Obviously, a reference sample of same material (where required also same geometry) and precisely known absorption is the most favored solution. However, this case is barely realistic for common "interesting" materials possessing very low absorption at the application wavelength. Alternatively, reference materials showing high initial absorption at the desired laser wavelength are applied [5]. In this case, spectrometers are used to create a "reference" by measuring the absorption at a particular wavelength and correction values are calculated for the materials under investigation bearing in mind the changes in the thermo-optical material properties. However, this approach fails or lacks in accuracy if scattering losses cannot be neglected in the spectrometer results and if no or differing thermo-optical material data is on-hand, respectively. Recently, a new calibration approach has been applied to optical materials that show a high absorption outside the application wavelength range—preferably in the infrared range to reduce scattering effects—but where other lasers sources are accessible. In that case, the high absorption is first measured in a spectrometer at an available laser wavelength and second by the corresponding laser in the photo-thermal setup. Thereby, the photo-thermal signature for the particular material is provided and finally, the photo-thermal measurement is repeated at the application laser wavelength.

For the Common Path Interferometry (PCI) technique the pump beam is focused to a very small spot in the sample and the absorption (heat) induced change in the optical path results in a change of the self-interference of a much larger probe beam [5]. Behind an aperture, this change is registered by a fast and sensitive detector (Fig. 15.4). For separating between substrate and coating absorption, the cross-point between pump and probe beams is moved through the sample while steadily monitoring the probe beam signal. Figure 15.5 shows the result of such a scan.

A different way to detect the properties of the probe beam due to the pump beam induced absorption is to use a Hartmann-Shack sensor. Here, the change of the wave front of a probe beam due to a pump beam of much smaller size is monitored by a CCD camera with high sensitivity and resolution [7, 8]. Figure 15.6 shows the scheme of the measuring principle of a Hartmann-Shack wave front sensor.

A representative experimental setup is shown in Fig. 15.7.

For the configuration in Fig. 15.7, the separation between substrate and coating absorption is achieved similar to the calorimetry, i.e. either by a sample set of different thicknesses or by comparing coated to uncoated samples. Recently, a 90°-configuration between pump and probe beam has been applied for easier separation between substrate and coating absorption. Predominantly, numerical simulation and analysis is applied for calibrating the experimental data from the Hartmann-Shack wave front sensor. Absorption values are calculated from the simulations by fitting to the experimental data [8, 9]. This approach requires both, sophisticated simulation models and—as mentioned above—a precise knowing of the multiplicity of required material parameters. In particular for materials possessing a non-negligible expansion coefficient, calculating the stress and its effect on the refractive index via elasto-optical coefficients is challenging. Consequently, for the majority of optical materials either only relative absorption data are provided or rather complex strategies are required [9].

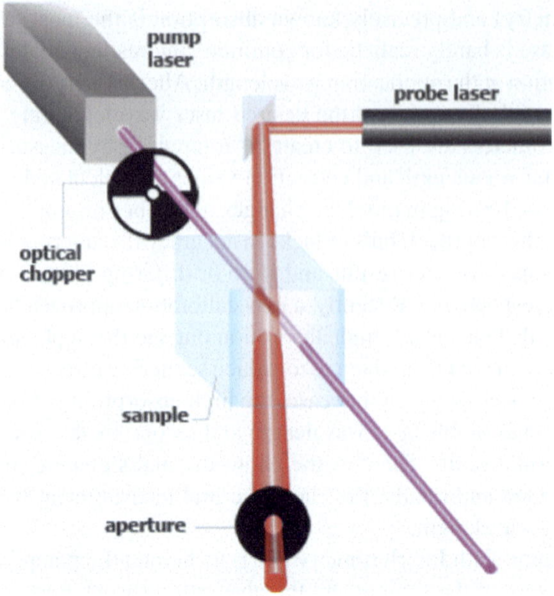

Fig. 15.4 Scheme of the common path interferometry (PCI) concept (taken from [6])

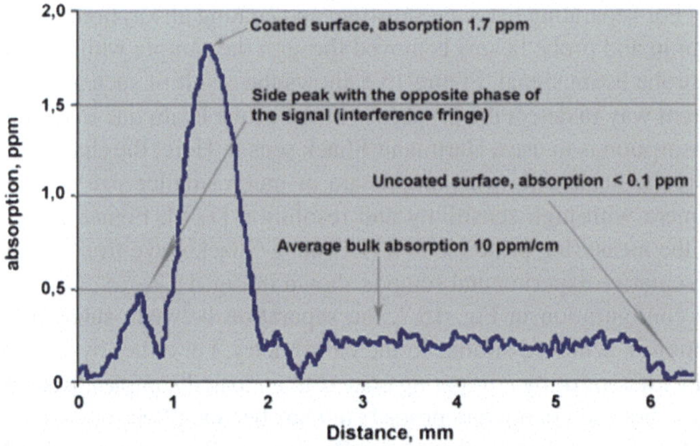

Fig. 15.5 A longitudinal scan through AR-coated fused silica substrate. Pump power 5 W, wavelength 1064 nm. Surface calibration was used for the vertical scale (taken from [5] with kind permission)

Taking the above into consideration, it is not unexpected that photo-thermal techniques show coinciding absolute absorption data mainly for silica or coatings on silica substrates due to the rather simple simulation models for silica as a result of its negligible expansion coefficient.

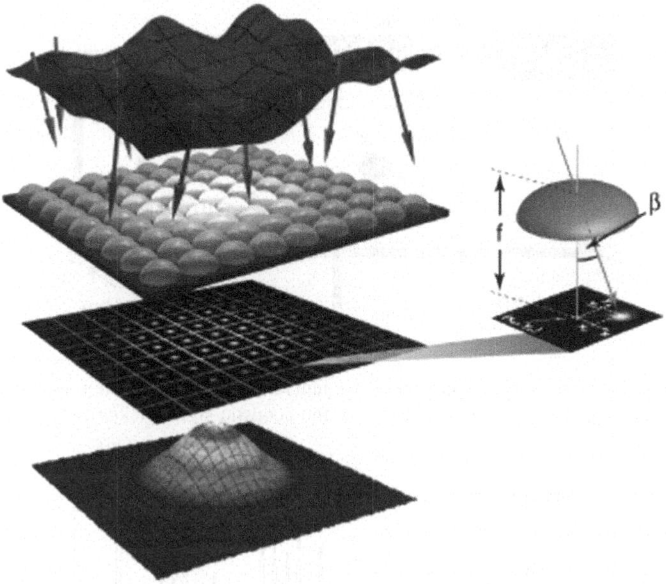

Fig. 15.6 Principle of the Hartmann-Shack wave front sensor (taken from [9] with kind permission)

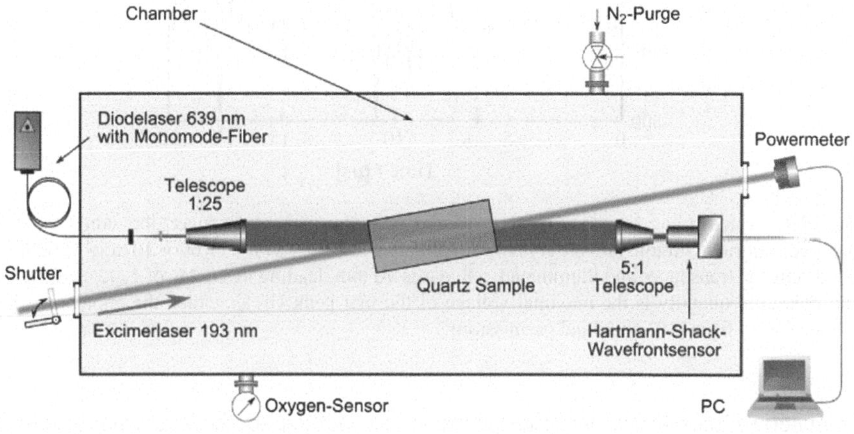

Fig. 15.7 Setup for measurement of the laser induced photo-thermal wave front deformation (taken from [9] with kind permission)

15.1.3 Photo-Acoustic Technique

Historically a technique long in use is the photo-acoustic spectroscopy (PAS). Here, acoustic waves are detected by (piezo) microphones which are generated by the absorption induced shock wave within the sample. PAS is in particular suited for materials showing high thermal expansion like nonlinear or laser crystals. In contrast,

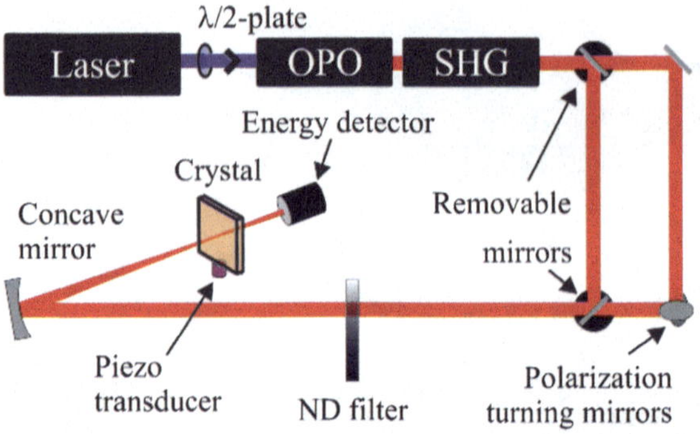

Fig. 15.8 Experimental setup: Light pulses are generated and focused onto a sample. A piezo transducer is attached to the sample to measure the acoustic signal (taken from [10] with kind permission)

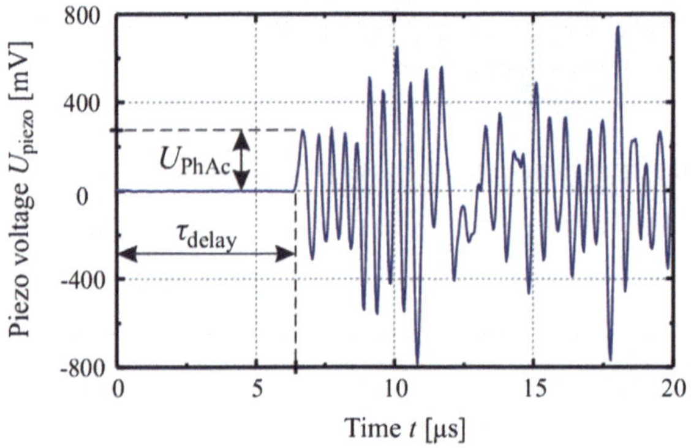

Fig. 15.9 Typical photoacoustic signal measured with the piezo transducer, the sample is an undoped congruent lithium niobate crystal with dimensions $(x \times y \times z) = 1 \times 60 \times 10\,mm^3$. The distance between transducer and illuminated cylinder is 40 mm, leading to a delay of $\tau_{delay} = 6.5\,\mu s$. The measured quantity is the maximal voltage of the first peak U_{PhAc}, called the photoacoustic voltage (taken from [10] with kind permission)

for materials like silica with negligible expansion coefficients, the accessible sensitivity is limited. Another challenge for PAS is the separation between surface and bulk absorption due to the associated shock waves and their reflections at sample surfaces.

After getting low attention in the past 1–2 decades due to a lack of sensitivity, recent developments show up with improved sensitivity for bulk absorption measurements [10]. Figures 15.8 and 15.9 show an example for an experimental PAS setup and a typical measurement signal, respectively.

Similar to the mentioned photo-thermal techniques, absolute calibration is a challenge and mostly done by using a reference sample of known absorption. Recently, the use of a tunable OPO laser source allowed to measure the sample of investigation at different wavelengths, e.g. also at wavelengths of high absorption that are accessible to spectrometric measurements. By this, an absolute calibration is achieved for the sample of investigation. In order to keep any systematic error due to scattering effects negligible, the comparing high absorption measurements have been done in the infrared wavelength region [10].

15.2 Laser Induced Deflection (LID) Technique

The laser induced deflection (LID) technique also belongs to the photo-thermal technique s with a pump-probe-configuration [11]. Within the sample, the absorbed pump laser power results in a temperature profile formation (Fig. 15.10). This temperature profile is turned into a refractive index profile (= thermal lens) by the thermal expansion and the temperature dependent refractive index. The refractive index gradient is proportional to the absorbed pump laser power and yields a probe beam deflection that is detected by position sensitive detectors (PSD). Combined with sophisticated electronics, probe beam deflections in the range of some Nanometers can be detected, corresponding to absorptions on the sub-ppm-level. In contrast to the photo-thermal techniques described in the previous section, the LID technique features some major differences. It uses a so-called transversal pump-probe-configuration, i.e. the probe beam is guided through the sample under 90° to the pump beam. Thereby, in most applications both beams do not cross each other but the probe beam passes the sample outside the pump beam area. Since the refractive index profile outside the pump beam area is only a function of the absorbed pump laser power, the LID technique measures independent on the actual pump laser beam geometry. Further, the pump beam is not focused into the sample and typically has a larger beam size than the probe beam. Finally, the measurement data is obtained in steady-state mode, i.e. the absorption induced refractive index profile does not change anymore.

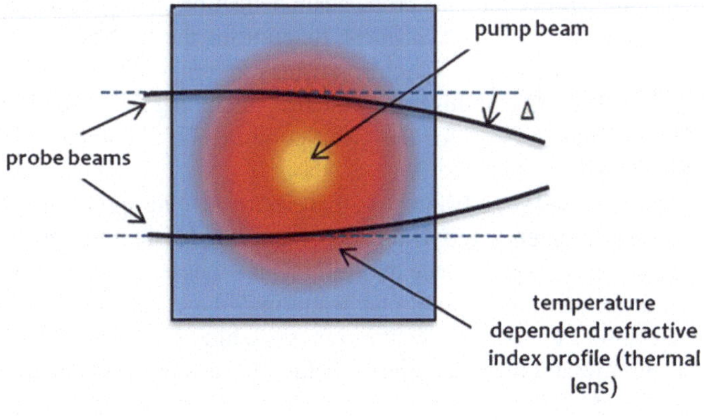

Fig. 15.10 Sketch of the LID measuring concept (taken from [12] with kind permission

Fig. 15.11 Calibration samples of different materials for bulk and surface/coating absorption, respectively

15.2.1 Absolute Calibration, Measurement Procedure and Absorption Calculation

As stated above, the absolute calibration is a critical issue for all photo-thermal absorption measurement techniques. For the LID technique, the calibration approach of the calorimetry has been adapted for the first time to photo-thermal techniques [13]. To derive absolute absorption data by the LID technique, the thermal lens is generated by electric heaters. This procedure needs to be applied separately for both, bulk and coating (surface) absorption as well as for each combination of sample material and geometry. For bulk absorption calibration, a hole in the center of the sample's aperture is required to put in a self-made electrical heater of same length as the sample (Fig. 15.11). For coating/surface absorption calibration, SMD heaters are mounted onto a very thin copper plate (thickness ~200 μm) which is attached centrally to the sample surface. The copper plate is essential to secure a high thermal conduction to the sample. It is obvious that these calibration approaches cannot be applied to a collinear pump-probe configuration but only to transversal pump-probe configurations where the probe beam is not hitting the irradiated/heated sample part.

For the calibration procedure, the probe beam deflection is measured as a function of the electric power. As a result, a linear function of the LID deflection signals versus electric power is obtained which spans over several orders of magnitude for the electric power. The slope of this linear function (including the zero-point) defines the calibration coefficient F_{CAL} for the given combination of sample material and geometry (Fig. 15.12).

The unique feature and key advantage of this electric heating approach is the ability to calibrate the measurement setup without any knowledge of thermo-optical material parameters. Recently, energy balance measurements—i.e. separate determination of reflectance, transmission, absorption and scattering—have been done for different materials and coatings to verify the calibration procedure. The results have given the proof that within measurement accuracy a value of 1 has been obtained in each of the investigations [14, 15].

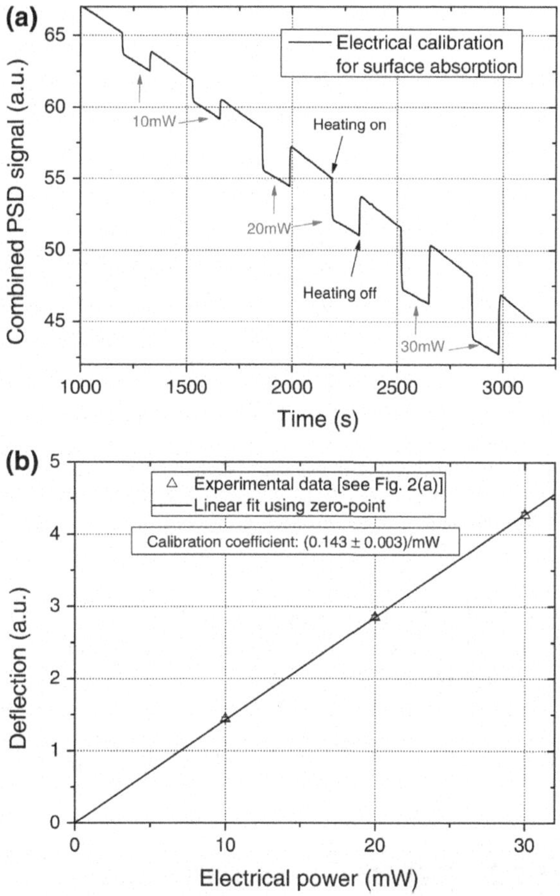

Fig. 15.12 **a** Combined signal from position sensitive detectors (PSD) during electrical calibration procedure for surface absorption measurement on CaF_2 samples of geometry 15 mm × 15 mm × 10 mm. **b** Deflection signal as function of the electrical power for the calibration procedure in **a** The slope of the linear fitting defines the calibration coefficient (taken from [12] with kind permission

Once the calibration has been performed, a sample of investigation can be investigated. An exemplary measurement cycle for an MgF_2 single layer at 193 nm laser irradiation is given in Fig. 15.13.

Each cycle in the measurement comprises the time between laser irradiation "on" and "off" events. Before any sample irradiation start, the "base line" of the LID signal is required to be constant or show a constant drift (Fig. 15.12a). The latter typically results from environmental changes in the setup, e.g. the temperature. Small temperature changes already affect optic mountings within the setup by thermal expansion resulting in small changes in the probe beam path and thus its location on the position sensitive detectors. Practically constant drifts over a measurement cycle of typically a few minutes are taken onto account during data analysis. For absorption analysis, the LID deflection signal is defined as the difference between the "base line" signal and the constant signal during irradiation. From the data analysis, the LID deflection signals for all measurement cycles are obtained. Their mean value

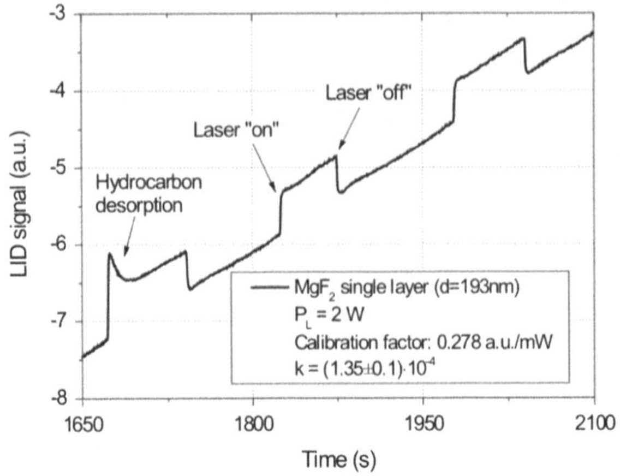

Fig. 15.13 Surface sensitive measurement concept: experimental results for the measurement of a single MgF_2 layer onto a CaF_2 substrate at the laser wavelength of 193 nm (taken from [16] with kind permission

is defined as the LID intensity I_{LID}. Knowing the constant mean pump laser power P_L for all cycles and the calibration coefficient F_{CAL}, the coating absorption A is calculated by

$$A = \frac{I_{LID}}{F_{CAL} \cdot P_L}. \tag{15.1}$$

15.2.2 LID Measurement Concepts

Is there a preference, whether two probe beams pass the sample outside the irradiated area or one probe beam passes inside the irradiated area? What deflection should be used—in direction of the pump laser (= horizontal) or perpendicular to it (= vertical)? The two different concepts are sketched in Fig. 15.14 and their application depends on the actual measurement tasks.

The vertical concept uses two probe beams above/beneath the irradiated spot as well as the probe beam deflection perpendicular to the pump beam direction. The high signal-to-noise ratio makes the vertical concept the best option for absorption measurements in bulks and highly reflective coatings. The only parameter which should be changed is the position of the probe beams along the sample length, i.e. for measuring coatings the probe beams pass the sample close to the coated surface while for bulk absorption measurement the probe beams pass the sample in the middle of the sample length.

For the horizontal concept, in contrast, one probe beam is guided centrally through the irradiated area at a position closest to one sample surface. Its deflection in direction

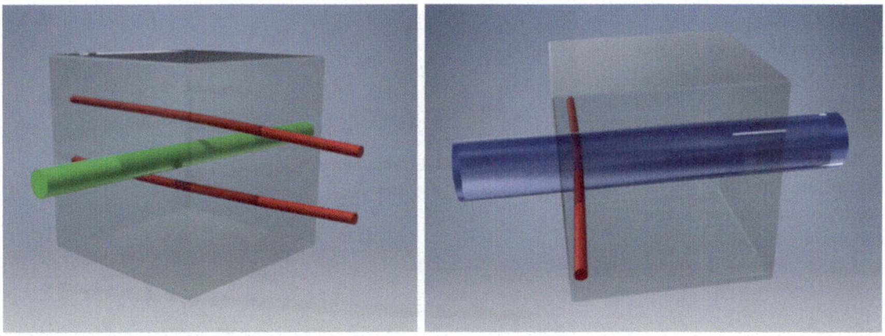

Fig. 15.14 (left) Vertical LID concept: two probe beams are guided outside the irradiated volume and deflection perpendicular to pump beam is measured; (right) Horizontal LID concept: one probe beam is guide through the irradiated volume and deflection in direction of the pump beam is measured

of the pump beam detects the surface absorption virtually free of the sample's bulk absorption. This makes the horizontal concept the ideal choice for investigating transparent optical coatings like anti-reflecting or partially reflecting coatings, where typically both, bulk and coating absorption, are present in the measurement signal and need to be separated.

The sensitivity of photo-thermal absorption measurement techniques depends on the thermo-optical properties of the investigated material. For many interesting materials like nonlinear optical or laser crystals, which favorably form only a small thermal lens, this often results in an insufficient sensitivity. In addition, the LID technique so far requires two additional polished side-faces for probe beam guiding due to the transversal pump-probe-configuration. Furthermore, the required sample aperture of at least 8×8 mm^2—the pump beam and two probe beams need to pass within the sample—prevents the use of the LID technique for investigating very small sized nonlinear optical and laser crystals.

The recently demonstrated Sandwich-LID concept (Fig. 15.15) solves these issues [17]. The basic principle is that a small sample is sandwiched between two larger tiles of an appropriate optical material. While the pump laser still hits the sample of investigation, the probe beams are guided through the sandwich tiles instead of the sample. The probe beam deflection now results from the thermal lens that is formed in the optical tiles by heat transfer from the irradiated sample. Now, the required sample aperture solely depends on the pump beam size and allows the investigation of typical sample dimensions of crystals like LBO or BBO.

Until now it was required to calibrate anisotropic or doped optical materials in dependence on the sample orientation and doping, respectively. For the Sandwich-LID concept, only the thermo-optical parameters of the optical tile material are of interest. Hence, it could be shown that calibration now is independent on the actual orientation or doping of the sample. Furthermore, it could be verified by measurements that the amount of heat transferred into the optical tiles does not depend on the polishing of the contact surfaces. Therefore, the requirement for additional

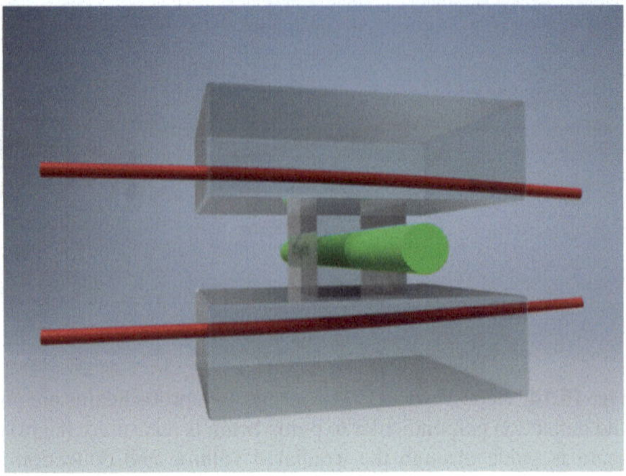

Fig. 15.15 Sketch of the Sandwich-LID concept: two probe beams are guided outside the pump laser beam and outside the sample through sandwich tiles of an appropriate optical material and deflection perpendicular to the pump beam is measured

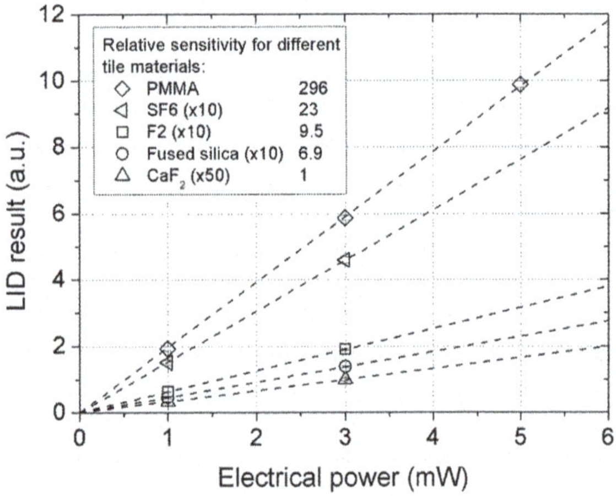

Fig. 15.16 LID results for a CaF_2 substrate (10 mm × 10 mm × 20 mm) with electrical surface heater (4 mm × 4 mm) placed between different sandwich tile materials (20 mm × 20 mm × 10 mm) (taken from [17] with kind permission)

side-face polishing is omitted. Moreover, the investigated material does not need to be transparent for the probe beam anymore. This clears the way for measuring materials that are nontransparent for visible wavelengths like some infrared materials or highly reflective mirrors on aluminum substrates.

The most outstanding feature of the new concept, however, is the ability to strongly increase the sensitivity for materials with a low photo-thermal response by choosing appropriate optical tiles (Fig. 15.16). This becomes possible since the probe beams uses now the thermo-optical properties of the optical tiles instead of those of the sample material. Separating the sites of pump beam absorption (sample of investigation) and probe beam deflection (optical tiles) allows a sensitivity increase for many optical materials by more than an order of magnitude. Thereby, much lower absorption is measured with a fixed pump laser power or a much lower pump laser power is required to measure a given absorption.

Table 15.1 Surface absorption values at 193 nm for all investigated CaF_2 samples in dependence on the ion beam figuring surface treatment (taken from [12] with kind permission)

Sample	Treatment	Absorption (% per surface @193 nm)
#1	*Pitch polishing (standard)*	***0.08*** *± 0.008*
#2	*Pitch polishing (standard)*	***0.083*** *± 0.01*
#5	IBF (standard parameters)	0.3 ± 0.035
#6	IBF (double neutralization)	0.39 ± 0.03
#7	IBF (double treatment time)	**0.445** ± 0.035
#8	IBF (step-like treatment)	0.355 ± 0.03
#9	IBF (optimized removal)	0.365 ± 0.035
#10	IBF (preserve treatment)	0.355 ± 0.03

15.2.3 Experimental Results

15.2.3.1 Surface and Interface Absorption at 193 nm

Assuring constant figures of merit—e.g. aberration control and contrast—for high-quality imaging optics for optical lithography or space applications, demands enormous efforts in all fabrication steps. To optimize the image of an optical system and to reduce aberrations, high-end optical surfaces are figured by ion beam polishing/sputtering. For many different optical materials like fused silica and low-expansion glasses and ceramics, this ion beam figuring (IBF) process has been well-established inside the optics fabrication chain. In contrast, IBF has not yet been studied and applied extensively for precise CaF_2 surface shaping, even though CaF_2 is the second material of choice for the optical lithography at 193 nm. A potential reason might be the fear to introduce extensive subsurface damage by the interaction of high energy particles with the CaF_2 lattice. Recently, the change in surface absorption of deep UV (DUV)-grade CaF_2 at 193 nm as a function of applied IBF parameters has been investigated using the horizontal LID concept [12]. Table 15.1 summarizes the IBF induced surface absorption at 193 nm compared to pitch-polished surfaces without IBF treatment. While the two pitch-polished samples show very comparable absorption slightly below 0.1%, all IBF treatments result in a significantly increased surface absorption. This subsurface damage is considered to result from ion-beam-induced generation of CaF_2 defects. It is worth to notice that the absorption at maximum is increased by more than a factor of 5 after applying double IBF treatment time.

It is supposed that two different mechanisms are responsible for the surface absorption increase, namely fluorine dislocation and depletion, respectively. The lack of fluorine within the CaF_2 lattice, give rise to the generation of F centers (depleted fluorine position in the CaF_2 lattice) and their agglomerates. Interestingly, after AR coating deposition, the measured surface absorption including the AR coating is

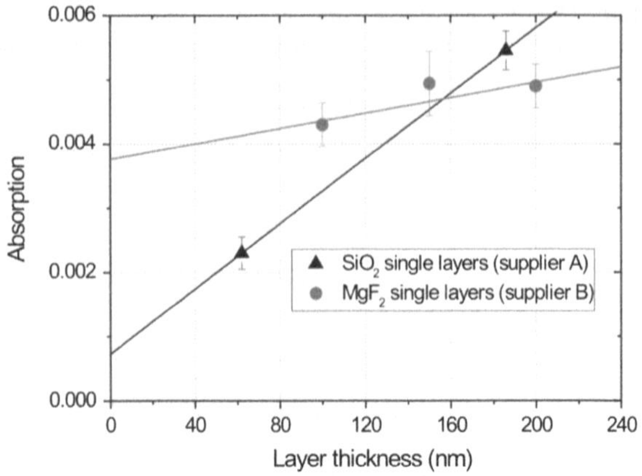

Fig. 15.17 Horizontal LID concept: comparison between absorption data at 193 nm of single SiO_2 and MgF_2 layers on CaF_2 substrates in dependence on their layer thickness. The linear fits are drawn to demonstrate the large differences in the interface absorptions (intersection with the y axis) obtained by different manufacturers (taken from [16] with kind permission)

very comparable to the data obtained for AR coated samples with conventionally surface treatment. Consolidated, the investigations are an indicator that the coating deposition provides a significant absorption annealing by two different processes. On one side, the elevated deposition temperature allows for a relocation of disordered fluorine atoms in the CaF_2 lattice. On the other side, fluorine-depleted sites in the subsurface region are partially-to-fully refilled by the fluorine from the coating materials. Comparing the effects of the two absorption annealing processes reveals that for the majority of the investigated IBF treatments, local fluorine depletion is the predominant source for the raised subsurface absorption in CaF_2.

Numerous studies e.g. by laser induced damage tests have indicated that interface engineering, i.e. surface polishing and cleaning, is the key to low absorbing transparent coatings. Inadequate surface preparation before coating and/or energetic particles at the start of ion-assisted processes can yield surface absorption not caused by the thin film itself. Consequently, strongly varying interface/surface absorption is likely to occur by different polishing/cleaning/coating technologies. Any inadequate interface quality may then yield insufficient laser induced damage threshold (LIDT) values. In order to investigate the interface quality it is required to distinguish between interface and thin film absorption. This can be allowed for by measuring a series of samples with different optical film thicknesses but identical surface preparation. Figure 15.17 gives the results for LID absorption measurements at 193 nm of single SiO_2 (thickness: 62 and 186 nm) and MgF_2 (thickness: 100, 150, and 200 nm) layers on CaF_2 substrates taken from different coating manufacturers. Although a quantitative evaluation is not done here due to the low number of measurements, the incorporated linear function in Fig. 15.17 proofs that the interface absorption for the MgF_2-coated samples is remarkably higher than for the SiO_2-coated samples. Contrary, the thin film absorption itself appears to be much lower for the MgF_2 than for the SiO_2 thin films.

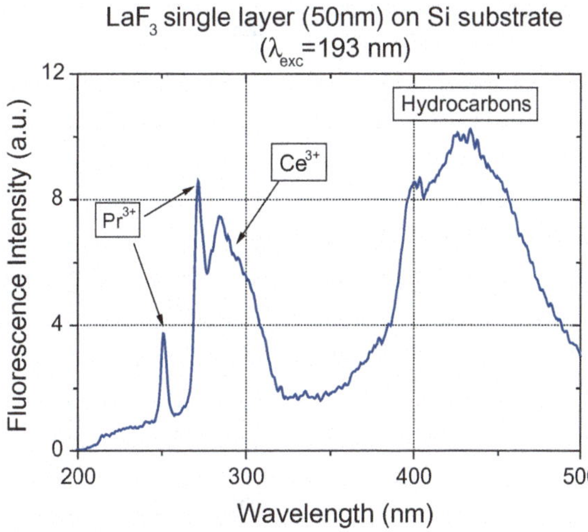

Fig. 15.18 LIF spectrum, excited at 193 nm with a fluence of H = 15 mJ/cm^2, for a 50 nm LaF_3 single layer deposited onto a silicon substrate. The spectrum was taken at a recording duration of 10 μs (taken from [13] with kind permission)

15.2.3.2 Effect of Hydrocarbons Absorption in Thin Films at 193 nm

Hydrocarbons incorporation is a familiar effect for optical coatings when stored under or close to ambient conditions. Compared to bulk materials, the rather porous structure of thin films results in a much higher affinity to store hydrocarbons and thus be affected by their strong absorption in the UV and IR wavelength regions. When irradiated with UV light, hydrocarbons contamination is visible by their broad blue fluorescence which also can be detected using laser induced fluorescence (Fig. 15.18). In addition, the LIF spectrum in Fig. 15.18 indicates residual impurities like Cerium and Praseodymium within the LaF_3 single layer. In order to avoid DUV laser induced fluorescence of common substrates (silica, CaF_2), the LIF investigations have been done using coated silicon substrates. It has been proven in advance that DUV irradiation of uncoated silicon substrates does not result in a detectable fluorescence signal.

When starting UV/DUV laser irradiation, the strong absorption of the hydrocarbons may yield damaging of the optics. Hence, it is strongly suggested to perform a pre-irradiation at low fluence/power level allowing for a light induced hydrocarbons removal without coating damage. This effect can be achieved either by direct laser irradiation (Fig. 15.19a) or by UV lamp illumination. However, due to the significantly lower power density, the latter requires a considerably longer time. One have to keep in mind though, that the laser induced removal is of temporal nature. Hydrocarbons adsorption repeats when laser irradiation is stopped. Figure 15.19b shows the development of hydrocarbons re-adsorption with time after stopping the laser irradiation at 193 nm, but still maintaining nitrogen purging. For each measurement, first a complete laser cleaning was obtained followed by the given storage time in nitrogen purged environment. The experimental data are fitted well by an exponential function indicating a re-adsorption time constant of about six hours. It is notice-

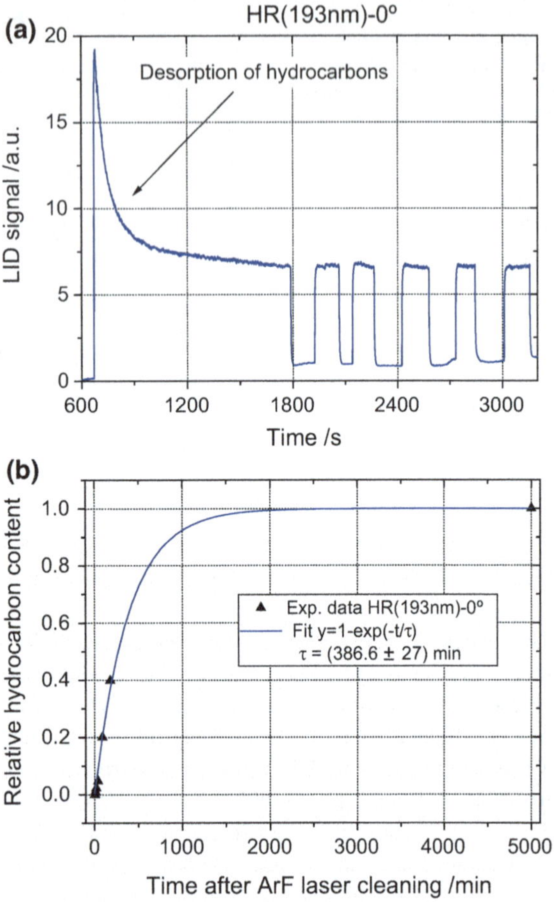

Fig. 15.19 Investigation of desorption and re-adsorption of hydrocarbons using an HR 0° coating at a laser wavelength of 193 nm in a nitrogen purged environment: **a** Experimental results for the laser induced desorption of hydrocarbons and **b** re-adsorption of hydrocarbons as a function of the time between consecutive laser irradiations (taken from [16] with kind permission)

able, that after only two hours the hydrocarbons can substantially raise the coating absorption again. For ambient air conditions, the time until a complete hydrocarbons re-adsorption will be significantly shorter. Since the hydrocarbons incorporation is a function of the thin film structure, the measurement of the desorption effect like in Fig. 15.19a can be used to qualitatively investigate the coating porosity (compare with the model introduced in Sects. 2.3.3 and 8.1.6 in this regard).

The hydrocarbons effect can not only be detected for relatively thick HR coatings but also for thin AR coatings as shown in Fig. 15.20. After an accumulated irradiation dose of almost 100 J/cm^2, the hydrocarbons are desorbed from the AR coating and the fluorescing Cerium impurity (280–310 nm) becomes detectable.

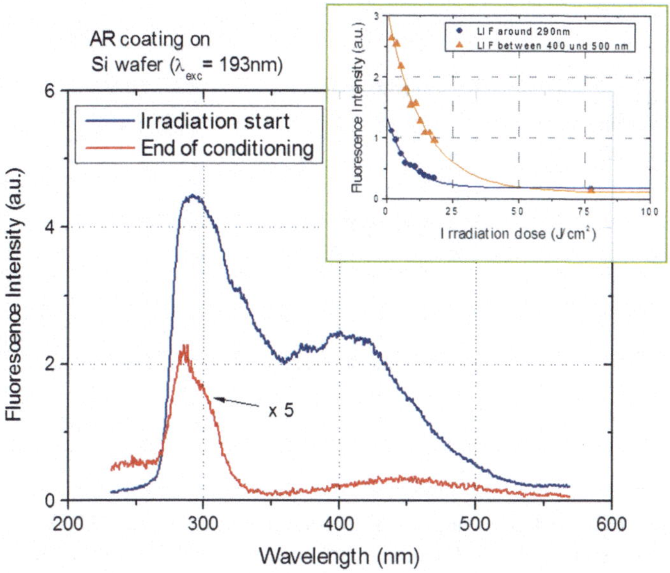

Fig. 15.20 LIF spectra of an AR coating on a silicon substrate showing the change in the hydrocarbon fluorescence peaks around 290 and 430 nm that is due to ArF laser irradiation at a fluence of H ~ 15 mJ/cm^2. The spectra were taken at a recording duration of 100 ns. The spectrum after ArF laser irradiation is rescaled for better viewing. Inlet: Integral fluorescence intensities for the two hydrocarbon peaks around 290 and 430 nm as a function of the ArF laser irradiation dose (taken from [13] with kind permission

15.2.3.3 Single Layer and Multilayer Investigations

When measuring the absorption of common optical coatings like anti-reflective (AR), partial-reflective (PR) or highly-reflective (HR), the obtained numbers are barely addressable to a particular coating material used in the coating stack or to the interface zone between substrate and first layer. Therefore, it is of great interest to investigate the absorption of single layers and thus *k* values for the individual coating materials. Using these numbers, total absorption data for the coating stacks can be calculated by means of the particular coating design. For high laser intensities and/or short laser wavelengths not only the linear but also a potential nonlinear absorption in the coatings needs to be considered.

In general, optical materials like SiO_2 or metal fluorides show very small intrinsic absorption due to their large bandgaps. For common DUV lasers, however, the combination of high photon energies and short pulse durations allows for intrinsic or defect related multi-photon absorption, which is not accessible to spectral transmission measurements. Therefore, direct absorption measurement techniques have been recently used to investigate the nonlinear absorption behavior in common DUV bulk materials, mainly at the important wavelength 193 nm of the ArF excimer laser. Contrary, there is a lack of experimental results with respect to the

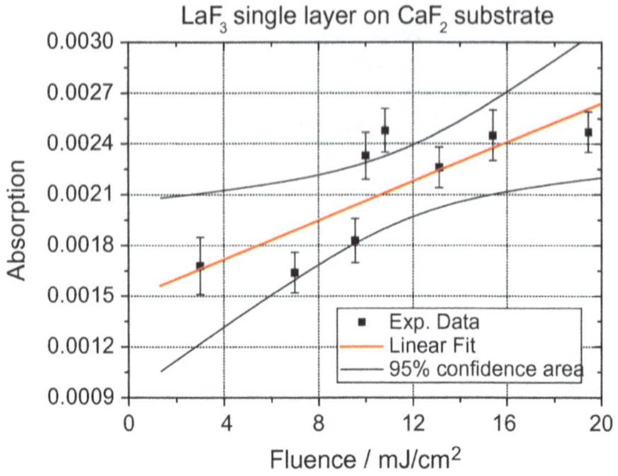

Fig. 15.21 Fluence dependent absorption data for the investigated LaF_3 thin film, measured by surface sensitive LID technique (taken from [21] with kind permission)

nonlinear properties of the common thin film materials for the DUV wavelength region. One reason might be the linear absorption of DUV light within the coatings, which is significantly higher than in the corresponding bulk materials. Thus, any intrinsic nonlinear absorption is potentially too small to be distinguished from the strong linear absorption background. Recent direct absorption measurements, however, have shown measurable nonlinear absorption in thin single layers [18–20]. The experimental results have indicated two-photon absorption (TPA) coefficients that are remarkably higher than the intrinsic TPA coefficients in the corresponding bulk materials. To a certain extent, these enlarged nonlinear absorption values are referred to sequential two-step absorption processes via intermediate defect energy levels [18].

For investigating the two-photon absorption, the samples of interest are measured as a function of the laser intensity. In the case of very weak absorption, i.e. $\alpha h \ll 1$ and $\beta I_0 h \ll 1$, the absorption A of a thin film with thickness h as a function of the incident laser intensity I_0 simplifies to

$$A = (\alpha + \beta \cdot I_0) \cdot h, \tag{15.2}$$

where α and β denote the one- and two-photon absorption coefficient, respectively. Accordingly, combining thin film absorption measurements as function of the laser intensity (or fluence) with a linear data extrapolation (15.2) allow for a calculation of the TPA coefficient β. Figure 15.21 shows exemplarily the absorption of a LaF_3 thin film as a function of the laser fluence at 193 nm wavelength. Table 15.2 summarizes the calculated extinction coefficients and TPA coefficients for the investigated MgF_2 and LaF_3 thin films. The measured TPA values provide the proof that potentially due to large defect concentrations in thin films, sequential two-step absorption occurs yielding several orders of magnitude higher TPA coefficients than typically obtained for fluoride single crystals.

Table 15.2 Summary of the linear absorption (k) and the two photon absorption (nonlinear absorption) coefficient β of MgF_2 and LaF_3 single layers obtained by the horizontal LID concept (taken from [21] with kind permission)

Thin film material	Small signal absorption k	Nonlinear absorption coefficient [cm/W]
MgF_2 (1)	$(1.8 \pm 0.1)*10^{-4}$	$(1.8 \pm 0.6)*10^{-5}$
MgF_2 (2)	$(6.9 \pm 0.35)*10^{-4}$	$(5.1 \pm 3.8)*10^{-5}$
LaF_3	$(2.0 \pm 0.3)*10^{-4}$	$(10 \pm 3)*10^{-5}$

In order to evaluate the physical origins of thin film absorption, the laser induced fluorescence has recently become a more and more sophisticated technique, in particular for the UV and DUV wavelength regions. The zero background LIF technique allows detecting concentrations of fluorescing defects or impurities down to the ppm level or even below. The separation of luminescence contributions from thin films and substrates is a key factor for thin film LIF investigations. Unfortunately, calcium fluoride and fused silica, being the most common UV substrate materials, exhibit strong (intrinsic and extrinsic) bulk emissions by e.g. self-trapped excitons and non-bridging oxygen hole centers. Due to the much larger excitation volume compared to the thin films, the substrate fluorescence commonly suppresses any chance to detect a potential thin film signal and therefore strongly affects the use of laser induced fluorescence for the thin film analysis. Interestingly, recent experiments have revealed that 193 nm laser irradiation of 80 nm thick aluminum films and standard silicon wafers does not result in any detectable photoluminescence. Consequently, it is strongly recommended to use these substrate/underlying layer materials for thin film fluorescence analysis. As example, Fig. 15.22 shows the recalculated LIF spectrum of an Al_2O_3 thin film onto silicon substrate showing characteristic UV emission between 3 and 4.5 eV which appear on nanosecond time scale after the 6.4-eV photon excitation. From the spectral positions and fluorescence lifetimes of the observed bands in Fig. 15.22, oxygen vacancies—typically introduced during the deposition process—are identified as the main fluorescence feature for 193 nm excitation of Al_2O_3 thin films. In particular, the observed spectral and temporal luminescence characteristics prove that—similar to crystalline Al_2O_3—F^+, F, and F^- center transitions between 3 and 4.5 eV are present in the emission.

An important field of interest for absorption and accompanied fluorescence measurements is the demand for deposition process optimization. Besides the evaluation of different coating raw materials, e.g. the effect of low-oxide containing fluorides, the multitude of deposition process parameters is a large playground. One of the important parameters is the deposition temperature, i.e. the temperature of the substrate surface that is to be coated. We have investigated the influence of the deposition temperature on the fluorescence and absorption properties of LaF_3 single layer, all deposited using identical raw material. Figures 15.23 and 15.24 show the LIF spectra with short and long living fluorescence lifetimes, respectively. The LIF recording

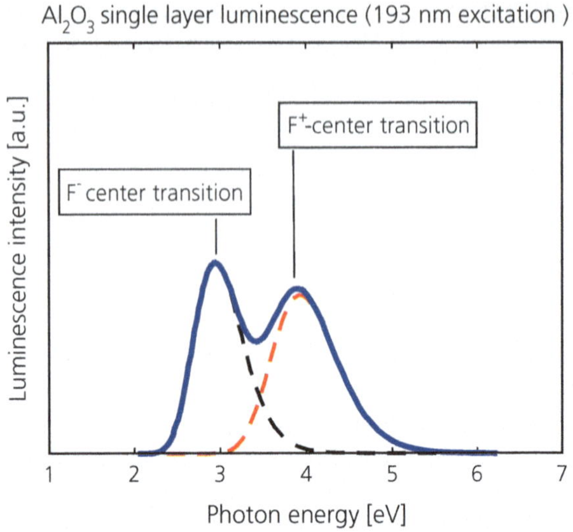

Fig. 15.22 Recalculated photoluminescence of Al_2O_3 thin films 20 ns after 6.4-eV laser excitation: the revised measurement data provide the thin-film luminescence intensity without interference overlay. The characteristic color-center emission between 3 and 4.5 eV can be decomposed into two Gaussian emission centers around 3 and 3.8 eV (taken from [22] with kind permission)

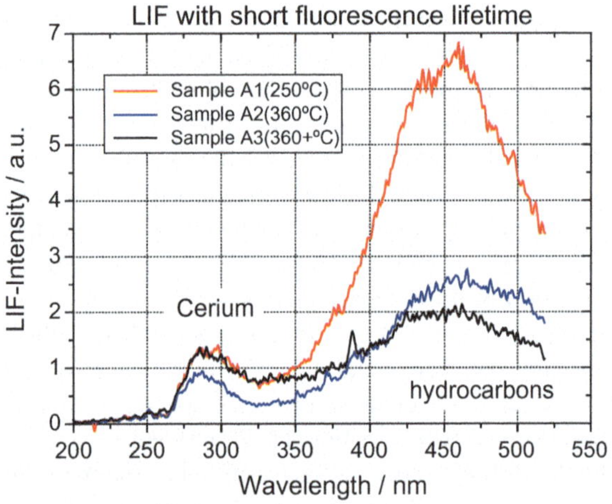

Fig. 15.23 Overview of LIF spectra with short fluorescence lifetime (<100 ns) for LaF_2 single layers deposited from the same raw material but different deposition temperatures

time is one main parameter that allows distinguishing between different fluorescing species overlaying in a particular wavelength region.

From the visible hydrocarbons fluorescence (Fig. 15.23) it has been expected and confirmed experimentally that the absorption of all LaF_3 layers decreases upon 193 nm laser irradiation (Fig. 15.25). It is clearly proved that the sample with the highest hydrocarbons incorporation possesses both, the highest initial absorption and the strongest absorption decrease upon laser irradiation. In addition, the highest stationary absorption (Table 15.3) is observed for the lowest deposition temperature sample (250 °C).Table 15.3 also shows the measured LaF_3 layer densities. There is a strong correlation between the layer density and the measured hydrocarbons fluorescence,

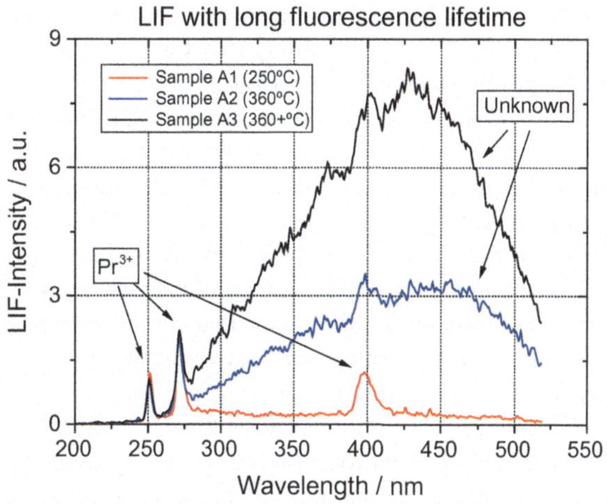

Fig. 15.24 Overview of LIF spectra with long fluorescence lifetime (≫100 ns) for LaF_3 single layers deposited from the same raw material but different deposition temperatures

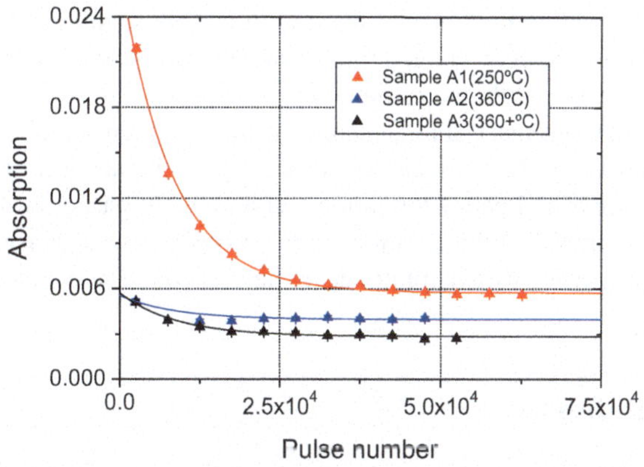

Fig. 15.25 Course of LaF_3 layer absorption upon ArF laser irradiation until reaching stationarity

Table 15.3 Summary of deposition temperature, stationary absorption and layer density for the investigated LaF_3 single layers

Deposition temperature	Absorption (10^{-3})	Layer density [g/cm^3]
250 °C (A1)	5.8 ± 0.3	4.58
360 °C (A2)	4.0 ± 0.2	5.77
360+ °C (A3)	2.8 ± 0.2	5.89

i.e. a decreasing layer density allows more effective hydrocarbons incorporation. Again, the experimental results give a proof that measuring the initial hydrocarbons fluorescence is a fast and sensitive tool to qualitatively investigate single layer densities.

Not only the initial absorption but also the stationary absorption after 193 nm laser irradiation strongly varies for the LaF_3 single layers. With increasing deposition temperature the stationary absorption decreases. The LaF_3 raw material is ruled out since the constant impurity concentrations (e.g. Pr^{3+} in Fig. 15.24) reveal identical raw materials for all layers. From Fig. 15.24 it is seen, that the samples strongly differ in a broad fluorescence of currently unknown origin (300…550 nm) with long fluorescence lifetime. Interestingly, this fluorescence is strongest for the highest deposition temperature (→lowest stationary absorption) and virtually absent for the lowest deposition temperature (→highest stationary absorption). Hence, the underlying absorption must be very weak compared to the measured stationary values and/or the fluorescence quantum yield is very high (photo-thermal absorption measurements only detect absorption that is transferred into sample heating!). However, the stationary absorption correlates with the hydrocarbons fluorescence and layer density. Two potential assumptions are suggested. First, despite the laser induced hydrocarbons desorption, not all of the incorporated hydrocarbons are removed from the layers. Consequently, a higher remaining hydrocarbon content is expected for a lower deposition temperature resulting in a higher stationary absorption. Second, the lower density for lower deposition temperature can be an indicator for a higher disorder in the LaF_3 lattice structure. This would yield a higher density of absorbing lattice defects which do not show fluorescence upon 193 nm laser irradiation, e.g. F centers and their agglomerates.

In summary, direct absorption measurements and laser induced fluorescence have proven to be valuable tools for state-of-the-art thin film analysis in complement to established characterization techniques. In particular for low loss optical coatings, the introduced techniques allow investigating extrinsic as well as intrinsic material properties which are not accessible to common characterization tools.

References

1. A.C. Tam, Overview of photothermal spectroscopy in *Photothermal Investigation of Solids and Fluids*, ed. by J.A. Sell (Academic, Boston, 1989) 1–33
2. B. Li, H. Blaschke, D. Ristau, Combined laser calorimetry and photothermal technique for absorption measurement of optical coatings. Appl. Opt. **45**, 5827–5831 (2006)
3. ISO 11551:2003: Optics and optical instruments/Lasers and laser-related equipment/Test method for absorptance of optical laser components, http://www.iso.org
4. S. Leonid, Light-induced absorption in materials studied by photothermal methods. Recent Pat. Eng. **3**, 129–145 (2009)
5. A. Alexandrovski, M. Fejer, A. Markosian, R. Route, Photothermal common-path interferometry (PCI): new developments. Proc. SPIE **7193**, 71930D (2009)
6. www.stan-pts.com/
7. S. Yoshida, D.H. Reitze, D.B. Tanner, J.D. Mansell, Method for measuring small optical absorption coefficients with use of a Shack-Hartmann wave-front detector. Appl. Opt. **42**(24), 4835–4840 (2003)
8. K. Mann, A. Bayer, U. Leinhos, M. Schöneck, B. Schäfer, Photothermal measurement of absorption and wavefront deformations in UV optics. J. Phys: Conf. Ser. **214**, 012015 (2010)
9. B. Schäfer, J. Gloger, U. Leinhos, K. Mann, Photo-thermal measurement of absorptance losses, temperature induced wavefront deformation and compaction in DUV-optics. Opt. Exp. **17**(25), 23025 (2009)

10. N. Waasem, S. Fieberg, J. Hauser, G. Gomes, D. Haertle, F. Kühnemann, K. Buse, Photoacustic absorption spectrometer for highly transparent dielectrics with parts-per-million sensitivity. Rev. Sci. Instrum. **84**, 023109 (2013)
11. M. Guntau, W. Triebel, A novel method to measure bulk absorption in optically transparent materials. Rev. Sci. Instr. **71**, 2279–2282 (2000)
12. C. Mühlig, S. Bublitz, R. Feldkamp, H. Bernitzki, Effect of ion beam figuring and subsequent AR coating deposition on the surface absorption of CaF_2 at 193nm. Appl. Opt. **56**(4), C91–C95 (2017)
13. C. Mühlig, W. Triebel, S. Kufert, S. Bublitz, Characterization of low losses in optical thin films and materials. Appl. Opt. **47**(13), C135–C142 (2008)
14. A. Burkert, C. Mühlig, W. Triebel, D. Keutel, U. Natura, L. Parthier, S. Gliech, S. Schröder, A. Duparre, Investigating the ArF laser stability of CaF2 at elevated fluences. Proc. SPIE **5878**, 58780E (2005)
15. C. Mühlig, G. Schmidl, J. Bergmann, W. Triebel, Characterization of high reflecting coatings and optical materials by direct absorption and cavity ring down measurements. Proc. SPIE **7102**, 71020T (2008)
16. C. Mühlig, S. Kufert, S. Bublitz, U. Speck, Laser induced deflection technique for absolute thin film absorption measurement: optimized concepts and experimental results. Appl. Opt. **50**(9), C449–C456 (2011)
17. C. Mühlig, S. Bublitz, W. Paa, Enhanced laser-induced deflection measurements for low absorbing highly reflecting mirrors. Appl. Opt. **53**(4), A16–A20 (2014)
18. O. Apel, K. Mann, A. Zoellner, R. Goetzelmann, E. Eva, Nonlinear absorption of thin Al_2O_3 films at 193 nm. Appl. Opt. **39**, 3165–3169 (2000)
19. B. Li, S. Xiong, Y. Zhang, S. Martin, E. Welsch, Nonlinear absorption measurement of UV dielectric components by pulsed top-hat beam thermal lens. Opt. Commun. **244**, 367–376 (2005)
20. L. Jensen, M. Mende, St. Schrameyer, M. Jupé, D. Ristau, Role of two-photon absorption in Ta_2O_5 thin films in nanosecond laser-induced damage. Opt. Exp. **37**(20), 4329–4331 (2012)
21. C. Mühlig, S. Bublitz, S. Kufert, Nonlinear absorption in single LaF_3 and MgF_2 layers at 193 nm measured by surface sensitive laser induced deflection technique. Appl. Opt. **48**(35), 6781–6787 (2009)
22. J. Heber, C. Mühlig, W. Triebel, N. Danz, R. Thielsch, N. Kaiser, Deep UV laser induced luminescence in oxide thin films. Appl. Phys. A **75**, 637–640 (2002)

Chapter 16
Cavity Ring-Down Technique for Optical Coating Characterization

Christian Karras

Abstract Mirror reflectance values above $R > 0.999$ can be accurately determined by measuring the life time of photons coupled into a cavity consisting of at least two of these mirrors (Cavity ring-down, CRD). This technique is superior compared to other methods such as laser ratiometric or spectrophotometric measurements. The accuracy of the measurement is determined by the cavity alignment and the reflectance value of the reference. When applying the CRD technique for high reflectance measurements, care has to be taken to minimize or estimate cavity losses (CRD-losses) resulting from scattering, absorption, mirror transmission and diffraction. The present chapter describes the fundamentals of the CRD technique for both, continuous wave or pulsed applications and provides experimental advises for setting up a CRD system capable of reliably detecting reflectances above $R > 0.999$.

16.1 Introduction

Using stacks of dielectric layers allows the design of high reflective (HR) optical coatings with reflectivity coefficients of $R > 0.99$ [1]. Those HR coatings are e.g. required for the design of beam routing or cavity mirrors for high power laser applications. An insufficient mirror reflectivity and consequently a larger transmission, absorptance or scattering might result in laser induced damage of the optical components or can be critical in terms of laser safety. Another possible field of application is the design of efficient optical filters for selected spectral bands e.g. to isolate a Raman signal from Rayleigh scattered light.

In order to achieve those large reflectivities, stacks of several hundred layers might be necessary and all of these have to be manufactured with sufficiently large accuracy. In the past decades the reflectivity value of the layer system was solely simulated. An

C. Karras (✉)
Leibniz-Institute of Photonic Technology IPHT, Albert-Einstein-Straße 9, 07745 Jena, Germany
e-mail: christian.karras@leibniz-ipht.de

O. Stenzel and M. Ohlídal (eds.), *Optical Characterization of Thin Solid Films*, Springer Series in Surface Sciences 64,
https://doi.org/10.1007/978-3-319-75325-6_16

accurate measurement of R was commonly not provided by the coating manufactures and remained the task of the end user.

Determining such large reflectivities is far from being simple. Hence it was announced as "2010 Measurement Problem" at the "Optical Interference Coatings" conference (OIC) in 2010 [2]. Several methods suitable for measuring the mirror reflectivity were compared in the framework of this task:

Spectrophotometric measurements: The coated sample is placed into a spectrophotometer in order to determine its transmission (or reflection) value. This method allows the determination of the reflection (or transmission) coefficient over a broad spectral range without the need of comprehensive subsequent data analysis in a fairly simple, commercially available setup. The maximum reflectivity that can be measured reliably is determined by the accuracy of the spectrograph. It is typically limited to $R < 0.995$ [3].

Laser ratiometric measurements : Applying this technique, the ratio of an incident and reflected or transmitted laser beam is determined and from this the coating reflectivity is deduced. Although reflectance measurements of $R = 0.9999$ were reported [4] laser ratiometric measurements failed to reliably determine those values in the OIC measurement contest [2]. Typically the limit of the achieved reliably determined mirror reflectivity amounted to $R < 0.999$. The accuracy of this technique is basically limited by fluctuations in the laser power and accuracy of its detection.

Cavity ring-down measurements: Applying the Cavity ring-down (CRD) technique the reliable determination of mirror reflectivities $R = 0.998$ and above [2] is possible.

A detailed explanation of the CRD technique for the determination of large coating reflectivities is the subject of this chapter. It majorly focuses on conveying the general technological aspects to the reader, in particular the discussion of important experimental design issues when accurately measuring mirror reflectivities above $R > 0.99$.

In the first section, the general CRD concept will be presented and differences between pulsed and continuous wave measurements will be outlined. Also the concept of retrieving the coating reflectivity from CRD raw data will be explained. In order to support the understanding of the experimental details this section also includes a brief summary dealing with optical resonators.

The second part provides an experimental guide for setting up a CRD system in order to determine reflection coefficients as high as $R = 0.9999$ and above. Also different possibilities for light generation and detection suitable for CRD measurements are compared and issues of possible cavity designs are analyzed. In addition a comprehensive analysis the system accuracy is presented.

The chapter closes with evaluating the limits of the technique.

16.2 The CRD Technique for Detecting Reflectivities

16.2.1 The General CRD Concept and Physical Basics

The CRD method was originally developed decades ago for the accurately measuring weak absorption coefficients in gases [5] and has evolved to a standard tool in absorption spectroscopy ever since [6–11].

In contrast to competing techniques for measuring the reflectivities such as laser ratiometric approaches or spectrophotometric methods, which aim for comparing the incident and transmitted light of a reflective surface, the CRD method follows an indirect approach. The light is coupled into an optical cavity in which it travels around. At each round trip, a small fraction is being coupled out at a cavity mirror. As the power inside the cavity decreases the absolutely outcoupled light power decreases as well with an increasing number of round trips. The outcoupled power is measured as a function of time. The detected time trace is directly related to the losses of the cavity (CRD-losses, in the following referred to as "losses"). Assuming negligible scattering and absorption inside the cavity, these losses arise from a nonzero mirror transmission $T = 1 - R > 0$. Consequently, the losses deduced from the time trace reveal the mirror reflectivity. The typical time constant, characterizing the decay is sometimes called "photon life time of the cavity". This approach enables highly accurate measurements, since the result is independent from typical noise sources, such as laser pulse fluctuations.

The CRD concept and a typical result of one high reflectance measurement is exemplary depicted in Fig. 16.1.

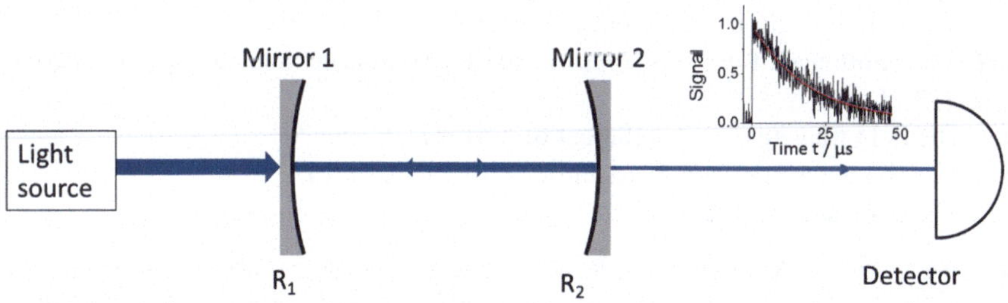

Fig. 16.1 Scheme of a Cavity ring-down setup. The light (trace illustrated by the blue line) is coupled into a cavity, consisting of mirrors 1 and 2 with reflectances R1 and R2. The power of the outcoupled light is measured as a function of time behind the cavity

Table 16.1 Selected characteristic values of an optical resonator

• $t_r = \frac{2L}{c}$	The time of one round trip inside the cavity
• $RM = \sqrt{R_1 R_2}$	The geometrical average of the mirror reflectivity
• $TM = \sqrt{T_1 T_2}$	The geometrical average of the mirror transmission
• $\mathcal{F} = \frac{\pi\sqrt{RM}}{1-RM}$	The "Finesse" of the resonator
• $g_i = 1 - \frac{L}{\rho_i}$	The "g-parameter" of mirror i
• $N_i = \frac{a_i^2}{\lambda\rho_i} \cdot \sqrt{\frac{(\rho_i - L)(\rho_i + \rho_j - L)}{L(\rho_j - L)}}$	The Fresnel number for a stable two mirror cavity with arbitrary mirrors (i, j $\in$ 1, 2; j $\neq$ i) [15], λ is the vacuum wavelength inside the cavity

16.2.2 *Fundamentals of Optical Resonators*

The center piece of a CRD system is the optical cavity. Thus it is worth to briefly revisit selected cavity properties which are important for further understanding. For a more detailed explanation of optical cavities the reader is referred to respective literature [12–14].

The optical characteristics of a resonator are dictated by its geometric parameters (a selection of derived parameters can be found in Table 16.1):

- ρ_i The radius of curvature for mirror i
- R_i The reflectance of mirror i
- r_i The field reflection coefficient of mirror i
- T_i Transmittance of mirror i. In the absence of scattering and absorption it equals to $T_i = 1 - R_i$
- t_i The field transmission coefficient of mirror i
- a_i The diameter of mirror i (assuming a round aperture)
- L The resonator length.

In order to apply Occam's razor, losses due to scattering and absorption inside the cavity are considered to be negligible for now, we assume a cavity consisting out of two mirrors only and the refractive index inside the cavity is considered to be 1.

For a Cavity ring-down experiment, one is interested in the temporal characteristics of the resonator. It can be expressed in terms of a Green's function $G(t)$ [16]. This is the impulse function of the resonator with respect to an electrical field and is determined by the round trip time and the geometrical average of the mirror reflectivity and transitivity:

$$G(t) = t_1 t_2 \sum_{n=0}^{\infty} (r_1 r_2)^n \, \delta\left(t - \left(n + \frac{1}{2}\right) t_r\right), \tag{16.1}$$

with δ being the Dirac function. The temporal output signal (intensity) of the cavity is the absolute square of the convolution of (16.1) with the input electrical field. It is noted, that in (16.1) the field reflection and transmission coefficients are considered to be a real values, meaning that the reflection at the mirrors is not creating an additional phase shift.

The Fourier transform of (16.1) leads to the spectral transmission characteristics of the cavity. Its square absolute value is the well-known Airy formula describing the spectral transmission of a Fabry–Perot like cavity:

$$T(\omega) = |FT(G(t))|^2 = \frac{TM^2}{\left(1 - RM^2\right)\left(1 + \left(\frac{2}{\pi}\mathcal{F}\right)^2 \sin^2\left(\frac{\omega t_r}{2}\right)\right)}, \tag{16.2}$$

with its typical resonance frequencies at $\nu_q = \frac{q}{t_r} = \frac{qc}{2L}$, and q being a positive integer. The spectral distance between two resonances $\Delta\nu = \frac{1}{t_r} = \frac{c}{2L}$ is known as "free spectral range" of the cavity.

The full width half maximum (FWHM) of the resonance peak is the ratio of the free spectral range and the Finesse:

$$\delta\nu = \frac{\Delta\nu}{\mathcal{F}} \tag{16.3}$$

Particularly if a light beam with a sufficiently large coherence length is coupled into the cavity (narrow spectral bandwidth), the spectral transmission behavior becomes crucial (c.f. Sect. 16.2.4).

Equations (16.1) and (16.2) do not take the geometrical properties of the mirrors such as the radius of curvature into account. This, however, determines the stability of the cavity. A stable resonator arrangement as prerequisite for a successful CRD measurement has to fulfill the stability criterion:

$$0 \leq g_1 g_2 \leq 1 \tag{16.4}$$

The supported field distribution of the resonator is obtained by finding the Eigensolution of the Kirchhoff diffraction integral [12, 17]. For sufficiently large and round apertures those are given by Gauss–Laguerre modes. They are characterized by the integer mode parameters p and l and represent an orthonormal basis for every field configuration. The complete set of cavity Eigenmodes are referred to as "transverse modes" (TEM_{plq}).

The resonance frequencies (and thus the free spectral range) of this cavity modes are different for different transverse modes. They are given by:

$$\nu_{plq} = \frac{1}{t_r}\left[q + \frac{2p+l+1}{\pi}\mathrm{acos}\sqrt{g1g2}\right] \tag{16.5}$$

Not that for $g_1 g_2 = 1$ (planar resonator), the free spectral range is $\Delta\nu = \frac{1}{t_r}$ as derived above, whereas for a $g_1 g_2 = 0$ arrangement (confocal resonator) it amounts to $\Delta\nu = \frac{1}{2t_r}$.

Furthermore when using aperture limited cavity mirrors diffraction on the mirror rims takes place. This introduces additional losses which are larger for higher modes as those possess a larger field diameter on the mirrors. The exact computation of diffraction losses requires the numerical solution of the Kirchhoff diffraction integral. A measure allowing the estimation of the diffraction losses is the Fresnel number N_i.

In case of large N_i value, the losses on one mirror may be estimated [15, 17]:

$$\kappa_i = \frac{2\pi\,(8\pi N_i)^{2p+l+1}}{p!\,(p+l+1)\,!}\mathrm{e}^{-4\pi N_i} \tag{16.6}$$

The loss per round trip due to diffraction is given by:

$$\kappa = \frac{1}{2}(\kappa_1 + \kappa_2) \tag{16.7}$$

16.2.3 CRD Using Pulsed Light Sources

The detected field behind the cavity is given by the convolution of this Greens function and the incident light field.

Let's first assume a short laser pulse to be coupled into the cavity with a pulse duration $\Delta\tau_{Pulse} < t_r$ and an arbitrary envelope function of the electrical field $\mathcal{E}(t)$ (c.f. Fig. 16.2a). The latter could for instance be a Gaussian function. For typically used cavities of several tens of cm, the pulse duration should be in the range of $\Delta\tau_{Pulse} \approx 1\ ns$ or below. Even very short pulses of several fs only, should not be a problem, since the dispersive pulse broadening introduced by the air in the cavity would not cause the pulse duration to exceed several ps. It is, however, noted that a too large dispersion introduced by the mirror surfaces has to be avoided, when very short pulses should be used.

The output signal behind the cavity will be given by:

$$\begin{aligned} |G(t) \otimes E_{in}(t)|^2 &= \left| t_1 t_2 \sum_{n=0}^{\infty} (r_1 r_2)^n\, \mathcal{E}\left(t - \left(n + \frac{1}{2}\right) t_r\right) \mathrm{e}^{-i\omega\left(t - \left(n+\frac{1}{2}\right)t_r\right)} \right|^2 \\ &= \underbrace{T_1 T_2}_{TM^2} \sum_{n=0}^{\infty} \left(\underbrace{R_1 R_2}_{RM^2} \right)^n \mathcal{E}^2\left(t - \left(n + \frac{1}{2}\right) t_r\right) + IF, \end{aligned} \tag{16.8}$$

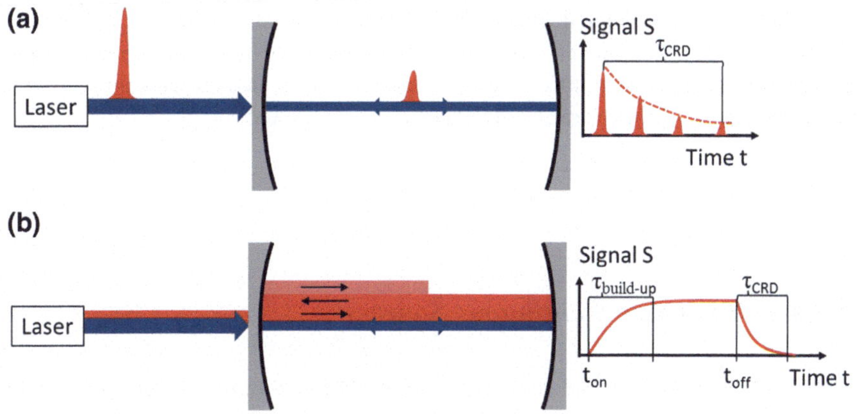

Fig. 16.2 Cavity ring-down measurements with **a** pulsed and **b** continuous wave laser sources: The blue arrows indicate the direction and position of the light beam. The pulse and the continuous light train including the field enhancement inside the cavity is illustrated in red in **a** and **b** respectively. The cavity output signal is depicted on the right hand side: τ_{CRD}: Cavity ring-down time characterizing the decrease of the outcoupled maximum pulse intensity **a** or the decrease after the cw source is switched off at t_{off}. $\tau_{build\text{-}up}$: characteristic time for the build-up of the field enhancement after the cw light is switched on at t_{on}

with IF being interference terms. Due to the short pulse duration the coherence length of the incoupled light is very short and the interference terms are thus negligible.

Let's further assume, that the exact temporal shape of the envelope cannot be resolved by the detector, which is an adequate assumption for $\Delta\tau_{Pulse} = 1$ ns. Thus after each round trip a time integrated signal S of the pulse is measured at the output of the cavity, which is proportional to the mirror transmission, the incident pulse intensity I_{in}, and the reflection after the respective number of roundtrips. For N round trips it is

$$S_n \propto I_{in} \cdot TM^2 \cdot RM^{2N} \tag{16.9}$$

The time for N round trips is

$$t_N = 2 \cdot N \cdot \frac{L}{c}, \tag{16.10}$$

Thus S_n is transformed to $S(t)$

$$\begin{aligned} S(t) &\propto I_{in} \cdot TM^2 \cdot RM^{\frac{tc}{L}} \\ &= I_{in} \cdot TM^2 \cdot \mathrm{e}^{\frac{tc}{L} \cdot \ln(RM)} \\ &\approx I_{in} \cdot TM^2 \cdot \mathrm{e}^{-\frac{tc}{L}(1-RM)} \\ &= I_{in} \cdot TM^2 \cdot \mathrm{e}^{-\frac{t}{\tau_{CRD}}}. \end{aligned} \tag{16.11}$$

Consequently, the outcoupled signal follows an exponential time behavior. The time constant τ_{CRD} of the exponential function is determined by fitting the measured time trace exponentially or by plotting it in a logarithmically scaled diagram and fitting it linearly. From τ_{CRD} the average reflectance can be computed:

$$\tau_{CRD} = \frac{L}{c\,(1 - RM)} \rightarrow RM = 1 - \frac{L}{\tau_{CRD} \cdot c} \tag{16.12}$$

16.2.4 CRD Using Continuous Sources

Applying a continuous wave laser source in CRD measurements (or a pulse with $\Delta\tau_{Pulse} > t_r$) the interference terms in (16.8) have to be taken into account and the outcoupled signal behind the cavity is not consisting out of distinct pulses anymore.

Let's first assume a continuous light wave with amplitude A which is switched on at a certain time $t_{on} = 0$. In addition its frequency should meet the resonance condition of the resonator (c.f. (16.2)).

$$\begin{aligned} |G(t) \otimes E_{in}(t)|^2 &= \left| t_1 t_2 \sum_{n=0}^{\infty} (r_1 r_2)^n \cdot \mathrm{A} \cdot \Theta\left(\mathrm{t} - \left(\mathrm{n} + \frac{1}{2}\right)\mathrm{t_r}\right) \mathrm{e}^{-i\omega\left(t - \left(n + \frac{1}{2}\right)t_r\right)} \right|^2 \\ &= TM^2 \cdot \tilde{A} \cdot \sum_{n=0}^{\infty} RM^{2n} \cdot \Theta\left(\mathrm{t} - \left(\mathrm{n} + \frac{1}{2}\right)\mathrm{t_r}\right), \end{aligned} \tag{16.13}$$

with Θ being the Heaviside step function. Since the incident light field is in resonance with the cavity, a standing wave is present in the resonator and the interference terms in (16.8) lead to a field enhancement inside the cavity. This effect and the square of the incident field amplitude are summarized in $\tilde{A}$ in (16.13).

The sum in (16.13) describes the overlay of newly incoupled light with the light which is traveling around in the resonator. It is illustrated in Fig. 16.2b. The signal behind the cavity after N roundtrips can be calculated by solving the sum in (16.13) for this case:

$$\begin{aligned} &\sum_{n=0}^{N} RM^{2n} \cdot \Theta\left(\mathrm{t} - \left(\mathrm{n} + \frac{1}{2}\right)\mathrm{t_r}\right) \\ &= \left[\sum_{n=0}^{N} RM^{2n} \cdot \Theta\left(\mathrm{t} - \left(\mathrm{n} + \frac{1}{2}\right)\mathrm{t_r}\right)\right] \cdot \frac{RM^2 - 1}{RM^2 - 1} \\ &= \frac{\left(RM^{2N} + RM^{2(N-1)} + \cdots + RM^2 + 1\right) \cdot \left(RM^2 - 1\right)}{RM^2 - 1} \\ &= \frac{\left(RM^{2(N+1)} + RM^{2N} + \cdots + RM^4 + RM^2\right) - \left(RM^{2N} + RM^{2(N-1)} + \cdots + RM^2 + 1\right)}{RM^2 - 1} \\ &= \frac{RM^{2(N+1)} - 1}{RM^2 - 1} = \frac{1 - RM^{2(N+1)}}{1 - RM^2} \end{aligned}$$

$$\rightarrow S(N) \propto TM^2 \cdot \tilde{A} \cdot \sum_{n=0}^{N} RM^{2n} \cdot \Theta\left(\mathrm{t} - \left(\mathrm{n} + \frac{1}{2}\right)\mathrm{t_r}\right) = TM^2 \cdot \tilde{A} \cdot \frac{RM^{2(N+1)} - 1}{RM^2 - 1} \tag{16.14}$$

Note that for many round trips the output signal becomes constant $S(N \rightarrow \infty) = \frac{TM^2 \cdot \tilde{A}}{1 - RM^2}$.

If the signal is switched off at a certain time $t_{off} > t_{on} = 0$, at which the incident beam has performed N^* round trips already the output signal of the cavity after the N'th roundtrip ($N > N^*$) is given by:

$$\begin{aligned} S(N) &\propto TM^2 \cdot \tilde{A} \cdot \sum_{n=0}^{N} RM^{2n} \cdot \left[\Theta\left(\mathrm{t} - \left(\mathrm{n} + \frac{1}{2}\right)\mathrm{t_r}\right) - \Theta\left(\mathrm{t} - \left(\mathrm{n} + \mathrm{N}^* + \frac{1}{2}\right)\mathrm{t_r}\right)\right] \\ &= TM^2 \cdot \tilde{A} \cdot \frac{RM^{2(N^*+1)} - 1}{RM^2 - 1} \cdot RM^{2(N-N^*)} \\ &= TM^2 \cdot \tilde{A} \cdot \frac{RM^{2(N^*+1)} - 1}{RM^2 - 1} \cdot RM^{2\tilde{N}}, \end{aligned} \tag{16.15}$$

with $\tilde{N}$ being the number of round trips after the light was switched off. With the exception of the proportional constant, which describes the field enhancement due to the signal overlay in the cavity $S(\tilde{N})$ can be treated analogous to the pulsed case in order to derive the correlation between the $S(\tilde{N})$ the time trace and the average mirror reflectance.

If the incident field and the resonator are not in resonance the situation is more complicated, since the interference terms have to be considered. In the comprehensive work of Lee et al. [16] the impact of spectral detuning between the cavity and the incident light was analyzed also as function of $\Delta\tau_{Pulse}$. The result is depicted in Fig. 16.3. In the presence of detuning the amplitude of the output signal $S(\tilde{N})$ decreases and only a small fraction of this output follows an exponential decay which can be used to determine the average mirror reflectance RM.

These results agree with what is expected when analyzing the spectral transmission behavior in the presence of detuning (Fig. 16.4).

The transmission of the cavity is only high in the range of the resonances. In this case the exponential decay can be observed behind the cavity, after the pulse or when the cw—signal is switched off. Due to the field enhancement, the signal is much stronger as if short pulses would be used. If, however, the frequency of the incoupled laser light and the resonance frequency of the cavity are detuned, the transmittance of the cavity drops. The light is not confined in the cavity and only part which is transmitted by the mirrors passes. In the case of a small detuning (or a broad pulse) there are always parts of the incident spectrum which are in resonance and parts which are not. The former contribute to the exponentially decaying signal and can be used to determine RM, the latter do not.

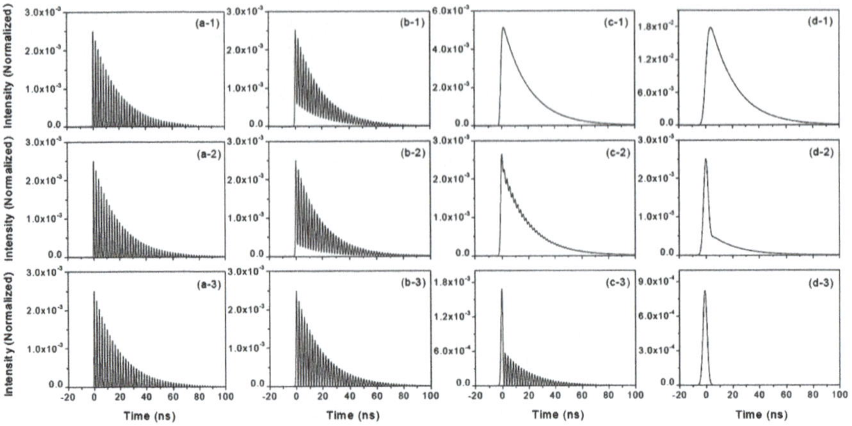

Fig. 16.3 Ring-down signals altering behaviors with the variation in pulse duration $\Delta\tau_{Pulse}$ and detuning between the detuning $\delta\omega_{det}$ between laser pulse central frequency and the cavity resonance: $\Delta\tau_{Pulse}$ were set to be **a** $0.25 \times t_r$, **b** $0.5 \times t_r$, **c** $1 \times t_r$, and **d** $2 \times t_r$ with t_r being the round trip time of the light in the resonator (depicted in separate columns). For each pulse $\delta\omega$ amounted to (1) $0 \times \Delta\omega$, (2) $0.25 \times \Delta\omega$, and (3) $0.5 \times \Delta\omega$ with $\Delta\omega$ being the free spectral range of the cavity in units of the angular frequency (depicted in separate rows). The low finesse cavity with mirror reflectivity of R = 0.95 was considered for the calculations and the origin of the abscissa, $t = 0$, is set as the moment when the transmitted peak of the input pulse component undergone no round-trip is about to escape from the cavity. This image was taken from [16] (Copyright (1999) The Japan Society of Applied Physics)

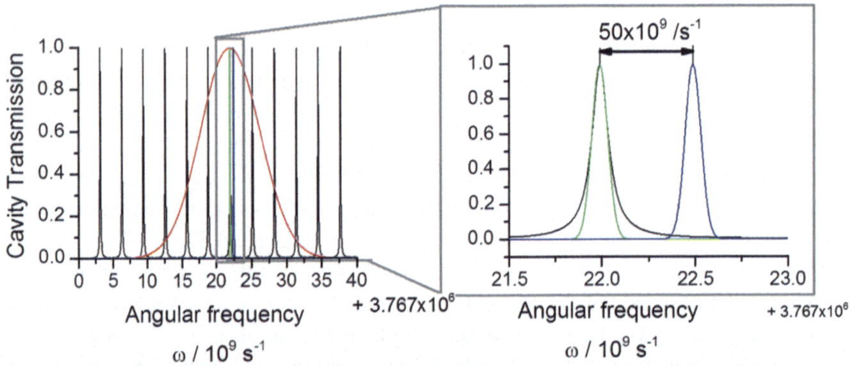

Fig. 16.4 Transmission of a cavity with a Finesse of F = 29 (black, R = 0.9) around a wavelengths of $\lambda = 500$ nm ($\omega \approx 3.767 \times 10^{15}\mathrm{s}^{-1}$) and the spectra of a broad band (red) pulse ($\Delta\lambda_{Pulse} = 5$ pm, $\Delta\omega_{Pulse} = 10^9\mathrm{s}^{-1}$) and a narrow band signal (green, blue, $\Delta\lambda = 0.5$ pm, $\Delta\omega = 10^8\mathrm{s}^{-1}$). Note that a detuning of $\delta\omega_{det} = 50 \times 10^9\mathrm{s}^{-1}$ of the narrow band signal causes the transmission to drop to less than 10%, whereas a detuning of the broad band signal (red) would not cause a significant change in transmission, as there are always resonance peaks covered by the light spectrum

Due to the large finesse of the cavity, detuning is very likely and definitely present, when the incoupled signal is spectrally broader than a resonance peak. E.g. the finesse of a two—mirror cavity with mirror reflectances of $R = 0.9999$ at each side is $F \approx 31000$. For a $L = 30\,\mathrm{cm}$ long cavity (free spectral range $\Delta\nu \approx 500\,\mathrm{MHz}$), the transmission bandwidth is than $\delta f \approx 16\,\mathrm{kHz}$. For a central wavelength of $\lambda = 500\,\mathrm{nm}$ this corresponds to a linewidth of $\delta\lambda \approx 0.013\,\mathrm{fm}$ only.

It is pointed out that the present consideration is only a simple one dimensional estimation. In case of spatial extension of the beam, please refer to more detailed literature regarding optical cavities [12, 18].

16.2.5 Calculating the Mirror Reflectivity

Until now, only the geometrical average of the reflectances RM was considered. It is directly related to the time constant of the exponential decay (c.f. (16.12)). When performing a CRD measurement one is, however, interested in the absolute individual reflectance R_i of each mirror.

Let's assume a linear cavity (two mirrors) where the reflectance of one mirror (reference mirror R_{ref}) is known with sufficient accuracy. In that case the unknown reflectivity of the sample can simply be calculated from the RM value and R_{ref}:

$$R = \frac{RM^2}{R_{ref}}. \tag{16.16}$$

In the more general case, that no reflectance is known, the mutual RM values of three mirrors with initially unknown reflectances R_1, R_2 and R_3 have to be measured:

$$\begin{aligned} RM_1 &= \sqrt{R_1 R_2} \\ RM_2 &= \sqrt{R_1 R_3} \\ RM_3 &= \sqrt{R_2 R_3} \end{aligned} \tag{16.17}$$

From those values, the absolute reflectance of each mirror can be calculated:

$$\begin{aligned} R_1 &= \frac{RM_1 RM_2}{RM_3} \\ R_2 &= \frac{RM_1 RM_3}{RM_2} \\ R_3 &= \frac{RM_2 RM_3}{RM_1} \end{aligned} \tag{16.18}$$

For reasons of accuracy (c.f. Sect. 16.3.6), the mirror with the largest reflectivity should be used as reference mirror.

Until now, only mirrors designed for an angle of incidence $\varphi = 0$ were considered. In case of $\varphi \neq 0$, a cavity consisting of three mirrors has to be set up, with the sample mirror in the center (c.f. Fig. 16.5).

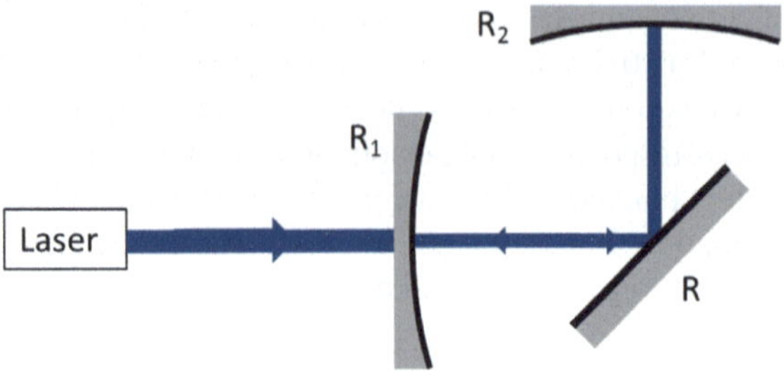

Fig. 16.5 Setup of a CRD measurement using three mirrors. The angle of incidence for the sample mirror with reflectance R is not zero

In case of a cavity consisting of three mirrors generally the center mirror (nonzero angle of incidence) is to be determined.

In this case it is sufficient to only measure the RM value of the linear cavity consisting of the cavity mirrors R_1 and R_2 (RM_{2M}) as well as the RM value of the folded cavity (RM_{3M}). R can be computed by the ratio:

$$R = \frac{RM_{3M}}{RM_{2M}}, \tag{16.19}$$

16.3 Making It Run! A Guide Towards a CRD System

If one aims for accurately measuring reflectivities as high as $R = 0.9999$ and above, the design of the CRD setup is far from being trivial. Care has to be taken in order to choose the proper light source, a suited detection unit and an appropriate cavity design. This section focusses on giving some advices for setting up the CRD system, which provides reliable results for high reflectivity measurements.

16.3.1 The Light Source

Basically both, pulsed as well as continuous wave laser sources may be used for the CRD experiment. In both cases, the laser power is the most critical parameter for a successful measurement.

If pulsed sources are used, assuming a pulse duration $\Delta\tau_{Pulse} \ll t_r$ field enhancement effects will be absent (c.f. Sects. 16.2.3 and 16.2.4). This is generally true for pulse durations $\Delta\tau_{Pulse} \approx 1$ns and shorter. In that case the attenuating effect of the cavity has to be taken into account. E.g. a cavity with a reflectance of $R = 0.99999$ for each mirror shows an optical density of $OD = 10$. Thus care has to be taken, to ensure a detectable transmission through the cavity, i.e. a sufficiently large laser power. The number of transmitted photons can easily be estimated by:

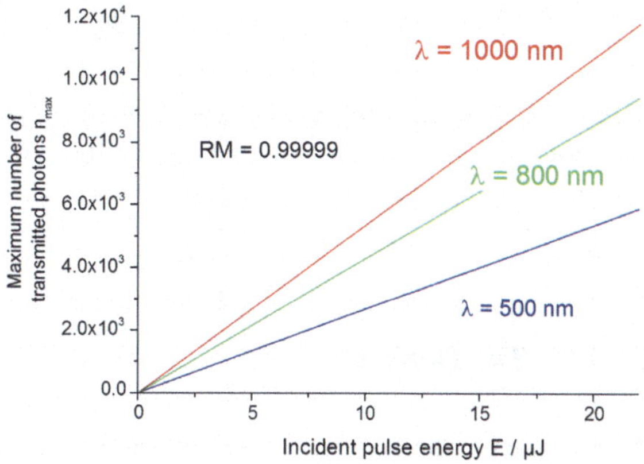

Fig. 16.6 Number of the maximum transmitted photons n_{max} as a function of the pulse energy E for a geometrically average reflectance of RM = 0.99999 for incident wavelength of λ = 500 nm (blue), λ = 800 nm (green) and λ = 1000 nm (red)

$$n_{max} = \frac{\lambda (1 - RM)^2}{hc} E, \tag{16.20}$$

with λ being the wavelength, h the Planck constant, c the speed of light and E the input pulse energy. For three different wavelengths, the transmitted number of photons is depicted exemplary in Fig. 16.6. This number of course refers to the number of photons that can maximally be detected after the first round trip. In order to get a reliably measure a CRD time trace one should be able to measure 10% of n_{max}.

If a continuous wave laser source is used, care has to be taken, to ensure a fast modulation of the laser in order to avoid a convolution between the ring down signal and the switch off characteristics. Hence the decay time of the laser after switching should be much smaller than the ring down time. Most modern laser systems support fast electronic switching down to a few nanoseconds, which is sufficient for characterizing reflectances of $R = 0.99$ ($\tau_{CRD} = 100$ ns, assuming a cavity length of $L = 30$ cm). If the laser system does not support properly fast switching an acousto-optical modulator (AOM) has to be applied to modulate the laser beam.

In case of a cw laser the transmitted signal S can be as large as the corresponding input due to the Fabry-Perot characteristics of the cavity. This, however, requires the light field to match the cavity resonances. This can be achieved by adjusting one mirror position using a piezo driven actuator and a feed-back loop (c.f. [19–21]). Additionally, the spectral width of the laser emission must be very narrow, as the CRD cavity is comparable with a Fabry-Perot-interferometer of high finesse (c.f. Sect. 16.2.4).

If broader sources are used, the cavity "picks" the respective components out of the emission spectrum (c.f. Fig. 16.4). Only those will be transmitted through the cavity and contribute to the CRD signal (c.f. Sect. 16.2.4). If the spectral width is large enough, no cavity locking is required. Though, the drawback is a tremendous decrease in transmission efficiency. Assuming a spectral width of the incoupled laser which is much larger than the free spectral range of the cavity, the transmitted power P_T can roughly be estimated by:

$$P_T \approx P_{in} \cdot \frac{1}{\mathcal{F}} = P_{in} \cdot \frac{1-R}{\pi\sqrt{R}}, \tag{16.21}$$

with P_{in} being the incident laser power. As a rule of thumb, the output power of the used laser should amount to at least $P_{Laser} = 100\,\text{mW}$ in order to achieve satisfying results, assuming a mirror reflectivity in the range of $R \approx 0.9999$. Continuous wave diode lasers or diode pumped solid state (DPSS) lasers are suited light sources.

16.3.2 The Detection Unit

As discussed in the previous section the amount of light to be detected is usually very small. E.g. the maximum transmitted power amounts only to $P_T = 10\,\text{pW}$ in case of a mirror reflectance of $R = 0.99999$ and a laser power of $P_{Laser} = 100\,\text{mW}$. Thus the requirement on the detectivity of the photo detector is very high. On the other hand the temporal resolution of the detector should be high, in particular for measuring small reflectivities around $R \approx 0.99$ or lower. The raise time τ_{rise} of the apparatus function should thus not exceed 10 ns.

Especially due to the requirement of high sensitivity the choice of the detector limits to either photo multiplier tubes (PMTs) or avalanche photo diodes (APDs). Both detectors are in principle capable of single photon counting.

16.3.2.1 Photo Multiplier Tubes

Photo multiplier tubes (PMT's) are based on multiplication of electrons emitted by a photocathode. The material of the photocathode basically determines the spectral response of the multiplier. For the visible spectral range silicon based photo cathodes are typically used. For longer wavelength their quantum efficiency (QE) is poor. Typically the QE for silicon based PMTs drops below $QE < 1\%$ for wavelengths $\lambda > 800\,\text{nm}$ [22, 23].

InGaAs or InP/InGaAsP based cathodes are more suited for the near or mid infrared spectral. Their detectivitiy ranges up to wavelengths of around $\lambda = 1.7\,\mu\text{m}$. They require, however, sufficient cooling and are fairly cost intensive [23, 24]. Their QE is still much lower than that of silicon based diodes in the visible spectral range.

PMT's typically show a large linear gain up to gain factors around 10^9, a low excess noise factor as well as a large aperture of up to several centimeters in diameter. In particular the latter makes them easy to use as detector for CRD setups, as the alignment of the detector is then very easy. No additional imaging optics behind the cavity is needed.

Their signal rise time typically lies in the range of $\tau_{rise} \approx 1\,\text{ns}$, which adequate even for high loss CRD measurements.

16.3.2.2 Avalanche Photo Diodes

Avalanche photo diodes (APDs) are generally a low cost alternative for measuring small signals. Especially for infrared applications APDs were proven to be superior compared to PMTs due to their larger quantum efficiency [22]. The major disadvantage of APDs is their larger excess noise and smaller linear gain. Thus especially in the visible region the signal-to-noise ratio of APDs is poor compared to that one of PMTs.

Furthermore, the aperture of APDs, which possess the required temporal resolution of $\tau_{rise} \approx 10\,\text{ns}$ is in general very small, amounting to an aperture diameter below 1 mm. Thus a sufficient imaging optic behind the resonator is required in order to detect the complete transmitted signal.

For measurements in the far infrared ($\lambda > 1.7\,\mu\text{m}$), HgCdZnTe/HgCdTe based photo diodes are the only sensor that show a sufficiently large specific detectivity for Cavity ring-down measurements [25, 26]. With those signals up to $\lambda = 10\,\mu\text{m}$ can be measured.

16.3.3 Broad Band Measurements

If the sample mirror reflectivity should not only be characterized at one particular wavelength, but over a broad range the laser light source and possibly the detection unit have to be extended in order to support broad band measurements.

In this case a tunable laser source such as an optical parametric amplifier (OPA) or oscillator (OPO) has to be used. Those systems are, however, very cost intensive and require circuitous tuning of the laser. In general, when changing the output laser wavelength in these systems the output beam position changes, which requires a realignment of the CRD cavity. Alternatively a short laser pulse (e.g. a nanosecond pulse from a Q-switched laser system) can be coupled into a photonic crystal fiber (PCF) generating a white light continuum.

Its major advantage compared to using a tunable light source is the simultaneous detection of the whole loss spectrum. This technique was e.g. applied to determine the mirror reflectance ($0.9976 < R < 0.9998$) spectrally resolved over a spectral range reaching from $\lambda = 530\,\text{nm}$ to $\lambda = 760\,\text{nm}$ [27]. The results are depicted in Fig. (16.7a)

In both cases a laser pulse ($\Delta\tau_{Pulse} \approx 1\,\text{ns}$) was spectrally broadened in a photonic crystal fiber before coupling it into the CRD cavity. The transmitted signal was recorded behind the cavity using a time gated CCD system in combination with a spectrograph.

Later a comparable setup was applied for characterizing photo damage in CaF_2 crystals induced by intense ultra violet radiation ([28], c.f. Fig. 16.7b). Within the framework of the latter study, the broad band reflectivity of the used mirrors was found to amount to $R > 0.9993$ in the spectral range of $480\,\text{nm} < \lambda < 650\,\text{nm}$ with a spectral resolution of $\Delta\lambda = 0.3\,\text{nm}$. The acquisition of the whole spectrally

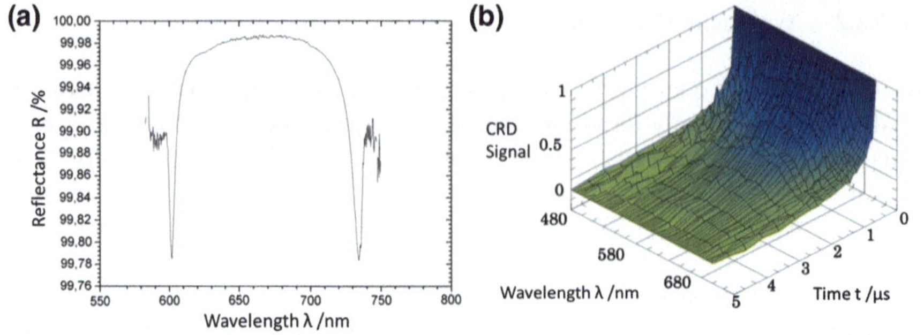

Fig. 16.7 Examples for broad band CRD measurements: **a** Spectrally resolved reflectivity curve for two cavity mirrors (reflectivity > 99.98% between 635 and 698 nm). Image taken from [27] (Reprinted from [27], with the permission of the OSA), **b** Normalized broad band transmission (480 nm $< \lambda <$ 650 nm) of a CRD cavity with a CaF_2 sample inside. Image taken from [28] (Reprinted from [28], with the permission of AIP Publishing)

resolved CRD time trace took less than 2 min. The spectrally resolved detection in this case was achieved by using a spectrograph (Shamrock 303i) in combination with a gated and intensified CCD camera (Andor iStar).

16.3.4 The Cavity Design

16.3.4.1 Stability

For a successful CRD measurement, the cavity has to be stable. Thus the constrain (16.4) has to be fulfilled for the cavity mirrors.

For the ratio

$$\alpha_i = \frac{L}{\rho_i}, \rightarrow L = \alpha_i \rho_i \; i = 1, 2, \tag{16.22}$$

for each cavity mirror i, the relation between α_1 and α_2 for a stable cavity is given by:

$$\begin{aligned} \frac{\alpha_1}{\alpha_1 - 1} &> \alpha_2 > 1 \quad for \quad \alpha_1 > 1 \\ \frac{\alpha_1}{\alpha_1 - 1} &< \alpha_2 < 1 \quad for \quad \alpha_1 < 1 \end{aligned} \tag{16.23}$$

It is pointed out that even for one convex mirror the cavity can be stable, if the second mirror is chosen appropriately. Thus the CRD technique can be used to determine the reflectivity of convex mirrors as well.

For the special case that both mirrors have comparable radii of curvature $\rho_1 \approx \rho_2 \approx \rho$, ($\alpha_1 \approx \alpha_2$) the cavity must be shorter than twice ρ:

$$L < 2\rho \tag{16.24}$$

It is pointed out that the particular case of $L = \rho$ (confocal resonator, $g_1 g_2 = 0$), $\rho = \infty$ (planar resonator $g_1 g_2 = 1$) or $L = 2\rho$ (concentric, $g_1 g_2 = 1$) are theoretically stable. These resonator configurations are in practice, however, not suited for CRD cavities, as they are very instable with respect to misalignment.

16.3.4.2 Effect of Diffraction Losses

Even in case of a stable cavity, the light inside will experience losses due to diffraction because of the finite mirror diameter. The diffraction losses are larger for smaller diameters a_i and can basically be estimated using the effective Fresnel numbers N_i in combination with (16.6) and (16.7). Although diffraction losses are typically very small for Fresnel numbers above 3.5 according to common literature [29] it is occasionally worth to estimate them for the CRD cavity since the losses to be determined are very small as well. For common cavity configurations with cavity lengths in the $L \approx 10$ cm region and mirror diameter in the $a_i \approx 1$ cm range the Fresnel numbers amount to several thousand, thus diffraction losses are generally negligible.

16.3.5 Coupling of Cavity and Light Source

Especially when coupling modulated continuous wave light or long pulses into the resonator it off resonant light will be suppressed effectively due to the high Finesse values (c.f. Sect. 16.2.4). This requires a fine tuning between the incoupled light field and the cavity.

16.3.5.1 Cavity Feed Back

As discussed in Sect. 16.2.4 an accurate locking of the cavity length to the laser frequency is necessary, particularly when using narrow band laser systems. One way to simplify this experimental challenge is to lock the laser frequency to the respective cavity length by exploiting optical feedback. Therefore diode lasers are required, which show an emission spectrum much broader than the free spectral range of the cavity. If the back reflected light of the optical cavity is coupled back into the laser, the emission in case of strong optical feedback will lock to the supported cavity modes. Thus the transmission of the resonator will be much larger, leading to a bigger

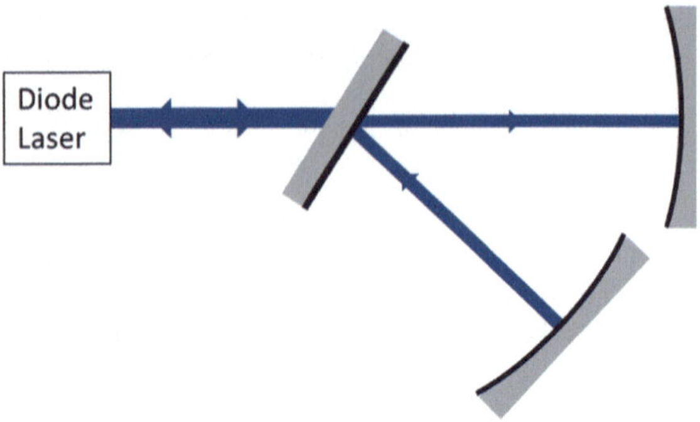

Fig. 16.8 Setup of a cavity feedback CRD setup: The light is coupled into the cavity. Back reflected light is coupled into a diode laser source causing it to lock to the cavity resonances

CRD signal and less noise [30]. A typical optical feedback CRD setup is depicted in Fig. (16.8).

A detailed explanation of the optical feedback mechanism is beyond the scope of this work. The reader is referred to respective literature for that [31–35].

16.3.5.2 Mode Matching

If an arbitrary light field is coupled into the cavity it will be decomposed in different transversal modes [12]. In case of a stable cavity conformation ($0 < g_1 g_2 < 1$) the resonance frequencies for different transverse modes are detuned for several MHz with respect to each other (c.f. (16.5)). Consequently all except of one TEM mode, for which the cavity might be locked (generally TEM_{00}) will be suppressed causing a decrease of coupling efficiency between the light source and the cavity.

In order to get the most efficient incoupling of light, the incident beam should be matched to one selected cavity mode (c.f. Fig. 16.9). Typically the TEM_{00} mode is chosen. Therefore the incident beam has to be shaped in a way that its radius of curvature at the cavity mirrors is identical with the radius of curvature of those mirrors. One should be aware that a nonplanar incoupling mirror changes the beam divergence. The mode matching can be performed by using only one lens between the laser source and the cavity in order to adapt the laser beam profile to the cavity mode. Using a telescope (2 lenses) or a zoom optic (3 lenses) will however simplify the optical setup, as the exact distance between laser source and cavity is not that critical anymore. Particularly when mirrors of varying radius of curvature shall be measured a multilens system will strongly simplify the adaption of mode matching to the respective radius of curvature.

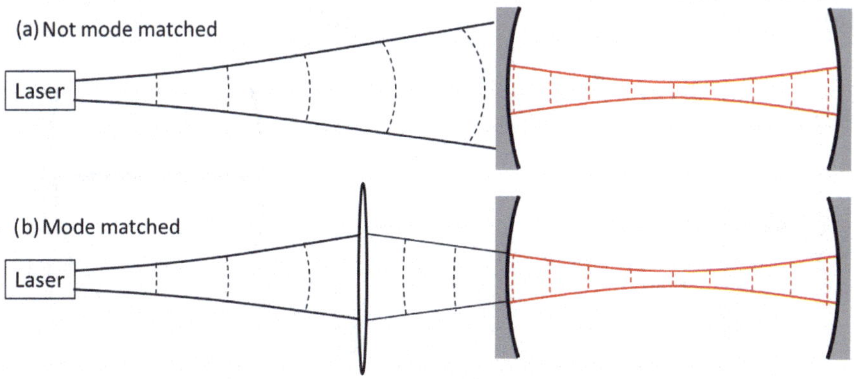

Fig. 16.9 **a** Not mode matched cavity setup: the field of the incident laser beam (black) and the TEM_{00} mode are not adapted. The cavity will the incident field cause to be decomposed in higher cavity modes (not depicted here). The dashed lines illustrate the phase fronts of the field distributions, **b** Mode matched case: The incident laser field (black) is adapted to the cavity TEM_{00} mode (red)

16.3.6 Accuracies

One crucial issue of performing a reliable CRD measurement is the proper analysis of the measurement accuracy. It is generally determined by the fitting accuracy and the alignment error.

16.3.6.1 The Fitting Accuracy of the Time Trace

Although the accuracy of fitting the decay time curve exponentially depends on the noise of the detector, it is generally very low, if the sample rate of the detector is sufficiently large. Typical sample rates of modern read out electronics are in the range of $f_{sample} = 1$ GHz. From those time traces a fitting accuracy of the time constant of $\frac{\Delta\tau_{CRD}}{\tau_{CRD}} \approx 0.5\%$ can easily be achieved, particularly if a PMT is used for read out (c.f. Sect. 16.3.2). In case of very noisy data, e.g. due to a very small transmitted power and/or the need of using APDs for signal detection the accuracies can be achieved by sufficient averaging of the time trace.

16.3.6.2 The Alignment Accuracy of the Optical Cavity

Even a slight misalignment will increase the diffraction losses of the cavity. As discussed above the diffraction losses are in general very low and for common cavity configurations negligible compared to reflection, scattering or absorption losses. The influence of improper cavity alignment onto the cavity stability[13] and losses has been analyzed theoretically and can be estimated by modifying the Fresnel numbers

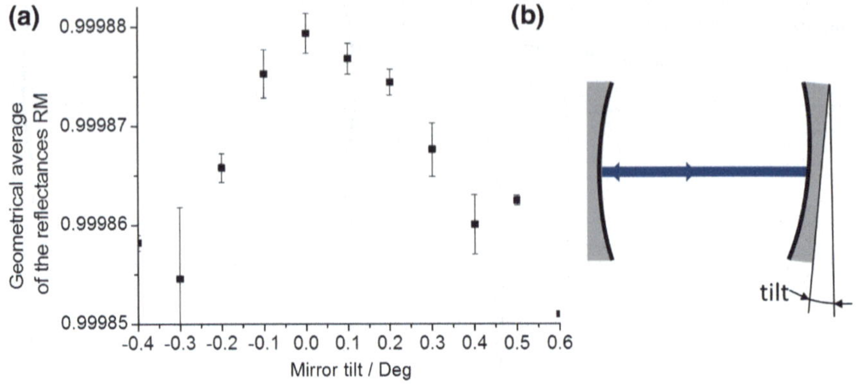

Fig. 16.10 **a** Effect of tilting on cavity mirror onto the geometrical average of the reflectances RM computed from the measured cavity ring-down time τ_{CRD}. A tilt of 0.1 degree causes a signal decrease of approximately 5%. The error bars arise from repeating the experiments for the fixed detuning angle 10 times. **b** scheme of the experiment

N_i by incorporating a tilt of the respective mirror [12]. For N_i values above $N_i > 1000$ the increase in losses is still negligible, especially, if the cavity configuration is close to cavities with g-values around 0.5–0.8 as those configurations are very stable with respect to misalignment.

What on the other hand might have a crucial influence on the losses is a smaller reflectivity of the mirrors for angles of incidence deviating from the designed value. In this case the impact of the misalignment depends not only on the cavity geometry but also on the particular layer design.

For one CRD configuration ($L = 30$ cm, $\rho_1 = \rho_2 = 100$ cm $\rightarrow g_1 = g_2 = 0.7$) it was estimated by conducting a systematic study. Therefore the system was repeatedly aligned and misaligned. Afterwards, one cavity mirror was titled by a defined angle. The results are summarized in Fig. 16.10. In this case the statistical deviation of the cavity losses (the signal decay time) due to misalignment amounts to 5%. This value is a good estimate as standard alignment error for the CRD measurement.

One important aspect of the CRD method is that systematic errors due to poor system alignment will only decrease the signal decay time τ_{CRD}. This leads to an over estimation of the cavity losses, i.e. an under estimation of the mirror reflectances. Thus a possible misalignment must only be taken into account at the lower error margin of the reflectance. The upper margin is solely determined by the accuracy of the exponential fit.

16.3.6.3 Accuracy of the Reflectivity Measurement

What in-fact was determined by the systematic misalignment study in the previous section is the accuracy of the losses ($1 - RM$ in the absence of scattering and absorption). The correlation between the relative and the absolute error of RM thus is given by:

$$\Delta RM_{+}^{rel} = \frac{\Delta RM_{+}}{1 - RM} = 5\%$$
$$\Delta RM_{-}^{rel} = \frac{\Delta RM_{-}}{1 - RM} = 0.5\% \quad (16.25)$$

If the reflectivity of one mirror in a linear cavity is retrieved using a geometrical averages RM_1 to RM_3 of three mirrors in all their combinations (c.f. (16.18)) the accuracy of the reflectivity is given by:

$$R_1 = \frac{RM_1 RM_2}{RM_3} \quad \rightarrow \quad \Delta R^{rel} = \frac{\Delta R}{1 - R} = 3\Delta RM^{rel} = \begin{cases} +15\% \\ -1.5\% \end{cases} \quad (16.26)$$

In case of a folded cavity consisting out of three mirrors (16.19) with the sample mirror in the center, the accuracy is given by:

$$R = \frac{RM_{2M}}{RM_{3M}} \rightarrow \quad \Delta R^{rel} = \frac{\Delta R}{1 - R} = \Delta RM_{2M}^{rel} + \Delta RM_{3M}^{rel} = \begin{cases} +10\% \\ -1.0\% \end{cases} \quad (16.27)$$

If a previously defined reference mirror is used in order to retrieve the reflectance of a sample mirror (c.f. (16.16)) the accuracy is given by:

$$R = \frac{RM^2}{R_{ref}} \rightarrow \quad \Delta R = \frac{2RM}{R_{ref}} (1 - RM) \cdot \Delta RM^{rel} + \frac{RM^2}{R_{ref}^2} \left(1 - R_{ref}\right) \times \Delta R_{ref}^{rel} \quad (16.28)$$

In this case the accuracy for the determination of the mirror reflectivity strongly depends on the reflectance of the reference mirror. If the reference reflectivity is very small compared to that of the sample mirror the former determines the cavity losses and the accuracy for determining R is decreasing tremendously. The effect is depicted in Fig. 16.11. On the other hand, the accuracy for determining R can be very high, if the reflectivity of the sample mirror is very large compared to that of the sample. Hence one should aim for using the mirrors with the highest reflectance values as references.

16.4 Limits of the Technique

The indirect determination of the reflectivity coefficient by measuring the losses, which makes the measurement independent of the exact knowledge of the input laser power and thus stable with respect to laser fluctuations shows a significant

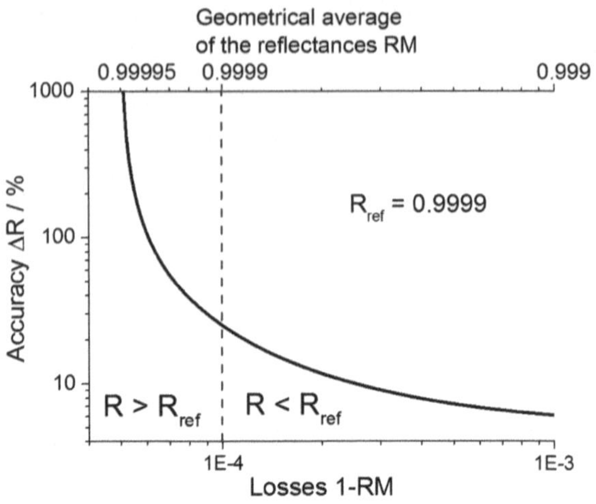

Fig. 16.11 Accuracy ΔR of determining the reflectance R of a sample mirror when a reference mirror $R_{ref} = 0.9999$ is used as a function of the geometrically averaged reflectance $RM = \sqrt{RR_{ref}}$ or the cavity losses $1 - RM$, respectively

drawback: The cavity losses are composed of absorption, scattering and reflection (besides losses due to inappropriate cavity design, c.f. Sect. 16.3.4). Hence, if absorption and/or scattering becomes dominant compared to reflection inadequacies, those will lead to an underestimation of R using the CRD method. In such cases, the CRD measurement has to be supported by accurate scattering and absorption measurements for determining the correct R values. This is particularly critical in case of very high mirror reflectivities $R > 0.99999$, poor surface qualities or when light in the short wavelength visible or ultraviolet spectrum is used for characterization.

Sophisticated techniques for evaluating scattering or absorption are subject of Chaps. 14 and 15 of this book.

In the case of small mirror reflectivities $R < 0.99$ corresponding to large losses, the decay time of the CRD signal is very short ($\tau_{CRD} < 100\,\text{ns}$ for $L = 30\,\text{cm}$). In this case the temporal resolution of the detector or the sampling of the read out electronic limits the accurate determination of R. Such low reflectances can, however, easily be determined by spectrophotometric or laser ratiometric measurements.

In this context CRD technique should not be considered to be superior with respect to spectrophotometric or laser ratiometric measurements but more as a complementing method.

Finally, using the CRD technique the reflectance can only be determined accurately in the center of the mirror. Although scanning of the mirror surface could be performed in a certain range, close to the rim of the mirror diffraction losses will exceed the reflectance losses.

16.5 Summary

In this chapter the possibility of applying the cavity ring-down (CRD) method in order to determine reflectances above $R > 0.99$ was discussed in detail. For such large reflectances CRD is superior compared to e.g. laser ratiometric or spectrophotometric measurements as the reflectivity is determined indirectly by measuring the losses of a cavity in which light is confined. Since this is done by determining the life time of the light inside the cavity the measurement accuracy is independent of laser power fluctuations and does not require a large dynamic range of the detector.

A detailed guideline of performing reliable CRD measurements is also fixed in the ISO standard 13142 [36] since 2015.

A successful CRD measurement does, however, require a carefully designed and aligned cavity. In particular the cavity length, the mirror diameters and their radius of curvature must be chosen appropriately in order to avoid losses due to cavity instabilities or diffraction.

The CRD method is a complementing technology with respect to laser ratiometry or spectrophotometric rather than a competing one for measuring reflectances. It is superior for $R > 0.99$ but inferior for $R < 0.99$.

References

1. R. Paschotta, *Encyclopedia of Laser Physics and Technology* (Wiley-VCH, 2008)
2. A. Duparre, D. Ristau, Optical Interference Coatings 2010 Measurement Problem. Appl. Opt. **50**(9), C172–C177 (2011)
3. P. Steve Upstone, Inc., Validating UV/VIS Spectrophotometers, Technical Note, (2012) https://www.perkinelmer.com/lab-solutions/resources/docs/TCH_Validating_UV_Visible.pdf (accessed: February 2018)
4. H. Nasibov, I. Mamedbeili, D. Riza, E. Balaban, F. Hacizade, High-precision measurements of reflectance, transmittance, and scattering at 632.8 nm. Laser Sources and Applications. *Proc. SPIE. 8433*, 843313 (2012)
5. A. O'Keefe, D.A.G. Deacon, Cavity ring-down optical spectrometer for absorption measurements using pulsed laser sources. Rev. Sci. Instrum. **59**(12), 2544–2551 (1988)
6. T. Zeuner, W. Paa, G. Schmidl et al., Optical readout of a nanoparticle based sensor by cavity ring-down spectroscopy. Sens. Actuators B: Chem. **195**, 352–358 (2014)
7. G. Rempe, R. Lalezari, R.J. Thompson et al., Measurement of ultralow losses in an optical interferometer. Opt. Lett. **17**(5), 363 (1992)
8. R.A. Washenfelder, A.R. Attwood, J.M. Flores et al., Broadband cavity-enhanced absorption spectroscopy in the ultraviolet spectral region for measurements of nitrogen dioxide and formaldehyde. Atmos. Measure. Tech. **9**(1), 41–52 (2016)
9. M. Goulette, S. Brown, H. Dinesan, W. Dubbé, G. Hübler, J. Orphal, A. Ruth, and A. Zahn, A New Cavity Ring-down Instrument for Airborne Monitoring of N2O5, NO3, NO2 and O3 in the Upper Troposphere Lower Stratosphere, in Imaging and Applied Optics 2016, OSA technical Digest (online) (Optical Society of America, 2016), LTh1G.2 (2016)
10. H. Waechter, F. Adler, M. Beels, et al., *Diverse industrial applications of Cavity Ring-Down Spectroscopy* (Photonics North (PN), Quebec City, QC, 2016), pp. 1–1 http://doi.org/10.1109/PN.2016.7537897

11. C. Dhiman, M.S. Khan, M.N. Reddy, Phase-shift cavity ring down spectroscopy set-up for NO2 sensing: design and fabrication. Defence Science Journal **65**(1), 25–30 (2015)
12. N. Hodgson, H. Weber, *Laser Resonators and Beam Propagation - Fundamentals, Advanced Concepts and Applications* (Springer, New York, 2005), p. 793
13. A.E. Siegman, *Lasers* (University Science Books, Sausalito, 1990), p. 1285
14. Joachim Herrman, B. Wilhelmi, *Laser für ultrakurze Lichtimpulse - Grundlagen und Anwendungen*, vol. 324 (Akademie-Verlag, Berlin, 1984)
15. W. Brunner, R. König, P. Nickles, L.W. Wieczorek, G. Wiederhold, *Der Laser*in W. Brunner, K. Junge, Wissensspeicher Lasertechnik, 1. uflage (Fachbuchverlag Leipzig, 1982)
16. J.Y. Lee, H.W. Lee, J.W. Hahn, Time domain study on cavity ring-down signals from a Fabry-Pérot cavity under pulsed laser excitation. Jpn. J. Appl. Phys. **38**(11), 6287–6297 (1999)
17. H. Kogelnik, T. Li, Laser beams and resonators. Appl. Opt. **5**(10), 1550–1567 (1966)
18. D. Romanini, I. Ventrillard, G. Méjean, et al., *Introduction to Cavity Enhanced Absorption Spectroscopy* (Springer, Berlin, Heidelberg, 2014)
19. B.A. Paldus, C.C. Harb, T.G. Spence et al., Cavity-locked ring-down spectroscopy. J. Appl. Phys. **83**(8), 3991–3997 (1998)
20. D. Romanini, A.A. Kachanov, N. Sadeghi et al., CW cavity ring down spectroscopy. Chem. Phys. Lett. **264**(3–4), 316–322 (1997)
21. T.G. Spence, C.C. Harb, B.A. Paldus et al., A laser-locked cavity ring-down spectrometer employing an analog detection scheme. Rev. Sci. Instrum. **71**(2), 347–353 (2000)
22. W.G. Lawrence, G. Varadi, G. Entine, E. Podniesinski, P.K. Wallace, A comparison of avalanche photodiode and photomultiplier tube detectors for flow cytometry, in imaging, manipulation, and analysis of biomolecules, cells, and tissues, p. 68590M (2008), http://doi.org/10.1117/12.758958
23. K.K. Hamamatsu Photonics, *Photomultiplier Tubes Basics and Applications*, 3rd edn. (2007)
24. S.O. Flyckt, C. Mamonier, *Photomultiplier Tubes: Principles and Applications* (Philips Photonics, Brive, 2002)
25. C. Bahrini, Y. Bénilan, A. Jolly et al., Pulsed cavity ring-down spectrometer at 3 μm based on difference frequency generation for high-sensitivity CH4 detection. Appl. Phys. B **121**(4), 533–539 (2015)
26. S. Zhou, Y. Han, B. Li, Simultaneous detection of ethanol, ether and acetone by mid-infrared cavity ring-down spectroscopy at 3.8 μm. Appl. Phys. B **122**(7) (2016)
27. G. Schmidl, W. Paa, W. Triebel et al., Spectrally resolved cavity ring down measurement of high reflectivity mirrors using a supercontinuum laser source. Appl. Opt. **48**(35), 6754–6759 (2009)
28. T. Zeuner, W. Paa, C. Muhlig et al., Note: Broadband cavity ring-down spectroscopy of an intra-cavity bulk sample. Rev. Sci. Instrum. **84**(3), 036104 (2013)
29. Y. Le Grand, J.P. Tache, A. Le Floch, Sensitive diffraction-loss measurements of transverse modes of optical cavities by the decay-time method. J. Opt. S B **7**(7), 1251–1253 (1990)
30. Y. Gong, B. Li, Y. Han, Optical feedback cavity ring-down technique for accurate measurement of ultra-high reflectivity. Appl. Phys. B **93**(2–3), 355–360 (2008)
31. K. Petermann, *Semiconductor Lasers with Optical Feedback* (Springer, Netherlands, 1988)
32. A.T. Ryan, G.P. Agrawal, G.R. Gray et al., Optical-feedback-induced chaos and its control in multimode semiconductor lasers. IEEE J. Quantum Electron. **30**(3), 668–679 (1994)
33. V.Z. Tronciu, H.J. Wunsche, M. Wolfrum et al., Semiconductor laser under resonant feedback from a Fabry-Perot resonator: Stability of continuous-wave operation. Phys. Rev. E: Stat., Nonlin, Soft Matter Phys. **73**(4 Pt 2), 046205 (2006)
34. W. Lewoczko-Adamczyk, C. Pyrlik, J. Hager et al., Ultra-narrow linewidth DFB-laser with optical feedback from a monolithic confocal Fabry-Perot cavity. Opt. Express **23**(8), 9705–9709 (2015)
35. B. Dahmani, L. Hollberg, R. Drullinger, Frequency stabilization of semiconductor lasers by resonant optical feedback. Opt. Lett. **12**(11), 876 (1987)
36. Electro-optical systems—Cavity ring-down technique for high-reflectance measurement. ISO 13142:2015. Geneva, Switzerland: ISO

Index

O. Stenzel and M. Ohlídal (eds.), *Optical Characterization of Thin Solid Films*, Springer Series in Surface Sciences 64,
https://doi.org/10.1007/978-3-319-75325-6

The manufacturer's authorised representative in the EU is Springer Nature Customer Service Centre GmbH, Europaplatz 3, 69115 Heidelberg, Germany. If you have any concerns regarding our products, please contact ProductSafety@springernature.com

Printed and bound by CPI Group (UK) Ltd, Croydon, CR0 4YY
05/10/2026
02228138-0020